SEMICONDUCTOR
Principles and Applications

In Partnership with the **NJATC**

Australia • Brazil • Japan • Korea • Mexico • Singapore • Spain • United Kingdom • United States

Semiconductor Principles and Applications
Second Edition
NJATC

Vice President, Technology Professional Business Unit:
Gregory L. Clayton

Product Development Manager:
Ed Francis

Product Manager:
Stephanie Kelly

Editorial Assistant:
Nobina Chakraborti

Director of Marketing:
Beth A. Lutz

Executive Marketing Manager:
Taryn Zlatin

Marketing Specialist:
Marissa Maiella

Director of Production:
Carolyn Miller

Production Manager:
Andrew Crouth

Content Project Manager:
Kara A. DiCaterino

Art Director:
Bethany Casey

Production Technology Analyst:
Thomas Stover

For product information and technology assistance, contact us at
Professional & Career Group Customer Support, 1-800-648-7450

For permission to use material from this text or product, submit all requests online at **www.cengage.com/permissions**
Further permissions questions can be emailed to
permissionrequest@cengage.com

Library of Congress Cataloging-in-Publication Data
Card Number: 2007028498

ISBN-13: 978-1-4180-7341-1
ISBN-10: 1-4180-7341-5

Delmar Cengage Learning
5 Maxwell Drive
Clifton Park, NY 12065-2919
USA

Cengage Learning products are represented in Canada by Nelson Education, Ltd.

For your lifelong learning solutions, visit **delmar.cengage.com**

Visit our corporate website at **www.cengage.com**

Printed in Canada
1 2 3 4 5 XX 10 09 08

In Loving Memory of

Stan Klein

CONTENTS

PREFACE

INTRODUCTION

Welcome to the second edition of *Semiconductor Principles and Applications,* which has been redesigned and updated to provide knowledge of the fundamentals to electrical technologists in apprenticeship programs, vocational-technical schools and colleges, and community colleges. The text emphasizes a solid foundation of classroom theory supported by on-the-job hands-on practice. Every project, every piece of knowledge, and every new task will be based on the experience and information acquired as each technician progresses through his or her career. This book, along with the others in this series, contains a significant portion of the material that will form the basis for success in an electrical career.

This text was developed by blending up-to-date practice with long-lived theories in an effort to help technicians learn how to better perform on the job. It is written at a level that invites further discussion beyond its pages while clearly and succinctly answering the questions of *how* and *why.* Improvements to this edition were made possible by the continued commitment by the National Joint Apprenticeship and Training Committee (NJATC) in partnership with Delmar Cengage Learning to deliver the very finest in training materials for the electrical profession.

For excellence in your electrical and telecommunications curriculum, look no further. The NJATC has been *the* source for superior electrical training for thousands of qualified men and women for over 65 years. Curriculum improvements are constant as the NJATC strives to continuously enhance the support it provides to its apprentices, journeymen, and instructors in over 285 training programs nationwide.

ABOUT THIS BOOK

The efforts for continuous enhancement have produced the product you see before you: this technically precise, academically superior edition of *Semiconductor Principles and Applications.* Using a distinctive blend of theory-based explanation partnered with hands-on accounts of what to do in the field and peppered with Technical Tips and Field Notes, this book will lead you through the study of semiconductor devices, practical application circuits, the importance of electronic circuits, and troubleshooting, as well as discussions on newer technologies in the field.

KEY FEATURES

This text has been strengthened from top to bottom with many new features and enhancements to existing content. All-new chapter features provide structure and guidance for learners. This formerly engineering-oriented book has been revamped with illustrations, diagrams, pictures, drawings, and real world applications to help readers apply the voluminous conceptual content. Enhanced and concrete Chapter Objectives are complemented by solid and reinforcing Chapter Summaries, Review Questions, and Practice Problems. Chapter contents are introduced at the beginning of each chapter, and then bolstered before moving on to the next chapter. Throughout each chapter, concepts are explained from their theoretical roots to their application principles, with reminders about safety, technology, professionalism, and more.

Semiconductor Principles and Applications, Second Edition, has been expanded to more fully explore a number of concepts through a major reorganization of the chapters. This enhancement leads to more logical discussions, with topics building on each other from one chapter to the next. Combining related topics and bolstering them with additional content makes the book even more reader-friendly. In addition, chapter pedagogical features have been boosted: objectives are more action-oriented, end-of-chapter summaries and review questions force readers to delve deeper into the content to prove a true understanding, and extra information has been added in sidebars to grab the reader's eye.

See the following pages for examples of these new features.

F E A T U R E S

AMPLIFIER GAIN

An **amplifier** provides gain—the ratio of the output to the input. An amplifier achieves gain by converting one thing to another. For example, a lever converts force to movement or movement to force. One pound of force on the end of a beam may lift 3 pounds on the other end, but it will not lift it far. An electronic amplifier uses circuit DC power and divides that power between the output terminal and a load resistor based on the strength of the input signal. Figure 6–1 shows a block diagram of an amplifier circuit.

Amplifier
A device that provides gain without much change in the original signal waveform.

Gain
The ratio of the output signal to the input signal of an **active component**.

Active component
Components of an electronic circuit that use a power source to process a signal. The processing usually involves amplification or some other change in the signal that requires additional power. BJTs, FETs, and UJTs are examples of active components.

A_P
Power gain.

A_I
Current gain.

A_V
Voltage gain.

Amplification
The process of increasing the voltage, current, or power of a signal.

FIGURE 6–1 Amplifier block diagram.

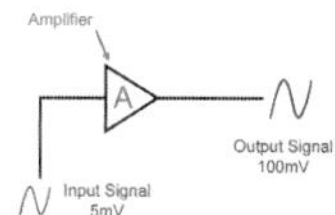

6.1 Types of Gain

If the objective for an amplifier is to produce gain, a measure of the gain of the amplifier is a measure of the amplifier's success. Gain is expressed as the ratio of output to input. Gain is shown in formulas using the symbol A with a subscript of $_P$, $_I$, or $_V$ to indicate power, current, or voltage, respectively.

Take care to write the result correctly; A_P, A_I, and A_V are unitless. This is because A is calculated by dividing power by power, current by current, or voltage by voltage. By not tacking units on the ends of **amplification** factors, statements such as "We have six times as many watts at the output than at the input" can be avoided. Normally, gain is expressed as a dimensionless number—"The amplifier has a voltage gain of 10, 20, 50, 110, . . ."

…put voltage lev-

FIGURE 6–2 Voltage gain.

V_{DC}
A
10V Output
Input

Gain = $\frac{\text{V output}}{\text{V input}}$

Gain = $\frac{10V}{.5V}$

Gain = 20

Running Glossaries are included in each chapter along with a comprehensive glossary at the end of the book.

FIGURE 13–6 Liquid service.

3-valve manifold
L
H
Low Pressure
Optional side-mounted drain/vent valve
High Pressure
Flow
Flow
Low Pressure
H
3-valve manifold
L
Drain/vent valve

FIGURE 13–7 Steam service.

Flow
Blocking valves
Plugged tee for steam service for sealing fluid
Low Pressure
High Pressure
Sufficient length for cooling
H
L
3-valve manifold

FIGURE 13–8 Gas service.

Vent drain valve
H
L
3-valve manifold
High Pressure
Low Pressure
Flow

vent. If measuring gases, the valves should be mounted to allow any liquid that has collected in the process tubing to drain. If measuring steam, the valves should be mounted the same as those for liquid measurement, because the lines should be filled with water. Filling the steam service lines with water is required to prevent steam from coming into contact with the transmitter sensor (Figure 13–8). When field calibration occurs, this liquid head on the impulse lines is accounted for naturally. When bench-calibrating these devices, you must allow for the liquid head to achieve accurate calibration.

The potential zero shift due to excessive pressure is provided to the sensor of the transmitter. Elevation and suppression are two conditions that an installer must consider when measuring liquids. For special cases of gas measurement, elevation and suppression may be considerations, but we now must consider a process that is lighter than the one we are used to dealing with. Mounting details should provide all necessary requirements for mounting height with respect to tap positions. Review and follow your site-specific mounting details.

It is possible that gas-measuring devices can be mounted to show an excess or lack of process due to the relationship of the process to the mounting position. Mounting details should be provided for special service devices to ensure that they are not located improperly.

High-Contrast Images give a clear and colorful view that 'pops' off the page.

TechTip!

As you have most likely observed at a carnival or fair with large portable truck-mounted generators, as the rides start and the demand for current increases, the generators have to work harder. This is usually apparent when the diesel engines work harder, creating more noise and more exhaust. As the electrical load or power requirement diminish, the engines settle back to the idle mode or throttle back. This is a direct indication that as there is more electrical work being done, more mechanical horsepower is required. Along with the mechanical changes are the voltage regulation controls, which are not as obvious but are constantly adjusting the magnetic field to maintain voltage.

ThinkSafe!

Many pneumatic devices are spring loaded. Always use extreme caution when removing mounting bolts, because great force can be exerted as the spring tension is relieved. This force can cause serious injury and equipment damage if not planned for and alleviated.

FieldNote!

"ELI the ICE man" is a phrase used in the industry to remind practitioners of the current and voltage relationships that exist in inductive circuits represented by L and in capacitive circuits represented by C. E leads I in an inductive circuit, hence the acronym ELI. This is the same circumstance as I lagging E as we have previously learned. The acronym ICE indicates that I leads E in a capacitive circuit. Thus, "ELI the ICE man" is a mnemonic used to help memorize the relationships of E and I in various circuits. Remember that in a purely resistive circuit, the E and I are in phase and have synchronous waveforms. In AC circuits with L and C and R the current with reference to the voltage may be anywhere from 90° leading to 90° lagging, or in between.

High-interest content is given in ThinkSafe!, FieldNote!, and TechTip! sidebars to make real-world connections to lessons learned in the chapters.

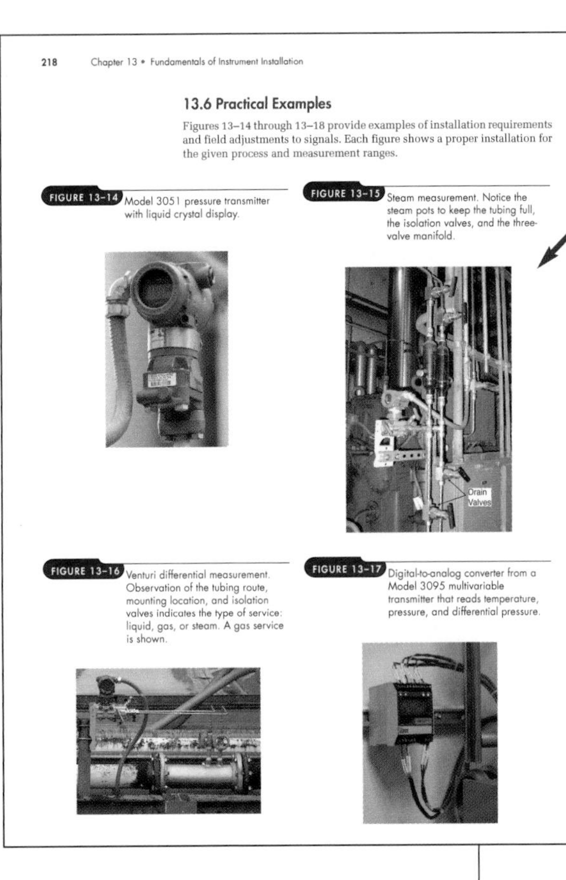
218 Chapter 13 • Fundamentals of Instrument Installation

13.6 Practical Examples

Figures 13–14 through 13–18 provide examples of installation requirements and field adjustments to signals. Each figure shows a proper installation for the given process and measurement ranges.

FIGURE 13–14 Model 3051 pressure transmitter with liquid crystal display.

FIGURE 13–15 Steam measurement. Notice the steam pots to keep the tubing full, the isolation valves, and the three-valve manifold.

FIGURE 13–16 Venturi differential measurement. Observation of the tubing route, mounting location, and isolation valves indicates the type of service: liquid, gas, or steam. A gas service is shown.

FIGURE 13–17 Digital-to-analog converter from a Model 3095 multivariable transmitter that reads temperature, pressure, and differential pressure.

New photos and illustrations are located near their text references and clarify explanations.

18 • Fiber Optics and Fiber-Optic Cable

Fibers usually come in bundles. Bundles are of two types: flexible or rigid. The flexible bundle is usually surrounded by a protective plastic coating, and at the ends of the cable the individual fibers are tied or joined together. In the rigid bundle, the individual fibers are melted together into a single rod and are shaped during the manufacturing process. Figure 18–10 shows a flexible fiber-optic bundle.

FIGURE 18–10 Fiber-optic bundles.

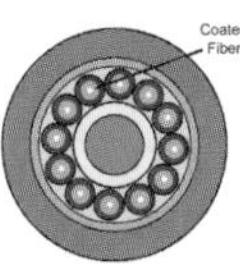

18.2 Optical Fiber Characteristics

To understand how a fiber-optic system operates, you need a fundamental knowledge of three areas: optics, electronics, and communications. In physics, light is treated as either electromagnetic waves or as photons (electromagnetic energy particles). For this discussion, we will concentrate on the electromagnetic wave characteristics of light. The light spectrum (light measured as a wave or electromagnetic frequency) is quite small when compared to the entire spectrum range. Figure 18–11 shows a chart of the electromagnetic spectrum. As you can see, there is only a small area of the spectrum we will consider when dealing with fiber optics, the optical spectrum from infrared to ultraviolet frequencies.

FIGURE 18–11 Electromagnetic spectrum.

The Optical Spectrum

Radio Microwaves X-Rays Gamma Rays

Infrared Ultraviolet

Frequency (Hz)

Photon Energy (ev)

Wavelength (m)

End-of-chapter problems reinforce critical concepts and relate to the worked-out examples in the chapter.

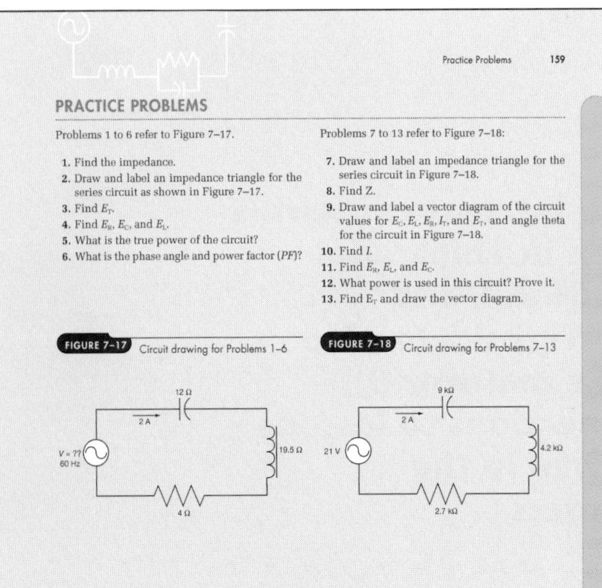
Practice Problems 159

PRACTICE PROBLEMS

Problems 1 to 6 refer to Figure 7–17.

1. Find the impedance.
2. Draw and label an impedance triangle for the series circuit as shown in Figure 7–17.
3. Find E_T.
4. Find E_R, E_C, and E_L.
5. What is the true power of the circuit?
6. What is the phase angle and power factor (*PF*)?

Problems 7 to 13 refer to Figure 7–18:

7. Draw and label an impedance triangle for the series circuit in Figure 7–18.
8. Find Z.
9. Draw and label a vector diagram of the circuit values for E_C, E_L, E_R, I_T, and E_T, and angle theta for the circuit in Figure 7–18.
10. Find *I*.
11. Find E_R, E_L, and E_C.
12. What power is used in this circuit? Prove it.
13. Find E_T and draw the vector diagram.

FIGURE 7–17 Circuit drawing for Problems 1–6

FIGURE 7–18 Circuit drawing for Problems 7–13

Step-by-step sample problems and solutions relate to the end of chapter exercises.

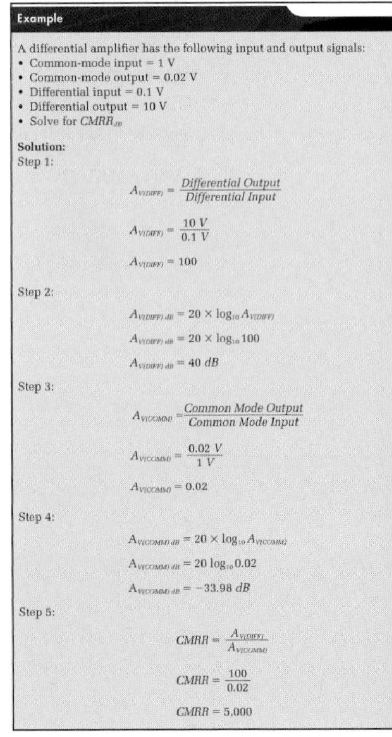
Example

A differential amplifier has the following input and output signals:

- Common-mode input = 1 V
- Common-mode output = 0.02 V
- Differential input = 0.1 V
- Differential output = 10 V
- Solve for $CMRR_{dB}$

Solution:

Step 1:

$$A_{V(DIFF)} = \frac{\text{Differential Output}}{\text{Differential Input}}$$

$$A_{V(DIFF)} = \frac{10\ V}{0.1\ V}$$

$$A_{V(DIFF)} = 100$$

Step 2:

$$A_{V(DIFF)\ dB} = 20 \times \log_{10} A_{V(DIFF)}$$

$$A_{V(DIFF)\ dB} = 20 \times \log_{10} 100$$

$$A_{V(DIFF)\ dB} = 40\ dB$$

Step 3:

$$A_{V(COMM)} = \frac{\text{Common Mode Output}}{\text{Common Mode Input}}$$

$$A_{V(COMM)} = \frac{0.02\ V}{1\ V}$$

$$A_{V(COMM)} = 0.02$$

Step 4:

$$A_{V(COMM)\ dB} = 20 \times \log_{10} A_{V(COMM)}$$

$$A_{V(COMM)\ dB} = 20 \log_{10} 0.02$$

$$A_{V(COMM)\ dB} = -33.98\ dB$$

Step 5:

$$CMRR = \frac{A_{V(DIFF)}}{A_{V(COMM)}}$$

$$CMRR = \frac{100}{0.02}$$

$$CMRR = 5{,}000$$

SUPPLEMENTAL PACKAGE

The Instructor Resource Kit is geared to provide instructors with all the tools they need in one convenient package. Instructors will find that this resource provides them with a far-reaching teaching partner that includes:

- PowerPoint® slides (electronic and hard-copy) for each book chapter that reinforce key points and feature illustrations and photos from the book,
- the Computerized Test Bank in ExamView format, which allows for test customization for evaluating student comprehension of noteworthy concepts,
- an electronic version of the Instructor's Manual, with supplemental lesson plans and support,
- the image library, which includes all drawings and photos from the book for the instructor's use to supplement class discussions, and
- a transition guide to help instructors map the changes from the previous edition to this new, stronger edition of the book.

ABOUT THE NJATC

Should you decide on a career in the electrical industry, training provided by the International Brotherhood of Electrical Workers and the National Electrical Contractors Association (IBEW-NECA) is the most comprehensive the industry has to offer. If you are accepted into one of their local apprenticeship programs, you'll be trained for one of four career specialties: journeyman lineman, residential wireman, journeyman wireman, or telecommunications installer/technician. Most importantly, you'll be paid while you learn. To learn more visit http://www.njatc.org.

NJATC ACKNOWLEDGMENTS

Technical Editors

William R. Ball, NJATC Staff
Jim Simpson, NJATC Staff

Contributing Writers

Ed Swearingen, Instructor, Alaska JATC
Chris MacCreery, Training Director, Battle Creek Elect. JATC
Paul LeVasseur, Training Director, Bay City JEATC
Gary Strouz, Training Director, Houston Electrical JATC

ADDITIONAL ACKNOWLEDGMENTS

This material is continually reviewed and evaluated by curriculum groups who are also members of the NJATC Inside Education Committee. The invaluable input provided by these individuals allows for the development of instructional material that is of the absolute highest quality. At the time of this printing the Inside Education Committee was comprised of the following members: Dennis Anthony, Chair; George Cunningham; Chris MacCreery; Ed Swearingen; and Neil Wilford.

The NJATC would also like to acknowledge the following Subject Matter Experts (SME's) for their hard work and dedication during the technical review process: Dale Goerge; Brett Hoffman; Chris MacCreery; and Ed Swearingen.

PUBLISHER'S ACKNOWLEDGMENTS

Delmar Learning and the author would also like to thank the following reviewers for their valuable suggestions and expertise:

John Fitzen, Instructor Electronics Department at Idaho State University, College of Technology, Electronics Department

Donald R. Montgomery, M.ED Administration & Supervision President of YTI Career Institute

David Hartle, Professor of Electrical Engineering Technology at SUNY Canton College

AUTHOR'S ACKNOWLEDGMENTS

I would like to acknowledge the technical and editorial expertise of my peer, friend, and colleague, Fred Rossnagel of Ferjr LLC, for his technical expertise and work. His technical ability is immeasurable and he is blessed with a tremendous talent. I would also like to acknowledge Mike Bearden of Mike Will Automations, for his push to make practical applications of this book to real-world problems. Thank you also to Jim Simpson of NJATC for his technical expertise and help on this book and to Richard Loyd for his technical expertise and guidance. Also, special thanks to Stephanie Kelly, her professional and coordination skills were a tremendous help and encouragement in this project . . . thanks, Stephanie, for the " Get 'er Done" approach as we say in the south.

ABOUT THE AUTHOR

David Carpenter is the owner of Integrity Company, a consulting company specializing in testing, design-build projects, inspection, codes, and standards. He serves as Chief Electrical Inspector with the City of Florence, AL. He has also worked as Engineer, Master and Journeyman Electrician for the Tennessee Valley Authority, Champion Paper, Inc., Occidental Chemical, Reynolds Aluminum, Amoco Chemicals, and as Project Manager for OMA, Inc. David holds a Ph.D. in Electrical Engineering.

David's professional certifications include I.A.E.I., S.B.C.C.I., and the U.S. Department of Labor as Inspector, Plan Reviewer, Master and Journeyman. He is the past chairman of the Alabama International Association of Electrical Inspectors and present Chairman of the Alabama Board of Electrical Contractors.

David has presented seminars for industry, government, educational institutions, and hospitals (e.g., NASA Arnold Engineering Development Center, TVA, Toshiba, Eastman Chemical, TOCCO Inc., University of Alabama Professional Development Engineering Department, University of Wisconsin, Professional Development, Wright State University, Clemson University, Southern Services and Alabama Power Company) and has performed research and testimony as a professional witness.

David is registered with the OSHA Training Institute as an instructor and has authored books on motors, motor controls, grounding and shielding of power systems and sensitive electronic equipment, instrumentation and process controls, hazardous locations, power quality, OSHA electrical standards-subpart S(<600 volts) & 269(High Voltage) and the NEC/ N.E.S.C.

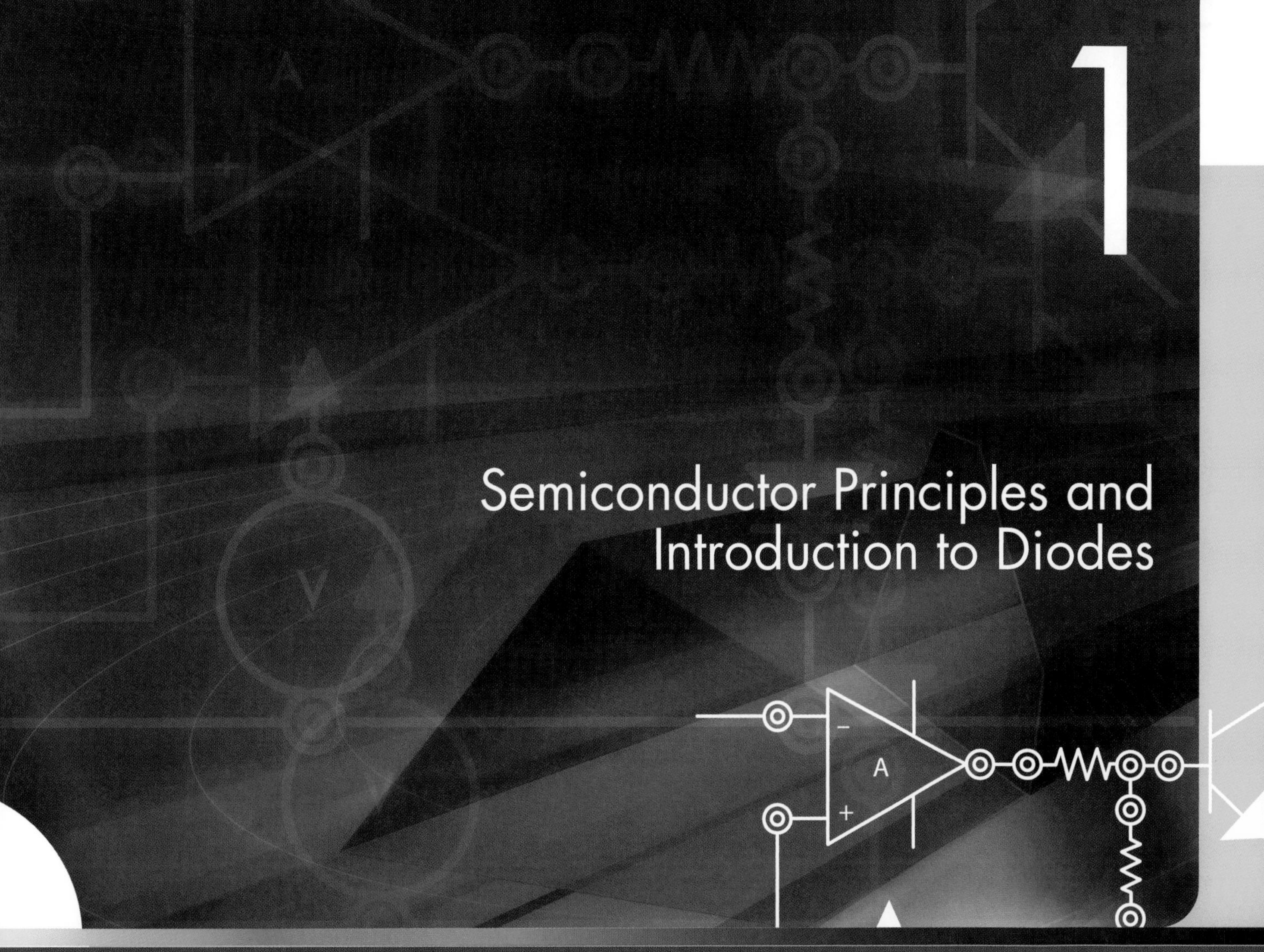

1 Semiconductor Principles and Introduction to Diodes

OUTLINE

OVERVIEW

Virtually all electrical and electronic circuits share the use of **passive circuit elements.** Prior to the 1960s, active devices such as transistors and vacuum tubes were used only in electronic circuitry. Since that time, improved semiconductors have found multiple applications in high-powered electric power circuits such as variable-frequency drives, uninterruptible power supply (UPS) systems, power conditioners, high-voltage DC transmission systems, and many others. In this chapter, you will learn the fundamental operating principles of semiconductors and will be introduced to the most basic of semiconductors—the diode.

OBJECTIVES

After completing this chapter, the student should be able to:

1. State in simple terms the difference between conventional and electron theories of current flow.
2. Explain the fundamental concepts of semiconductor materials.
3. Describe the basic operating principles of the PN junction.
4. Describe electron flow through a diode.
5. Identify diode symbols to indicate forward and reverse bias.
6. Interpret characteristic curves for semiconductor diodes.
7. Determine if diodes are operating properly.
8. Predict the output waveforms for diode circuits.
9. Calculate component and circuit values for current and voltage in circuits that have diodes.

Passive circuit element
An electric device that does not add energy to an electric circuit. Examples include resistors, inductors, and capacitors.

INTRODUCTION

1.1. Historical Background

The history of semiconductors is based on the work of Michael Faraday (1831) and James Clerk Maxwell (1864). These men, among others, showed that electric current could be produced by magnets in motion and that this electric current had electromagnetic wave properties.

In 1883, Thomas Edison discovered that electron flow can be generated between a hot filament and a positively charged plate when they are isolated in a vacuum chamber. In 1896, Guglielmo Marconi transmitted and received radio signals over a distance of two miles. These two events provided the foundation for modern-day tubes and later semiconductors.

1.2 Vacuum Tubes

Tubes have been used in radio and television for many years. **Amplitude modulation** (AM) radio was being used in the early 1900s, and **frequency modulation** (FM) was introduced in the 1930s. Electronic black-and-white television systems were developed by the early 1940s, and color TV was commercially available by the mid 1950s. High-power applications for tubes continued to dominate the communications field until full development of the transistor and integrated circuits.

Amplitude modulation
A system of attaching information to a single-frequency carrier wave. The information is included by varying the amplitude (magnitude) of the carrier wave.

Frequency modulation
A system of attaching information to a single-frequency carrier wave. The information is included by varying the frequency of the carrier wave.

1.3 Transistors

The invention of the transistor in 1948 permitted machinery and devices requiring electronic components to become smaller and more portable. The rows and rows of vacuum tubes and mechanical relays that once made computers room-size installations have now been replaced by transistors and integrated circuits, which have enabled some computers to be reduced in size to less than 3 inches by 5 inches. In addition, the use of semiconductors has dramatically reduced the power requirements and heat dissipation of electronic equipment. What used to require 240 V (volts) and 15 A (amperes, or amps) now takes less than 5 V and 250 mA (milliamperes).

KEY REVIEW ELEMENTS

1.4 Current Flow

The electrons in the outer ring of the atom are called valence electrons. The valence ring can contain a maximum of eight electrons. These electrons are more or less unstable and can be moved from atom to atom. The number of valence electrons determines how well the material conducts electric current.

There are two equally valid theories to describe electric current within a circuit. Understanding semiconductor operation requires appreciation of both theories.

- The **conventional theory of current flow** describes the movement of electrons from areas of positive charges to areas of negative charges. As electrons move to fill holes in one direction, they leave vacancies behind them.

Conventional theory of current flow
The theory of current flow in which current flows from a positive charge to a negative charge; in other words current flows from positive to negative through a circuit.

Electron theory of current flow The theory of current flow in which current flows from a negative charge to a positive charge; in other words current flows from negative to positive through a circuit.

Therefore, it appears that although electrons move in one direction holes move in the opposite direction. Many references can be found that utilize this concept.

- The **electron theory of current flow** defines electric current as movement of electrons from negative toward positive. This text utilizes the electron theory of current flow to describe the operation of circuits.

FieldNote!

An electrical malfunction was observed and reported by a plant operator one hot, sunny, summer afternoon. The technician was dispatched first thing the following morning only to find no evidence of the problem. The technician was called again that same afternoon because the problem had reappeared. It was obvious to the technician that the problem was intermittent and only occurs after the machinery has been run for a period of time. Stumped by what could be causing the problem, the technician decided to ask an experienced electrician his opinion. The experienced electrician explained to him that the problem was likely related to an electromechanical relay. A relay coil, after working through the cool morning hours may become too hot in the mid afternoon hours. When this occurs the coil resistance increases, reducing the current which in turn reduces the magnetic field needed to pull-in the armature contacts. An important concept to understand is; the resistance of most conductors, including copper coil windings rises as the temperature increases (**positive temperature coefficient**).

Positive temperature coefficient A condition whereby resistance, capacitance, length, or other physical characteristic of a material increases as its temperature rises.

1.5 Conductors, Semiconductors, and Insulators

Electrical Conductivity Principles

All materials can be classified as conductors, insulators, or semiconductors. Material classification depends on the ability of a particular material to conduct an electric current. The ability to conduct electricity depends on the number of free or relatively loose electrons in the material. All **semiconductor** devices—such as diodes, transistors, and integrated circuits (ICs)—are made of semiconductor materials.

Conductors

Conductors consist of materials whose atoms have only one or two electrons in the outer valence ring. These atoms (of copper, silver, gold, and so on) are good conductors because their outer electrons can be "pulled away" very easily. Electrical conductivity of conductors tends to go down as temperature goes up.

Insulators

Insulators are materials with very high resistance to current (electron flow). This means that their electrons strongly resist being "pulled away" from the outer valence ring of the atom. Insulators have almost full valence rings with seven or eight electrons. Wood, paper, and glass are good examples of insulators. Electrical conductivity of insulators tends to go up as temperature goes up.

Semiconductors

Semiconductors have four electrons in the outer valence ring. They are better conductors than insulators, but they do not conduct as well as conductors. The electrical conductivity of semiconductors reacts to temperature in the same way it does in insulators; that is, conductivity goes up as temperature goes up.

SEMICONDUCTOR CHARACTERISTICS

Semiconductor materials have characteristics of both insulators and conductors. Two of the most common semiconductors are silicon and germanium. These materials form crystals. In their pure (intrinsic) form, they do not have enough free electrons to be useful. Therefore, they are modified by adding (doping) an impurity to help the conducting process.

After the intrinsic crystal is doped, it is called an extrinsic semiconductor. **Doping material** comes in two types. The first has an excess of free electrons, and when it is added to a pure semiconductor crystal the resulting crystal is called an N-type semiconductor material. The second has a deficiency of free electrons (an excess of free holes), and when it is added to a pure semiconductor crystal the resulting crystal is called a P-type semiconductor. Figure 1–1 shows the resulting crystal effect with the different N-type and P-type doping materials.

Semiconductor
Any of various solid crystalline substances, such as germanium or silicon, having electric conductivity greater than insulators but less than good conductors. (Excerpted from the *American Heritage Talking Dictionary.* Copyright © 1997 The Learning Company, Inc. All rights reserved.)

Doping material
A material added to a semiconductor to cause either an N-type material (electron excess) or a P-type material (electron deficiency).

FIGURE 1–1 N-type and P-type materials.

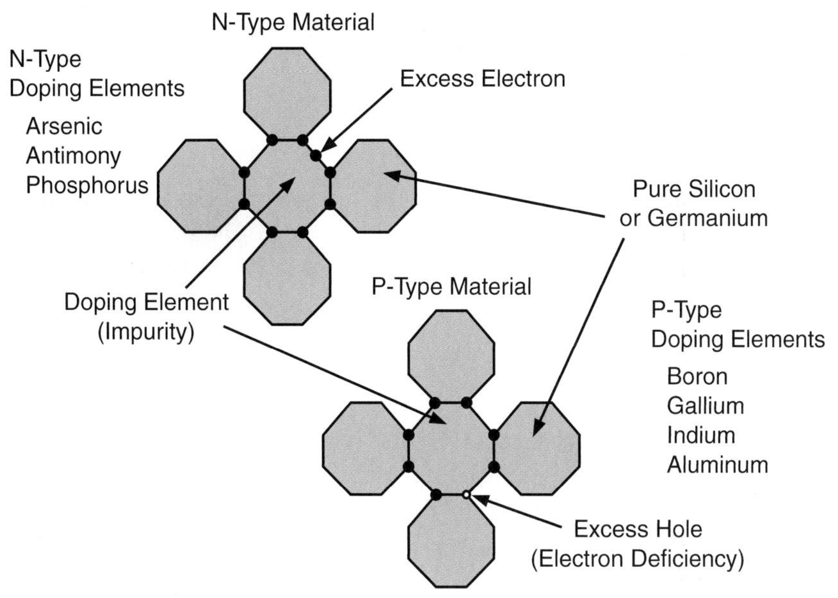

THE PN JUNCTION

1.6 Forming the PN Junction

As you can see from Figure 1–1, current will flow in either P-type (hole flow) or N-type (electron flow) material. When these two types of materials are joined, they form the enormously useful PN junction. The **PN junction** is vital to diode and semiconductor operation. During the manufacturing process of the PN junction, an interaction takes place between the two types of materials.

PN junction
The interface formed when a P-type material is conjoined with an N-type material.

Some of the excess electrons move into the P material and combine with the excess holes, and some of the excess holes move into the N material and combine with the excess electrons. This combination creates negative and positive ions—negative ions in the P material and positive ions in the N material. Figure 1–2 shows this PN junction effect. The area affected by the combining holes and electrons is called the depletion region or **depletion layer.**

Depletion layer
The layer that forms between the P and N material in a PN junction. Also called the depletion region.

FIGURE 1–2 The PN junctions.

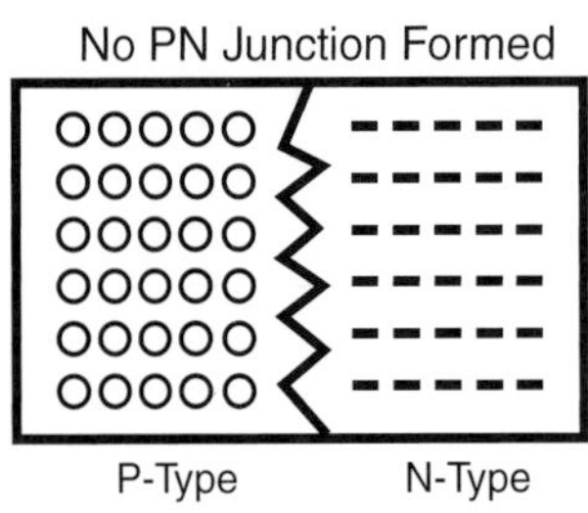

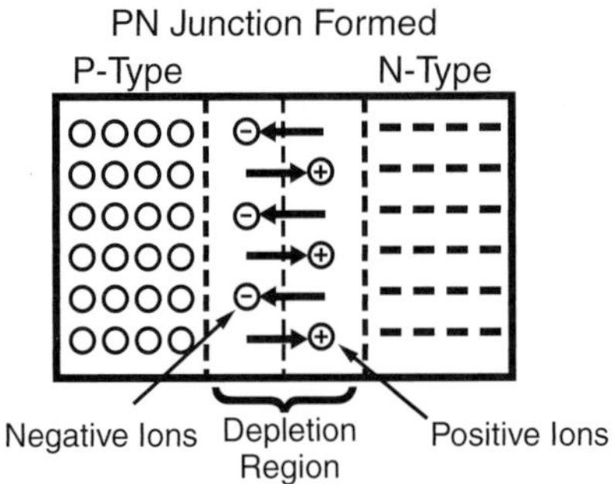

The negative and positive ions in the depletion region create an internal barrier voltage of 0.5 V to 0.7 V (silicon) and 0.25 V to 0.3 V (germanium) that prevents further current flow from the N-type or P-type material. For the sake of consistency, this text uses 0.7 V and 0.3 V as junction voltages. This internal voltage barrier is a constant that comes from the semiconductor material itself. This means that it is present in all PN junctions of diodes and transistors. As with any voltage, this barrier voltage has a polarity.

The negative ions (in the P material) are negatively charged and thus repel any further movement from the excess electrons in the N material. The positive ions (in the N material) are positively charged and thus repel any further movement from the excess holes in the P material. Figure 1–3 shows how this barrier voltage works.

FIGURE 1–3 The ion barrier voltage.

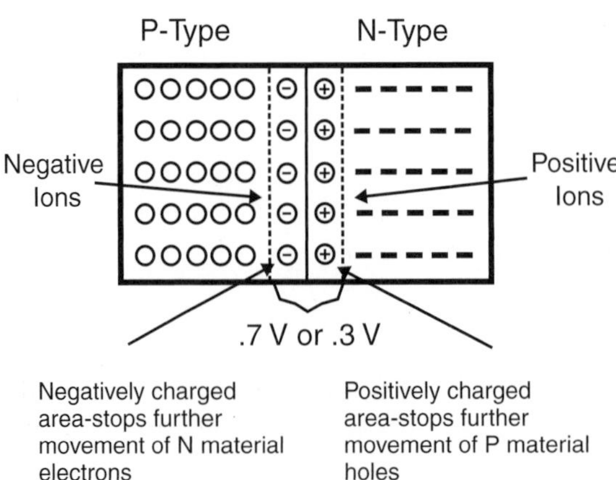

1.7 Biasing the PN Junction

The behavior of the PN junction can be changed by applying a **bias** voltage. The bias voltage changes the width of the PN junction's depletion region and therefore changes its resistance. Figure 1–4 shows the result of forward bias and reverse bias on the PN junction. The PN junction conducts when it is forward biased and does not conduct when it is reverse biased.

Bias
A current or voltage applied to a semiconductor device to obtain a specific result, such as conduction.

FIGURE 1–4 Biasing the PN junction.

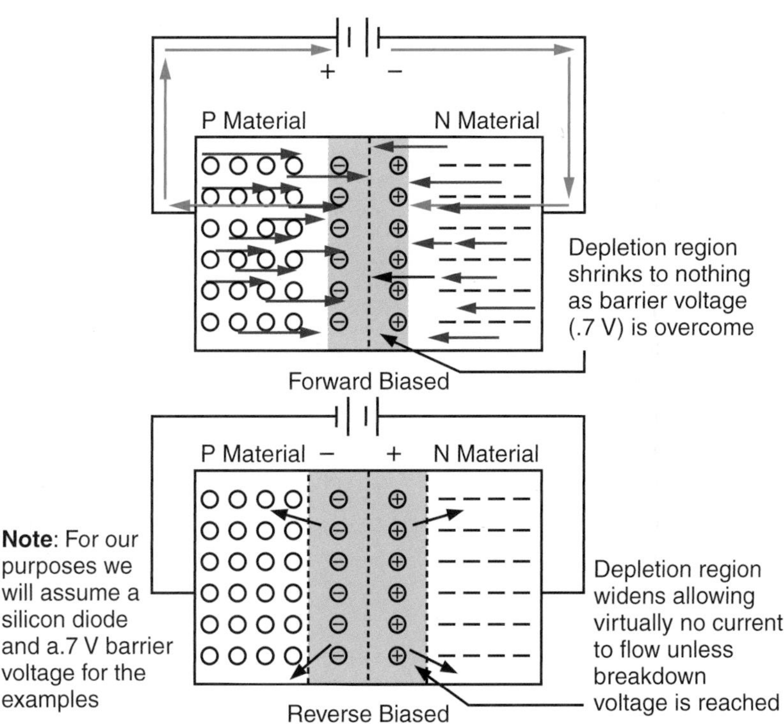

Forward Bias

A PN junction is forward biased by making the P material positive with respect to the N material. Depending on whether the PN junction is made of silicon (Si) or germanium (Ge), the typical amount of external voltage required to forward bias the junction is 0.7 V (Si) and 0.3 V (Ge). Power semiconductors may have greater forward voltage drops, depending on current magnitude. When the junction is forward biased, the depletion layer decreases and the PN junction conducts.

Reverse Bias

Reverse biasing occurs when the N material is made positive with respect to the P material. In this condition, the depletion layer widens and essentially no current flows.

1.8 Reverse Breakdown Voltage

Reverse breakdown voltage The reverse bias voltage required to cause a PN junction to fail.

Peak inverse voltage See *Reverse breakdown voltage.*

Avalanche voltage See *Reverse breakdown voltage.*

Excessive reverse bias will break down and possibly destroy the PN junction. When the voltage is large enough, the PN junction's resistance drops rapidly, a very high current develops, and the PN junction is destroyed. The voltage at which this happens is called **reverse breakdown voltage, peak inverse voltage,** or **avalanche voltage.**

There are certain types of semiconductor devices designed to operate in the reverse breakdown area. The Zener diode is the most common of these devices. The Zener diode is used to regulate voltage in a circuit, even when there is a widely varying current. You will learn more about these devices later on in your study. Normally, however, all semiconductor devices that have PN junctions operate at less than peak inverse or avalanche voltages. Figure 1–5 shows normal conduction and reverse breakdown (avalanche) voltage and current relationships during forward and reverse bias conditions.

FIGURE 1–5 PN junction voltage–current characteristic curve.

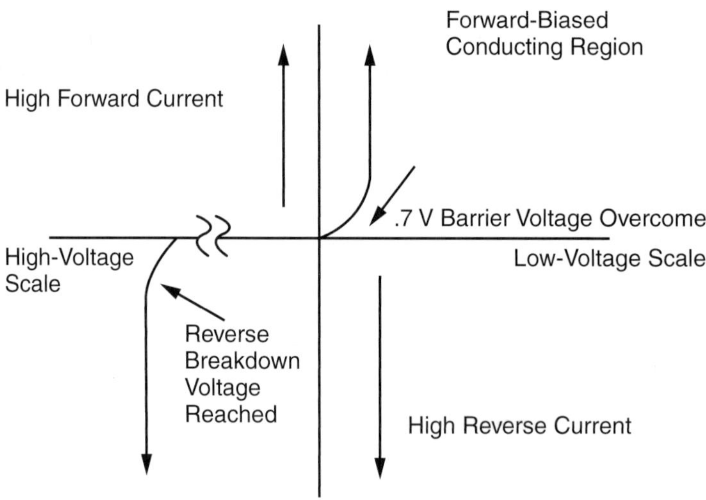

THE DIODE

1.9 Basic Principles

Diode A two-terminal semiconductor device that passes current of one polarity and blocks current of the opposite polarity.

A **diode** is a two-terminal (or two-lead) device that has a PN junction and acts as a one-way conductor. When forward biased, the diode conducts. When reverse biased, the diode conduction is so small that it is usually considered zero. There are other types of diodes (Zener, light-emitting, and so on). For now, however, when we use the term we are referring to a simple PN junction two-lead device.

1.10 Diode Construction and Characteristics

The Anode and the Cathode

Two examples of diodes and the schematic symbol are shown in Figure 1–6. The N material is called the cathode and the P material is called the anode. The diode will conduct when it is forward biased. Remember, electrons flow against the arrow (from the negative to the positive potential).

FIGURE 1–6 The diode: (a) schematic, (b) typical examples.

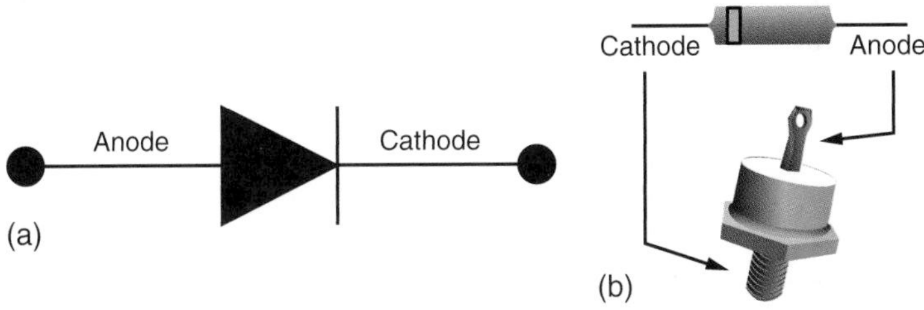

Forward and Reverse Biasing

Figure 1–7 shows a diode with forward bias connections in a simple series circuit. Note that the circuit is drawn in three different ways. The first way shows the diode connection with respect to the negative and positive terminals of the voltage supply. The second and third ways show the diode connection with respect to a voltage potential and ground. Note that in each case the anode is positive with respect to the cathode. As your career progresses, you will probably see all of these types of circuit connections.

FIGURE 1–7 A forward-biased diode allows current flow.

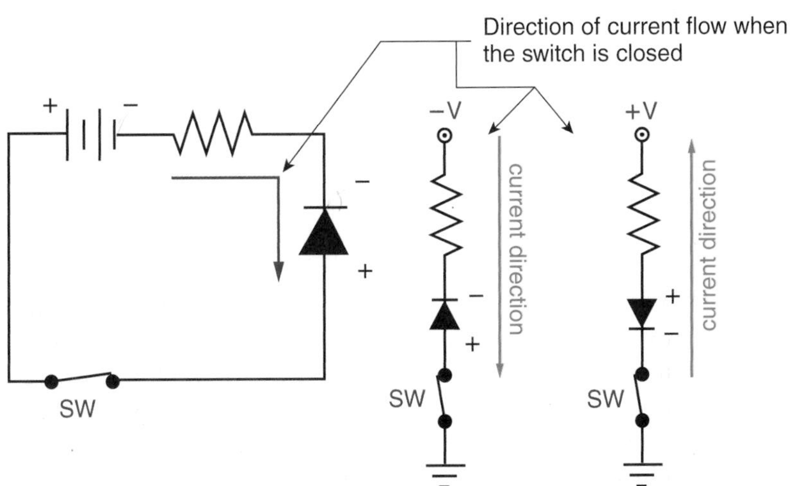

Figure 1–8 shows a diode with reverse bias connections in a simple series circuit. Again, note that the circuit is drawn in three different ways. The first way shows the diode connection with respect to the negative and positive terminals of the voltage supply. The second and third ways show the diode connection with respect to a voltage potential and ground. In all three diagrams, the anode is negative with respect to the cathode and therefore the diode will not conduct.

FIGURE 1–8 A reverse biased diode blocks current flow.

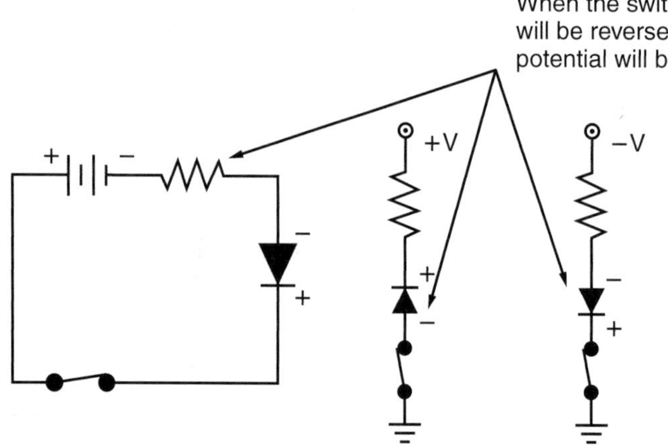

Rating Characteristics

Diodes have several characteristics that must be considered when using them in circuits. The most common are outlined in Table 1–1. Diodes have other characteristics. However, these three are among the most important.

TABLE 1–1 Diode Characteristics (Ratings)

Characteristic (Rating)	Description
Peak Inverse Voltage (PIV)	The maximum recurrent (or repetitive) reverse voltage the diode is rated to withstand. Exceeding this voltage will destroy the diode. Sometimes called the peak inverse voltage or the avalanche voltage.
Average forward current (I_O)	Maximum allowed DC forward current. Exceeding this current will destroy the diode.
Forward power dissipation ($P_{D(max)}$)	Maximum power that can be dissipated by the diode when it is forward biased. Exceeding this power limit will destroy the diode.

Diode Characteristic Curves

All diodes also have characteristic curves. These are plotted on charts to show how the diode will react at different temperatures and with different circuit voltages and currents applied. Figure 1–9 shows examples of typical diode characteristics.

FIGURE 1–9 Diode characteristic curves.

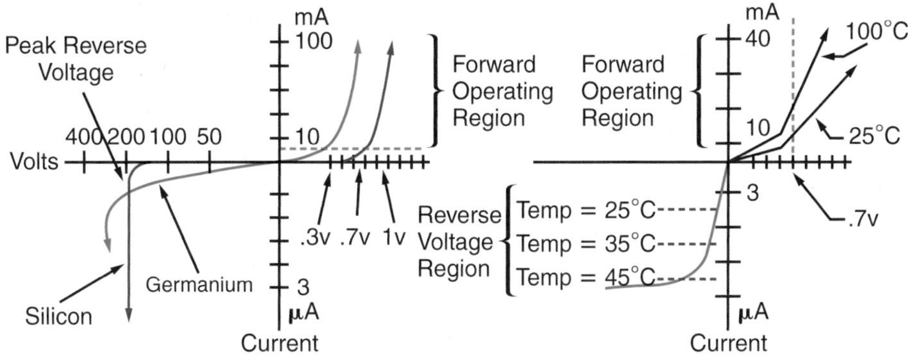

TechTip!

The steps that follow employ a DMM to test a diode.

1. Although it is often possible to test diodes while in-circuit, it is safer and more reliable (with power off and capacitors discharged) to remove the suspect diode from the circuit before testing.
2. Set the DMM function switch to its Diode Test position (diode symbol).
3. Connect the positive DMM lead to the diode anode lead, and negative DMM lead to the diode cathode lead. Expect the DMM to read at the barrier voltage. If the DMM reads lower, the diode is shorted. If the DMM reads higher, the diode is open. Figure 1–10a shows a DMM connected to test a PN forward junction.
4. Reverse the test leads. Expect the DMM to read "open," just as with unconnected test leads. Figure 1–10b illustrates connection of a DMM to measure a reverse junction.

1.11 Diode Testing

When silicon and germanium diodes fail, they usually fail in one of two ways (or modes).

- *Open failure mode:* When forward bias is exceeded, current no longer flows. This often happens when I_{output} is greatly exceeded. The diode may appear to be burned or cracked.
- *Short failure mode:* When reverse breakdown voltage is found to have decreased dramatically. The same physical symptoms as for open failure mode may be present.

The characteristics of silicon and germanium diodes may be determined with a semiconductor curve tracer. However, most modern digital multimeters (DMM) will detect these two failure modes with their diode testing function. A typical DMM is shown in Figures 1–10a and 1–10b.

FIGURE 1–10 (a) Testing a PN junction (forward biased), (b) Testing a PN junction (reverse biased).

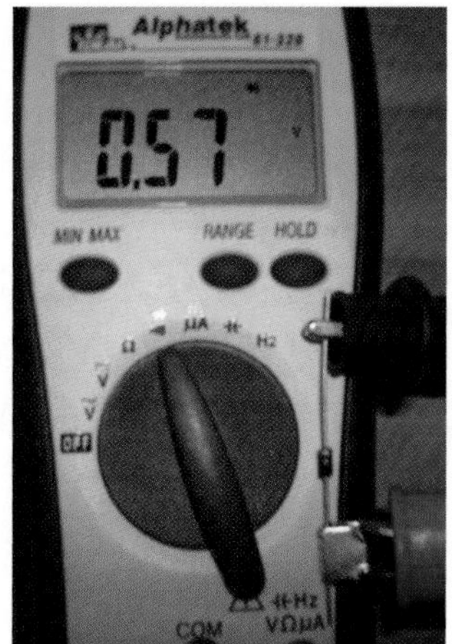

(a)

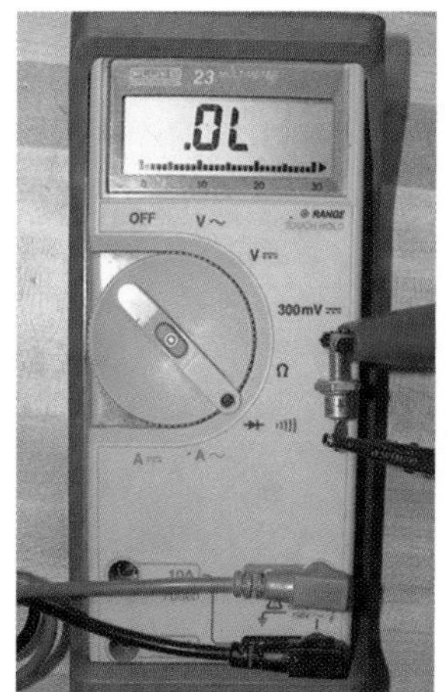

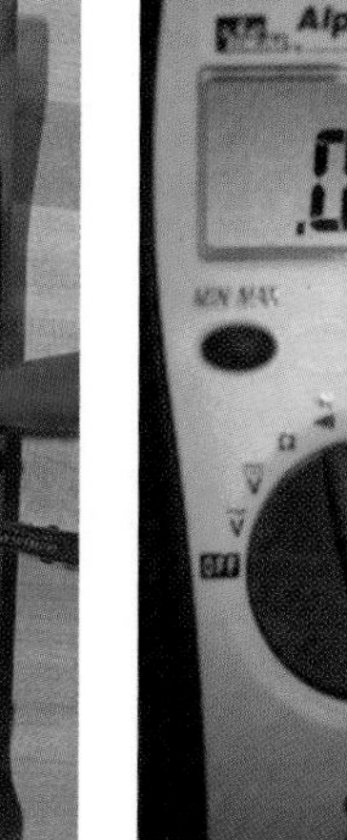

(b)

DIODE CIRCUIT ANALYSIS

Diode circuit analysis is more difficult than it might appear at first. Remember that many of the circuit analysis tools you have used previously depend on the circuit being bilateral; that is, all of the elements must work the same both ways. Clearly, diodes do not behave in this way. Therefore, tools such as superposition and Kirchhoff's laws must be used very carefully.

1.12 Calculating DC Voltage and Current

The following illustrates the effect of the 0.7-V barrier voltage in a simple series circuit. Referring to Figure 1–11, calculate the voltage drops for R_1 and D_1 (silicon) and the total circuit current. Using Ohm's law:

$$V_T = V_{D1} + V_{R1}$$

Where

$$V_{D1} = 0.7 \text{ volts (silicon)}$$

$$V_{R1} = V_T - V_{D1} = 12 - 0.7 = 11.3 \text{ V}$$

Ohm's law:

$$I_T = \frac{V_{R1}}{R_1} = \frac{11.3 \text{ V}}{1{,}200\ \Omega} = 9.42 \text{ mA}$$

NOTE: This same example can be calculated using a germanium diode with a barrier voltage of 0.3 V.

FIGURE 1–11 Simple DC diode circuit.

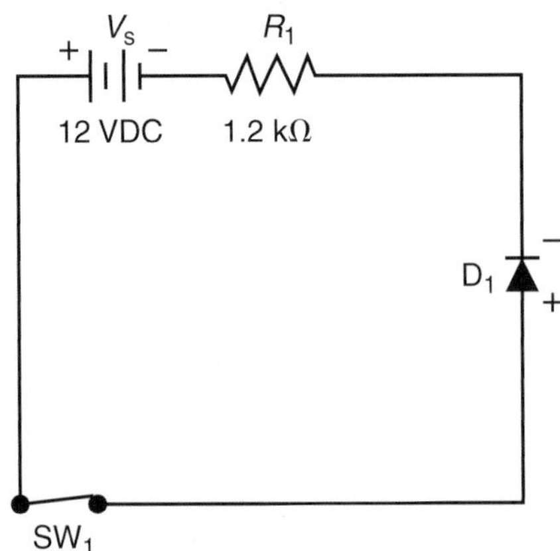

1.13 Diodes and AC Circuit Analysis

As you have seen, the diode is polarity sensitive. When forward biased, the cathode is negative, and the anode is positive. Likewise, when reverese biased, the cathode is positive and the anode is negative. This polarity can be used in an AC circuit to **rectify** the output signal of the circuit. This happens because a diode placed in a circuit with an AC power supply will conduct (be forward biased) only half the time. Figure 1–12 shows a diode connected to an AC power source in a series circuit and the resulting output waveform.

Rectify
To change alternating current (AC) to direct current (DC).

FIGURE 1–12 Diode used to rectify AC.

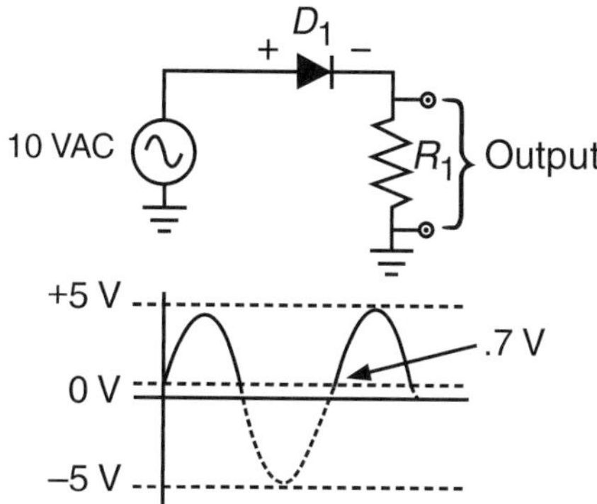

Practical Applications

A practical application of this rectifying effect can be found in meters. In some meters, diodes are placed in series to allow current in only one direction (see Figure 1–13). In other meters, diodes are placed in parallel (shunted) to the meter movement. This limits the current through the movement. (See Figure 1–14 for an example of this application.)

FIGURE 1–13 Diode and meter movement in series.

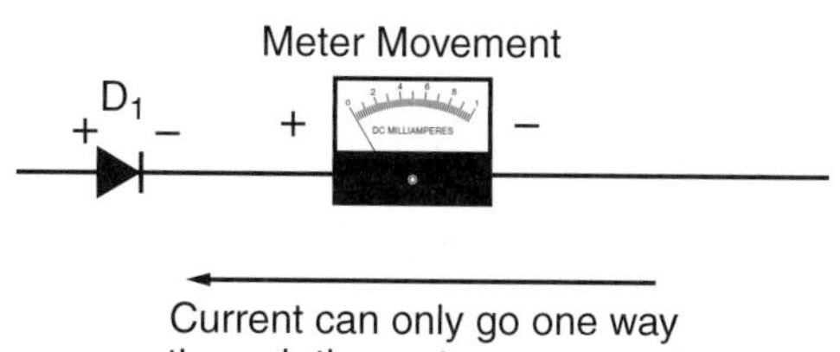

FIGURE 1–14 Diodes and meter movement in parallel.

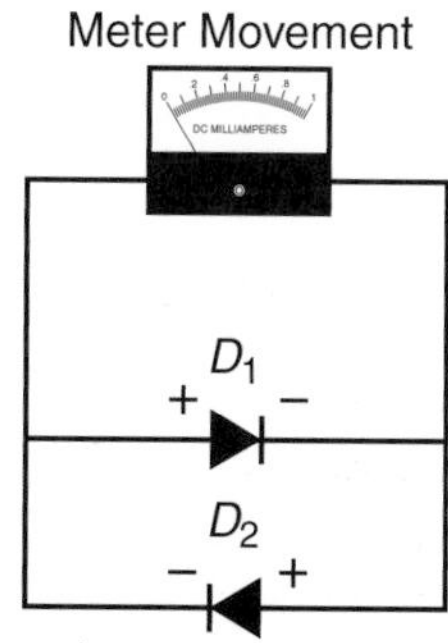

Figure 1–15 shows a half-wave rectifier being used as a voltage reduction circuit. With the diode in the circuit (switch up), only half the voltage waveform is let through. Although the current has been rectified, it still has some reduced magnitude. When the switch is thrown down, the load receives the entire waveform and thus the full system voltage. You can combine diodes to vary the circuit's output. Refer to Figure 1–16 for the following example.

FIGURE 1–15 Half-wave rectifier application.

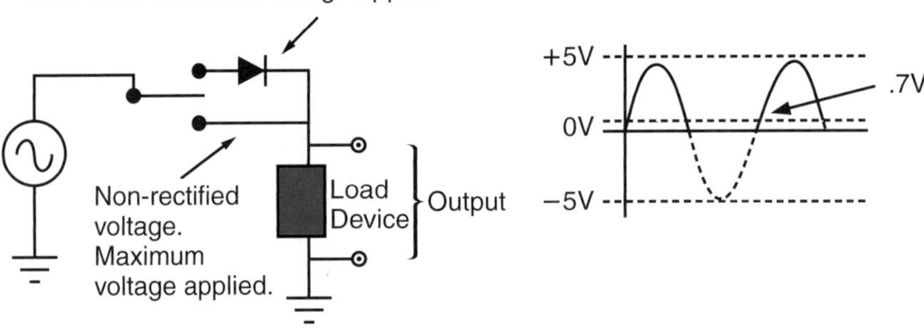

FIGURE 1–16 Multiple diode circuit.

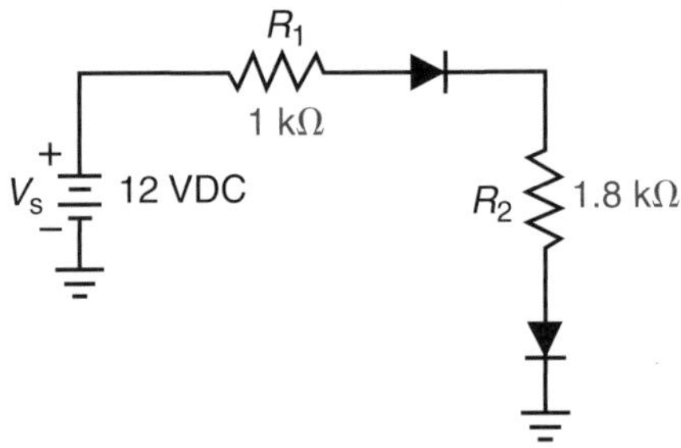

Example

Calculate the voltage drops across R_1 and R_2. Calculate I_T for the circuit. Using Ohm's law:

$$I_T = \frac{[V_s - (0.7\text{ V} + 0.7\text{ V})]}{R_T} = \frac{(12\text{ V} - 1.4\text{ V})}{2{,}800\ \Omega} = 3.79\text{ mA}$$

and

$$V_{R1} = I_T \times R_1 = 3.79\text{ mA} \times 1.0\text{ k}\Omega = 3.79\text{ V}$$

$$V_{R2} = I_T \times R_2 = 3.79\text{ mA} \times 1.8\text{ k}\Omega = 6.82\text{ V}$$

When checking to see if your calculations are correct, add all voltage drops. Their sum should equal the voltage applied.

$$V_T = 6.82\text{ V} + 3.79\text{ V} + 0.7\text{ V} + 0.7\text{ V} = 12.0\text{ V}$$

The following example uses multiple power sources in a circuit with diodes. Refer to Figure 1–17 when calculating the following example.

Example

Calculate the total circuit current and voltage drop across R_1.

$$I_T = \frac{V_{R1}}{R_1} = \frac{12\text{ V} - 0.7\text{ V} - 4\text{ V} - 0.7\text{ V}}{2{,}200\ \Omega} = 3.0\text{ mA}$$

$$V_{R1} = I_T \times R_1 = 3.0\text{ mA} \times 2.2\text{ k}\Omega = 6.6\text{ V}$$

FIGURE 1–17 Diode and multiple power sources.

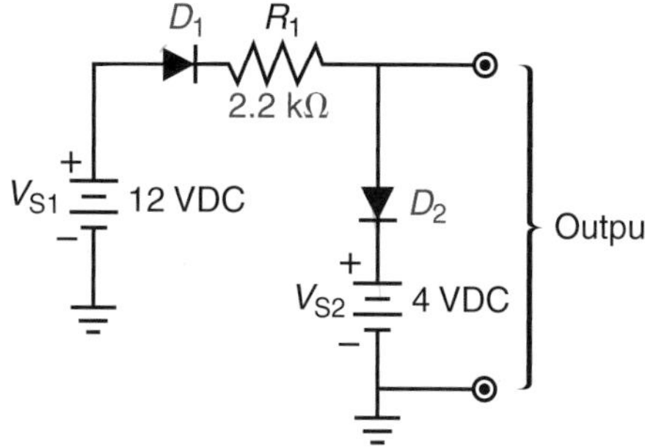

ELECTROSTATIC DISCHARGE

Electrostatic discharge (ESD) is the sudden, rapid, momentary movement of charges from one object to another. Lightning is an excellent example of ESD on a gigantic scale. At the other extreme, a small spark of less than 400 V is possible in an air environment. ESD is a serious threat to semiconductors. When working with semiconductors, the technician can take some simple precautions to avoid harming these devices.

1.14 The Electrostatic Discharge Threat

ESD can destroy or damage semiconductors. Damage can occur in several ways. Instantaneous ESD current across a PN junction causes heating which may be sufficient to literally blow a hole through the junction. The consequence of this will be a change in the semiconductor operating characteristics, or a short circuit. Another common result of ESD current is an open-circuit condition which occurs when a pin terminal separates from the semiconductor, or its weld is weakened. So, ESD damage may produce a wide range of undesirable semiconductor performance, some of which may not be obvious without detailed analysis. Some semiconductors, though damaged, may seem to function as expected, until well after an ESD event has occurred. These devices are said to have latent damage, and may be called "Walking Wounded." We associate static discharge with the finger-to-doorknob spark and shock. Semiconductor damage can, however, occur with as little as a few volts of ESD exposure; a spark need not occur. Semiconductor manufacturers may label their products as ESDS, or **E**lectro**S**tatic **D**ischarge **S**ensitive components.

1.15 Causes of Electrostatic Discharge

Some dissimilar materials, when separated (or as the result of friction) may develop large static charges. Persons or tools in contact with one of these materials represent a significant threat to the health of semiconductors. Charges transferred by induction, from one object to another, are also a hazard to nearby semiconductors.

1.16 Electrostatic Discharge Protection

Many substances are especially dangerous when in close proximity to ESDS components. Among these materials are styrofoam, wool clothing, carpeting, paper, and most plastics. Even most antistatic materials are only coated, and therefore over time lose those protective properties. Some practices adopted by the electronics industry to avoid ESD damage include the following.

- ESD bags, when properly employed, allow for safe transport of electronic components.
- ESD workstation mats, when properly grounded, keep components at ground potential.
- ESD wrist and heel straps, when properly worn and grounded, prevent static charges from accumulating on individuals in sensitive work areas.
- Proper grounding of workbench equipment such as soldering irons prevent static charges from damaging semiconductor components.
- Humidity control limits the build-up of static charges in an enclosed area.

Figure 1–18 shows an ESD wrist strap and an ESD-protected temperature-controlled solder station. Figure 1–19 shows an electronics workbench covered with an ESD-protective mat, a protective floor mat, wrist strap, and monitor conveniently mounted beneath the benchtop. Note the building ground point, behind the bench, where the mats and monitor may be easily wired. Figure 1–20 shows common ESDS device packaging and symbols.

FIGURE 1–18 Grounded soldering iron and wrist strap.

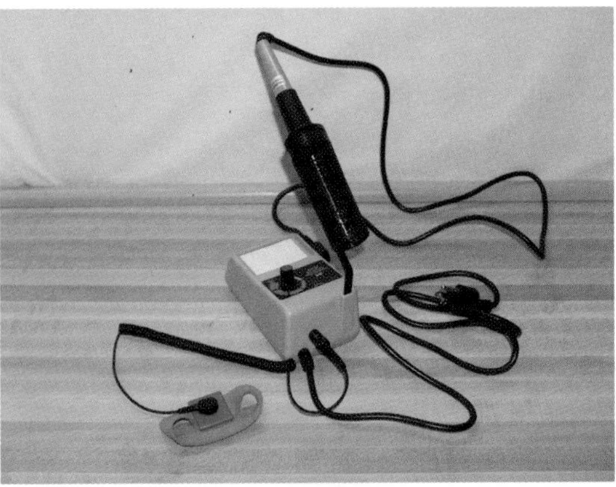

FIGURE 1–19 Typical ESD-protective workbench.

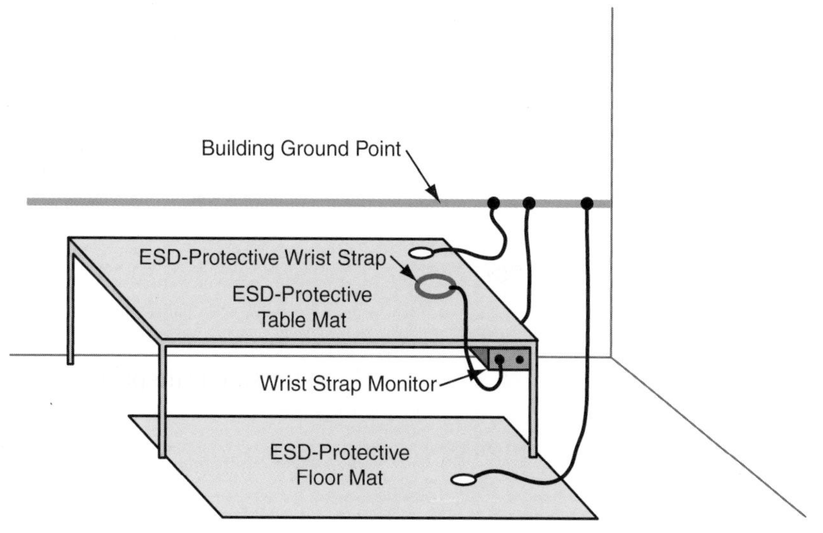

FIGURE 1–20 ESDS packaging and symbols.

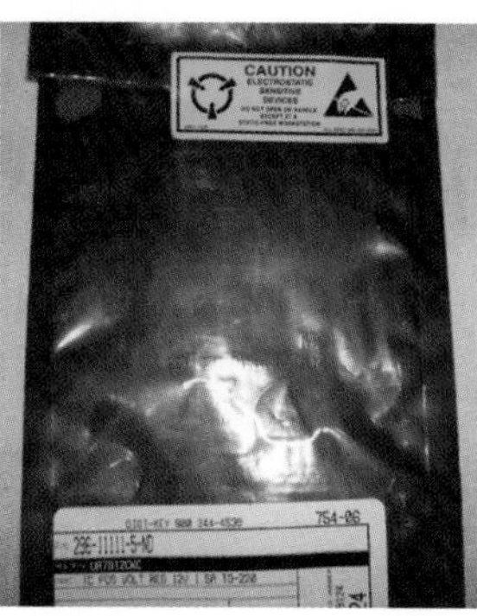

SUMMARY

Semiconductor devices such as diodes and transistors utilize the elements Silicon (Si) and Germanium (Ge). These elements are usually doped to achieve specific conduction characteristics. Doping creates P-type or N-type materials.

Joining P-type and N-type materials creates a barrier (or depletion) region at the junction, across which a barrier voltage exists. Barrier (or junction) voltage depends upon the semiconductor materials and doping used. Silicon junction voltage ranges from 0.5 V to 0.7 V; Germanium junction voltage ranges from 0.25 V to 0.3 V. Devices of this nature are called diodes.

Current can be made to flow from N-type to P-type material if the barrier voltage is exceeded. This biasing condition is called forward-biasing. Current cannot flow from P-type to N-type material unless the reverse breakdown voltage is reached. This biasing condition is called Reverse-Biasing. These characteristics enable the diode to convert, or rectify ac current to dc current.

Diodes are most often field-tested using the common digital multimeter.

Electrostatic discharge (ESD) can damage or destroy semiconductors. Electrostatic discharge sensitive (ESDS) devices, including diodes, are often (though not always) labeled. Protective measures should be employed to help assure the safety of ESDS components. These measures include properly grounded wriststraps, table mats, and test equipment.

REVIEW QUESTIONS

1. Discuss the depletion layer in a PN junction.
   a. What is its origin?
   b. How does it affect the operation of the junction?
2. What are three critical characteristics of diodes that must be considered to avoid destroying the diode?
3. Describe the characteristics of a P-type material and an N-type material.
4. Draw a diode circuit with a resistor, diode, and battery.
   a. Forward biased
   b. Reverse biased
5. What are the forward voltage drops for silicon and germanium PN junctions?
6. Examine Figure 1–21. What is the output voltage?
7. If you wish to analyze the circuit shown in Figure 1–21 using the superposition theorem, you must assume a small current flow through the diodes when they are reverse biased. Why?
8. The load device shown in Figure 1–22 sees less voltage when the switch is up than when it is down. Why?

FIGURE 1–21 Diode and multiple power sources.

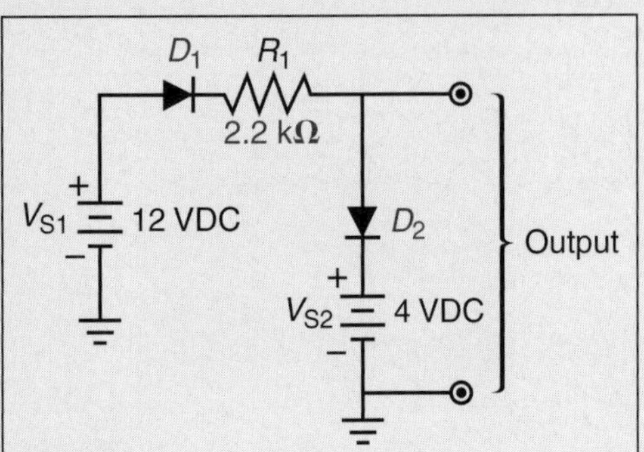

FIGURE 1–22 Half-wave rectifier application.

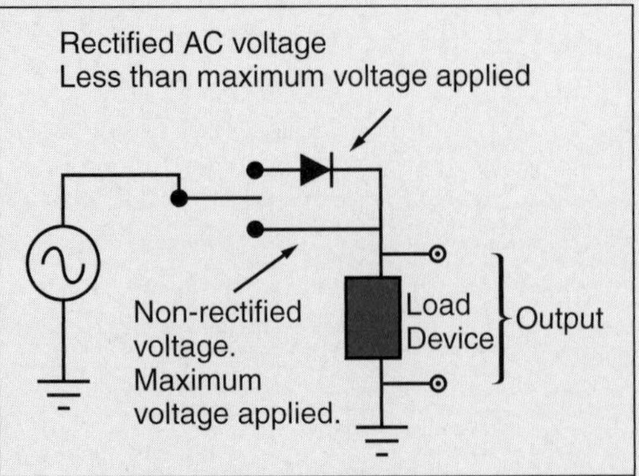

9. Examine Figure 1–23. How much current will flow if the battery voltage is 0.5 V and the diode is a silicon diode?

10. If a technician repaired a piece of equipment with ESDS components without using ESD-protective measures, could that equipment contain "walking wounded" (damaged) components?

FIGURE 1–23 A forward-biased diode allows current flow.

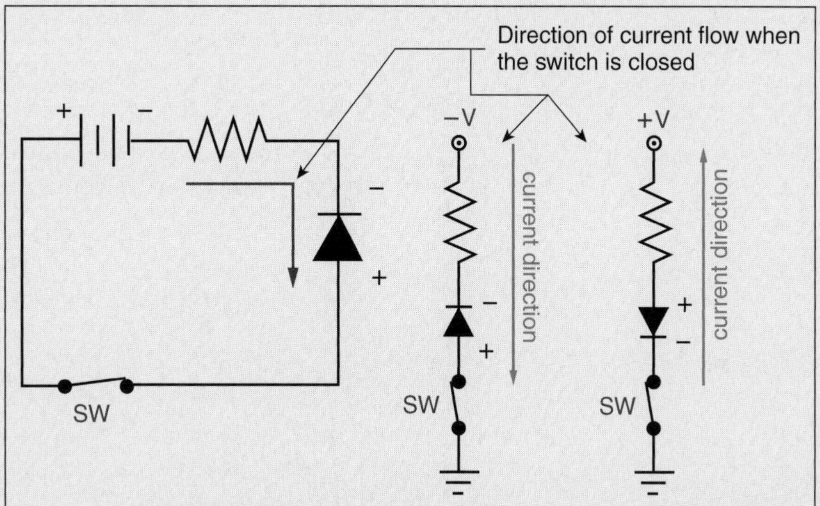

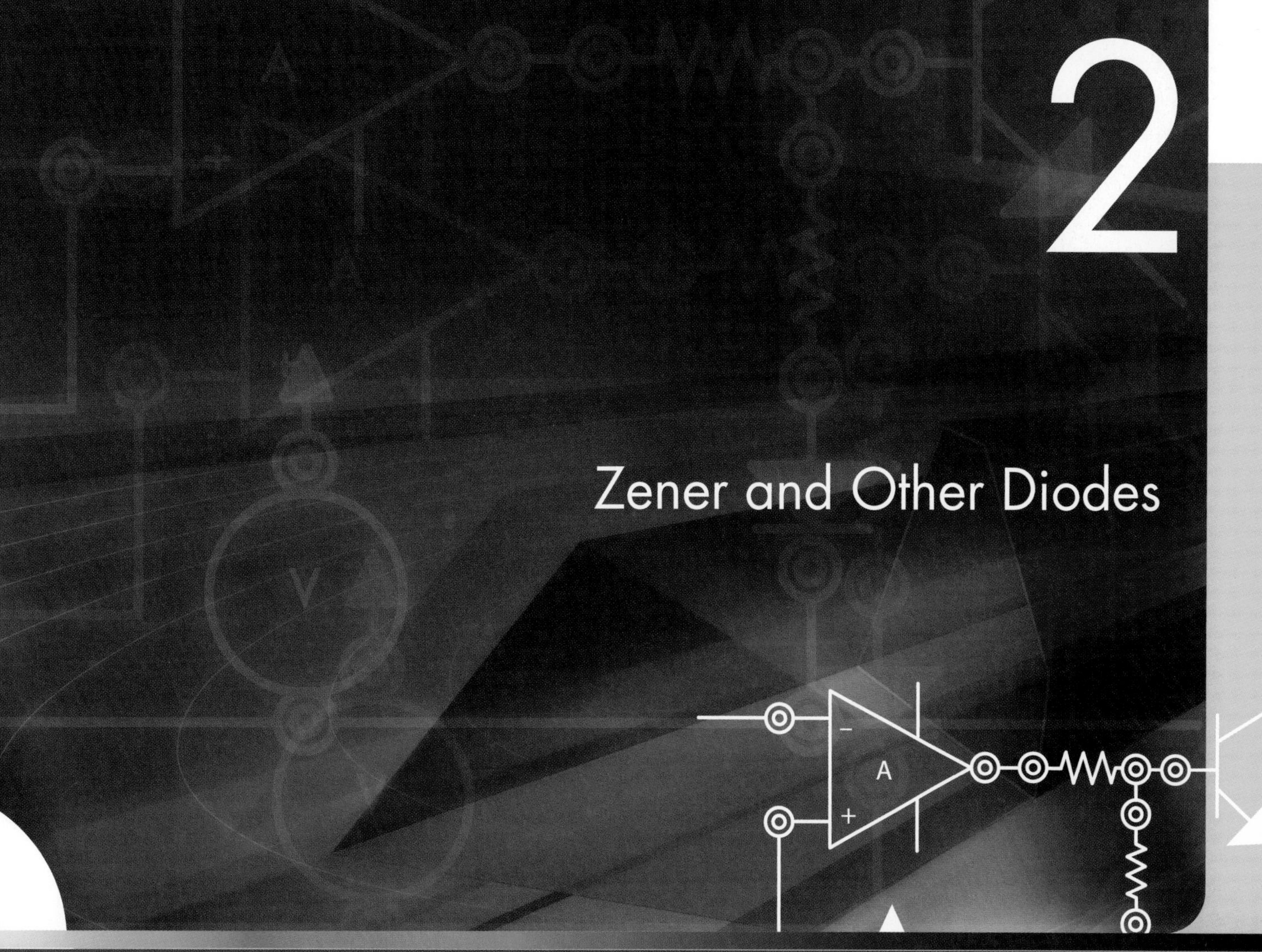

2 Zener and Other Diodes

OUTLINE

OVERVIEW

There are many variations of the simple diode. In this chapter, you will learn about five types. Some of these types are used primarily in electronic circuitry, whereas others (such as the Zener diode) are used in both electronic and power applications. The information in this chapter builds on your previous training and serves as a background for subsequent chapters.

OBJECTIVES

After completing this chapter, the student should be able to:

1. Describe the operation of Zener diodes.
2. Explain current flow in a Zener diode voltage regulator circuit.
3. Draw a characteristic curve for a hypothetical Zener diode.
4. Draw a schematic diagram of a simple Zener voltage regulator, and label all components and expected voltage drops.
5. Calculate component and circuit values for current, voltage, and power in circuits that have Zener diodes.
6. Draw schematic symbols for Varactor, tunnel, PIN, and Schottky diodes.
7. Give application examples for Zener, Schottky, tunnel, PIN, and Varactor diodes.

INTRODUCTION

The Zener diode is a special and very useful type of diode and is one of the more prevalent and useful forms of diode made today. Zener diodes are unique in that they normally operate in the reverse bias mode. When forward biased, Zener diodes operate as normal semiconductor (rectifier) diodes. Figure 2–1a shows the schematic symbol for a Zener diode. Figure 2–1b shows low- and high-power Zener diodes.

FIGURE 2–1 Zener diode. (a) Schematic symbol. (b) Typical examples.

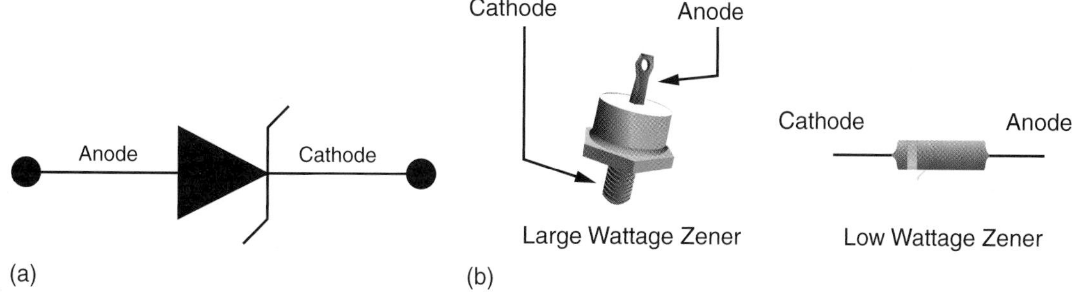

2.1 Zener Diode Characteristics

Figure 2–2 shows a typical characteristic curve for a 15 volt (V) Zener diode. As you can see, the Zener diode voltage–current (VI) curve is somewhat different than for diodes you studied previously. When the reverse bias voltage reaches a level called the **Zener voltage**, current flows in the diode and the voltage drop remains essentially constant. This characteristic makes Zener diodes particularly useful in circuits such as voltage stabilizers or regulators.

Zener voltage
The Zener diode's rated reverse breakdown voltage. The regulating voltage of a zener diode.

FIGURE 2–2 Zener diode characteristic curve.

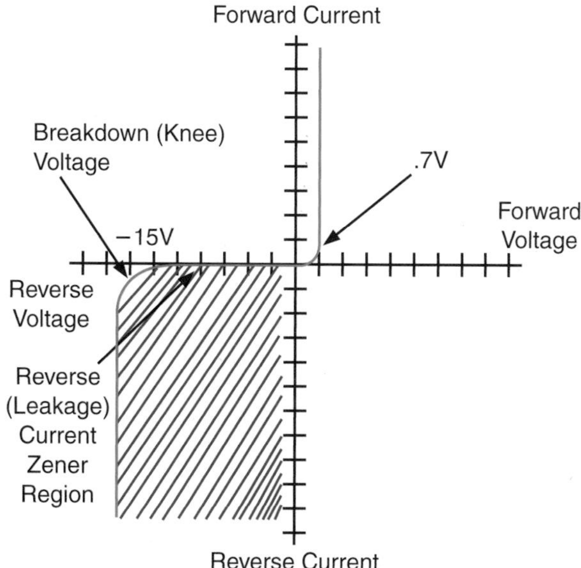

Zener region
Voltages equal to, or greater than the Zener voltage.

The area to the left of the breakdown or Zener voltage is called the **Zener region**. Depending on the specific diode, Zener voltages range from 1.8 V at ¼ watt to 200 V at 50 watts. The wattage ratings refer to the maximum power dissipation allowed for the Zener before it will be damaged. Zener diodes rated at greater than 1 watt usually come in stud-mount packages as shown in Figure 2–1b. Heat sinks can be used to dissipate heat and increase the power capability of the diode. The most common Zener values are 3.3 V to 75 V.

Note that as the reverse voltage increases the (reverse) leakage current remains almost constant until the breakdown voltage is reached. The current then increases dramatically. (Recall that this is called the *avalanche point.*) This breakdown voltage is the Zener voltage for Zener diodes. In practical terms, before the Zener (breakdown) voltage is reached the Zener diode performs like an open circuit. Although it is imperative for the normal PN junction diode or rectifier to operate below this voltage to prevent diode damage, the Zener diode is intended to operate at this voltage.

Knee voltage
The voltage at which current (other than leakage) begins to flow across a diode junction.

The V_{BR} (breakdown voltage) is a manufactured characteristic of the Zener diode. It is also called the **knee voltage.** Zener voltage is usually indicated as V_Z. Other critical characteristics are outlined in Table 2-1 (see also Figure 2–3). The tolerance of Zener diodes is typically 5%.

TABLE 2–1 Zener Diode Characteristics

Rating	Description
I_{ZM}	The maximum Zener current allowed before damage will occur.
I_{ZT}	The Zener current at which the diode was tested to determine its rating. For example, a 6.2 V Zener diode has its voltage (6.2 V) tested at the factory at a value of 10 milliamperes (mA).
$P_{D(max)}$	The maximum power dissipation allowed before damage will occur.

FIGURE 2–3 Zener diode current region.

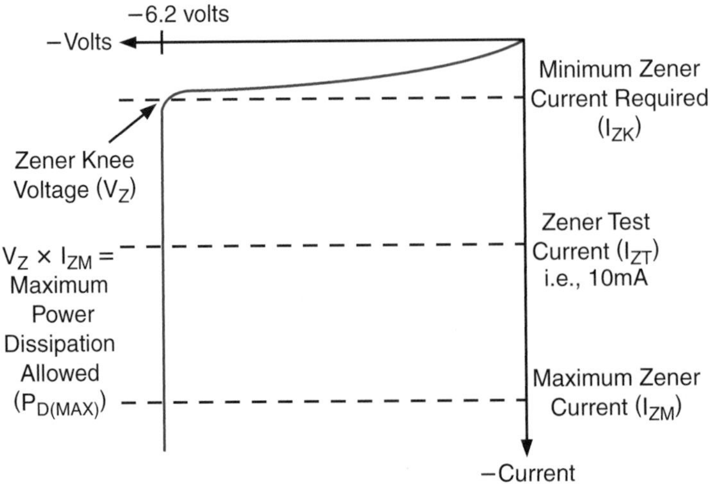

CIRCUIT ANALYSIS

Diodes are not bilateral devices. This means that you must be very careful in trying to analyze such circuits using standard methods. Methods such as superposition, Kirchhoff's voltage law, and others will work reliably only if the system is evaluated for current flow in the forward direction or for voltages across the diode in excess of V_Z.

2.2 Zener Diode Circuit Operation

Figure 2–4 shows a photograph of an oscilloscope waveform showing voltage taken across a 6.2 V grounded-anode Zener diode. Voltage applied through a current-limiting resistor R_s varies from −30 V at the left side of the trace to +30 V at the right side of the trace. The vertical scale on the oscilloscope is 2 V per division. Note that maximum voltage in the forward direction is approximately −0.7 V, whereas when reverse biased Zener voltage levels off at +6.2 V (as explained in material following).

FIGURE 2–4 (a) Oscilloscope photo of a 6.2 V Zener operation. (b, c) Operation of a typical Zener circuit.

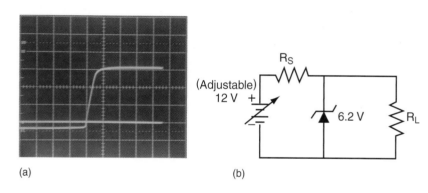

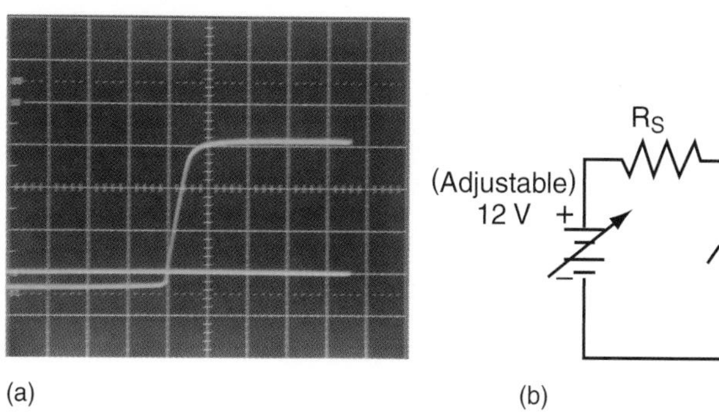

RS
(Adjustable)
12 V
6.2 V
RL

(a)

(b)

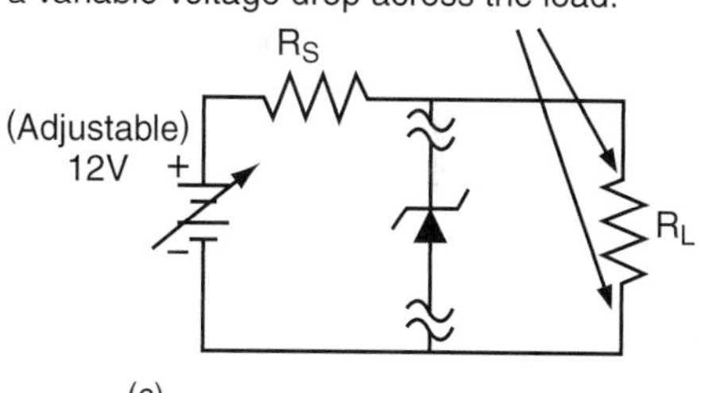

Until 6.2 volts is reached by the DC source, the Zener will act like an open in the circuit supplying a variable voltage drop across the load.
RS
(Adjustable)
12V
RL

(c)

Figure 2–4b shows a simple Zener diode circuit with a variable DC power supply. Note that the Zener is in parallel with the load and reverse biased. This is the normal configuration for the Zener diode. Voltage regulation performed by the Zener operates per the following parameters (see Figure 2–4c).

- As the DC voltage is increased from 0 V, the Zener will act like an open circuit.
- When the Zener voltage reaches 6.2 V or greater, the Zener avalanches and draws current. It also maintains a constant 6.2 V across the load.
- Operation of this circuit is highly dependent on the values of R_S and R_L.

For the following explanation of this process, refer to Figure 2–5. Start by calculating the current flow through the Zener diode with switch SW_2 open. This is a simple application of Kirchhoff's voltage law (KVL). KVL states: **The algebraic sum ($\sum$) of voltages around a series circuit is equal to zero volts.** Therefore the mathematical equation for this theory is: $V_{Source} + V_{RS} + V_Z = 0$.

FIGURE 2–5 Zener circuit analysis. (a) KVL example, (b) 1.2kΩ load, (c) 2.2kΩ load, (d) 4.4kΩ load.

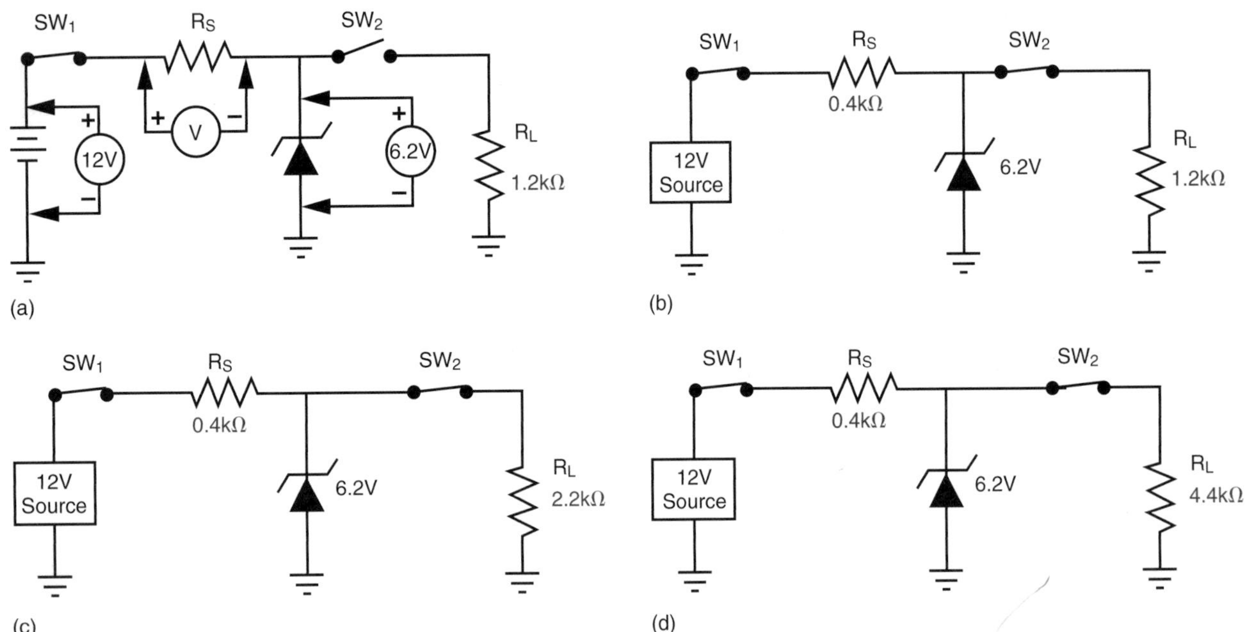

To test this theory refer to Figure 2-5a, trace the path through the circuit in a clockwise direction starting with the source voltage and record the voltage value of each component, pay close attention of the polarity as it enters each of the voltmeters in the circuit. Record each value as you trace the circuit. After tracing the circuit the equation will look like this:

$$-12\text{ V} + V_{RS} + 6.2\text{ V} = 0\text{ V}$$

Always record voltages across each component (either clockwise or counterclockwise, but *be consistent* with your choice).
This equation could also be expressed as:

$$12\text{ V} - (V_{RS} + 6.2\text{ V}) = 0\text{ V}$$

Another way to express this concept is to state: **The sum of voltages around a series circuit is equal to the applied voltage.** Therefore:

$$V_{Source} = V_{RS} + V_Z$$

$$12\text{ V} = V_{RS} + 6.2\text{ V}$$

To find the value of V_{RS} the formula can be rearranged:

$$V_{RS} = 12\text{ V} - 6.2\text{ V}$$

$$V_{RS} = 5.8\text{ V}$$

Now that you know the value of V_{RS} the final (original) equation will look like this:

$$-12\text{ V} + 5.8\text{ V} + 6.2\text{ V} = 0$$

Or:

$$12\text{ V} - (5.8\text{ V} + 6.2\text{ V}) = 0$$

Using Ohm's law, the current through R_2 can be calculated as follows:

$$I_{RS} = \frac{V_{RS}}{R_S}$$

$$I_{RS} = \frac{5.8\text{ V}}{400\Omega}$$

$$I_{RS} = 14.5\text{mA}$$

Note that these values will stay the same for each of the following three examples as long as enough current is supplied to the Zener diode to allow it to stay in its Zener region. This can be checked later. First, consider the following three examples.

Example

In Figure 2–5b, both switches are shown in their closed positions. Again, because the Zener and the load resistance are in parallel you know that $V_Z = V_{RL}$.
Therefore:

$$I_{RL} = \frac{V_Z}{R_L}$$

$$I_{RL} = \frac{6.2\text{ V}}{1.2\text{ k}\Omega}$$

$$I_{RL} = 5.17\text{ mA}$$

By KCL, the Zener current is calculated as follows.

$$I_Z = I_{RS} - I_{RL}$$

$$I_Z = 14.5\text{ mA} - 5.17\text{ mA}$$

$$I_Z = 9.33\text{ mA}$$

Example

In Figure 2–5c, both switches are shown in their closed positions. Because the Zener and the load resistance are in parallel, you know that $V_Z = V_{RL}$.
Therefore:

$$I_{RL} = \frac{V_Z}{R_L}$$

$$I_{RL} = \frac{6.2\text{ V}}{2.2\text{ k}\Omega}$$

$$I_{RL} = 2.82\text{ mA}$$

By Kirchhoff's current law (KCL), the Zener current is calculated as follows:

$$I_Z = I_{RS} - I_{RL}$$

$$I_Z = 14.5\text{ mA} - 2.82\text{ mA}$$

$$I_Z = 11.7\text{ mA}$$

Example

In Figure 2–5d, both switches are shown in their closed positions. As previously, because the Zener and the load resistance are in parallel you know that $V_Z = V_{RL}$.
Therefore:

$$I_{RL} = \frac{V_Z}{R_L}$$

$$I_{RL} = \frac{6.2 \text{ V}}{4.4 \text{ k}\Omega}$$

$$I_{RL} = 1.41 \text{ mA}$$

By KCL, the Zener current is calculated as follows:

$$I_Z = I_{RS} - I_{RL}$$

$$I_Z = 14.5 \text{ mA} - 1.41 \text{ mA}$$

$$I_Z = 13.1 \text{ mA}$$

Note two very important points.

- As the load resistance increases, the Zener current increases.
- The maximum Zener current is 14.5 mA.

2.3 Checking an Installation

For any Zener circuit, the Zener current must stay between two limits.

- The minimum Zener current must be sufficient to bias the Zener into its Zener region.
- The maximum Zener current must be low enough that the Zener power does not exceed its maximum.

Assume that the minimum Zener current for the diode used in the previous examples is 8 mA. This means that the minimum load that can keep the Zener operating properly is:

$$R_L = \frac{V_Z}{I_{RS} - I_Z}$$

$$R_L = \frac{6.2 \text{ V}}{14.5 \text{ mA} - 8.0 \text{ mA}}$$

$$R_L = 954 \ \Omega$$

Any value of load resistance less than 954 Ω will result in a Zener current that is less than the required value and the Zener will not regulate. At the other end, you need to calculate as follows the total power dissipation of the Zener at its highest current.

$$P_Z = V_Z \times I_Z$$

$$P_Z = 6.2 \text{ V} \times 14.5 \text{ mA}$$

$$P_Z = 89.9 \text{ mW}$$

In this particular circuit, even a ¼ watt Zener [250 milliwatts (mW)] will be sufficient.

2.4 Other Zener Circuit Considerations

Insufficient Reverse Bias

A Zener diode may be taken out of its Zener range by either of two conditions.

- Insufficient current (as discussed previously)
- Insufficient supply voltage (as indicated in Figure 2–6)

Actually, these conditions amount to the same thing. The circuit conditions cause the available voltage and current to be below the Zener level. In this condition, the Zener behaves like any reverse biased diode. In Figure 2–6, the voltage of the source is not high enough to properly reverse bias the Zener. Therefore, the Zener behaves like any diode and the voltage drop across the resistors is easily calculated using the voltage divider theorem.

In a previous example (see Figure 2–5), you saw that when the load resistance drops below 954 Ω the Zener will drop out of its Zener range. This is because at that load level too much voltage is dropped across the R_S resistor. This makes the voltage available for the Zener too low to properly bias it.

FIGURE 2–6 A Zener with reverse bias less than its Zener voltage.

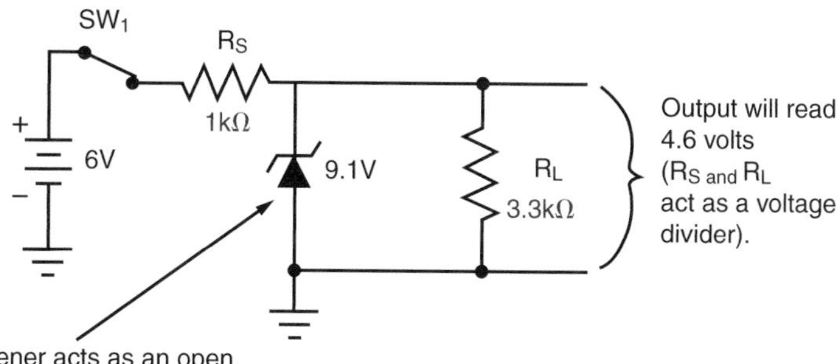

Forward Biased Zener

A forward biased Zener diode behaves exactly the same way a normal diode behaves, as indicated in Figure 2–7. This circuit can be solved by a simple application of KVL:

$$V_{Source} + V_{RS} + V_{Z(forward)} = 0\text{ V}$$

Then:

$$V_{Source} - (V_{RS} + V_{Z(forward)}) = 0\text{ V}$$

Then:

$$V_{RS} = V_{Source} - V_{Z(forward)}$$

$$V_{RS} = 6\text{ V} - 0.7\text{ V}$$

$$V_{RS} = 5.3\text{ V}$$

Then

$$I_{RS} = \frac{V_{RS}}{R_S}$$

$$I_{RS} = \frac{5.3\text{ V}}{1000\ \Omega}$$

$$I_{RS} = 5.3\text{ mA}$$

FIGURE 2–7 A forward biased Zener diode.

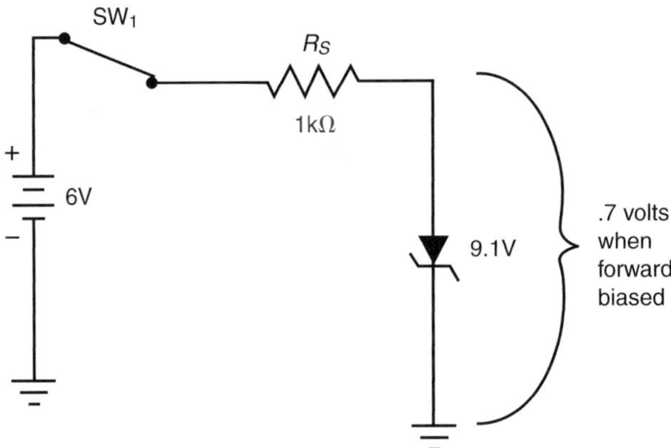

Reverse Biased Zener Diodes in Series

Figure 2–8 shows an example of how Zener diodes connected in series provide an additive constant voltage drop across the circuit load. Recall from earlier in Figure 2–4. The Zener diode acts as a constant voltage source to the parallel load. The same principle applies in Figure 2–8, except that the series Zener voltage drops are additive. All voltage drops summed provide a constant voltage to the parallel load.

FIGURE 2–8 Reverse biased Zener diodes in series.

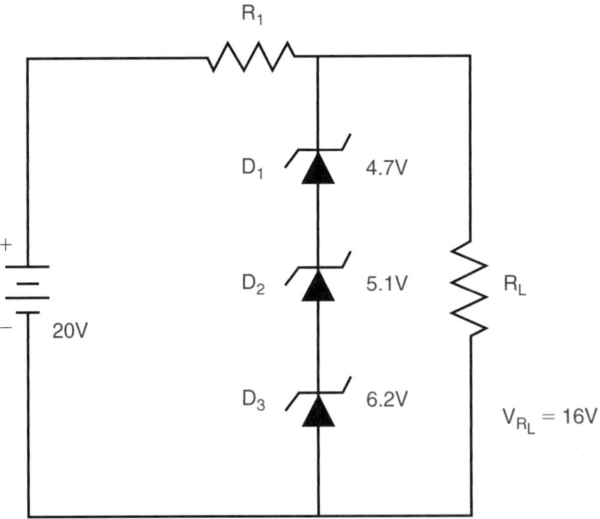

2.5 Diodes in an AC Circuit

Non-Zener

Antiparallel
Two circuit elements connected in parallel with opposite polarities.

Figure 2–9 shows a circuit and waveform for **antiparallel** diodes connected across the output of an AC supply.

FIGURE 2–9 Regular diodes with an AC source.

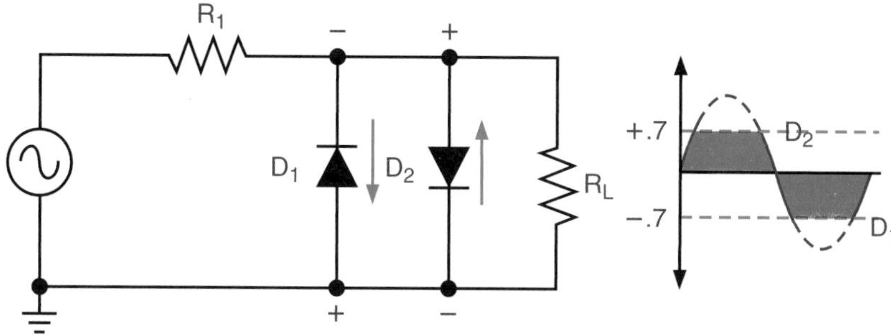

Note that because of the internal 0.7 V drop of the diodes, both alternations of the AC cycle are "clipped" or blocked. The diode conducts only after its forward biasing voltage is reached (approximately 0.7 V). This means that until the forward bias is greater than 0.7 V the diode acts as an open circuit. After 0.7 V is reached, the diode holds its voltage and therefore the voltage across R_L is 0.7 V. When D_1 is conducting, D_2 is reverse biased (open)—and vice versa.

An example of multiple diodes in series and their clipping effect on the parallel load is shown in Figure 2–10. Note that the difference in this circuit from that of Figure 2–9 is that there are two diodes in series in the first parallel branch. These diodes' internal voltage drops are additive and provide a total drop of 1.4 V before they are forward biased and conduct. This is shown in the effect of the output signal voltage across the load.

The waveform of Figure 2–10 shows that in the positive direction 0.7 V is developed across the load. In the negative direction, 1.4 V (or 2 diode voltage drops) is developed across the load.

FIGURE 2–10 Clipping circuit with two diodes in one parallel leg.

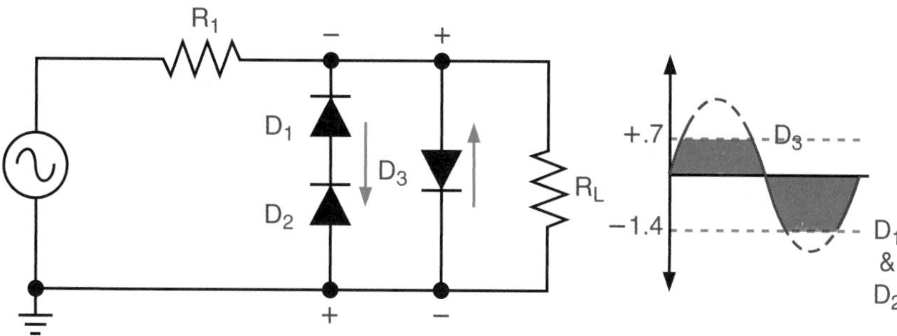

Zener Diodes

Figure 2–11 shows an interesting variation on clipping. Note that in Figure 2–11a the Zener diode D_2 and regular diode D_3 are in series and are the reverse of each other. In this connection, when the Zener is reverse biased the diode will be forward biased (and vice versa).

FIGURE 2–11 Clipping with a regular diode and a Zener diode.

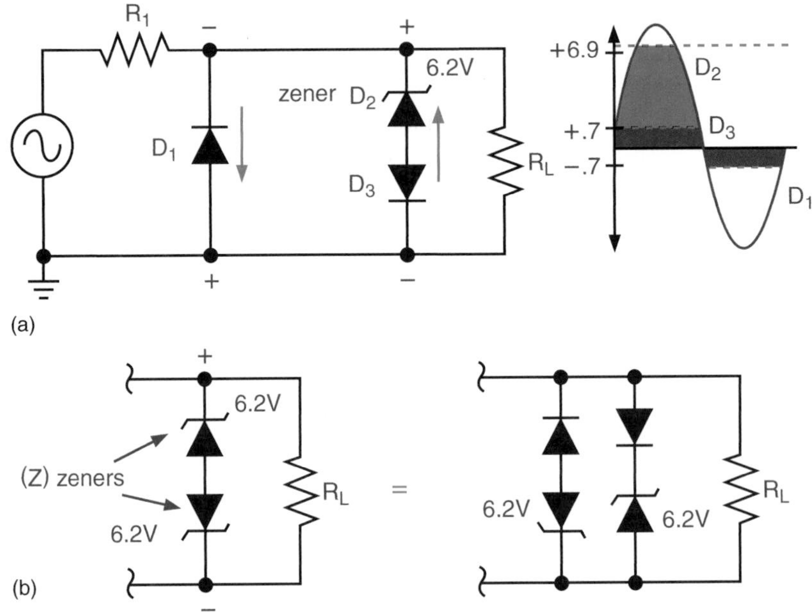

On the negative half-cycle, the D_2D_3 combination will not conduct. Current will go through D_1 because it is forward biased, thus providing a 0.7 V output. During the positive half-cycle, the D_2D_3 combination will not conduct until the voltage has reached the Zener voltage (6.2 V) plus the diode forward voltage (0.7 V). The resulting waveform is shown in Figure 2–11a.

Figure 2–11b shows two circuits that are functionally equivalent in regard to the output voltage to R_L when an AC voltage is applied. The back-to-back Zener diodes in the left circuit function the same in both the positive and negative cycles. This is because in the positive or negative cycle one Zener will be forward biased (0.7 V) and the other Zener will be reverse biased (Zener voltage of 6.2 V). The final output will be +6.9 volts to −6.9 volts across R_L.

In the circuit on the right in Figure 2–11b, the Zeners are in series with the rectifier diodes. The diodes create a voltage drop of 0.7 V, as the opposing Zener diode did in the left circuit. In the positive or negative cycle, the Zeners will operate at the Zener voltage of 6.2 V. Therefore, the sum of the Zener and diode is 6.9 V. Again, the final output will be +6.9 V to −6.9 V across R_L. This example assumes that the Zeners and non-Zener diodes have internal voltage drops of 0.7 V.

OTHER TYPES OF DIODES

This section explores how heavier doping of the PN junction with impurities can change the way certain diodes respond to a forward biasing current and voltage. This section also discusses another special type of diode—the PIN diode. This diode does not have a PN junction. Other special diodes (such as photodiodes, light-emitting diodes, and laser diodes) are covered in a later semiconductor lesson.

2.6 Tunnel Diodes

Figure 2–12a shows a schematic diagram indicating the symbol for a tunnel diode utilized as an oscillator. This application of the tunnel diode is commonly found in communications receivers for which low-power high-frequency signals are required. Figure 2–12b shows a characteristic curve of a tunnel diode; Tunnel diodes are normally operated in the **negative resistance** region. They can also operate at higher temperatures than silicon or germanium diodes. Another use of tunnel diodes is in high-frequency oscillators.

Negative resistance
A special purpose diode in which current reaches a peak. Further voltage increase results in a rapid current decrease.

FIGURE 2–12 (a) Tunnel diode oscillator circuit. (b) Tunnel diode characteristic curve.

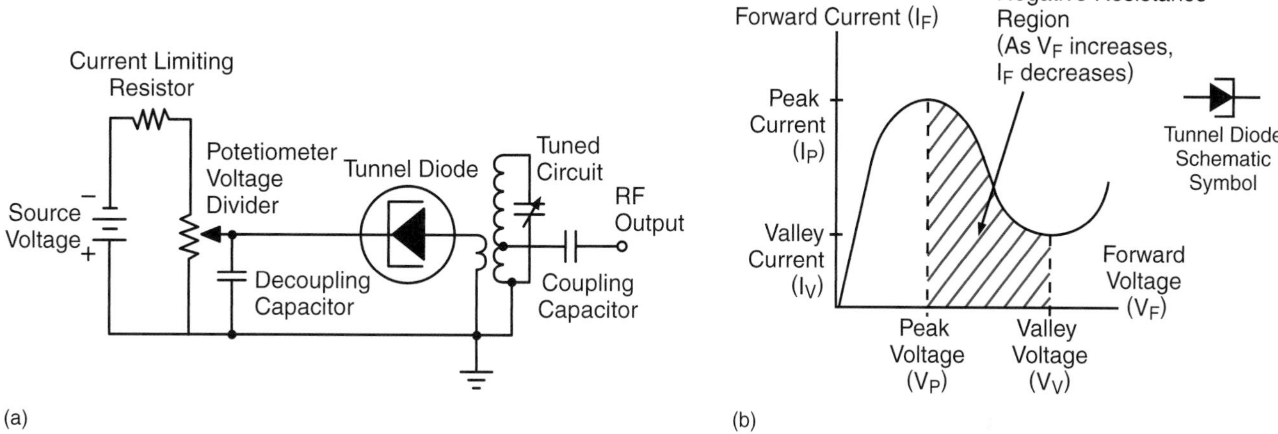

2.7 Schottky or Shockley Diodes

These diodes are also known as **hot-carrier** or **surface-barrier diodes**, and they are used in switching circuits and microwaves. They are also frequently used as a trigger device for SCRs (silicon-controlled rectifiers). When forward biased, they have a low breakover voltage of 0.3 V. Their main advantage is that they reverse polarity (forward to reverse bias) almost instantaneously, allowing the diode to be used in VHF ranges.

Hot-carrier diode
A diode exhibiting a thermal difference across the PN junction, having a very fast off-to-on and on-to-off transition.

Surface-barrier diode
See Hot-carrier diode.

This high-frequency response is made possible by one of this diode's unique characteristics. Namely, it uses metallic flakes (sort of a doping) on top of the silicon semiconductor-type material to produce its "junction." Figure 2–13 shows the characteristic curve of a Schottky diode.

The Schottky can also have two PN junctions (PNPN), but only two electrodes (anode and cathode). See Figure 2–14 for the schematic symbology of typical diode packaging. The usual forward biased resistance for the Schottky is small, just a few ohms. This is considerably less than the forward biased resistance of normal rectifier diodes. The Schottky is rated by its threshold (trigger) voltage. This is the voltage required to forward bias or turn on the diode. A Schottky can have a third electrode. If a third electrode, called a gate, is used to "fire" or turn on the diode the Schottky is called an SCR. SCRs are discussed in a later lesson. Figure 2–15 is a schematic diagram of an AM radio detector circuit. Schottky diodes are especially suited to these circuits because of the low threshold voltage, which allows detection of weaker signals than would be possible with conventional pn diodes.

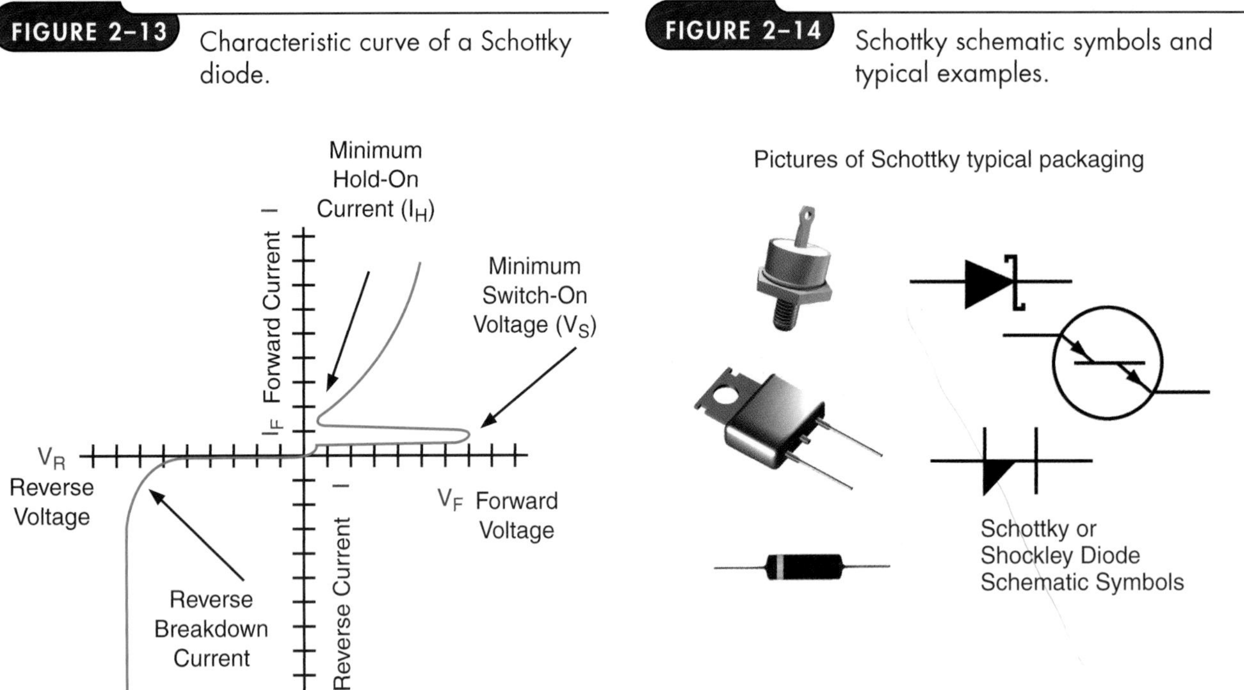

FIGURE 2–13 Characteristic curve of a Schottky diode.

FIGURE 2–14 Schottky schematic symbols and typical examples.

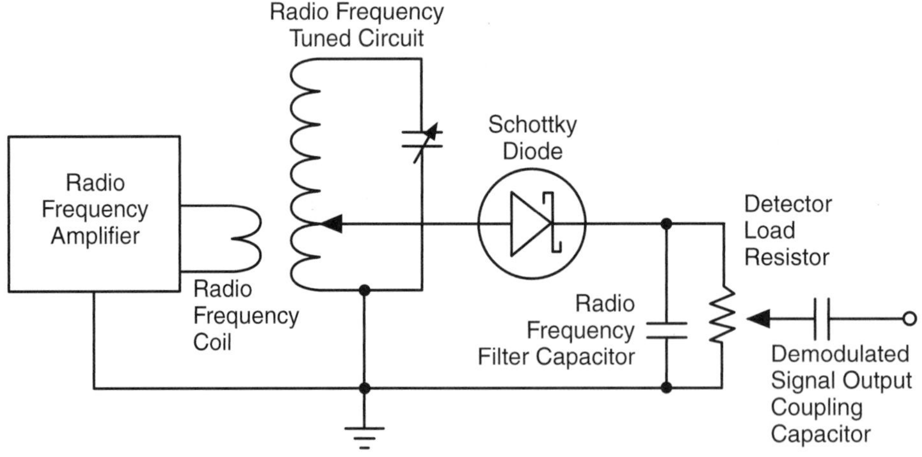

FIGURE 2–15 AM radio detector circuit.

2.8 PIN Diodes

The most common uses of the PIN diode is in radio frequency (RF), microwave, and modulator circuits and as a switch. The PIN diode differs from the PN junction diode because it does not have a PN junction. Instead, it consists of three materials: P-type, N-type, and a pure (intrinsic) silicon slice. The intrinsic silicon slice is between the P-type and N-type materials, and thus there is no PN junction. In fact, the *I* in PIN comes from the "intrinsic" material slice or layer between the P-type and N-type materials.

The intrinsic silicon has a very high resistance (recall that intrinsic silicon or germanium acts much like an insulator unless doped with impurities). The thicker the intrinsic material layer the higher the resistance and the greater the breakdown voltage. See Figure 2–16 for an example of the construction of a PIN diode.

FIGURE 2–16 Construction of a PIN diode.

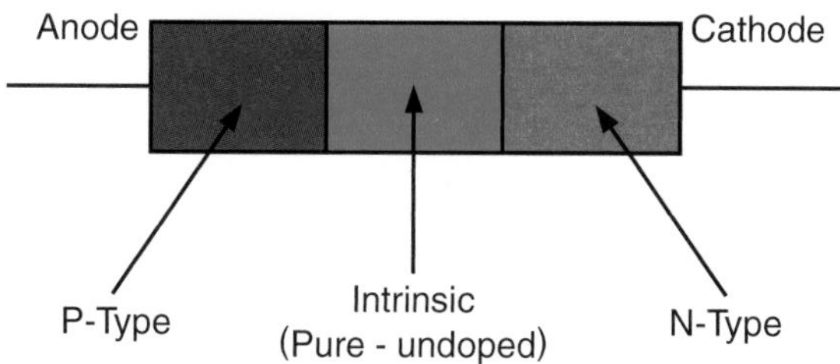

Another distinct characteristic of the PIN diode is that it does not have an actual knee voltage. Unlike the PN junction diode, there is no abrupt "turn-on" point but rather a gradual increase in forward bias voltage. This causes a corresponding increase in forward bias current. See Figure 2–17 for a PIN diode characteristic curve.

FIGURE 2–17 PIN diode characteristic curves.

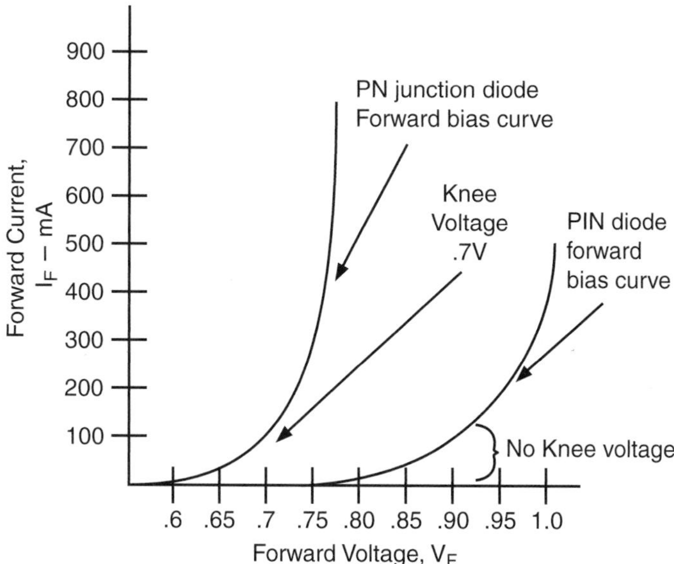

The PIN family of diodes has two basic uses. One type is used in high-frequency (can be greater than 300 MHz) circuits as signal carriers and switches. The second PIN-type diode is used in a range of high-power diodes. This is possible because of the intrinsic layer that supports a high breakdown voltage needed for high-power applications. Light-emitting diodes (LEDs) and laser diodes are discussed later in the book.

Figure 2–18 shows an elementary schematic diagram of a solid-state RF switch. Solid-state switching circuits are exceptionally small and lightweight, consume very little power, and operate at very high frequencies. They are often found in handheld radios and cell phones. When bias 1 is biased positive, PIN diode 1 is forward biased—allowing signals from the antenna to pass to the receiver input. When forward bias is applied to PIN diode 2, the transmitter output passes through PIN diode 2 to the antenna. The receiver cannot be damaged by the high-power transmitter output because the PIN diodes are never biased-on simultaneously.

FIGURE 2–18 PIN diode RF switch.

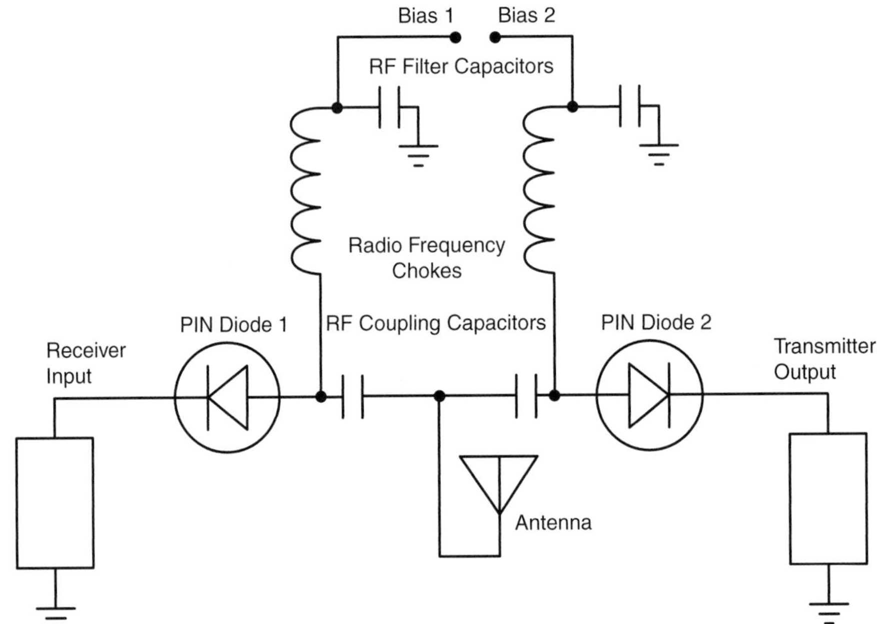

FIGURE 2–19

Varactor diode schematic symbol.

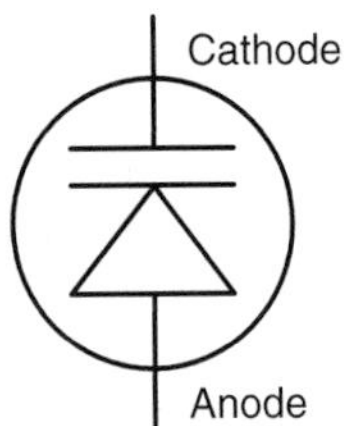

Voltage-capacitance curve A graph upon which Voltage on the x-axis, is plotted against capacitance on the y-axis (or reverse).

Reverse breakdown voltage The reverse-bias voltage beyond which current in excess of leakage current begins to flow. This is usually a destructive condition.

2.9 Varactor Diodes

See Figure 2–19 for the schematic symbology of the Varactor diode. Recall that capacitance of a capacitor is dependent on three factors: area of the plates, distance between those plates, and the dielectric material between those plates. PN junctions have similar factors. The area of the PN junction is obvious. However, the other two criteria come into play. As you know, an unbiased PN junction permits very little current.

A barrier region exists between the two PN terminals. The doped areas of the PN junction can be considered the dielectric material. Therefore, capacitance exists across the PN junction. Neither the area of the junction nor the makeup of the doped materials can be modified. However, the effective thickness of the barrier region can be controlled. Barrier region thickness can be increased by increasing reverse bias of the PN junction. Figure 2–20 depicts a Varactor diode **voltage-capacitance curve.**

As you may expect, increasing negative bias results in decreased capacitance. As with most diodes, exceeding **reverse breakdown voltage** will likely result in destruction of Varactor diodes.

FIGURE 2–20 Typical Varactor diode voltage-capacitance curve.

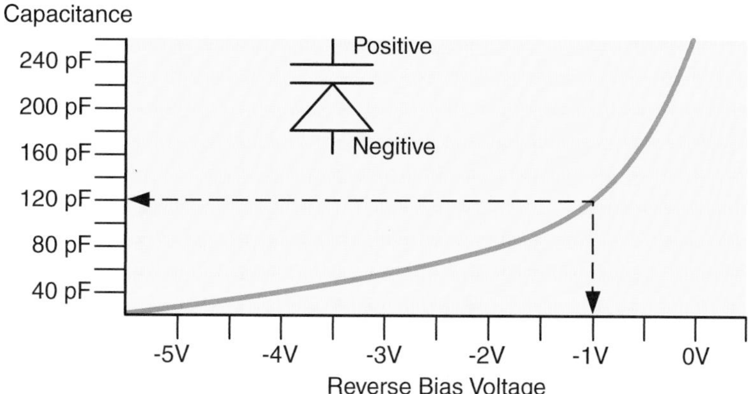

Varactor diodes are used primarily as voltage-controlled capacitors in high-frequency tuned circuits. Varying reverse bias of the Varactor results in changes to the resonant frequency of those circuits.

Figure 2–21 shows a schematic diagram of a typical Varactor tuning circuit. Note the DC path from the Varactor anode to ground. The positive bias at the Varactor cathode is furnished from the potentiometer armature by way of the RF choke. The remaining capacitor decouples the tuned circuit from the variable DC tuning voltage. The Varactor remains reverse biased throughout the range of potentiometer travel.

One advantage of this arrangement is that the tuning potentiometer may be located quite distant from the tuned circuit. Should forward bias occur, the Varactor will not function as a capacitor but as a low-resistance device. Varactor front-to-back resistance may be tested in the same manner as a conventional diode. Be aware that all PN junctions exhibit the Varactor characteristic to some extent.

FIGURE 2–21 Varactor diode utilized as a tank circuit tuning capacitor.

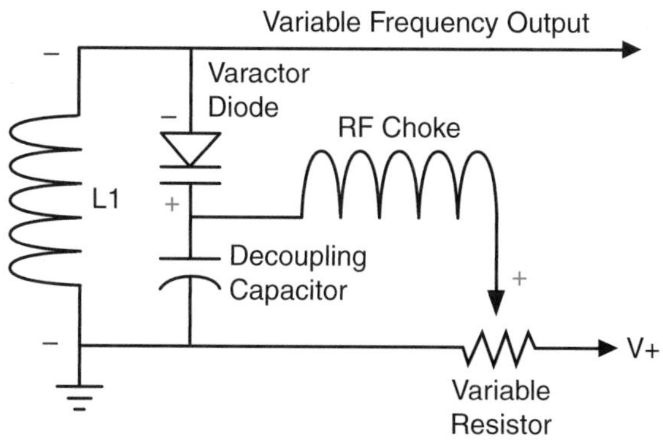

TechTip!

A maintenance technician was called upon to check on a problem with a Varacter-tuned tank circuit that would not stay on frequency. The technician verified the frequency was indeed shifting so he started considering possible causes. After considering each component in the circuit (similar to that of Figure 2–21) a hunch led him to believe that the DC control voltage was probably unstable at the Varactor cathode. Following his hunch, the technician used his oscilloscope to observe the DC potential at the Varacter cathode. The result of the test showed that the DC potential was indeed unstable. The technician started looking for obvious possible causes within the circuit such as bad solder joints and moisture or corrosion on and around the circuit card connection. All of which seemed to be ok. He then went on to check the wiring connections to the remotely located potentiometer, this also checked out ok.

Having checked all of the obvious possibilities, the technician needed to focus on the likelihood of a failed component. With a solid background in electronics, the technician started contemplating likely components that could be creating the problem. Knowing that semiconductors are highly reliable components, especially those not subject to high voltages and currents, led him to believe that some other component was causing the problem. The next possibility was the potentiometer; potentiometers are not extremely reliable devices especially in adverse conditions as well as in situations where they are not regularly exercised. The technician replaced the potentiometer and the tank circuit was restored to proper operation.

SUMMARY

Zener Diodes

The Zener is one of the special types of PN junction diodes. It is designed to operate in the reverse biased condition. When reverse biased, the Zener is normally used as a voltage regulator. Zeners are designed to maintain a specified voltage drop over a wide range of currents. Zeners come in sizes (specified voltages) ranging from 1.8 V at ¼ watt to 200 V at 50 watts. Multiple Zeners of various breakdown voltages can be placed in series to increase the designed regulated voltage (voltage drop). In the forward biased condition, the Zener acts much like a regular PN junction diode and has a forward voltage drop of approximately 0.7 V. Critical Zener characteristics and ratings are outlined in Table 2-1.

Other Diode Types

Four other common non-PN junction diodes are the tunnel, Schottky, Varactor, and PIN. All of these can be used with high-frequency and microwave applications. The tunnel diode has a unique characteristic called negative resistance. This means that voltage increases beyond peak forward voltage cause the forward current to decrease. Negative resistance allows the tunnel diode to be used as an oscillation device in high-frequency applications.

The Schottky diode is also known as the Shockley or hot-carrier diode. This family of diodes is used as a switching (on/off) device in high-frequency applications, especially microwaves, and frequently as trigger devices for SCRs. Some Schottky diodes use a metal doping on top of the silicon semiconductor material to produce a "junction." A Schottky diode that has a third electrode or gate is called an SCR.

The PIN diode is used extensively in modulator and RF circuits. The PIN diode does not have a PN junction but uses a third intrinsic or pure silicon layer between the P-type and N-type materials. The "intrinsic" silicon slice is represented by the *I* in *PIN*. This slice also acts as an insulator and sets up a high-resistance barrier that provides for greater breakdown voltage ratings. The thicker the intrinsic layer the higher the breakdown voltage. PIN diodes do not have specific knee voltages but rather increase forward conduction more gradually than PN junction diodes.

The Varactor diode is used principally as a voltage-controlled capacitor. It operates in the reverse bias mode only. Increasing the reverse bias voltage results in lower PN junction capacitance. The junction of a Varactor may be tested in the same way as a conventional diode.

REVIEW QUESTIONS

1. Describe the most common uses for the types of diodes discussed in this chapter.
2. Examine Figure 2–22. If the battery is increased to 12 V, what value of load resistors will allow the Zener to operate in its Zener range? Assume that the Zener requires a minimum of 5 mA to operate correctly.

FIGURE 2–22 A forward biased Zener diode.

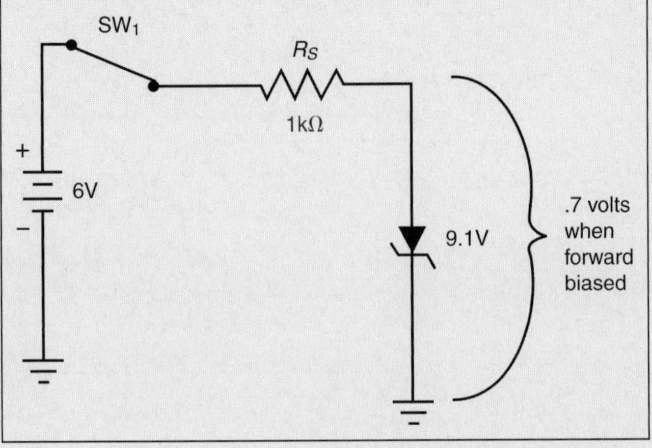

3. In Figure 2–23, what precautions would have to be observed to keep the two diodes from burning out due to forward current flow?

FIGURE 2–23 Clipping circuit with two diodes in one parallel leg.

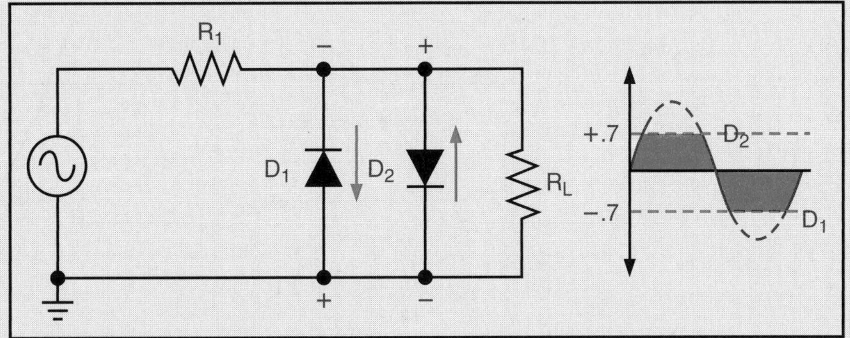

4. List at least four practical applications of a Zener diode regulator.
5. How is a PIN diode made? What is the doping characteristic of the intrinsic layer?
6. The tunnel diode is normally biased to operate in its negative resistance region. How does this differ from regular diodes or Zener diodes?
7. You have a need for a regulated DC 11.3 V voltage. You have four Zener diodes with ratings of 11.2, 6.2, 5.1, and 4.7 V. What is the best combination for your regulator?
8. In Figure 2–24, explain what the differences might be in the application of the two types of Zener diodes shown.
9. Draw a circuit that will limit (clip) an AC voltage to a maximum of +/−4.7 V.
10. Examine Figure 2–25. Assume the Zener $P_{D(MAX)}$ is 2 watts. What is the smallest size of R_S that could be used in this circuit?

FIGURE 2–24 Zener diode.

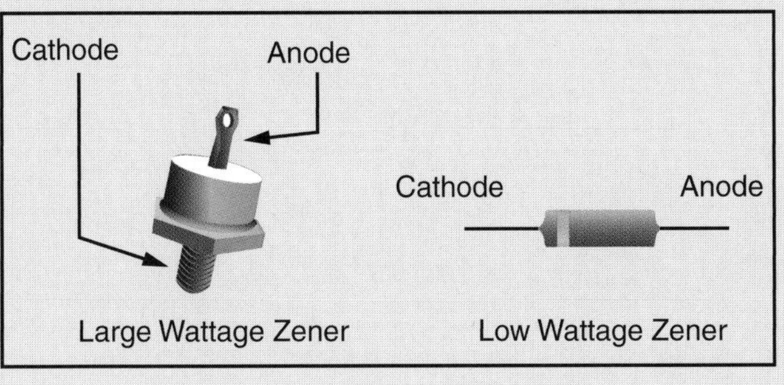

FIGURE 2–25 Operation of a typical Zener circuit.

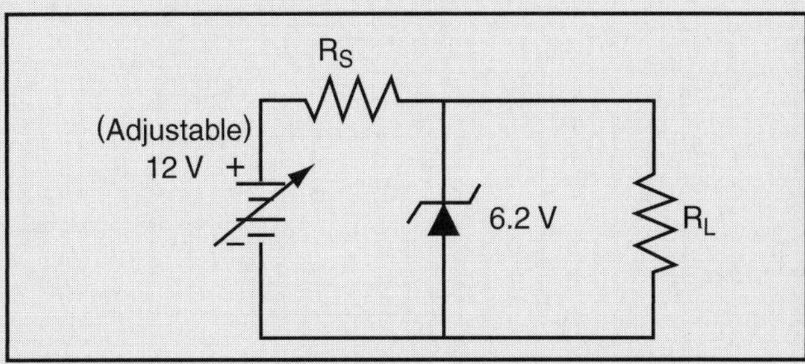

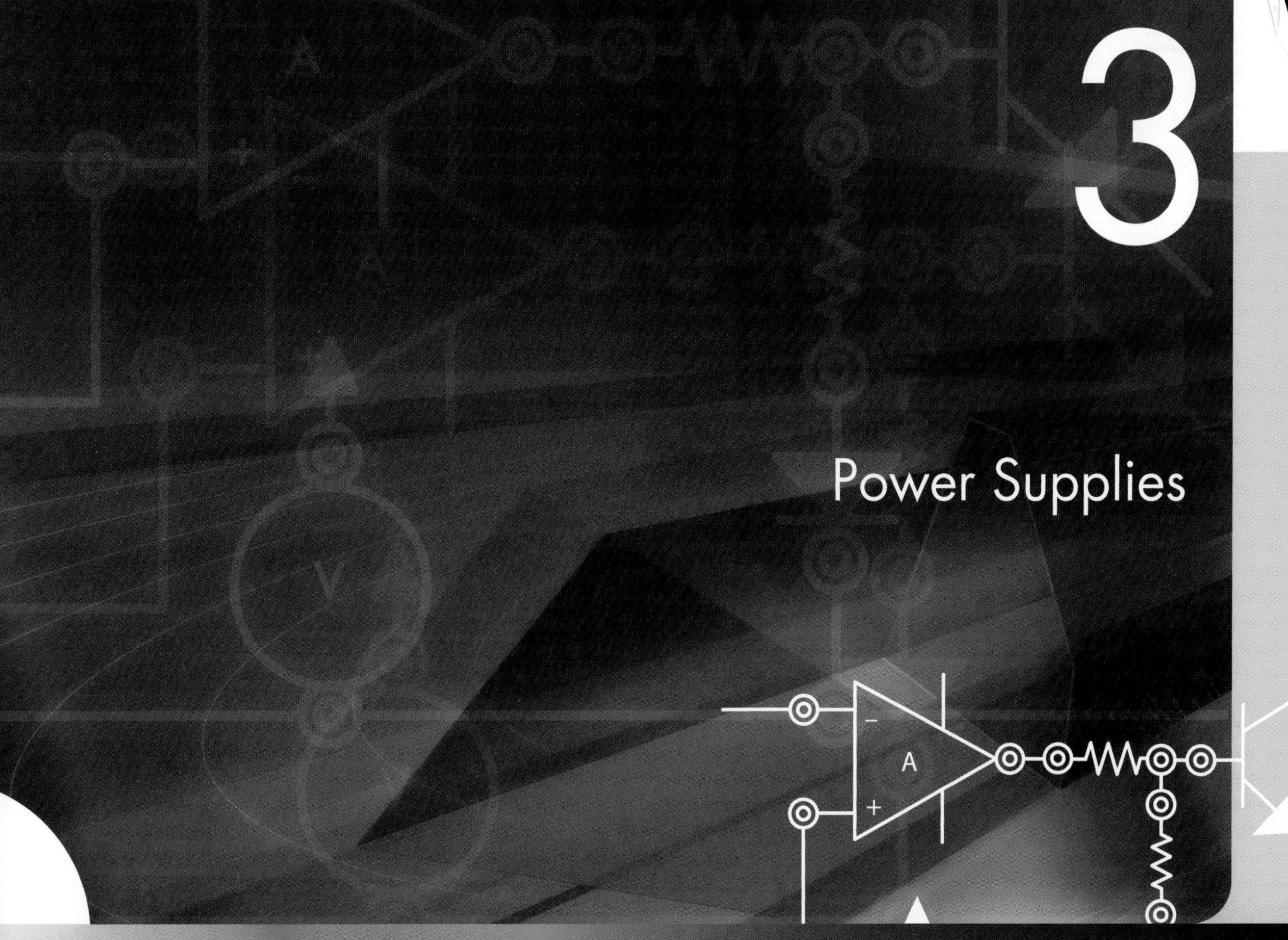

3 Power Supplies

OUTLINE

OVERVIEW

All physical processes require energy. The rate at which this energy is expended is called power. In our society, most electric power is distributed using alternating current (AC). Many electric and electronic circuits require a direct current (DC) form of power to operate properly.

In this chapter, you will learn the fundamentals of turning AC into DC. You learned something about this earlier when you saw that a diode will pass current only when it is forward biased. Thus, when an AC voltage is applied to a diode a pulsating DC is produced. From this simple concept, by using ever more complex circuitry AC power can be rectified into a DC that is virtually pure—almost as pure as the output from a battery.

OBJECTIVES

After completing this chapter, the student should be able to:

1. Draw schematic symbols for half-wave, full-wave center tap, and full-wave bridge rectifiers.
2. Explain the operation of capacitors as filters.
3. Explain the operation of chokes as filters.
4. Describe the operation of voltage regulators and dividers.
5. Predict the output waveforms of various rectifier circuits.
6. Discuss the operation of voltage doublers.

INTRODUCTION

Earlier you learned that semiconductor devices can be forward biased by a very small voltage, typically 0.3 V or 0.7 V (volts). Other semiconductor devices are "turned on" with voltages ranging from 1.8 V to 50 V. These low DC voltages are used to control the semiconductor circuits in the form of forward or reverse biases on the semiconductor devices. Other applications (such as oscillators, receivers, modulators, and transmitters) use AC voltages that "ride on" the DC control voltage to produce the "signals" (waveforms) processed by the various semiconductor devices.

Consider Figure 3–1, for example. In this circuit, a 2 $V_{P\text{-}P}$ signal is fed into a signal amplifier. The semiconductor circuits, possibly transistors, in the amplifier are biased with a DC supply voltage. The bias allows the transistors to control their output voltages by exactly reproducing the shape of the input magnified or amplified by a factor of 10. Note that (as shown in Figure 3–1) the output also has a 6 VDC component.

FIGURE 3–1 AC signal processing with DC control voltage.

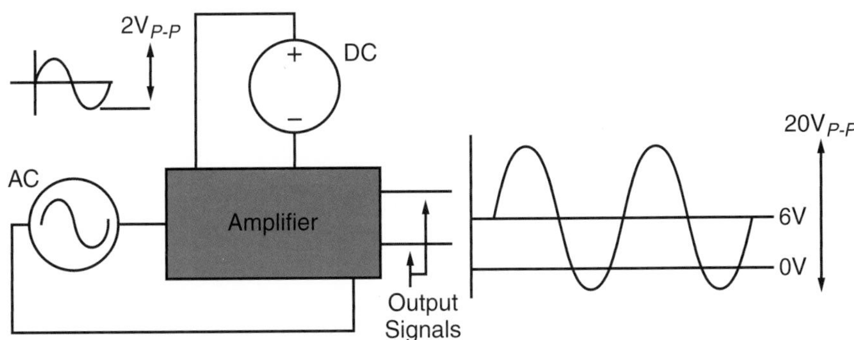

POWER SUPPLY SYSTEM CHARACTERISTICS

3.1 Series Voltage Sources

Different power supplies have many different output voltages and currents, but most of them use the same basic operating principles. Figure 3–2 shows a simple battery power supply. This configuration uses batteries in series, with each cell (V_{C1} through V_{C4}) supplying 25% of the total supply voltage.

FIGURE 3–2

Battery power supply with batteries in series.

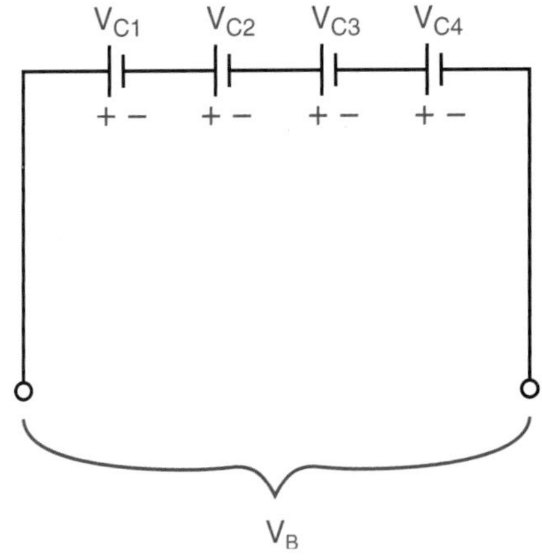

3.2 Parallel Voltage Sources

Figure 3–3 shows a power supply with the batteries connected in parallel. Recall that in this type of circuit the current is additive instead of the voltage. Each cell (V_{C1} through V_{C4}) supplies 25% of the total source current.

3.3 Dual-Voltage Power Supply

By placing a reference ground between cells V_{C2} and V_{C3} (indicated in Figure 3–4), point A becomes negative to the ground reference and point B is positive to the same reference point. This allows the supply source to provide two polarities to a circuit. This is necessary with some semiconductor circuits to properly bias transistors and diodes. A dual-voltage power supply is often called a bipolar supply.

FIGURE 3–3 Battery power supply with batteries in parallel.

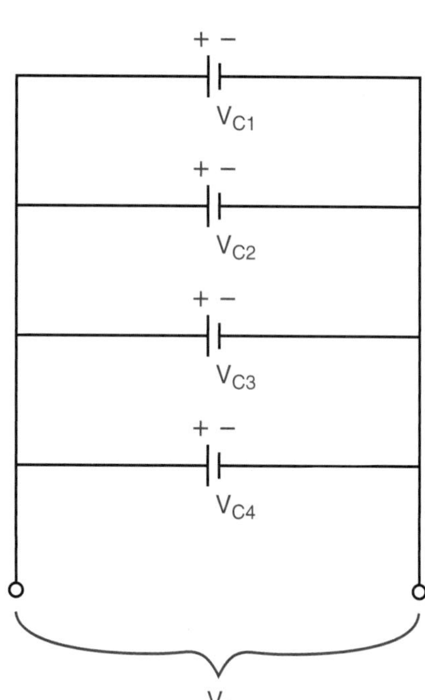

FIGURE 3–4 Dual-voltage power supply (also called a bipolar power supply).

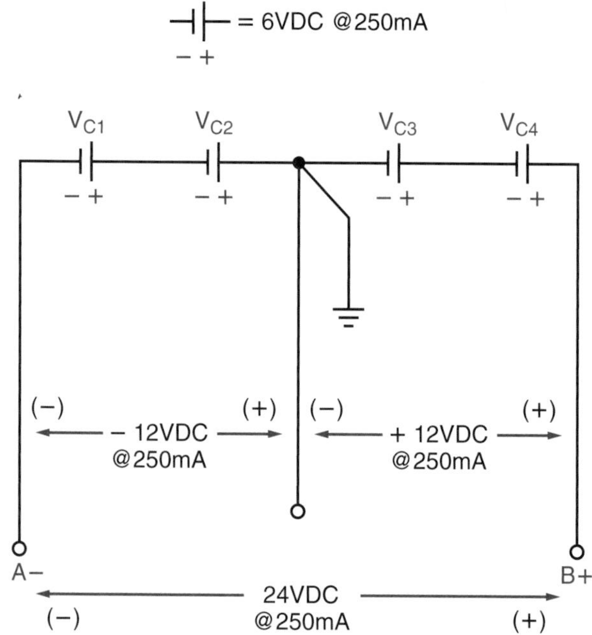

3.4 Power Supply Block Diagram

The power supplies discussed in this chapter use 120 VAC 60 Hz input (common household electric power in North America) and produce a range of DC voltage forms. The most common power supplies have three major operational sections: the transformer, the rectifier, and the filter. In other lessons on DC and AC theory, you learned how transformers and filters operate.

In the last two lessons, you analyzed the rectifier capabilities of diodes. In this lesson, all three operations are combined to produce a power supply.

This lesson builds on the five steps involved in a complete power supply system. These steps are:

1. Transform the AC voltage.
2. Rectify the AC to pulsating DC.
3. Filter the pulsating DC output.
4. Regulate the output.
5. Divide the output.

Figure 3–5 shows a simple block diagram of the first four steps listed previously, and the following material describes each process in more detail.

FIGURE 3–5 Basic power supply block diagram.

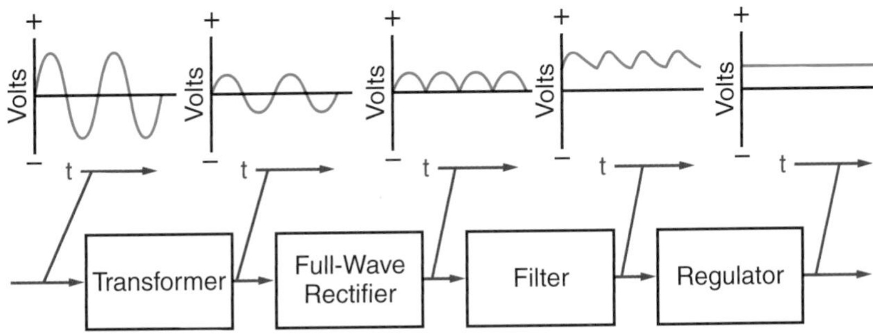

Transform the AC Voltage

This process is done by the transformer shown in Figure 3–5. The maximum DC output voltage will depend on the amount of AC supplied to the rectifier and the filter. On the other hand, excessive pulsing DC voltage into the filter will cause the filter to run hot and will probably result in excessive heat loss. By using a transformer, the voltage can be changed to an appropriate level to produce the desired DC output.

Rectify the AC to Pulsating DC

The box labeled "Full-Wave Rectifier" in Figure 3–5 has this job. In fact, you recall that a rectifier is a circuit that changes AC into pulsating DC. Not all rectifiers are necessarily full wave. However, the full-wave rectifier is more efficient and thus most large power supplies use full-wave rectifiers.

Filter the Pulsating DC Output

Most electric and electronic circuits that use DC require a smooth DC similar to that produced by a battery. By methods described in detail later in this chapter, the filter changes the pulsating DC to a purer DC by removing at least part of the AC component.

Regulate the Output

As the load current changes, the output voltage of a power supply tends to change. This is caused by the increased internal voltage drop within the power supply.

A regulator, such as a Zener diode, will maintain the output voltage at some predetermined value so that the load is always supplied at the needed voltage level. Note that the regulator shown in Figure 3–5 also helps with the filtering process.

Divide the Output

Some circuits require multiple DC voltage levels and/or multiple DC voltage polarities. A voltage divider circuit can be used to provide different voltages from the same supply. Such a circuit may involve a simple resistance voltage divider, a capacitor-type voltage doubler, or even multiple Zener diodes from which regulated voltages are taken. The voltage divider may be located before or after the regulator in a power supply. No divider is shown in Figure 3–5.

HALF-WAVE RECTIFIER

3.5 Rectifying Characteristics

Recall from the past lessons that one of the main purposes of a diode is to rectify AC voltage. Also recall that there are different forms of measurements for these rectified voltages, including **peak-to-peak**, **peak average**, and **RMS** (root mean square). The RMS value is also called the **effective value.**

Peak-to-peak
The peak-to-peak magnitude of an AC signal.

Peak average
The magnitude of a waveform as measured from the zero value to the peak value.

RMS value
The same as effective value. RMS stands for "root mean square."

Effective value
The magnitude of a DC waveform that generates as much power as a measured AC waveform.

The RMS rating is used to determine the effective output of an AC machine. We use this because the machine produces the peak only for an instant and then returns through its cycle. When the RMS values of voltage and current are used, the values identify the same amount of power as a like value of DC.

An AC waveform is rectified by turning it into pulsating DC. Figure 3–6a shows an AC waveform before and after it is rectified by a PN junction diode. Also recall that for the AC signal to be processed the diode must be forward biased. This forward bias voltage is typically 0.7 V. Figure 3–6b shows a typical half-wave rectifier circuit that would produce the output waveform shown in Figure 3–6a. Note that in this case the power is being supplied by a transformer.

As stated earlier, the voltage may need to be increased or decreased. This is done by step-up and step-down transformers. The two types are shown schematically in Figure 3–6c. The transformer in the schematic of Figure 3–6b is called a step-down transformer because the input is 120 VAC and the output is 15 VAC.

The voltage is reduced by the induction of the primary (input) coil into the secondary (output) coil of the transformer. Transformers may either step up (increase) or step down (decrease) the primary side voltage of the transformer to the secondary side output. This is done by increasing or decreasing the amount of wire coils (number of turns in the coil) of the transformer output (secondary) compared to the transformer's input (primary) side windings.

There is one class of transformers (called **isolation transformers**) that have the same number of turns on primary and secondary. The voltage does not change, but the transformer does allow "isolation" from the primary to the secondary. Figure 3–7a shows the polarity of the output voltage and path of current flow through the rectifier circuit.

Isolation transformers
A power transformer having a turns ratio of 1:1.

FIGURE 3–6 Elements of a half-wave rectifier. (a) Input and output waveforms. (b) Simple circuit. (c) Step-up and step-down transformers.

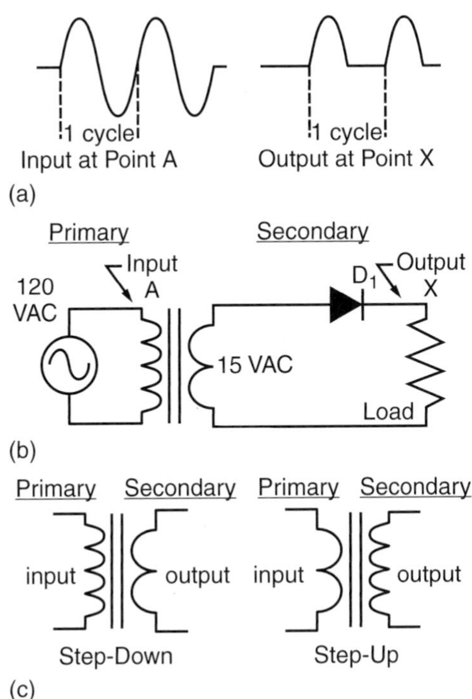

Note that by reversing the diode (as shown in Figure 3–7b) the current through the load is reversed. This changes the voltage polarity across the load. In Figure 3–7a, only the positive voltage peaks are passed to the load, and in Figure 3–7b only the negative voltage peaks are passed.

FIGURE 3–7 Half-wave rectifier. (a) Positive output. (b) Negative output.

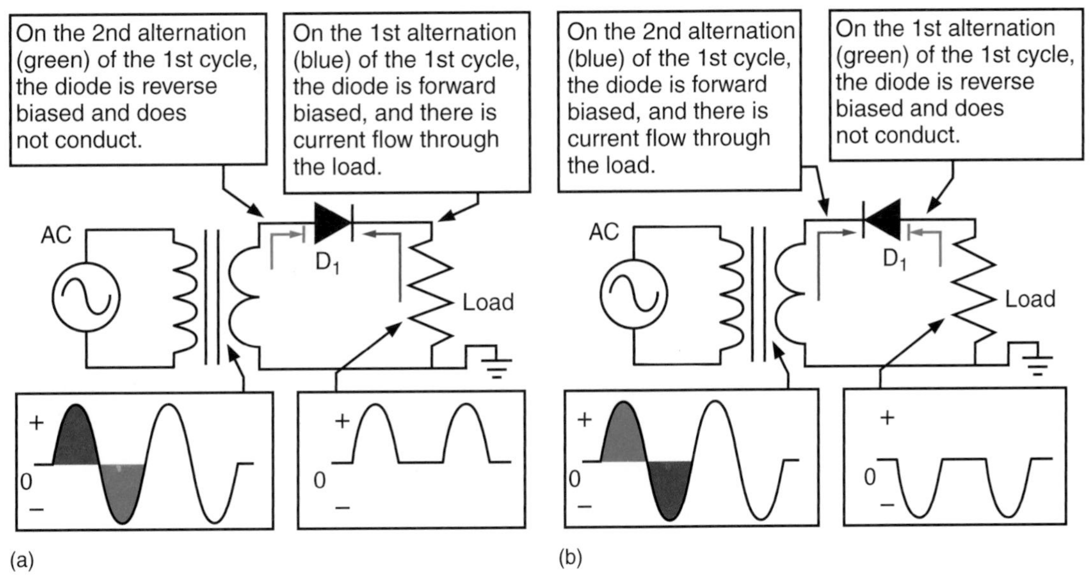

3.6 Circuit Analysis

Review of AC Measurements

The various values for AC voltage and current levels are almost always given in terms of the RMS value. This includes transformer output voltages. There are, in fact, several values for voltage and current that can be specified for an AC or a pulsating DC voltage or current. The values and their meanings are outlined in Table 3–1. Their location and labels for an AC sine wave are shown in Figure 3–8. Using the information, the values for the typical 120 V house supply are as follows:

FIGURE 3–8 A sine wave and its various voltage values.

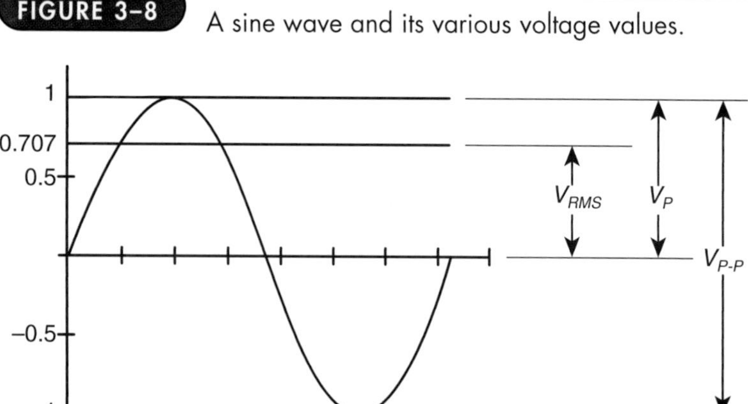

$V_{RMS} = 120$ V

$V_{P\text{-}P} = 120\ V_{RMS} \times 2 \times \sqrt{2} = 339.4$ V

$V_P = 120\ V_{RMS} \times \sqrt{2} = 169.7$ V

$V_{AVE} = 0$ V

Remember that voltages and currents in AC power systems are almost always measured and discussed based on their RMS value.

TABLE 3–1 Key Values for Various Types of Waveforms

		Unfiltered Voltage Across the Load		
Symbol	Description	Unrectified Sine Wave	Half-wave Rectified[1]	Full-wave Rectified
$V_{P\text{-}P}$	Peak-to-peak value. The distance from the highest peak to the lowest peak.	$2 \times V_P$	V_P	V_P
V_P	The distance from the zero line to the highest peak value, usually measured up.	$V_P = \frac{V_{P\text{-}P}}{2}$ or $V_{RMS} \times \sqrt{2}$	V_P	V_P
V_{AVE}	The arithmetic average of all values. V_{AVE} is also the DC (not RMS) value that would be measured by a non-RMS DC-responding meter.	0	$V_P \div \pi$	$2 \times V_P \div \pi$
V_{RMS} or V_{EFF}	The effective value or heating value of the waveform. A DC value that will produce the same amount of heat in a resistor.	$V_P \div \sqrt{2}$	$V_P \div 2$	$V_P \div \sqrt{2}$

[1] A sine wave has no average value because a voltage waveform above the zero line is a mirror image of the voltage waveform below the zero line.
[2] $\pi = 3.14159$

Example

Refer to Figure 3–9 and solve for the load V_{EFF}, V_{AVE}, and I_P and for the output waveform configuration. The voltage out of the secondary side of the transformer is 24 VAC peak to peak. From Table 3-1, the effective voltage of the transformer's secondary side is:

$$V_p = \frac{V_{P\text{-}P}}{2}$$

$$V_{EFF} = \frac{V_P}{\sqrt{2}}$$

$$V_{EFF} = \frac{12\text{ V}}{1.414} = 8.49\text{ V}$$

The half-wave output to the load has a peak voltage equal to the peak of the transformer output minus the forward drop across the diode:

$$V_{P(Load)} = V_{P(XFMR)} - V_D$$

$$V_{P(Load)} = 12\text{ V} - 0.7\text{ V} = 11.3\text{ V}$$

From Table 3-1, the average and RMS voltage for the load are:

$$V_{AVE} = \frac{V_{P(Load)}}{\pi}$$

$$V_{AVE} = \frac{11.3\text{ V}}{3.14} = 3.60\text{ V}$$

$$V_{RMS} = \frac{V_{P(Load)}}{2}$$

$$V_{RMS} = \frac{11.3\text{ V}}{2} = 5.65\text{ V}$$

The peak current through the load can now be calculated as follows:

$$I_P = \frac{V_{P(Load)}}{R_L}$$

$$I_P = \frac{11.3\text{ V}}{200\ \Omega} = 56.5\text{ mA}$$

The average and the RMS value of the current are:

$$I_{AVE} = \frac{I_P}{\pi}$$

$$I_{AVE} = \frac{56.5\text{ mA}}{3.14} = 17.99\text{ mA}$$

$$I_{RMS} = \frac{I_P}{2}$$

$$I_{RMS} = \frac{56.5\text{ mA}}{2} = 28.25\text{ mA}$$

The voltage waveform is shown in Figure 3–10.

FIGURE 3–9 Half-wave rectifier circuit analysis.

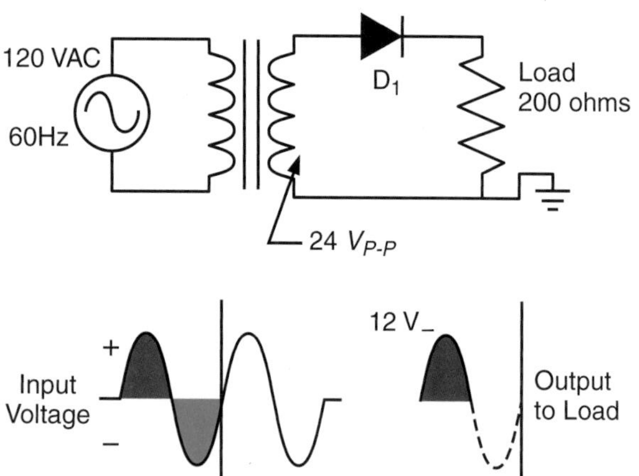

FIGURE 3–10 Half-wave load voltage waveform.

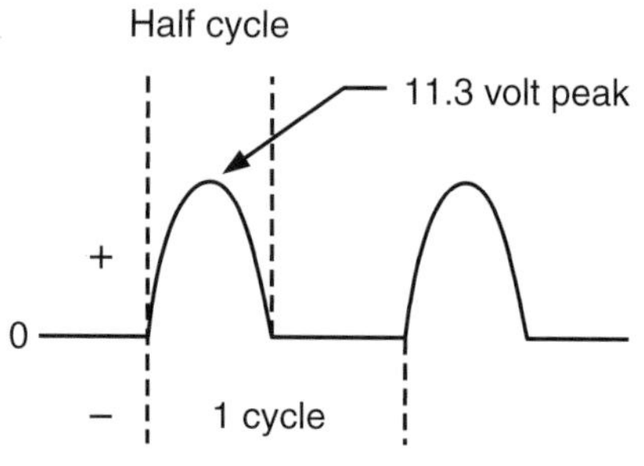

FULL-WAVE CENTER TAP RECTIFIER

3.7 Rectifying Characteristics

A full-wave center tap (CT) rectifier uses two diodes and a CT on the secondary of the transformer to produce a positive output voltage on both positive and negative alternations of the AC input voltage cycle. This waveform is pulsating DC with half the peak voltage of the secondary (less the diode forward voltage drop) because of the CT. Figure 3–11 shows a simple full-wave CT rectifier and accompanying output waveform.

FIGURE 3–11 Simple full-wave rectifier circuit.

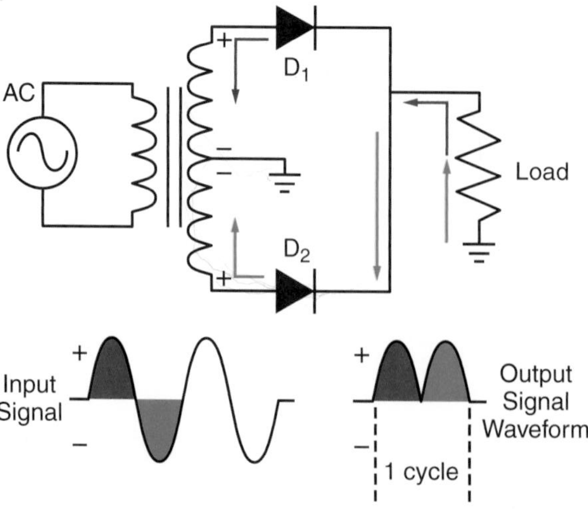

Note that by using a CT referenced to ground each half cycle produces an equal but opposite voltage across the diodes. When D_1 is conducting, D_2 is reverse biased with respect to ground and the load. The opposite is true when D_2 conducts: D_1 is reverse biased with respect to ground and the load. With this arrangement, each diode conducts and allows a voltage drop across the load on alternating half cycles. An output waveform is produced that has a positive peak for each half cycle of the AC input voltage or current.

Figure 3–12 shows various rectifier case styles. Case TO-220AB is a half-wave two-diode rectifier. The remaining cases are full-wave bridge four-diode rectifiers. The full-wave bridge is discussed later in the chapter.

FIGURE 3–12 Solid-state rectifier case styles.

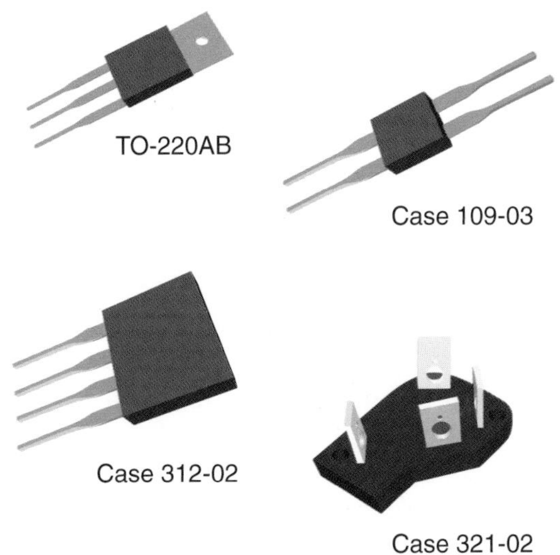

FIGURE 3–13 Full-wave rectifier circuit.

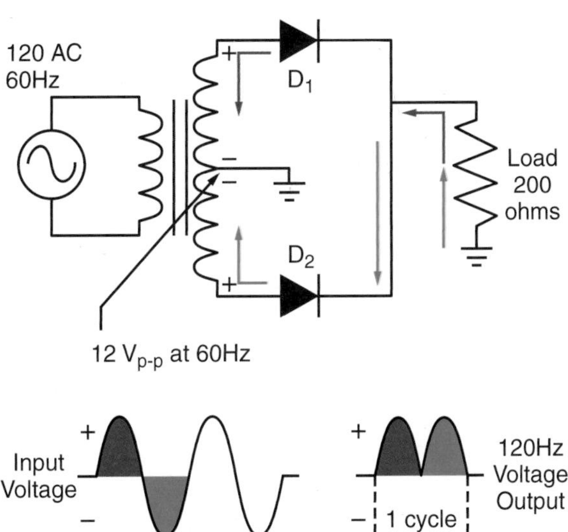

3.8 Circuit Analysis

Example

Using Figure 3–13, solve for the load voltages V_P, V_{EFF}, V_{AVE}, and I_P. Solution: The secondary of the transformer is 24 $V_{P\text{-}P}$ and it is center tapped. Note that the peak voltage supplied to the load from the secondary is only half value because of the center tap reference, therefore, the peak load voltage is:

$$V_{P(Load)} = \frac{V_{P\text{-}P}}{2}$$

$$V_{P(Load)} = V_P - D_1$$

$$V_{P(Load)} = 6\text{V} - 0.7\text{ V}$$

$$V_{P(Load)} = 5.3\text{ V}$$

Recall that a full-wave rectifier has twice as many DC output pulses as the half-wave rectifier. From Table 3–1, the average and RMS load voltages are:

$$V_{AVE} = \frac{2 \times V_{P(Load)}}{\pi}$$

$$V_{AVE} = \frac{2 \times 5.3\text{ V}}{\pi}$$

$$V_{AVE} = \frac{10.6\text{ V}}{3.141}$$

$$V_{AVE} = 3.37\text{ V}$$

$$V_{RMS} = \frac{V_{P(Load)}}{\sqrt{2}}$$

$$V_{RMS} = \frac{5.3\text{ V}}{1.414}$$

$$V_{RMS} = 3.75\text{ V}$$

The peak load current through the load can now be calculated.

$$I_P = \frac{V_{P(Load)}}{R_L}$$

$$I_P = \frac{5.3\text{ V}}{200\ \Omega}$$

$$I_P = 26.5\text{ mA}$$

To find the values of the average and RMS currents the formulas are:

$$I_{AVE} = \frac{2 \times I_P}{\pi}$$

$$I_{AVE} = \frac{2 \times 26.5\text{ mA}}{\pi}$$

$$I_{AVE} = \frac{53\text{ mA}}{3.141}$$

$$I_{AVE} = 16.87\text{ mA}$$

(continued)

$$I_{RMS} = \frac{I_P}{\sqrt{2}}$$

$$I_{RMS} = \frac{26.5 \text{ mA}}{1.414}$$

$$I_{RMS} = 18.74 \text{ mA}$$

Note that with half the voltage the full-wave CT rectifier supplies almost the same average and RMS current to the load. Moreover, the full-wave pulsating DC is at twice the input or line frequency, has less ripple, and will allow better filtering. In addition, the full-wave "center-tap (CT) transformer" rectifier peak voltage and current are 50% less than the half-wave rectifier.

This allows both the voltage and current rating of the full-wave CT rectifier diodes to be 50% less in comparison to the half-wave rectifier diode. Because the full-wave rectifier ripple frequency is twice that of the half-wave rectifier, you may expect the filter capacitor(s) to be of lower capacitance and of smaller physical dimensions.

Of course, this analysis does not mean that you can arbitrarily use half the voltage when designing a full-wave CT rectifier compared to a half-wave rectifier. It does, however, show that the extra pulse provided for each cycle makes a full-wave rectifier a more efficient and generally better approach.

FULL-WAVE BRIDGE RECTIFIER

3.9 Rectifying Characteristics

One of the most useful and common rectifiers is the full-wave bridge rectifier (bridge rectifier). It does not use a CT transformer but instead four diodes to rectify the incoming AC voltage or current. Figure 3–14 shows a simple bridge rectifier circuit. Notice that like the full-wave CT rectifier just discussed the pulsating DC output frequency is double that of the AC input or line frequency.

FIGURE 3–14 Bridge rectifier circuit.

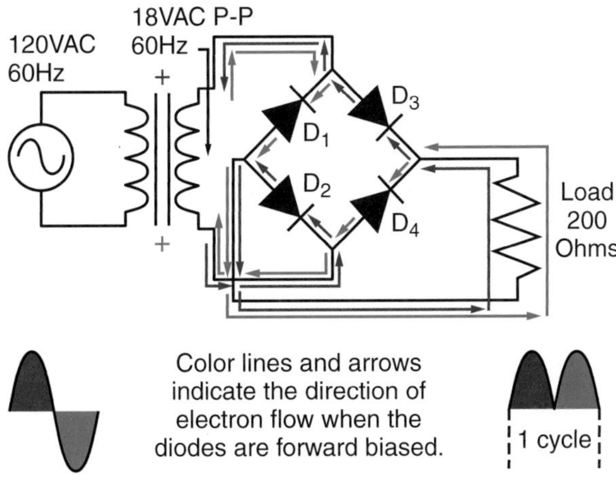

Another characteristic of the bridge rectifier is that the voltage drop across the diodes will be approximately double that of the CT full-wave rectifier because there are two diodes in each conducting path.

3.10 Circuit Analysis

The current path in Figure 3–14 is shown in blue for the first half cycle of the AC input voltage. The second half-cycle current path is shown in green. As the input AC builds to a positive (blue + symbol), the electrons flow from the negative terminal of the secondary through D_2 to the load. Electrons continue to flow through the load and on through D_3 back to the positive terminal. This completes the first half cycle of the AC input voltage.

The second half cycle builds a positive potential on the secondary side of the transformer (green + symbol). Electrons now reverse direction and flow from the negative secondary terminal through D_1 to the load. The electrons continue through the load and on through D_4 back to the positive terminal. This completes the second half cycle of the AC input. Note that the voltage or current always flows the same direction through the load no matter which input half cycle (negative or positive) is generating the signal.

Example

Refer to Figure 3–14 and solve for the following load voltages: V_P, V_{EFF}, V_{AVE}, and I_P.

Solution:

The peak voltage across the load is:

$$V_{P(Load)} = \frac{V_{P\text{-}P(XFMR)}}{2} - (V_{D1} + V_{D4})$$

$$V_{P(Load)} = \frac{18\text{ V}}{2} - (0.7\text{ V} + 0.7\text{ V})$$

$$V_{P(Load)} = 9\text{ V} - 1.4\text{ V}$$

$$V_{P(Load)} = 7.6\text{ V}$$

From Table 3-1, the average and effective output voltages are:

$$V_{AVE} = \frac{2 \times [V_{P(XFMR)} - (V_{D1} + V_{D4})]}{\pi}$$

$$V_{AVE} = \frac{2 \times (9.0\text{ V} - 1.4\text{ V})}{3.14}$$

$$V_{AVE} = 4.84\text{ V}$$

$$V_{RMS} = \frac{[V_{P(XFMR)} - (V_{D1} + V_{D4})]}{\sqrt{2}}$$

$$V_{RMS} = \frac{(9.0\text{ V} - 1.4\text{ V})}{1.414}$$

$$V_{RMS} = 5.37\text{ V}$$

Using Ohm's law, the peak, average, and RMS load currents are:

$$I_P = \frac{[V_{P(XFMR)} - (V_{D1} + V_{D4})]}{R_L}$$

(continued)

$$I_P = \frac{(9.0V - 1.4\text{ V})}{200\ \Omega}$$

$$I_P = 38\text{ mA}$$

$$I_{AVE} = \frac{V_{AVE}}{R_L}$$

$$I_{AVE} = \frac{4.84\text{ V}}{200\ \Omega}$$

$$I_{AVE} = 24.2\text{ mA}$$

$$I_{AVE} = \frac{V_{RMS}}{R_L}$$

$$I_{AVE} = \frac{5.37\text{ V}}{200\ \Omega}$$

$$I_{AVE} = 26.85\text{ mA}$$

Figure 3–15 summarizes some of what you have learned so far. There are three basic types of rectifiers: the half-wave, the full-wave CT, and the full-wave bridge. Each schematic and output waveform is shown in the three views presented.

Regulated power supplies using transistors are discussed in later chapters. Finally, there are other types of common power supplies (such as three-phase rectification and phase inverters) outside the scope of this text.

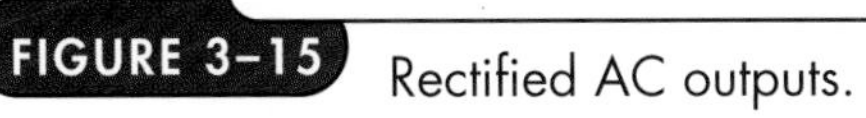

FIGURE 3–15 Rectified AC outputs.

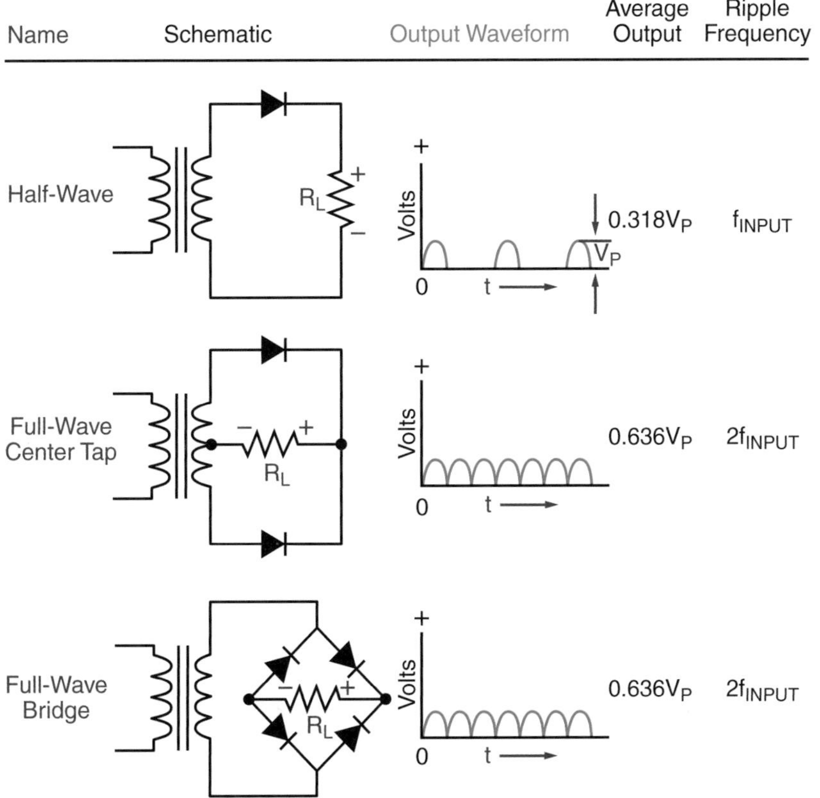

FILTERS

A filter is a device that helps to change the pulsating DC into a smoother DC output. It is called a filter because it filters or eliminates much of the ripple from the DC waveform. Virtually all of the so-called passive filters use either capacitors or inductors. Although their actual connections vary, the basic principle of both is the same. The capacitor or inductor absorbs energy during the peak of the voltage pulses and then releases it as the voltage falls off. Thus, the filter supports the voltage during its discharge cycle and the pulsations are decreased.

3.11 Capacitors

Dielectric
The insulating material separating capacitor plates, where electric energy is stored.

Power supply
An electric circuit used to convert electric voltages and currents from one form to another form suitable for a particular application. Example: change the AC main voltage into a DC voltage suitable for powering an electronic computer.

Capacitors perform a variety of jobs such as storing an electrical charge to produce a large current pulse. Two conductors, usually plates, separated by some type of insulating material (called the **dielectric**) make up a capacitor. The most common type of capacitor you will use in **power supply** work is the electrolytic capacitor. See Figure 3–16 for images of actual capacitors.

FIGURE 3–16 Types of capacitors.

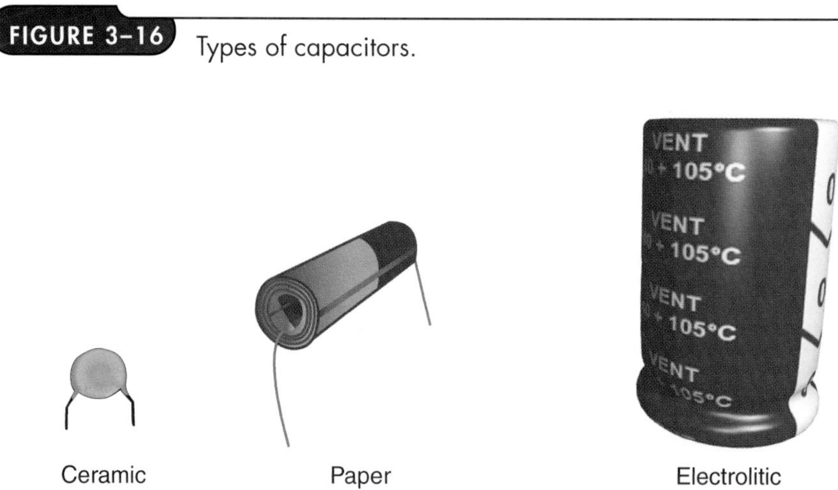

Voltage Ratings

The voltage rating of a capacitor is actually the working rating of the dielectric. The voltage rating is extremely important, and the life of the capacitor will be greatly reduced if it is exceeded. The voltage rating indicates the maximum amount of voltage the dielectric is intended to withstand without breaking down. If the voltage becomes too great, the dielectric will break down—allowing current to flow between the plates. At this point, the capacitor is shorted. For electrolytic capacitors, the voltage rating (also called the **DC working voltage**) is the DC or average voltage applied to the capacitor.

DC working voltage
The voltage rating of a capacitor based upon average applied voltage.

Capacitors as Filters

Capacitors oppose a change in voltage. When connected to alternating current, current will appear to flow through the capacitor. The reason is that in an AC circuit the polarity is continually changing, causing the current to change direction. This flow of current in and out of the capacitor constitutes current flow through a load connected in the circuit.

As noted, the change in voltage in the circuit is opposed by the charge stored in the capacitor. As the voltage potential goes negative from its peak value, the charge stored in the capacitor from the peak voltage now discharges into the circuit. This discharging voltage helps prevent the peak voltage from going negative and thus reduces the ripple in the rectified AC output.

Figure 3–17 shows the voltage opposition effect of a "filter" capacitor in a rectifier circuit. The term *filter* refers to the capacitor's ability to reduce the pulsating DC output ripple. Recall that a capacitor requires five time constants to fully charge or fully discharge. To effectively reduce the output ripple, the capacitor should not be allowed to completely discharge between half cycles. The following section of this lesson explores this concept in detail.

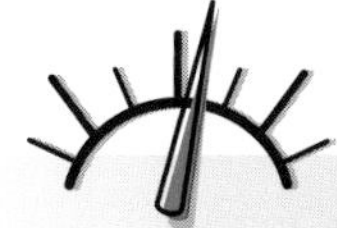

TechTip!

Replacement capacitors should be chosen having DC working voltages (DCWVs) at least equal to the average DC voltage across the capacitor plates. It is also desirable (because electrolytics can deteriorate from disuse or undervoltage) to select replacement capacitors with rated working voltages as close to but not less than the DCWV required.

FIGURE 3–17 Bridge rectifier with filter capacitor.

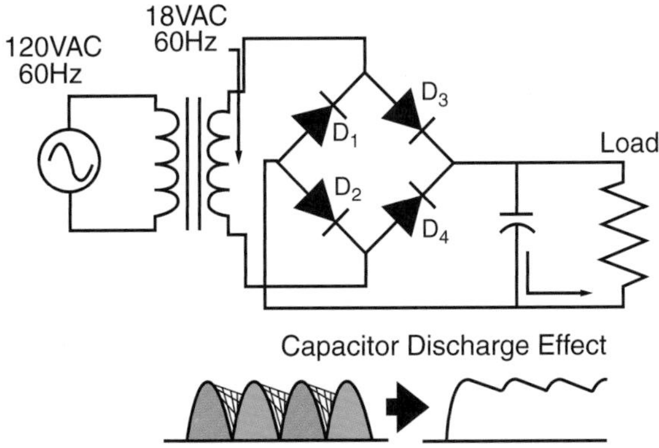

3.12 Rectifying Characteristic Time Constants

Theory

Capacitors charge and discharge at an exponential rate. The curve is divided into five time constants, and each time constant is equal to 63.2% of the remaining value. This pattern continues until the capacitor is fully charged. See Figures 3-18a and b for charge and discharge times. This rate can be changed by changing the resistance placed in parallel with the capacitor. Together they form what is called an RC time constant. R stands for resistance and C stands for capacitance. The formula for calculating the required time constant is

$$T = RC.$$

Here:

T = time in seconds

R = resistance in ohms

C = capacitance in farads

FIGURE 3–18 Capacitor time curves. (a) Charging. (b) Discharging.

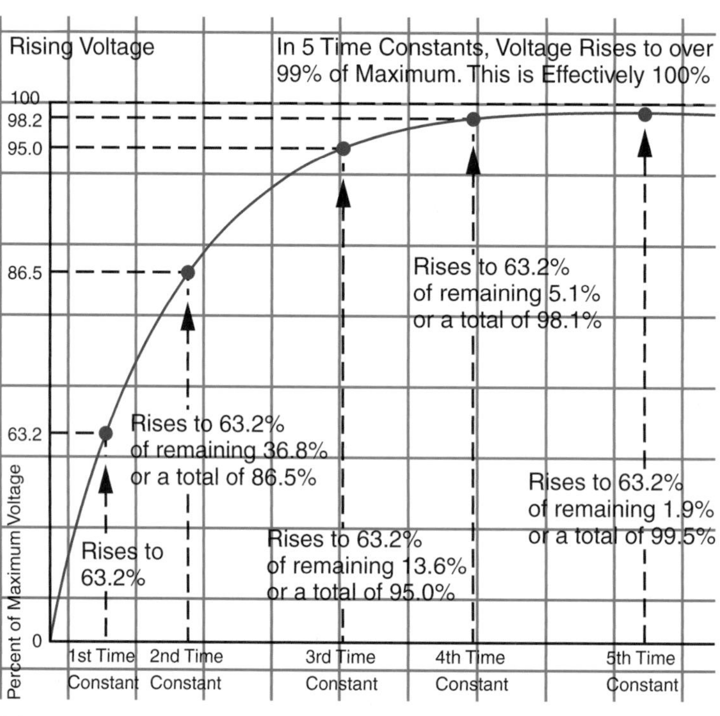

(a)

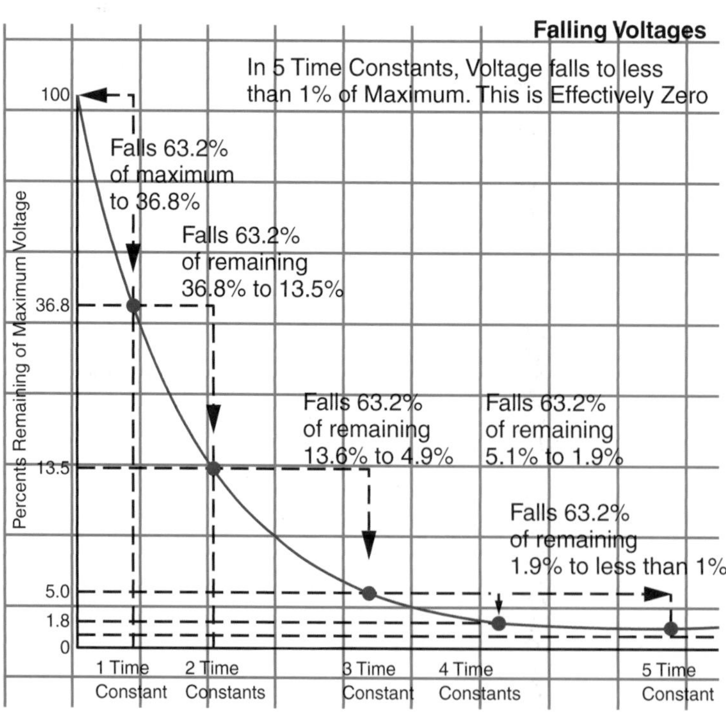

(b)

The actual formulas that govern the charge and discharge of the capacitor through a resistor can be analyzed by looking at the circuits shown in Figure 3–19.

FIGURE 3–19 Capacitor circuits. (a) Charging. (b) Discharging.

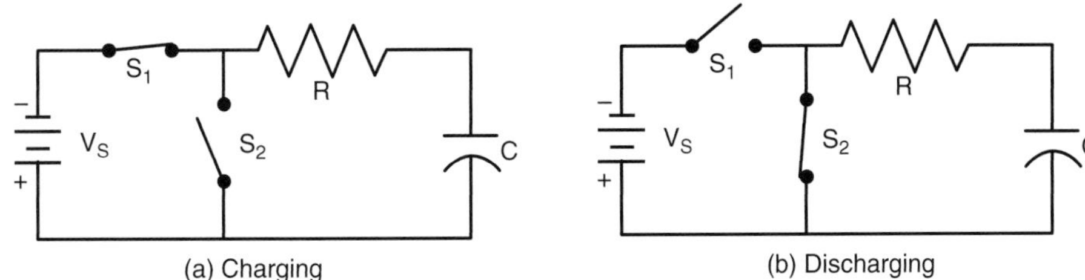

Charging

When switch S_1 is closed, the capacitor will start to charge. The voltage across the capacitor at any time after the switch closes is given by $V_C = V_S\left(1 - e^{-\left(\frac{t}{RC}\right)}\right)$. Although this formula may seem formidable, it is easily calculated using a scientific calculator, electronic spreadsheet, or other such tool.

Table 3–2 outlines the values for V_C as calculated from the previous formula for $t = 0, 1, 2, 3, 4$, and 5 seconds, assuming the $T = RC = 1$ second. Note that the values calculated from the formula agree with the values shown in Figure 3–18a. In fact, that is where the values came from.

TABLE 3–2 Capacitor Voltages for Charge and Discharge

	Time (*t*) in Seconds					
Condition	0	1	2	3	4	5
V_C charging • S_1 closed • S_2 open	0	0.632 V_s	0.865 V_s	0.950 V_s	0.982 V_s	0.993 V_s
V_C discharging • S_1 open • S_2 closed	V_s	0.368 V_s	0.135 V_s	0.0498 V_s	0.0183 V_s	0.0067 V_s

Discharging

If you open S_1 and immediately close S_2, the capacitor will discharge through the resistor. The values of voltage on the capacitor at $t = 0, 1, 2, 3, 4$, and 5 seconds are shown in Table 3-2. These values are calculated from the discharge formula, which is $V_C = V_S \times e^{-\left(\frac{t}{RC}\right)}$.

TechTip!

The following outlines how to troubleshoot a full-wave rectifier.

1. Make a visual inspection of the power supply. Look for broken or damaged parts.
2. Are you sure the power supply is malfunctioning? Be sure. Power up and verify that power supply (PS) DC output voltage is missing, out of tolerance, or exhibits excessive ripple.
3. If voltage is zero or near zero:
   a. Power down. Disconnect the load from the PS. Measure resistance across the load. If it is zero or near zero, the problem is *not* the PS. It is the load!
   b. Power up. Measure AC voltage present at the transformer secondary. If there is none, verify primary transformer voltage is present and correct. If it is, suspect the transformer. Otherwise, trace wiring back to the source to find an "open."
   c. Power down. Remove and replace the filter capacitor. While the capacitor is out of circuit, test the rectifier diode PN junction voltages to be sure they have not been damaged. If the diodes are inexpensive and available, it may be wise to replace them anyway because a shorted capacitor would have caused excessive diode current and possible failure.

continued

Example

You can calculate the effect of different sizes of filter capacitors by using the $T = RC$ formula. Assume a load of 200 ohms and a filter capacitor of 50 μF (50×10^{-6} farads). The time required for 1 time constant = 0.01 seconds ($T = RC = (200 \times (50 \times 10^{-6}) = 0.01$ second.). This is about the same as a half cycle for a 60-Hz line voltage. The capacitor will discharge about 63% of its charge, causing quite a bit of ripple.

Increasing the capacitor's size will increase the time constant proportionally. With an increase in time constant, the length of discharge decreases and the amount of ripple decreases. Increasing capacitance of a filter is not the complete solution to ripple problems, however. Increasing capacitance excessively will result in high initial surge current, perhaps causing damage to other circuit components.

Using the same formula ($T = RC$), assume a load of 200 ohms and increase the capacitor to 500 microfarads. One time constant would be 0.1 second. This means that there is a long time required for capacitor discharge compared to the time allowed by the circuit. Little charge will be lost by the capacitor, and the DC output ripple will be greatly reduced.

See Figure 3–20 for an example of ripple outputs using different filter capacitors.

FIGURE 3–20 Filter capacitor effect on ripple.

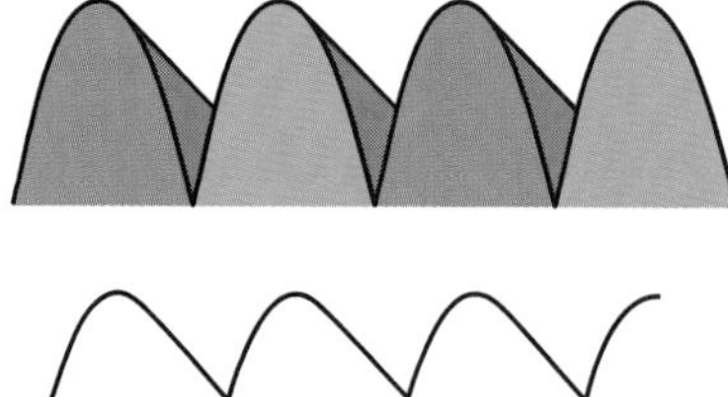

50μF Capacitor Discharge- loses about 63% of its charge from half cycle to half cycle.

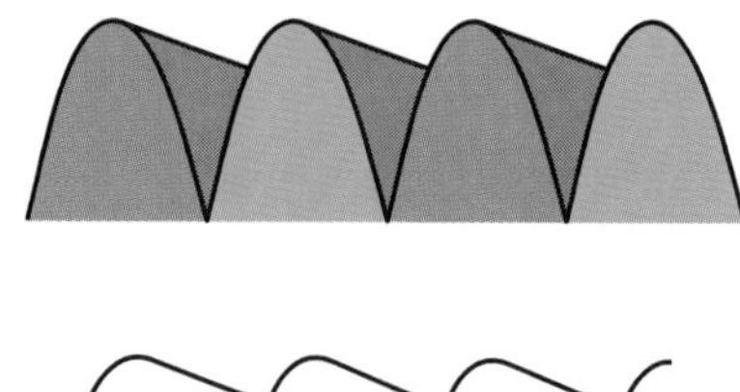

500μF Capacitor Discharge- loses about 30% of its charge from half cycle to half cycle.
Notice reduced ripple from larger capacitor

Note that the overall DC voltage output of a filtered rectifier will be higher than a nonfiltered one. This is because the charge on the capacitor keeps the voltage across the load near peak value. As you look at the time constant formula, $T = RC$, you will notice that if resistance decreases capacitance has to increase for time to remain the same.

A practical application of this is that when the load starts to draw a lot of current (resistance decreases) the capacitor must have a higher value to maintain the same filtering (ripple reduction) capabilities. This is one of the reasons capacitors for DC machinery usually have very large values.

3.13 Chokes

Coils (inductors) used as rectifier filtering devices are called **chokes.** Chokes get their name from the function they perform in the rectifier circuit. They "choke," or reduce, the current ripple to the load and thus help reduce the voltage ripple. Inductors (chokes) oppose a change in circuit current. The opposition to change in current by a choke is at the same rate as the opposition to voltage change for a capacitor.

It takes five time constants for an inductor to reach maximum current induction. The time constant for an inductor is $T = LR$. Each time constant is at 63.2% of the total current value remaining. Figure 3–21 shows an example of a choke used as an additional filter with a capacitor.

FIGURE 3–21 Choke used as a filter.

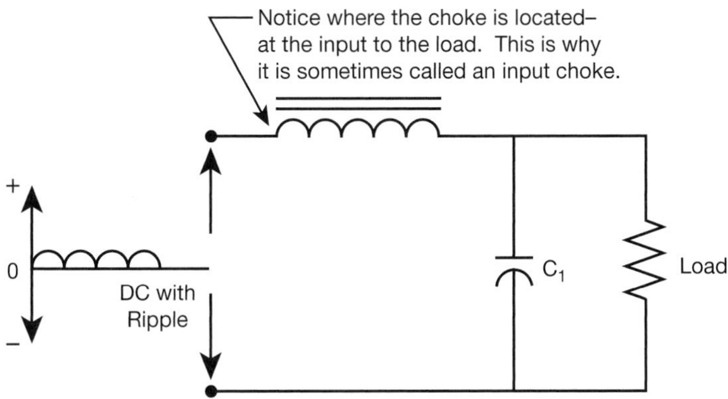
Notice where the choke is located–
at the input to the load. This is why
it is sometimes called an input choke.
+
0
–
DC with
Ripple
C1
Load
At low frequencies the choke's X_L will be extremely small and the current ripple will be felt on the load and across the capacitor. At high frequencies the X_L will increase and the ripple will be blocked from the load.

Another common type of choke filter is the power-line filter. Power lines are normally only 60 Hz and carry our common household currents and voltages. The lines do, however, provide a conduction source for RF (radio frequency) currents. These current signals cause interference with motors, fluorescent lighting controls, radios, computers, and other equipment. To filter out or trap these RF currents, an input choke is used. Figure 3–22 shows a balanced low-pass filter.

The term **low-pass** comes from a choke's ability to allow lower frequency current to pass, while attenuating unwanted frequency currents. At high frequencies, the choke's high impedance drops most of the high-frequency current ripple across the choke. Two chokes are used to balance the line's AC with respect to ground. Chokes are best used with unregulated power supplies that have high current demands. For smaller semiconductor power supplies, voltage regulation is used. This has two advantages over the choke: cost and size.

TechTip! (cont'd)

4. If the voltage is out of tolerance:
   a. Measure peak voltage. If the peak is too high, suspect overvoltage on the transformer primary, or shorted transformer primary winding (expect an overheated transformer). If peak voltage is too low, suspect the AC input voltage of being too low or a shorted transformer secondary (expect an overheated transformer).
   b. Inspect the ripple waveform. If peaks do not appear at twice the AC input frequency, a diode is open. Disconnect diodes, and measure their PN junction voltages.
   c. Check the ripple voltage. If it is too large, it is likely the filter capacitor has lost capacitance or is "open circuited." Replace the capacitor. Otherwise, the load may be heavier than normal.

Choke
A coil or inductor used as a filter, often in an AC-to-DC power supply.

Low-pass
The filter type which passes low frequencies more readily than higher frequencies.

ThinkSafe!

Although an oscilloscope is a useful tool for analyzing power supply (and other) circuits, the technician should be aware of limitations of his oscilloscope. Some oscilloscopes are severely limited in the maximum peak-to-peak voltage they can measure. Exceeding that limitation can endanger the oscilloscope and the operator.

Another major concern is electrical isolation of the vertical inputs from the oscilloscope chassis. Some oscilloscopes are not isolated, or are inadequately isolated from their AC power inputs. If not isolated, the circuit under test may become damaged during testing, the oscilloscope may be damaged, the AC power line may be overloaded, and most importantly the technician may be seriously injured by electrical shock. Do not be afraid to use your oscilloscope, but *know its specifications*.

TechTip!

Sometimes a rectifier power supply component becomes short circuited, and yet the primary transformer fuse may not open. Why not? Remember that *every* component in the power supply has at least some resistance. That includes power transformer primary and secondary windings. A transformer may seem excessively hot due to the overly high current drawn by a leaky or shorted component further downstream.

FIGURE 3–22 Balanced choke low-pass filter.

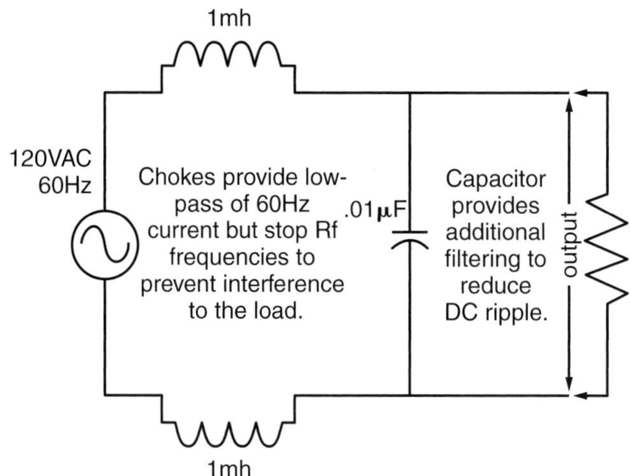

Figures 3–23a through d show a number of common low-pass filter configurations for power supplies. In the AC chapter on filters, we covered these and others as low-pass and high-pass filters.

FIGURE 3–23 Typical low-pass filters for power supplies. (a) Choke acts as a voltage divider with the load resistor. (b) Adding a bypass capacitor aids filtering by providing sharper cutoff at low frequencies. (c) This T-type filter uses an additional choke (L2) to further reduce ripple. (d) Pi (π)-type filters use an initial capacitor (C_1) to provide additional filtering.

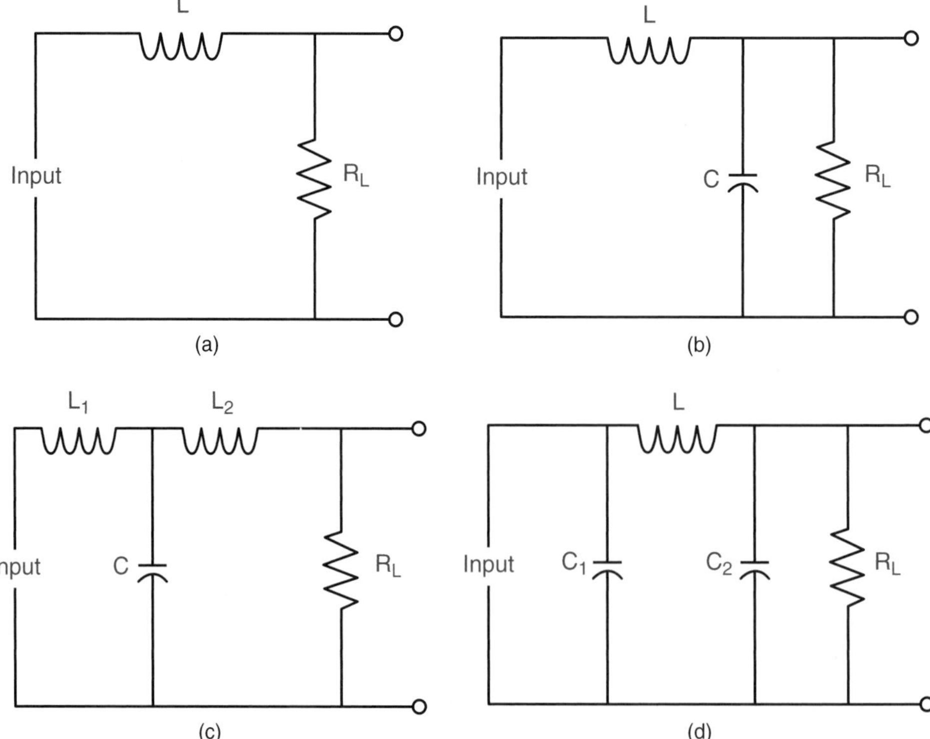

VOLTAGE REGULATION AND OUTPUT

So far in this chapter you have learned about the types of rectifiers: half-wave, full-wave CT, and bridge. You have seen how each of these takes an incoming AC signal and rectifies it to an output pulsating DC (DC with ripple). Finally, you have learned how capacitors and chokes are used to filter or reduce the DC ripple. The last major sections of a power supply are voltage regulation and output.

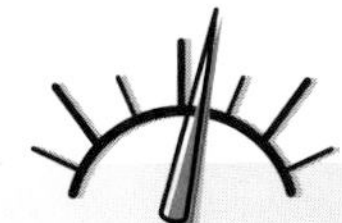

TechTip!

In regard to spike suppression, it is commonplace to find small nonpolarized capacitors wired in parallel with rectifier diodes. Do not let it confuse you. The purpose of these capacitors is to prevent reverse-voltage spikes from damaging the diodes. They are most often found in high-voltage power supplies. These capacitors rarely short circuit.

3.14 Voltage Regulation

Figure 3–24 shows an example of a regulated power supply. The Zener diode is rated at 9 V, and thus the output across the load will remain relatively constant. The capacitor will provide ripple reduction from the pulsating DC output of the bridge. Reducing the ripple in the DC will ensure that the Zener stays in reverse bias condition and operates properly even under varying load conditions.

FIGURE 3–24 Regulated power supply.

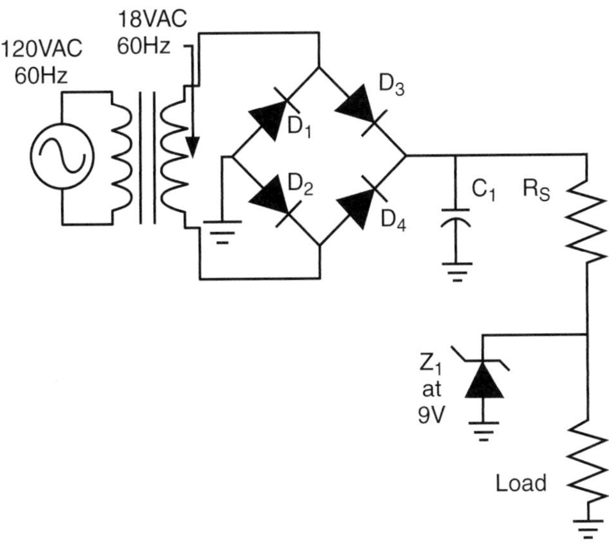

3.15 Power Supply Output

An alternative to Zener diode regulation for a power supply is to use an adjustable resistor in series with the output. Figure 3–25 shows the alternative schematic for a regulated power supply. Note that the outputs can be combined or used individually. This type of variable output uses a voltage divider network. For example, the voltage output between points *B* and *A* is 5 V, between *C* and *A* is 10 V, between *D* and *A* is 15 V, between *E* and *A* is 20 V.

Note that such a power supply is not normally used for circuits with widely varying current requirements. The reason is that the resistor has to be adjusted manually to keep the voltage output constant. The use of a 20 V Zener diode in parallel with the output resistors would be a more efficient circuit. The adjustable resistor may be replaced with an appropriate fixed resistor.

TechTip!

In regard to bleeder resistors, better-quality power supplies come with resistors connected across their output voltage points. They serve two purposes. One purpose is to drain current from the power supply filter capacitors once the power supply has been powered down. This ensures that the technician will not be harmed inadvertently. It also allows the technician to perform power-off testing of the equipment without worrying about stray currents.

The second purpose of bleeder resistors is to assist with regulation of the output voltage. If the normal load should become unusually light or disconnected, and without a bleeder, power supply voltage could climb as high as the peak power supply voltage. The potential exists for some components, especially solid-state components, to be damaged. When used, bleeder resistors are typically selected to draw approximately 10% of the power supply's rated output current.

The voltage divider regulation technique has the added advantage of acting as a bleeder resistor. However, should a voltage divider become open circuited huge overvoltages may appear at the load(s)!

FIGURE 3–25 Power supply with voltage regulation and multiple output voltages.

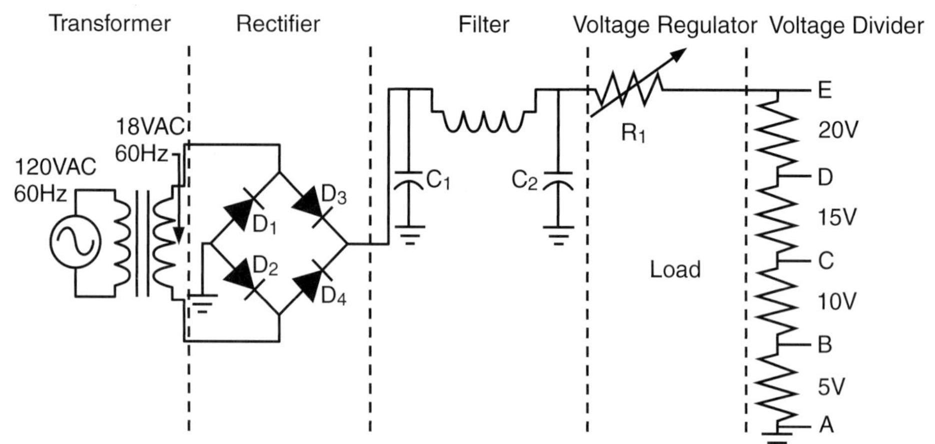

3.16 Voltage Doublers

Voltage doublers and other multiplier circuits provide a DC output that can be double, triple, and quadruple the peak input signal. Because the voltage doubler and multiplier circuits are not capable of generating voltage, the increased input voltage is offset by the loss of circuit current. For example, in a voltage doubler circuit 10 V at 50 mA input would yield slightly less than 20 V at 25 mA output. This is necessary to maintain power balance in the circuit (i.e., power in = power out + losses).

Figure 3–26 shows an example of a full-wave voltage doubler illustrating representative input and output voltages. There are two basic configurations for the voltage doubler. The half-wave is shown in Figure 3–27 and the full-wave is shown in Figure 3–28.

FIGURE 3–26 Voltage doubler.

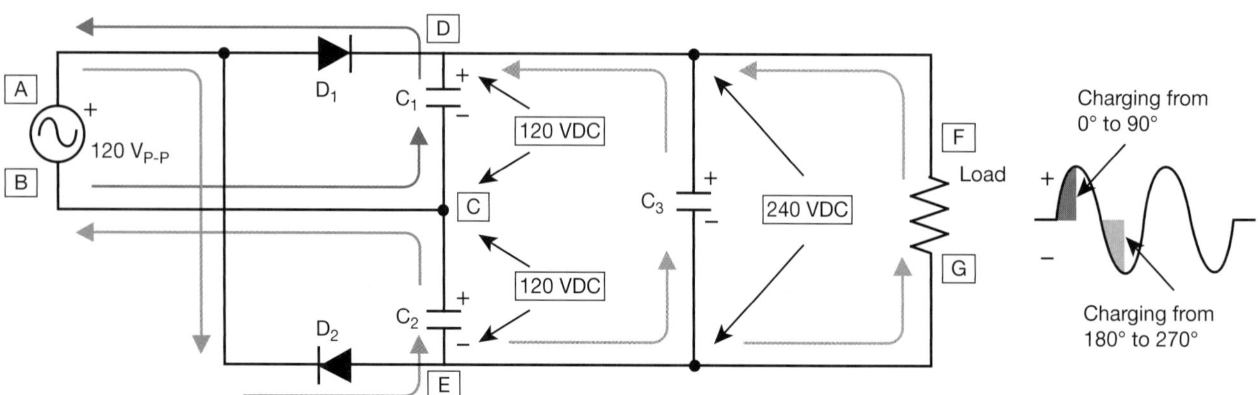

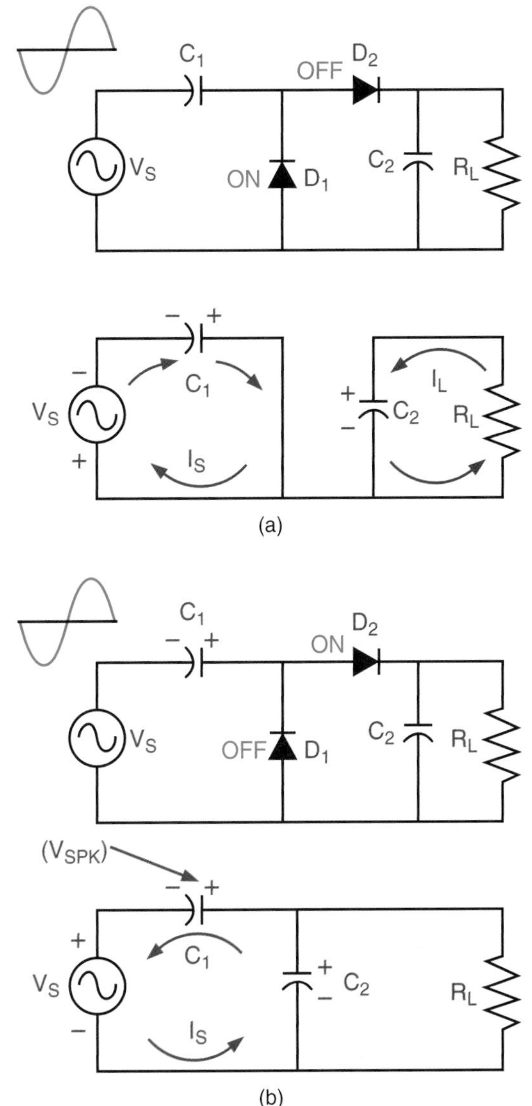

FIGURE 3–27

Half-wave voltage doubler.
(a) Negative half cycle.
(b) Positive half cycle.

Half-Wave Voltage Doubler

During the negative alternation of the input AC signal (Figure 3–27a), D_1 is forward biased and acts as a short and D_2 is reverse biased and acts as an open. The equivalent circuit shows that C_1 charges to the value of the source voltage and that C_2 discharges through the load resistor. As the input AC signal goes positive (Figure 3–27b), the diodes switch operation. D_2 becomes forward biased and D_1 is reverse biased. C_1 now discharges and acts as a series, aiding voltage to the AC signal source.

Full-Wave Voltage Doubler

A full-wave voltage doubler circuit is shown in Figure 3–28. During the positive alternation of the input (Figure 3–28b) signal, D_1 is forward biased and D_2 is reverse biased. This allows C_1 to charge to $V_{S(Peak)}$. As the input signal goes negative (Figure 3–28c), the diodes reverse their condition and C_2 now charges to $V_{S(Peak)}$. The two voltages in series provide 2 $V_{S(Peak)}$ across C_3 and the load. C_3 does not add to the charge, but has a value designed to reduce the ripple output from the charge–discharge cycle of the input signal and C_1 and C_2.

FIGURE 3–28 Full-wave voltage doubler. (a) Circuit. (b) Positive half cycle. (c) Negative half cycle.

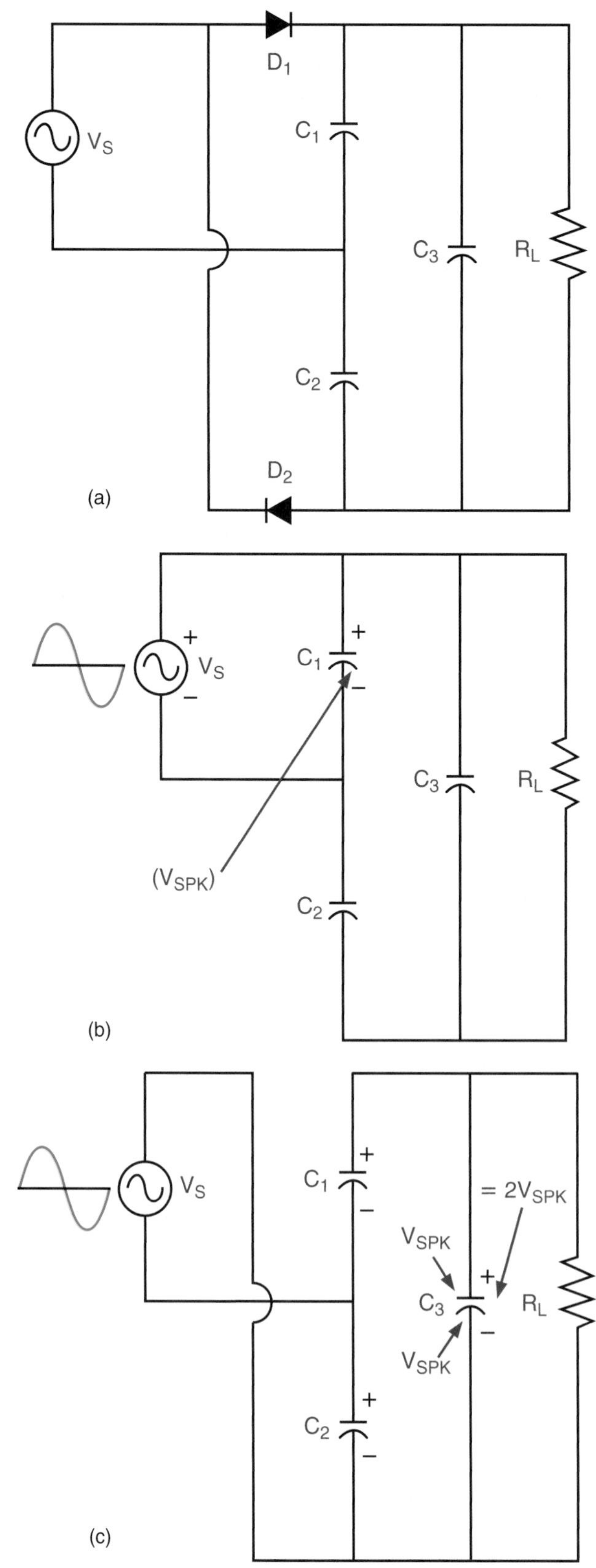

Principles of Operation

Examine Figure 3–26. The circuit operates as follows. During the first 90° of the sine-wave cycle, D_1 is forward biased and D_2 is reverse biased. Charging current will flow as indicated by the blue arrows.

C_1 will charge to the peak input voltage, less only the small voltage drop across the PN junction of D_1. Because the input voltage is 170 V_{Peak} (120 V_{RMS} $\times \sqrt{2}$), C_1 will charge to 169 V at 90° into the sine-wave cycle. No charging current will flow between 90° and 180° because C_1, now charged, causes D_1 to become reverse biased. The voltage at point D is positive with respect to point C.

Follow the green arrows when the input sine wave passes 180°. Point B becomes more positive than point A. Current can now flow from point A, through D_2 and C_2, and back to point B. Because current flows in the direction of the green arrows, C_2 will reach 169 V at 270° into the first sine-wave cycle.

Between 270° and 360°, D_2 is reverse biased. Both C_1 and C_2 are charged. Observe that the voltages across C_1 and C_2 are connected so that voltage across points D and E are additive. Therefore, voltage across filter capacitor C_3 and the load at points F and G will approach the sum of the voltage across C_1 and C_2 (or ideally, 338 VDC). C_3 is an additional filtering capacitor used to reduce the ripple of the DC output from C_1 and C_2.

In practice, ripple voltage in voltage doubler circuits is high, and with continuous loading DC output will never reach the peak output voltage. Because C_1 charges during the positive half cycle of input, and C_2 charges during the negative half cycle of input, the DC output voltage will have a ripple frequency of twice the sine-wave input frequency. For example, if the frequency of the input is 60 Hz output ripple frequency is 120 Hz. (Figures 3–29 through 3–31 show various types of power supplies.)

FIGURE 3–29 24 VDC to 5 VDC regulated power supply block.

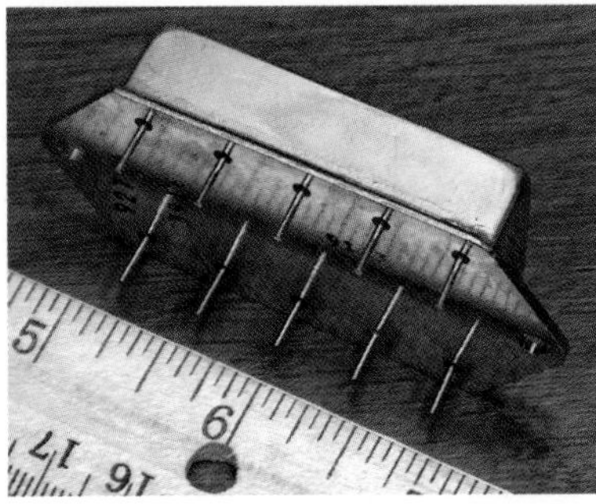

FIGURE 3–30 Multi-voltage switching power supply.

FIGURE 3–31 Series-regulated 12 VDC power supply.

SUMMARY

Power supplies have the following main components:

- Transformer
- Rectifier
- Filter
- Voltage regulator
- Voltage output or variable outputs

The transformer usually provides a stepped-up (increased) or stepped-down (decreased) AC voltage to the rectifier. The rectifier converts the AC signal into pulsating DC (DC with ripple). The filter removes much of the DC ripple through the combined actions of capacitors and inductors or chokes. The voltage regulator is constructed of either a variable resistor connected in series with the load or a Zener diode connected in parallel with the load. Output voltages can be controlled through voltage divider resistors connected in series. These resistors offer different "tap" points for varying output voltages.

REVIEW QUESTIONS

1. Explain in your own words the purpose of a DC power supply.
2. How do capacitors and chokes work to reduce the AC ripple in the output of a power supply?
3. Refer to Figure 3–32. What would happen to the circuit if diode D_3 were reversed?

FIGURE 3–32 Bridge rectifier circuit.

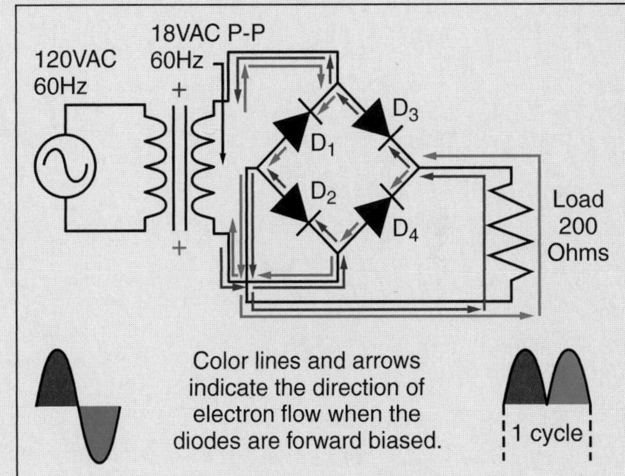

4. A certain power supply has a capacitor filter. What would likely happen to the output ripple if the capacitor were reduced to half its size?
5. Examine the half-wave and full-wave voltage doublers in Figures 3–33 and 3–35, respectively. Read and memorize the operation description given in the text, and then explain their operation without the book.

FIGURE 3–33 Half-wave voltage doubler. (a) Negative half cycle. (b) Positive half cycle.

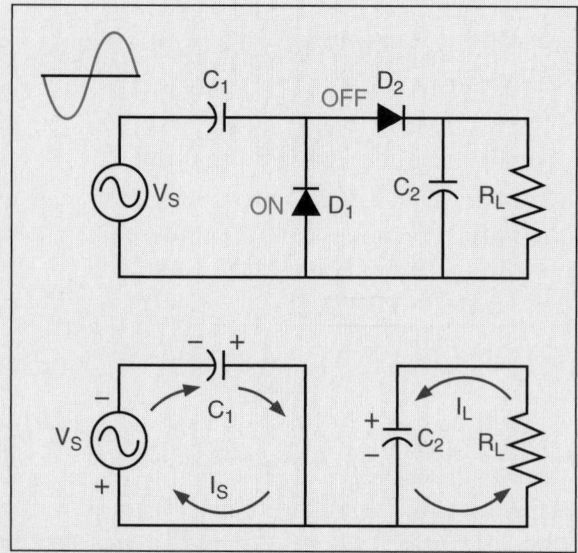

(a)

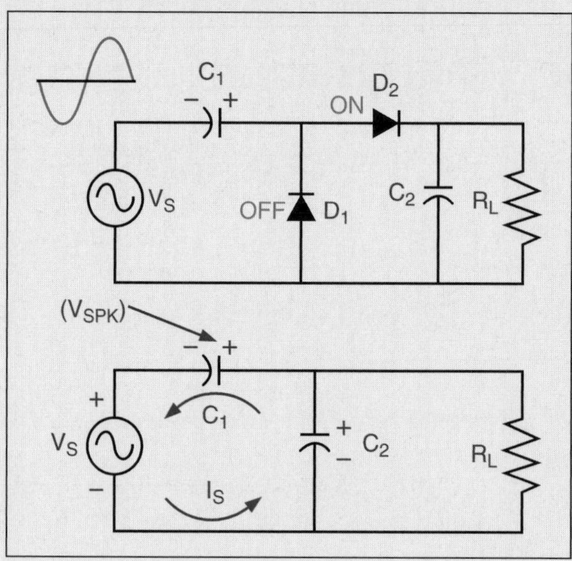

(b)

FIGURE 3–34 Full-wave voltage doubler. (a) Circuit. (b) Positive half cycle. (c) Negative half cycle.

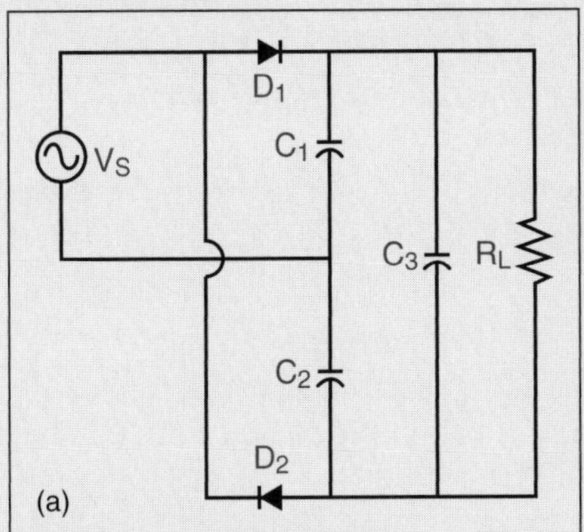

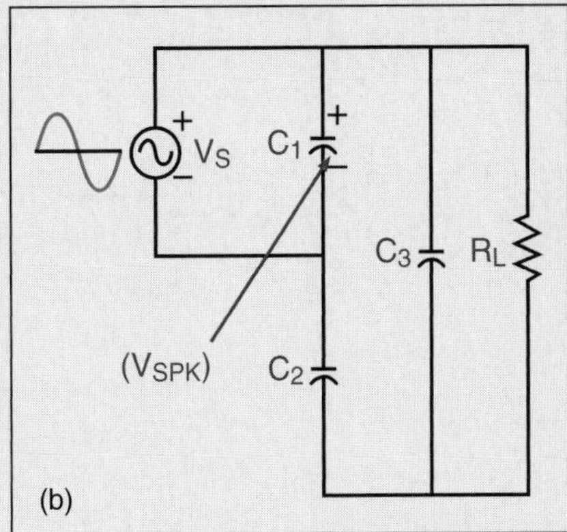

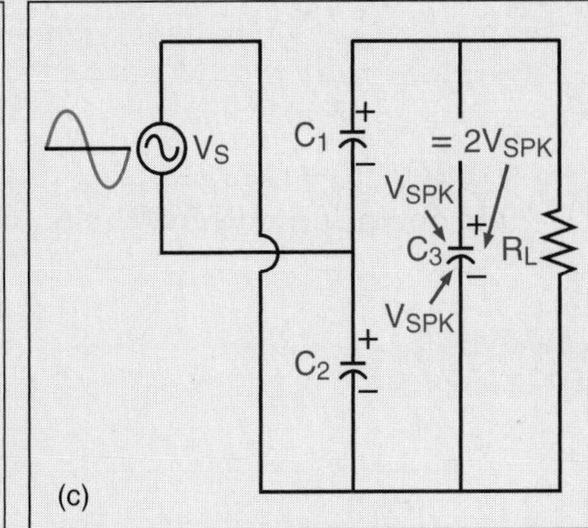

FIGURE 3–35 Regulated power supply.

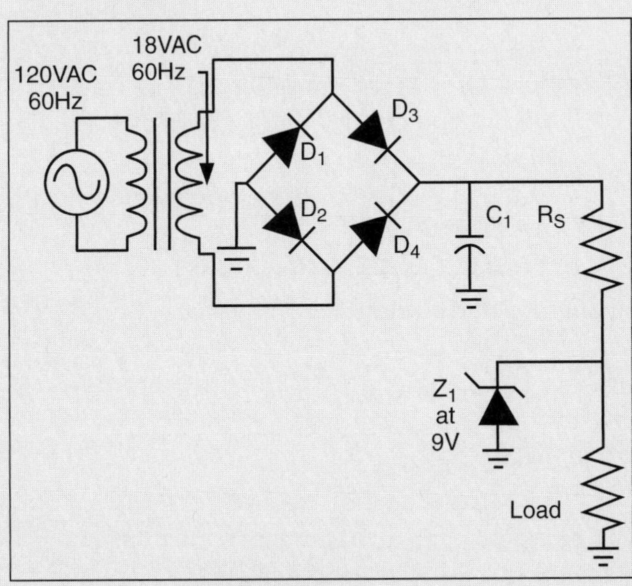

6. What is the purpose of the Zener diode shown in Figure 3–35?
7. In Figure 3–36, what changes would you make to obtain four equal voltages from the output V_{AB}, V_{BC}, V_{CD}, and V_{DE}?
8. Discuss the advantages you might expect by using a full-wave voltage doubler as opposed to a half-wave voltage doubler.
9. Using what you have learned so far, describe the operation of a capacitor input π filter.

FIGURE 3–36 Power supply with voltage regulation and multiple output voltages.

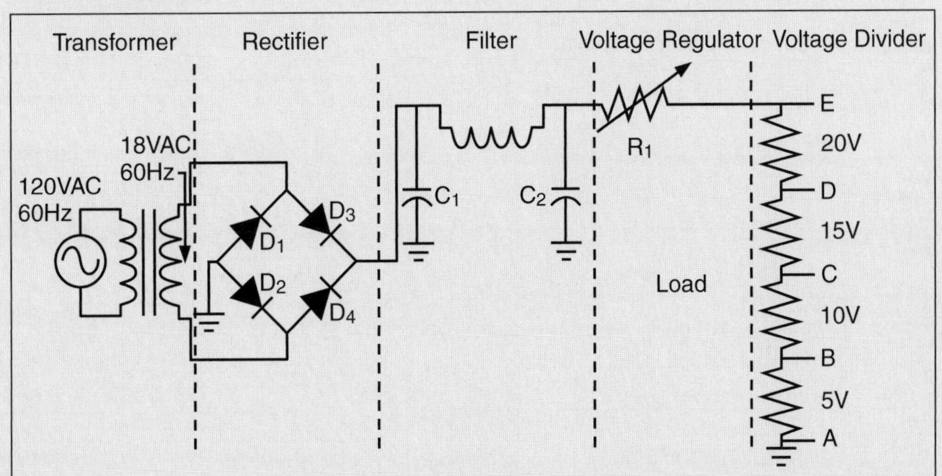

PRACTICE PROBLEMS

1. Carefully examine Figure 3–37 and determine the most probable cause of the following problems.
   a. When the power supply is turned on, the DC output voltage is about half its rated value and shows a large amount of ripple in the output signal.
   b. When turned on, the power supply operates normally for a few minutes and then the output voltage drops to almost zero and the transformer fuse blows.
2. Examine Figure 3–38. A certain RC circuit is charged to 100 VDC. If the circuit is discharged through a short circuit, what will the capacitor voltage be in 1.5 time constants?
3. A certain half-wave rectifier is being fed a 240-V_{RMS} 60 Hz signal. The load is 500 ohms. What is the load voltage and current (RMS, average, peak)?
4. A certain full-wave rectifier is being fed a 240-V_{RMS} 60 Hz signal. The load is 500 ohms.
   a. What is the load voltage and current (RMS, average, peak)?
   b. What is the ripple frequency?
5. A certain full-wave rectifier, having one open diode, is being fed a 240 V_{RMS} 60 Hz signal. The load is 500 ohms.
   a. What is the load voltage and current (RMS, average, peak)?
   b. What is the ripple frequency?
   c. It is performing exactly as a ____________ rectifier circuit.

FIGURE 3–37 Regulated power supply.

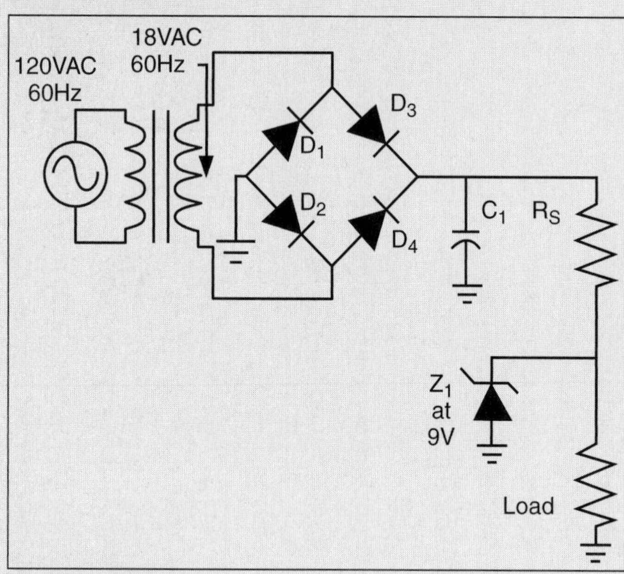

FIGURE 3–38 Capacitor time curves. (a) Charging. (b) Discharging.

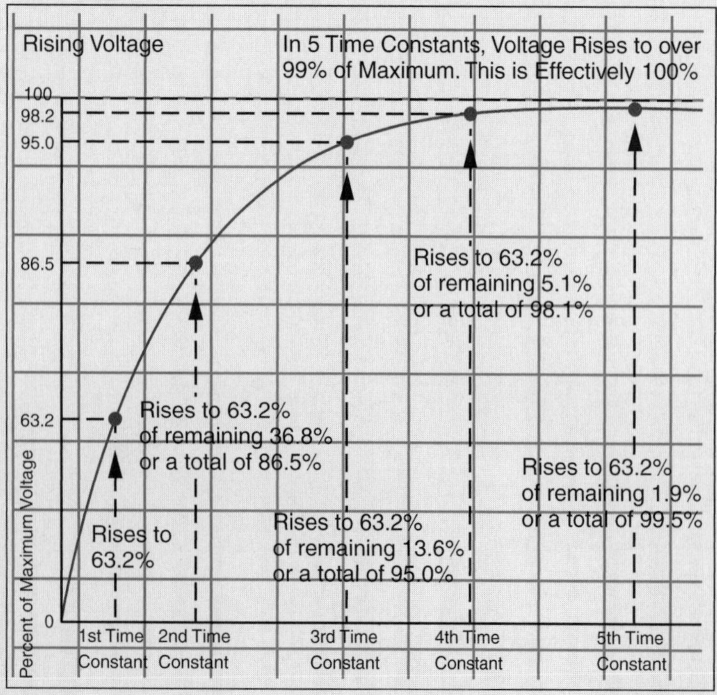

(a)

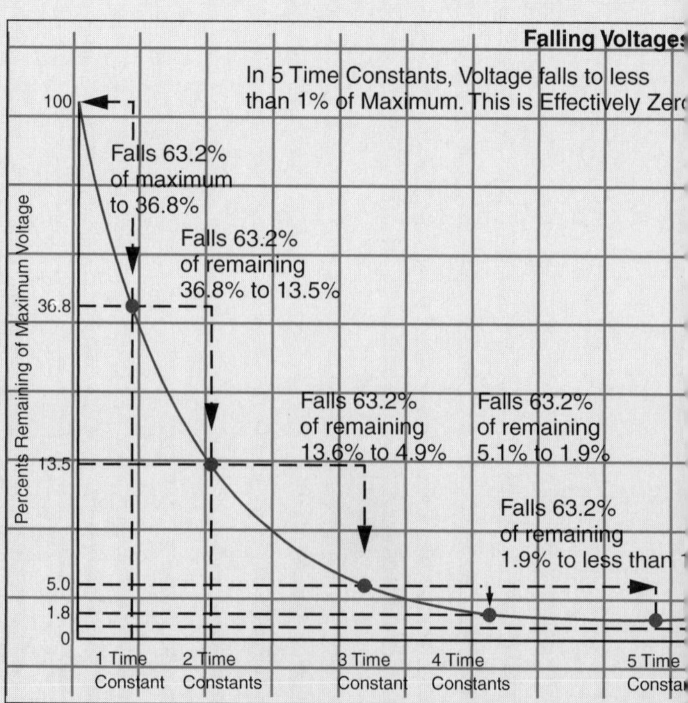

(b)

4 Transistors

O U T L I N E

OVERVIEW

The basic component of all modern electronics is the bipolar junction transistor (BJT). Integrated circuits are often nothing more than huge collections of bipolar transistors all on the same piece of semiconductor material. The transistor is very similar to a PN junction diode, but it has an additional PN junction. The output of a transistor provides amplification by taking the input signal and increasing its power.

The fundamental principle underlying the BJT was discovered in 1951 by the Bell Labs research team of John Bardeen and Walter Brattain. These two researchers noticed that when a signal was applied to contacts on a germanium chip the output taken from one contact was amplified compared to the input.

Dr. William Shockley, the supervisor of the team, took this information and within a few weeks developed the theory of the BJT. Using these brilliant theories, Shockley's team developed the first BJT by the end of 1951. The first commercial use of the BJT was in 1952 when it was used in electronic switching for telephones by Bell Telephone in Englewood, New Jersey.

OBJECTIVES

After completing this chapter, the student should be able to:

1. Draw and correctly label schematic symbols for NPN and PNP transistors.
2. Show biasing to establish and control current flow through a transistor.
3. Define the terms *amplification* and *power gain*.
4. Establish transistor characteristic curves.
5. Calculate gain and develop load lines for transistors in a given circuit.
6. Describe the steps involved in testing transistors.

BIPOLAR JUNCTION TRANSISTOR OPERATING CHARACTERISTICS

4.1 Bipolar Junction Transistor Construction and Symbols

Figure 4–1 shows the construction and schematic symbols of a **BJT**, and Table 4–1 outlines the names and purposes of each of the elements. The collector and emitter are always the same material (P or N), and the base material is always the opposite (N or P).

For instance, an NPN BJT has a collector and emitter made of N-type material and a base of P-type. The PNP has a collector and emitter made of P-type material and a base of N-type. Note that the physical construction may change (NPN or PNP), but there are always two PN junctions (see Figure 4–1). An easy way to remember whether the transistor is an NPN or PNP is by looking at the arrow on the symbol—the NPN transistor's arrow is "**N**ot **P**ointed i**N**."

Bipolar junction transistor (BJT) A three-terminal semiconductor device used for the control and amplification of signals in electronic circuitry. The most common of all types of semiconductors.

FIGURE 4–1 BJT construction and symbols.

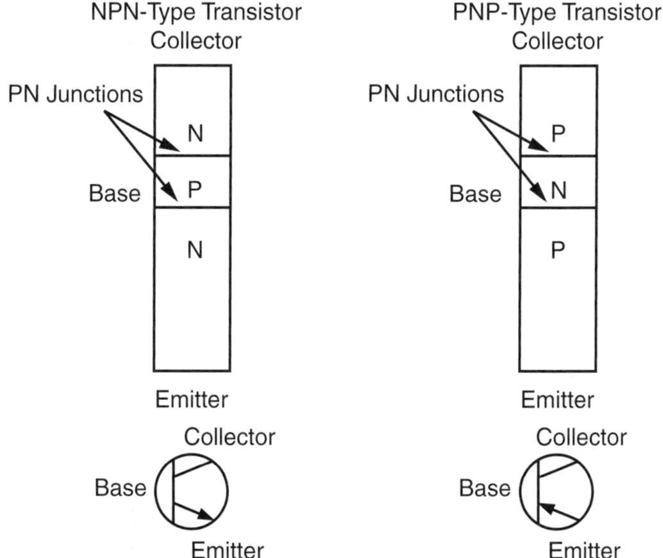

TABLE 4–1 BJT Elements and Their Purposes

Element	Purpose	Comments
Emitter	Serves as a source and "emits" free electrons or holes to the base and collector	The first transistor was of NPN construction. Thus, the emitter was a source of electrons. It is the largest and most heavily doped.
Collector	Receives most of the emitter's electrons	The collector is a "hole" receptor in PNP transistors. It is the second largest and second most heavily doped.
Base	Used as a control and a biasing source for the PN junctions	Small changes between the base-emitter junction create large changes in current between the emitter and collector. The base bias can also be adjusted to block conduction completely. It is the smallest and most lightly doped element.

4.2 Transistor Currents and Voltages

In normal operation, the base-emitter (B-E) PN junction is forward biased, and the base-collector (B-C) PN junction is reverse biased (Figure 4–2). Because the base is so lightly doped, not much of the emitter current flows through the base region but passes on to the collector region. The standard labels for the various currents are outlined in Table 4–2.

FIGURE 4–2 BJT normal biasing conditions.

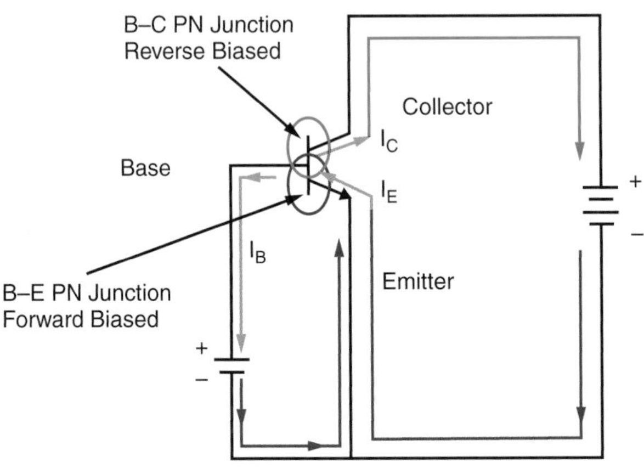

NPN is shown. Reverse polarities of power sources and direction of arrows for PNP.

TABLE 4–2 Lexicon and Symbology for Currents

Name	Symbol
Base current	I_B
Collector current	I_C
Emitter current	I_E

Alpha
A non-dimensional number representing the ratio of collector current to emitter current.

Remember that emitter current (electron flow) is always against the arrow when the B-E junction is forward biased. Figure 4–3 illustrates the relationship between the three currents. Note that the emitter current splits between the base and collector current. The base current is quite small compared to the collector current. The ratio of the collector to emitter current is called **alpha** (α). The formula for calculating alpha is:

$$\alpha = \frac{I_C}{I_E} = \frac{I_C}{I_B + I_C}$$

FIGURE 4–3 NPN transistor currents.

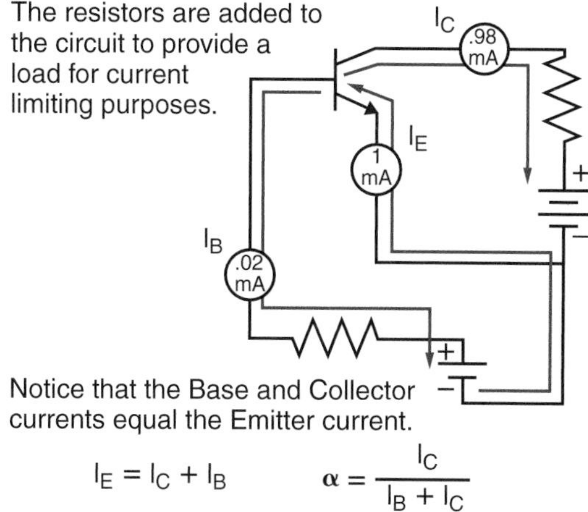

Alpha will always be less than 1 because the emitter current splits between the collector and the base. Using the same type of NPN transistor, voltages for the base, collector, and emitter can be identified. Figure 4–4 shows the voltages for each. Note that supply voltages use a double letter to indicate the voltage source for collector, base, or emitter voltages. For example, V_{EE} indicates the emitter supply voltage and V_{BB} indicates the base supply voltage.

FIGURE 4–4 NPN transistor voltages. (a) With respect to the earth. (b) Directly across transistor.

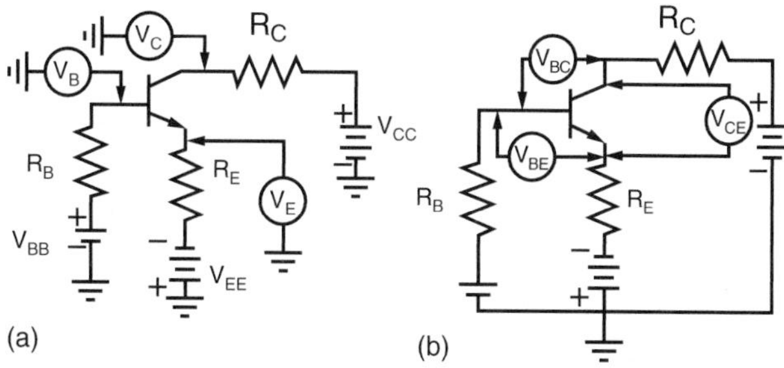

The biasing voltages may come from separate power supplies (as shown in Figure 4–4) or from a system of resistor voltage dividers. For example, the base and emitter voltages are often supplied by dividing V_{CC} across a set of biasing resistors. Figure 4–4b shows the labeling of the voltage drops across each region of the transistor.

4.3 Transistor Bias and Gain

Transistor Bias

Two additional terms are important to an understanding of the basic operation of a BJT. They are *cutoff* and *saturation*. Cutoff occurs when both PN junctions (B-E and B-C) are reverse biased.

This condition yields almost full V_{CC} voltage across the emitter-collector, as shown in Figure 4–5. In other words, the E-C junction is like an open switch. Recall that under normal operating conditions the B-E junction is forward biased and the B-C junction is reverse biased.

FIGURE 4–5 Transistor biased to cutoff.

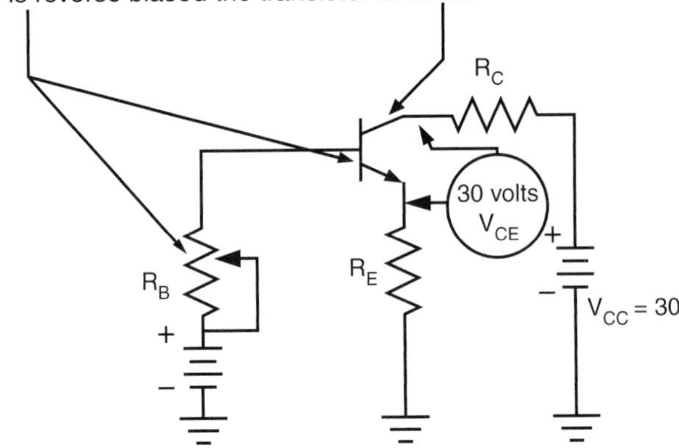
When R_B is set to V_B so that the
B-E junction is reverse biased and the B-C junction
is reverse biased the transistor is cut off
R_C
30 volts
V_{CE}
R_E
R_B
+
–
V_{CC} = 30

Saturation is the condition opposite that of cutoff. As I_B is increased, I_C increases. Saturation is the maximum current condition of I_C. This value is determined by V_{CC} and the resistance of R_C and R_E. At saturation, any increase in I_B will not affect I_C. Using these relationships, the saturation value is calculated as:

$$I_C \approx \frac{V_{CC}}{R_C + R_E} \qquad \text{Note: } \approx \text{ means "almost equal to"}$$

Figure 4–6 illustrates when a transistor us in saturation.

FIGURE 4–6 Transistor biased to saturation.

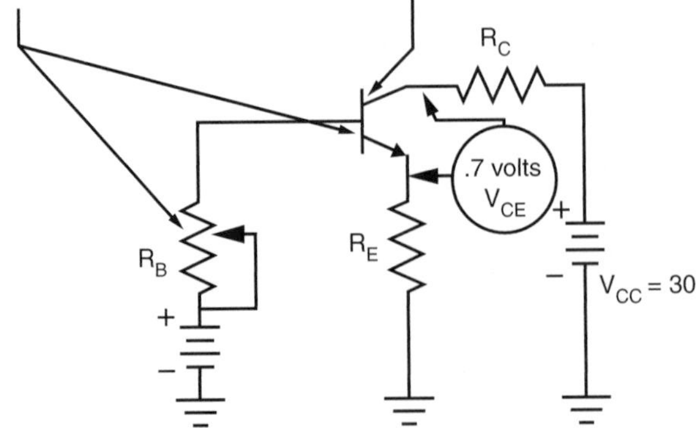
When R_B is set to reduce V_B so that the B-E junction is
forward biased and the B-C junction is conducting at
maximum IC the transistor is in saturation.
R_C
.7 volts
V_{CE}
R_E
R_B
+
–
V_{CC} = 30

Figure 4–7 illustrates the biasing and operational state of an NPN transistor when in cutoff, saturation, and normal operation. Table 4–3 identifies key points about each condition. Note especially the biasing differences among cutoff, saturation, and normal operation.

FIGURE 4–7 Transistor biasing configurations. (a) Cutoff. (b) Saturation. (c) Normal.

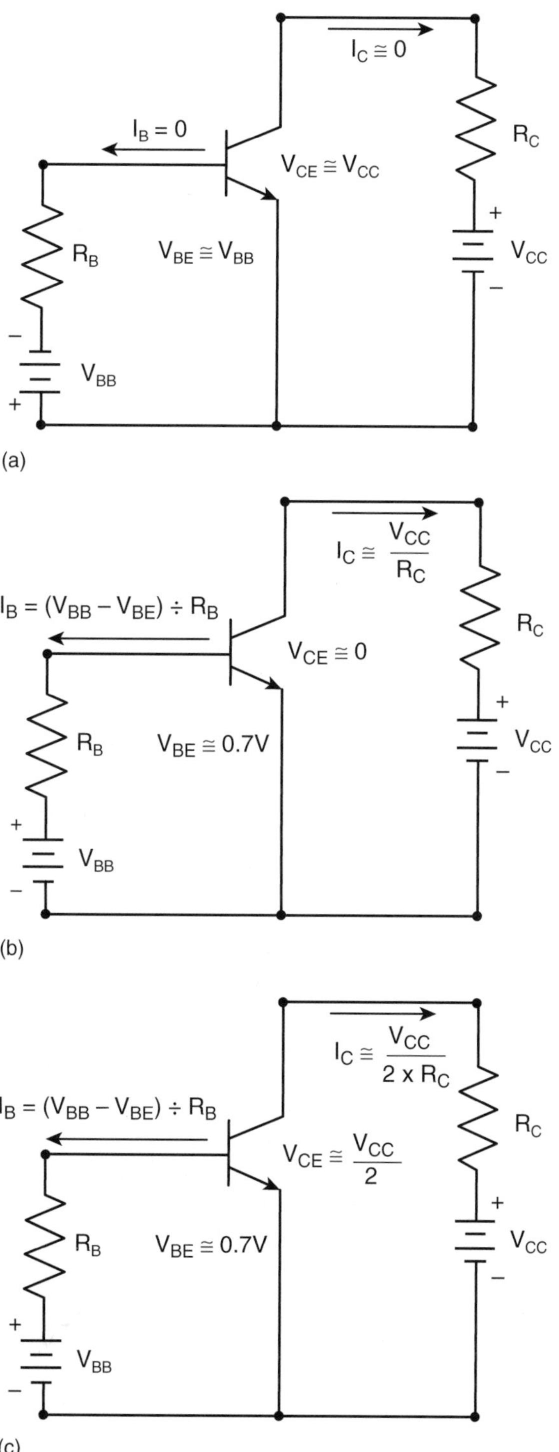

TABLE 4–3 Bias Conditions for Figure 4–7

Transistor Operation	Cutoff	Saturation	Normal
B-E junction	Reverse biased	Forward biased	Forward biased
C-E junction	Reverse biased	Reverse biased	Reverse biased

Transistor Gain

Amplify
To make larger or more powerful.

Gain
The amplification of current, voltage, or power by a transistor per selected circuit values.

Beta
The ratio of DC collector current to base current.

A transistor can **amplify** current, voltage, or power, depending on the circuit values selected. This amplification is called **gain.** The amount of DC current gain in a transistor is called **beta** (β). Beta is the ratio of DC collector current to DC base current. In practical application, a small amount of change in the base current will cause a large change in the emitter and collector currents. The formula for β is:

$$\beta = \frac{I_C}{I_B} \Rightarrow I_B = \frac{I_C}{\beta}$$

For example, if the β of a transistor is 75 and $I_B = 100\ \mu A$, then $I_C = 75 \times 100\ \mu A = 7.5$ mA.

Note: The symbol, ⇒,can be interpreted as "therefore."

FIGURE 4–8

Transistor gain example.

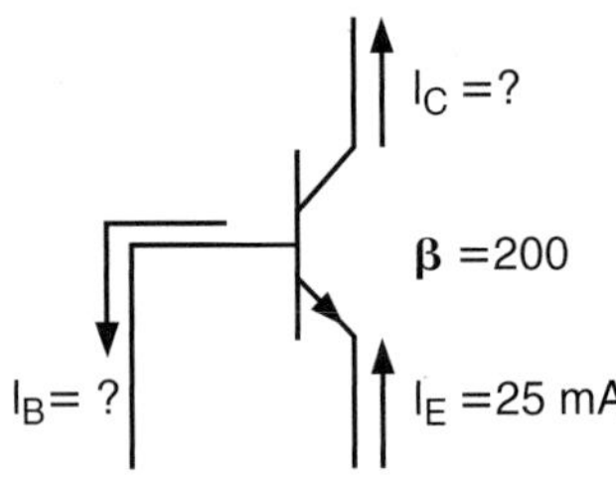

Example

With the information given in Figure 4–8, solve for I_B and I_C.

Solution:
Recall that α of a transistor is the ratio between the collector and emitter current and is expressed as:

$$\alpha = \frac{I_C}{I_E} \Rightarrow I_E = \frac{I_C}{\alpha}$$

Also recall that:

$$I_E = I_C + I_B$$

Substituting from the above equations yields:

$$\frac{I_C}{\alpha} = I_C + \frac{I_C}{\beta}$$

Dividing this equation results in:

$$\alpha = \frac{\beta}{1 + \beta} \text{ and } \beta = \frac{\alpha}{1 - \alpha}$$

And also:

$$I_E = I_E (1 - \alpha)$$

The first step is to find α.

$$\alpha = \frac{200}{200 + 1} = 0.995$$

Next, find I_B:

$$I_B = I_E (1 - \alpha)$$

$$I_B = 25 \text{ mA} \times (1 - 0.995)$$

$$I_B = 125\ \mu A$$

(continued)

Now you can solve for I_C:

$$I_C = I_E - I_B$$

$$I_C = 25 \text{ mA} - 125 \text{ µA}$$

$$I_C = 24.875 \text{ mA}$$

Note: Beta is also equal to the change in the value of I_B to I_C. Therefore:

$$\beta = \frac{\text{Change in } I_C}{\text{Change in } I_B} = \frac{\Delta I_C}{\Delta I_B} = \frac{I_C}{I_B}$$

The Greek symbol "Δ" can be interpreted as "A CHANGE IN". Remember that beta has no units.

TRANSISTOR CIRCUIT CONFIGURATIONS

There are three ways a transistor can be wired (configured) to produce different types of gain. The following sections discuss each of these unique configurations—common emitter, common base, and common collector (see Figure 4–9). Note the location of the input and output connections for all three configurations. The name of each configuration is derived from the element common to both the input and the output.

FIGURE 4–9 Three types of transistor circuits.

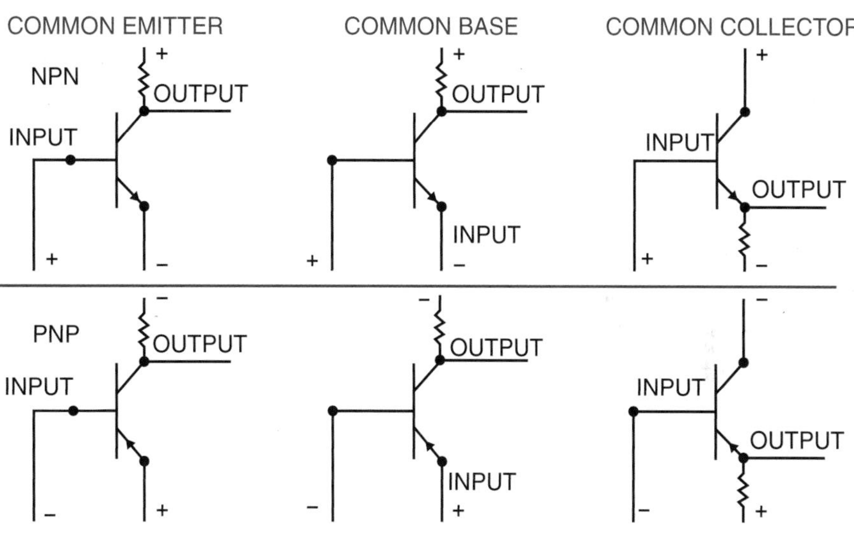

4.4 Common Emitter

Recall that the flow of the current from the emitter through the base to the collector is controlled by the amount of forward bias on the base-emitter junction. A slight change in I_B can cause a great change in I_C. The current flow in the base is a function of the base-emitter junction internal resistance (R_b) and the external resistance in the base circuit.

The external resistance is either a series-limiting resistor (R_s) or a voltage divider network. The common emitter has a voltage phase reversal due to the collector voltage being 180° out of phase with the input base voltage. The common emitter is the most common of the three configurations. Figure 4–10 shows a simple common-emitter circuit.

FIGURE 4–10 Simple common-emitter circuit.

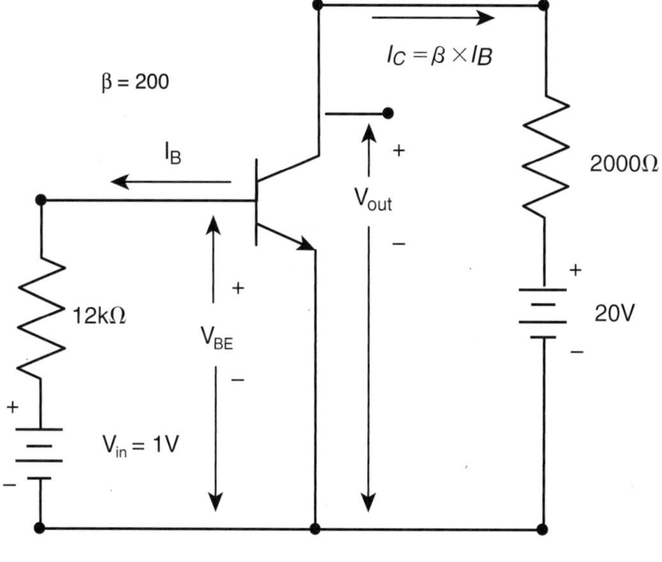

Example

Solve for the voltage gain of the common-emitter circuit in Figure 4–10.
Solution:

$$V_{GAIN} = \frac{V_{OUT}}{V_{IN}}$$

Kirchhoff's voltage law (KVL) may be used to find base current as follows. The algebraic sum of the voltage sources and the voltage drops in a closed circuit must equal zero. We know that the forward biased B-E voltage is about 0.7 V.

$$V_{BE} + V_{Resistor} - V_{Applied} = 0$$

$$V_{Resistor} = V_{Applied} - V_{BE}$$

$$V_{Resistor} = 1\text{ V} - 0.7\text{ V}$$

$$V_{Resistor} = 0.3\text{ V}$$

To find I_B:

$$I_B = \frac{V_{Resistor}}{Resistor}$$

$$I_B = \frac{0.3\text{ V}}{12{,}000\ \Omega}$$

$$I_B = 25\ \mu\text{A}$$

From the definition of β, we have:

$$I_C = \beta \times I_B$$

$$I_C = 200 \times 25\ \mu\text{A}$$

$$I_C = 5\text{ mA}$$

A KVL around the C-E circuit gives:

$$20\ V - V_{Out} + (I_C \times R_C) = 0$$

Therefore:

$$V_{Out} = 20\text{ V} - (I_C \times R_C)$$

$$V_{Out} = 20\text{ V} - (5\text{ mA} \times 2{,}000\ \Omega)$$

$$V_{Out} = 10\text{ V}$$

Then:

$$V_{Gain} = \frac{V_{Out}}{V_{In}}$$

$$V_{Gain} = \frac{10\text{ V}}{1\text{ V}}$$

$$V_{Gain} = 10$$

Note also that the common-emitter circuit creates power gain.

(continued)

The input power is:

$$P_{In} = V_{In} \times I_B$$

$$P_{In} = 1\ \text{V} \times 25\ \mu\text{A}$$

$$P_{In} = 25\ \mu\text{W}$$

The output power is:

$$P_{Out} = V_{Out} \times I_C$$

$$P_{Out} = 10\ \text{V} \times 5\ \text{mA}$$

$$P_{Out} = 50\ \text{mW}$$

To find the power gain:

$$P_{Gain} = \frac{P_{Out}}{P_{In}}$$

$$P_{Gain} = \frac{50\ \text{mW}}{25\ \mu\text{W}}$$

$$P_{Gain} = 2{,}000$$

Note the interesting result if V_{In} is decreased to 0.9 V:

$$V_{BE} + V_{Resistor} - V_{Applied} = 0$$

$$V_{Resistor} = V_{Applied} - V_{BE}$$

$$V_{Resistor} = 0.9\ \text{V} - 0.7\ \text{V}$$

$$V_{Resistor} = 0.2\ \text{V}$$

To find the value of I_B:

$$I_B = \frac{V_{Resistor}}{Resistor}$$

$$I_B = \frac{0.2\ \text{V}}{12{,}000\ \Omega}$$

$$I_B = 16.7\ \mu\text{A}$$

Then:

$$I_C = \beta \times I_B$$

$$I_C = 200 \times 16.7\ \mu\text{A}$$

$$I_C = 3.33\ \text{mA}$$

To find V_{Out} a KVL can be expressed as:

$$V_{Out} + (2{,}000\ \Omega \times 3.33\ \text{mA}) - 20\ \text{V} = 0$$

$$V_{Out} = 20\ \text{V} - (2{,}000\ \Omega \times 3.33\ \text{mA})$$

$$V_{Out} = 20\ \text{V} - 6.66\ \text{V}$$

$$V_{Out} = 13.34\ \text{V}$$

FIGURE 4–11

Common-emitter phase reversal.

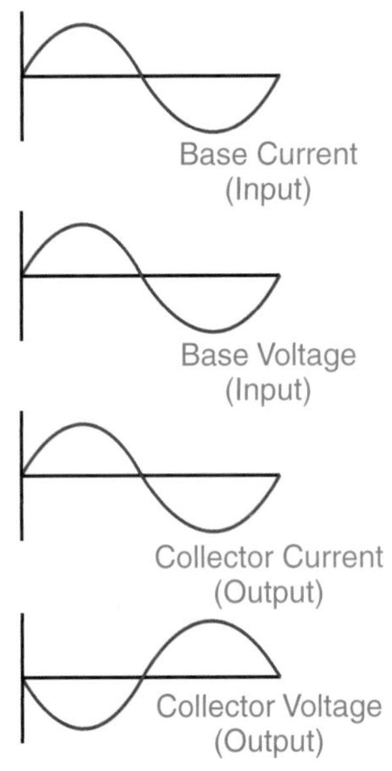

(continued)

A negative change in the input voltage created a positive change in the output voltage. A common-emitter amplifier inverts changes in the input signal voltage. This relationship is shown in Figure 4–11. As seen in Figure 4–11, the common emitter has a voltage phase reversal due to the collector voltage being 180° out of phase with the input base voltage.

4.5 Common Base

The common-base circuit does not exhibit any current gain (Figure 4–12) because the input is into the emitter and the output is from the collector. Remember that typically only 99% of the emitter current is seen in the collector. There are two advantages to the common-base circuit.

- The common base exhibits both voltage gain and power gain.
- The **input impedance** of the common base is very useful in certain types of power amplifier circuits.

Input impedance
The total of all electrical opposition, both resistive, and reactive, to current flow offered by a circuit input element.

Example

Solve for the voltage and power gain of the common-base circuit in Figure 4–12.

Solution:

$$P_{GAIN} = \frac{P_{OUT}}{P_{IN}} = \frac{I_{OUT}^{2} \times R_{OUT}}{I_{in}^{2} \times R_{IN}} = \alpha^2 \times \frac{R_{OUT}}{R_{IN}}$$

This is true because:

$$\alpha = \frac{I_C}{I_E} \Rightarrow \alpha^2 = \frac{I_C^2}{I_E^2} = \frac{I_{OUT}^{2}}{I_{IN}^{2}}$$

Combining these equations and substituting the values in Figure 4–12 gives:

$$P_{GAIN} = \frac{P_{OUT}}{P_{IN}} = \alpha^2 \times \frac{R_{OUT}}{R_{IN}} = 0.99^2 \times \frac{48{,}000\ \Omega}{200\ \Omega}$$

$$P_{GAIN} = 235.2$$

$$V_{GAIN} = \frac{I_{OUT} \times R_{OUT}}{I_{IN} \times R_{IN}} = \alpha \times \frac{48{,}000\ \Omega}{200\ \Omega} = 0.99 \times 240$$

$$V_{GAIN} = 237.6$$

FIGURE 4–12

Simple common-base circuit.

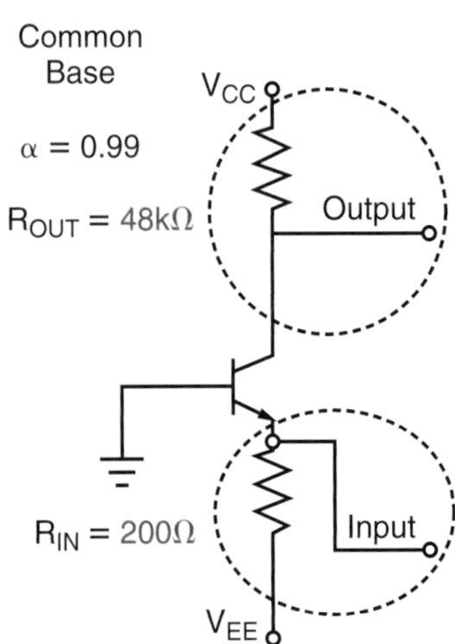

Note that the common-base voltage and power gains are dependent on input versus output resistance because the *I* in the formulas for voltage or power remains relatively constant for a value of α that is close to unity. In contrast, in a common-emitter circuit β (the relationship between input and output current) is the determining factor.

The common base is not a current amplifier because the collector current equals α times the emitter current and α is always less than 1. This configuration is sometimes used to match and amplify radio signals being fed from an antenna. The phase relationships for the common-base circuit are shown in Figure 4–13.

FIGURE 4–13 Phase relationships to the common-base circuit.

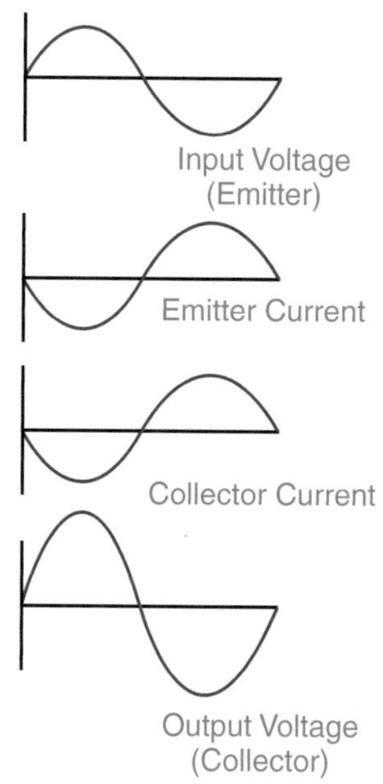

4.6 Common Collector

The third transistor configuration is the common collector. In this configuration, a small change in the base current causes a large change in the emitter-collector current. The voltage difference between the emitter and base is very small and almost constant because of the forward bias on the PN junction. The difference in current is then explained because of input versus output resistance.

This relationship can be expressed as follows. Assume that 1 mA of current is flowing in the emitter-collector circuit. Using what you know about β, you can calculate the base current to be $\frac{1}{\beta}$. Using Ohm's law, we knowthat $R = \frac{V}{I}$, and because V is relatively constant the resistance seen at thebase is equal to the emitter resistance times β. This (low) output resistance of the emitter compared to the (high) input resistance on the base is very useful in coupling a high-resistance (impedance) input to a low-resistance (impedance) output without loading the input circuit.

Figure 4–14 shows an example of a simple common-collector circuit. The common collector has a high input resistance and a low output resistance. The output voltage is always slightly less than the input voltage. This configuration is often used as an isolation or buffer amplifier, which explains why it is often called an **emitter follower.**

Emitter follower
Another term for a common collector amplifier.

FIGURE 4–14 Simple common-collector circuit.

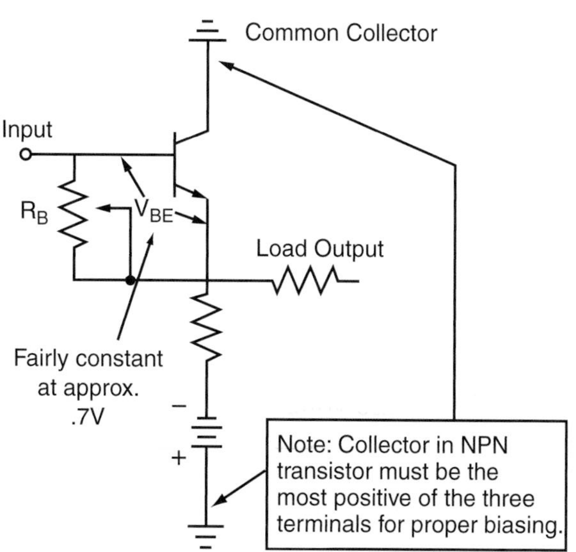

4.7 Comparison of the Three Types of Connections

Figures 4–15a through 4–15c show the features of each of the three types of BJT circuits.

FIGURE 4–15 Summary of circuit characteristics. (a) Common base. (b) Common collector. (c) Common emitter.

	Common Base
Input Impedance	Lowest (approx. 50Ω)
Output Impedance	Highest (approx. 1MΩ)
Current Gain	No (less than 1)
Voltage Gain	Yes (500-800)
Power Gain	Yes
Input-Output Phase Voltage	In Phase
Application	RF Amplifier
Configuration Diagram * The configuration is named after the transistor connection that is common to both the input (source) and the output (load) Another explanation is: The transistor connection that does not provide an input or output to the circuit. (a)	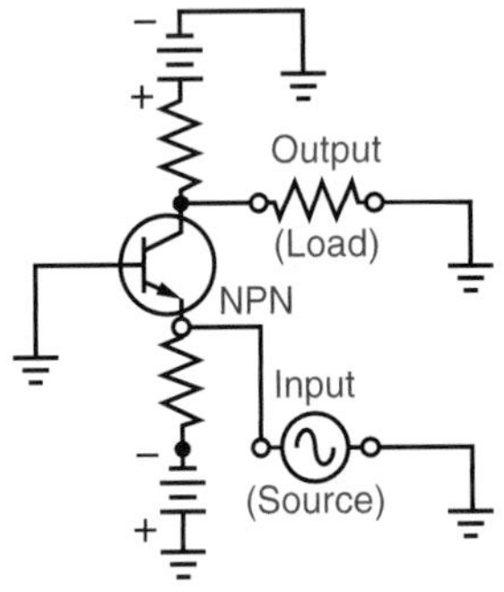 Output (Load) NPN Input (Source) Common Base

	Common Collector
Input Impedance	Highest (approx. 300kΩ)
Output Impedance	Lowest (approx. 300kΩ)
Current Gain	Yes (30-100)
Voltage Gain	No (less than 1)
Power Gain	Yes
Input-Output Phase Voltage	In Phase
Application	Isolation Amplifier Emitter Follower
Configuration Diagram * The configuration is named after the transistor connection that is common to both the input (source) and the output (load) Another explanation is: The transistor connection that does not provide an input or output to the circuit. (b)	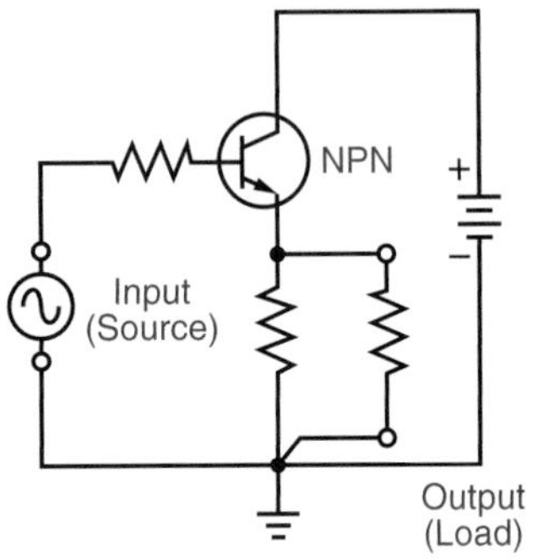 NPN Input (Source) Output (Load) Common Collector

	Common Emitter
Input Impedance	Medium (approx. 1kΩ)
Output Impedance	Medium (approx. 5kΩ)
Current Gain	Yes (30-100)
Voltage Gain	Yes (300-600)
Power Gain	Yes (Highest)
Input-Output Phase Voltage	180° Out of Phase
Application	Universal
Configuration Diagram * The configuration is named after the transistor connection that is common to both the input (source) and the output (load) Another explanation is: The transistor connection that does not provide an input or output to the circuit. (c)	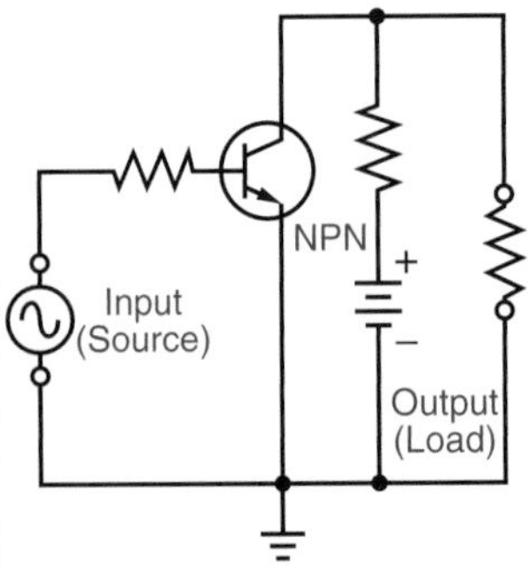 NPN Input (Source) Output (Load) Common Emitter

CHARACTERISTIC CURVES

4.8 Introduction

There are many types of characteristic curves that can be used to determine current and voltage relationships in the various types of transistor configurations. Some of these are the collector curves, the base curves, and the beta curves.

Collector curves plot the relationship among I_C, I_B, and V_{CE}. The base curves plot the relationship of I_B and V_{BE}. Beta (β) curves show how the value of β varies with temperature and I_C. Probably the most useful of these characteristic curves are the collector curves, shown in Figure 4–16.

FIGURE 4–16 Typical collector curves with load lines.

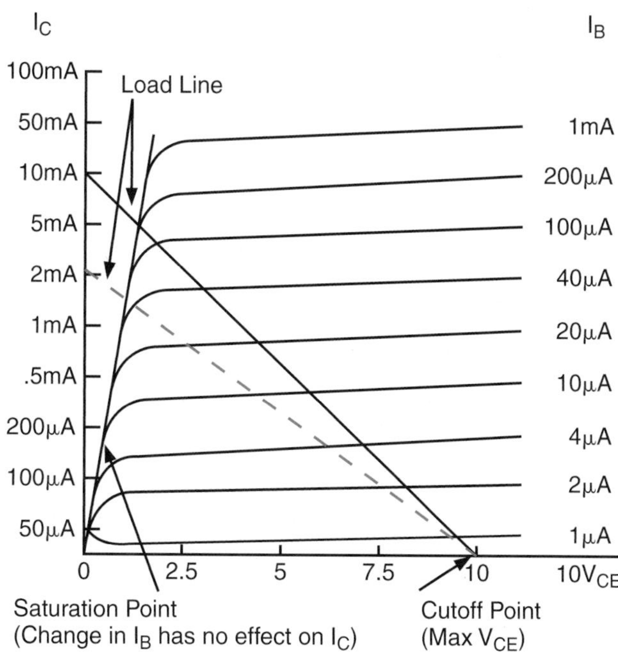

4.9 Collector Curves

Refer to the circuit in Figure 4–17 for the following discussion. To plot collector curves, the base current must be fixed and the collector-emitter voltage varied over a predetermined range. For each value of collector-emitter voltage, I_C is recorded. When the collector current begins to level off (becomes saturated), another base current is set and the process is repeated.

Setting the I_B to 0 assumes that the transistor is in cutoff. An example set of collector-emitter curves is shown in Figure 4–16. Note that the load line shown in Figure 4–16 has its cutoff at maximum V_{CE}, which is typically V_{CC}. For the same load line, saturation is at 10 mA for the collector current—with 1 mA of base current.

FIGURE 4–17 Typical load line circuit.

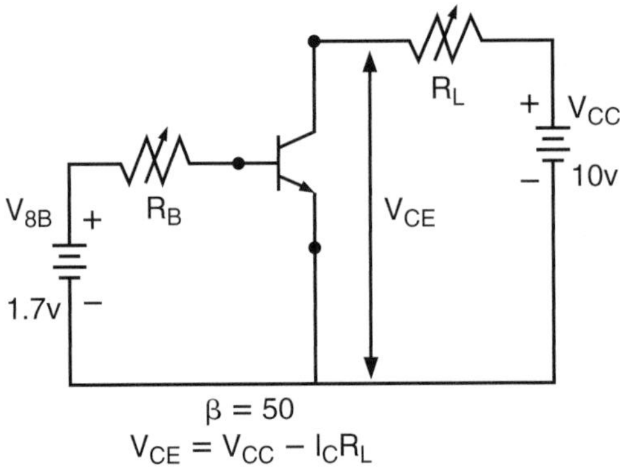

Example

The load line will shift if the base current is lowered. The dotted red line shows a load line for a base current of 40 μA, with an $I_{C(SAT)}$ of 2 mA. **Solve** for R_B and R_L:

$$R_B = \frac{1.7 \text{ V} - 0.7 \text{ V}}{40\ \mu\text{A}}$$

$$R_B = 25 \text{ k}\Omega$$

The −0.7 V drop comes from the base-collector junction of a silicon transistor. The voltage drop of a germanium transistor is about half that value. At saturation and cutoff, the load can be calculated by:

$$R_L = \frac{10 \text{ V}}{2 \text{ mA}}$$

$$R_L = 5 \text{ k}\Omega$$

4.10 Temperature Effects

The bias current of a transistor controls the amount of output the transistor will deliver. Not only do the DC source and biasing resistors set the bias current, but as mentioned previously the temperature contributes to the bias level. Theoretically, if the collector-base temperature of the transistor is kept constant and I_C is increased β will increase. A further increase in I_C beyond the maximum point would cause β to decrease.

In reality, when I_C is held constant and the collector-base temperature is varied the β changes directly with the temperature. Figure 4–18 illustrates how a wide range of temperatures affects β at different I_C for a typical transistor. A transistor data sheet usually includes a temperature-effect curve for that particular transistor. However, the temperature effect on several of the same model transistors can vary from transistor to transistor.

FIGURE 4–18 Temperature effects on operating characteristics.

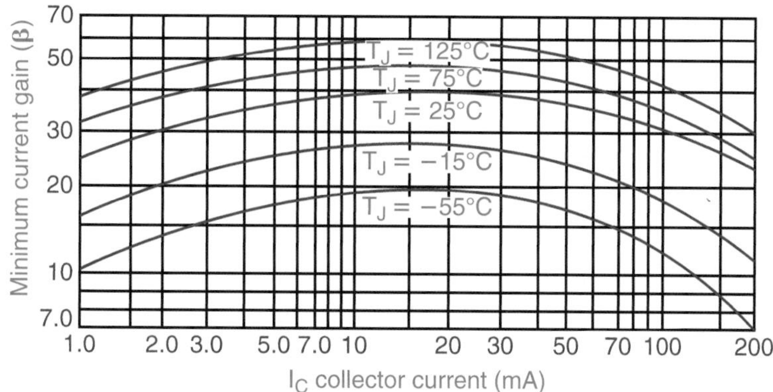

When the ambient temperature changes, the junction resistance changes along with the bias and β. This change of bias is called **thermal instability.** The designer must be aware of the range of ambient temperatures in which a circuit may be physically located. Several items can be implemented to help stabilize a circuit from thermal instability. The use of heat sinks to pull the heat out of the transistor through conductive heat dissipation will lower the junction temperature.

Thermal instability
Change in bias as the result of a change in ambient temperature.

Another approach to thermal protection is within the circuit design itself. Returning a portion of the output back into the input by means of an emitter feedback, collector feedback, or combination emitter/collector feedback circuit will also aid in the stabilization of the biasing.

Finally, the use of the voltage divider with emitter bias circuit can be utilized by including a resistor in the emitter that reacts to the ambient temperature. Resistors also increase in resistance when the temperature is increased. Therefore, the voltage on the emitter would decrease and cause an opposition to the base bias. By adding a **thermistor** or diode in the grounded portion of the voltage divider circuit, additional stabilization can be achieved. The diode or thermistor's resistance value changes as ambient temperature changes and alters the base biasing of the circuit.

Thermistor
A two-wire, temperature sensitive semiconducting device that exhibits resistance that varies inversely with temperature.

NPN VERSUS PNP TRANSISTORS

Both the NPN and PNP transistors are widely used throughout the electronics industry. The two transistor structures are forward biased differently, preventing the replacement of an NPN transistor with a PNP transistor (and vice versa). An NPN transistor's collector must be positive with respect to the emitter. Conversely, with a PNP transistor the collector junction must be negative with respect to the emitter.

The PNP transistor is based on hole current, and the NPN transistor is based on electron current. Electrons have better mobility than holes and can move faster through the crystal structure. Therefore, the NPN transistor is better for high-frequency circuits because it tends to operate faster.

TechTip!

An electronics manufacturer thoroughly tests every circuit card as it leaves the production line, and before stocking. A new version of a proven circuit card assembly began coming off the line, with a shocking 19% failure rate. The technician assigned to isolate the problem found that for those failing circuit cards output from Q17 was always 1.8 V. With power off, the technician measured resistance from E to B, B to C, and C to E and determined that all were 0 ohms.

Microscopic examination of the transistor solder joints appeared free from shorts. The technician began to remove one of the problem transistors, and found a shorting band wrapped around the transistor legs. This band, intended to minimize the possibility of transistor damage due to static discharge, had not been removed prior to installation by the manufacturing department. Removal of the shorting bands brought the faulty circuit cards back to life. No components needed to be replaced.

The NPN transistor is more widely used because manufacturers have more NPN types to offer. This fact allows circuit designers better flexibility with respect to the exact parameters their designs require. The NPN is also based on a negative ground design, which is more common than a positive ground system. Because the NPN transistor is implemented in more designs, the cost of the NPN transistor is usually less than the PNP.

An increased performance and more efficient design are often achieved by using both types of transistors in a circuit. Figure 4–19 shows a design utilizing both an NPN and a PNP transistor. In later chapters, you will discover the advantages of this type of design with respect to efficiency and output gain.

FIGURE 4–19 Complementary class B amplifier.

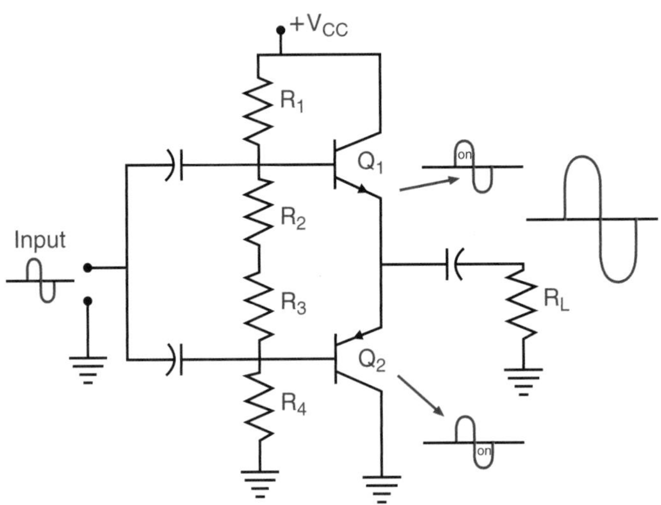

TESTING METHODS

4.11 Transistor Checker

There are various methods of testing transistors. A transistor checker is a special piece of equipment used to test most NPN and PNP transistors. It not only tests for proper operation but can check to determine β. Figure 4–20 shows a photograph of a typical transistor checker.

Other transistor-checking methods are used in testing and design laboratories. This type of equipment is called a **curve tracer**. Some laboratories set up special circuits to test the transistor under actual load and operating conditions. This is called **dynamic testing**.

Curve tracer
An item of test equipment capable of producing a family of transistor characteristic curves for a transistor under test.

Dynamic testing
A procedure for determining operation characteristics of an operating circuit.

4.12 Transistor Testing with a Digital Multimeter

The digital multimeter (DMM) is probably the most common piece of test equipment used to test transistors. Figure 4–21 shows the process for checking an NPN transistor with a DMM.

Because there is no physical PN junction between the collector and the emitter, the resistance will always be very high regardless of the ohmmeter's polarity.

FIGURE 4–20 Typical transistor checker.

FIGURE 4–21 Transistor checking with a digital multimeter. (a) E-B forward. (b) E-B reverse. (c) C-B forward. (d) C-B reverse.

Emitter-Base Forward Biased: 0.5 to 0.8 Volts as read using Diode Function on a DMM.

(a)

Emitter-Base Reverse Biased: "OL" should be read using Diode Function on a DMM.

(b)

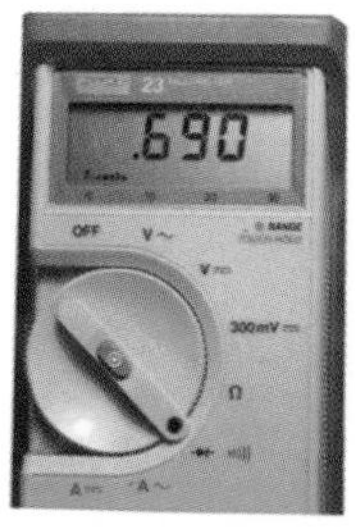

Collector-Base Forward Biased: 0.5 to 0.8 Volts as read using Diode Function on a DMM.

(c)

Collector-Base Reverse Biased: "OL" should be read using Diode Function on a DMM.

(d)

SUMMARY

In this chapter, you learned about bipolar junction transistors (BJTs). The BJT is a three-terminal device constructed of three semiconductor materials forming two PN junctions. The BJT's output current, voltage, and/or power are controlled by its input current. The output of a transistor provides amplification as defined by power = current × voltage.

The BJT has three terminals: the base, emitter, and collector. There are two basic types of BJTs, the NPN and PNP. Under normal operating conditions, a small change in base current produces a large change in collector and emitter current. This change ratio is called beta (β) and is the gain (h_{FE}) of the transistor. Beta is also called h_{FE} in some transistor models. Another important ratio is that of collector current to emitter current. This is called alpha (α).

There are three common configurations for BJTs. They are the common emitter, common base, and common collector. The term *common* refers to the terminal common to both the input (source) and the output (load). For example, the common-emitter circuit has the input at the base and the output from the collector of the transistor.

The load line is another useful characteristic of the BJT. The load line is a function of the collector at saturation and the collector-emitter voltage at cutoff. At saturation, any increase in base current will have little or no effect on the collector current (the collector current is saturated). At cutoff, the collector-to-emitter voltage is effectively that of the applied collector voltage (V_{CC}). This means that the transistor is not conducting.

Even though both the NPN and PNP transistors are widely used in the electronics industry, they cannot be interchanged because of their different bias polarities. The NPN transistor is the most common because it is based on a negative ground design, manufacturers offer it in a larger variety of types, and it costs less. PNP transistors are based on conventional current, whereas NPN transistors are based on electron current (usually NPN transistors operate slightly faster, which can make them better in high-speed applications). It is common to find both NPN and PNP transistors utilized in the same circuit to offer better efficiency and output gain.

Temperature affects the bias and β of a transistor circuit and is called thermal instability. Heat sinks can dissipate the heat from the transistor, but usually additional design issues are implemented to help control the increasing β. Feedback circuits such as the emitter feedback, collector feedback, or combination emitter/collector feedback circuit will aid in the stabilization of biasing. The voltage divider with emitter bias circuit also stabilizes the transistor temperature, especially when a diode or thermistor is placed in the circuit.

A digital multimeter (DMM) can be used to test a BJT for proper operation. If working properly, the forward bias condition of the base-emitter and base-collector junctions should read between 0.5 and 0.8 V on the DIODE TEST position of the DMM. The reverse bias condition of the same junctions should read Over Limit (OL).

If the transistor base-emitter junction checks good in both directions, check the base- collector junction in exactly the same way. All readings should be comparable to those shown in Figure 4–21. If the transistor is a PNP, reverse the DMM leads and look for the same readings.

REVIEW QUESTIONS

1. What are the two types of transistors? What is the advantage/disadvantage of each?
2. Which of a transistor's three semiconductor layers has the heaviest doping? Which is the largest? Which controls the current between the other two?
3. Describe the differences among the common-emitter, common-base, and common-collector circuits.
4. What is α of a transistor? What is β?
5. You need an amplifier circuit that has a large voltage gain. Which of the three circuit types would give you the best voltage gain? Which one is the worst?
6. Look at the red dashed load line in Figure 4–22. How much collector current will flow when the base current is 4 μA?
7. You are testing a transistor with a DMM. The B-E junction reads "OL" regardless of which polarity lead you attach to the base. What is probably wrong?
8. You are selecting a radio-frequency amplifier for the first stage of a radio receiver. The first stage connects directly to the antenna, which has an impedance of 73 Ω. Which of the three types of circuit connections would probably be the best, and why?
9. Explain how current can flow between the emitter and base of a transistor when it has to go backward across the C-B junction.
10. Describe the conditions you would find in a transistor circuit during cutoff and saturation.

FIGURE 4–22 Simple common-collector circuit.

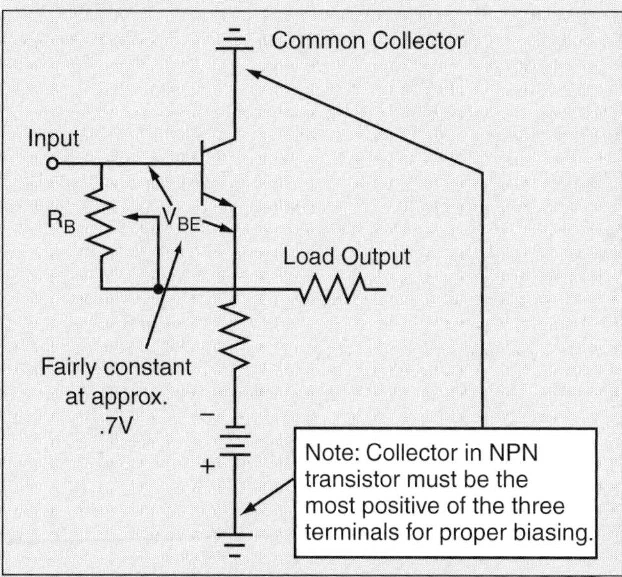

PRACTICE PROBLEMS

1. In Figure 4–23, assume the following values: $\beta = 250$, $R_B = 10{,}000\ \Omega$, and $R_C = 2{,}500\ \Omega$. All other values are as shown. Calculate V_{GAIN}, I_{GAIN}, and P_{GAIN}.

FIGURE 4–23 Simple common-emitter circuit.

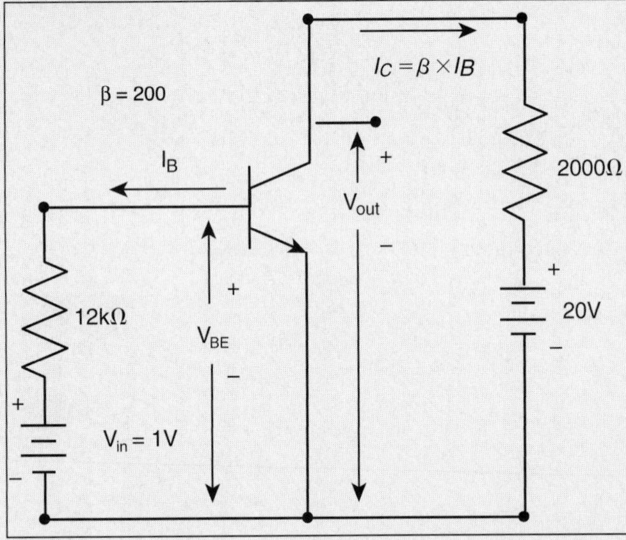

2. What is the voltage and power gain of the circuit of Figure 4–24 with the following values: $\alpha = 0.95$, $R_{OUT} = 25{,}000\ \Omega$, and $R_{IN} = 400\ \Omega$?

FIGURE 4–24 Simple common-base circuit.

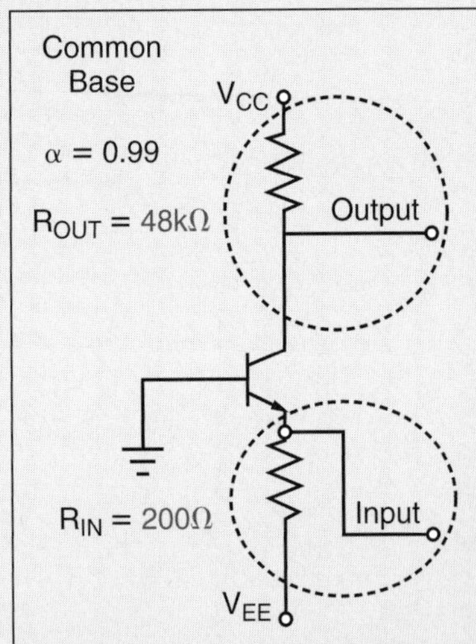

Quiescent point (Q-point)
Collector current present when no input signal is applied.

3. A transistor amplifier has the black load line shown in Figure 4–25. With no input signal, the base current I_B is 10 μA. This point is called the **quiescent point**. Answer the following questions.
   a. What is the collector current at the quiescent point?
   b. What is the collector voltage at the quiescent point?
   c. If an input signal (I_B) is applied with a peak-to-peak value of 20 μA, what is the peak-to-peak collector current?
   d. What is the peak-to-peak collector voltage for the 20 μA peak-to-peak base current?
4. In regard to the example for question 3, how much base current is required to:
   a. Drive the transistor to cutoff?
   b. Drive the transistor to saturation?

FIGURE 4–25 Typical collector curves with load lines.

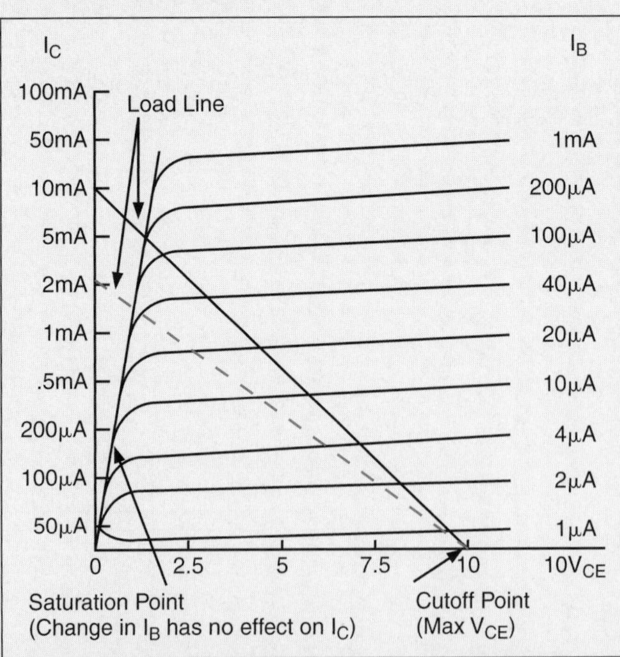

5. In the example shown in Figure 4–23, what is the efficiency of this amplifier? Do not forget to include the power being input by the base voltage supply.

5

JFETs, MOSFETs, and UJTs

O U T L I N E

OVERVIEW

In this chapter, you will learn about transistors that do not have bipolar junctions (BJTs). These transistors are classified as junction field effect (JFET), metal-oxide semiconductor field effect (MOSFET), and the unijunction (UJT). The phototransistor is covered in a later lesson on optical devices. The insulated gate bipolar junction transistor, a hybrid of the field effect transistor (FET) and the BJT, is explained in a later lesson (along with silicon-controlled rectifiers and triacs).

Recall from the past lesson that the BJT is a current-controlled device. In general, FETs are voltage-controlled devices. This means that the output characteristics of the FET are controlled by input voltages, not input currents.

OBJECTIVES

After completing this chapter, the student should be able to:

1. Draw schematic symbols for JFET, MOSFET, UJT, and IGBT transistors.
2. List and identify the characteristics of a JFET, MOSFET, UJT and IGBT.

JUNCTION FIELD EFFECT TRANSISTORS

5.1 Construction

The JFET has three terminals that are similar to the BJT. They are the source, drain, and gate. Figure 5–1 compares the JFET terminals with the equivalent BJT terminals. The BJT counterparts are emitter for the source, collector for the drain, and base for the gate. A fourth JFET component is the channel. The channel is the material that connects the source to the drain. Note that the gate surrounds the channel (see Figure 5–2). Also note that in the JFET schematic symbol the arrow points in toward the material for an n-channel JFET.

FIGURE 5–1 Comparison of BJT and JFET schematic symbols.

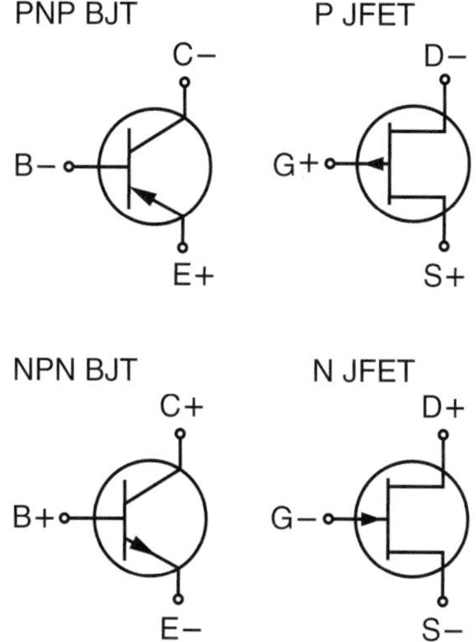

FIGURE 5–2 JFET construction.

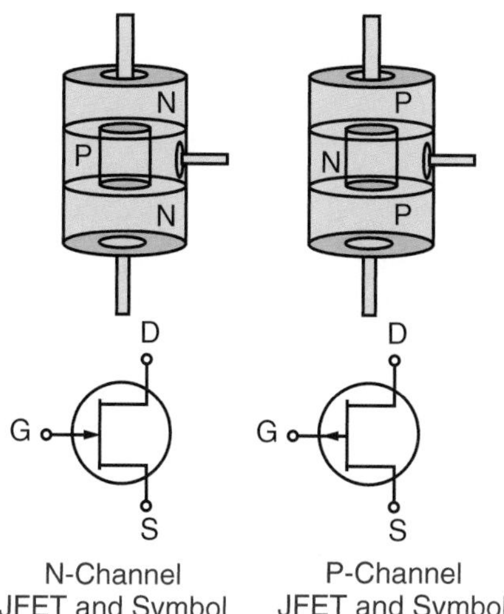

5.2 JFET Operating Characteristics

Similar to an NPN transistor that requires positive supply voltages (V_{CC} and V_{BB}), the N-channel JFET also requires positive supply voltages. The reverse is true for a PNP transistor and for a P-channel JFET. Figure 5–3 shows N-channel and P-channel supply voltage configurations.

FIGURE 5–3 N-channel and P-channel JFET circuits.

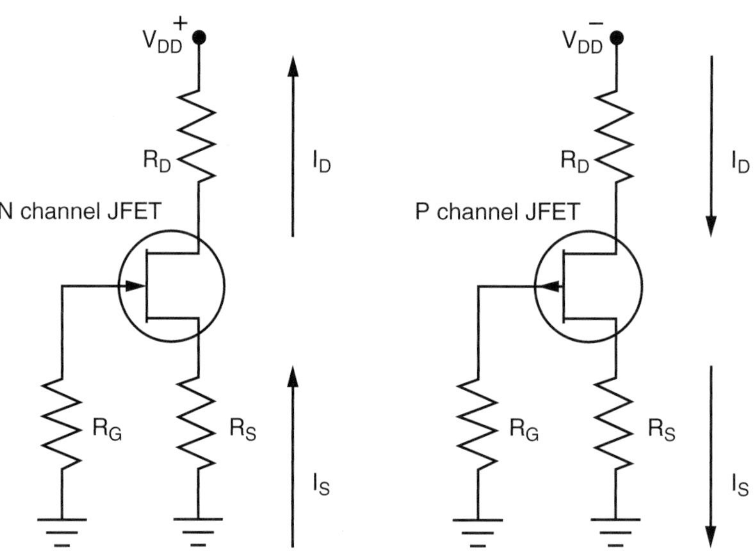

The supply voltage is labeled V_{DD} (similar to the BJT's V_{CC}). Unlike a BJT, which requires holes and electrons for conduction across the collector emitter, the JFET requires only electrons. As an interesting historical note, a patent covering the basic operation of a JFET was issued in the late 1930s. Unfortunately for the patent holder, semiconductor material technology at that time was not sufficient to build one.

Refer to Figure 5–4 for the following discussion. Because the gate surrounds the channel, a change in the width of the gate will control the current flow through the channel at any given voltage. To control the gate width, a reverse bias is applied to the gate-source junction. The reverse bias increases the depletion area (increases the P-material size) and reduces the N-channel. The reduced N-channel restricts the current flow from the drain to the source.

Pinchoff Voltage (V_P)
The drain-to-source voltage, with a given gate voltage, beyond which drain current will increase no further.

Pinchoff is another concept applicable to the JFET. If gate bias is set to a standard value (0 VDC is usually chosen), increasing drain-to-source voltage (V_{DS}) causes drain current (I_D) to rise. However, the channel voltage in the vicinity of the gate becomes progressively more positive. Therefore, reverse bias (gate-to-channel) effectively increases. The channel continues to narrow, thus limiting the increase of drain current. A point is eventually reached where V_{DS} will produce enough bias to completely choke-off channel current. This V_{DS} is called the **pinchoff voltage** (V_P). Increases of V_{DS} beyond V_P, up to and beyond breakdown voltage (V_{BR}) will not produce any increase in I_D.

FIGURE 5–4 JFET gate control.

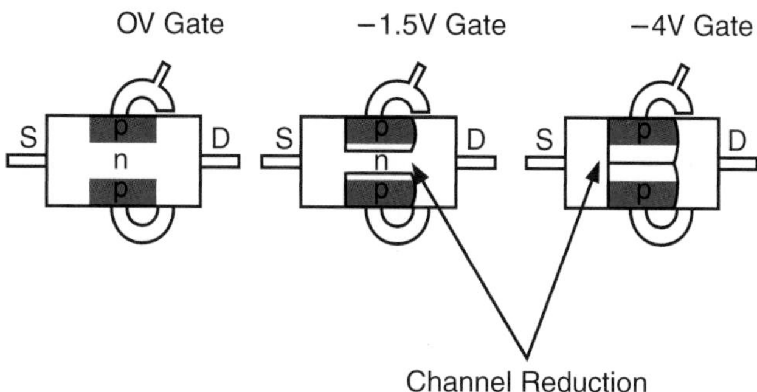

5.3 JFET Biasing

There are four basic biasing circuits for the JFET. They are gate biasing (very similar to BJT base biasing), self-biasing, voltage divider biasing (very similar to BJT voltage divider biasing), and current source biasing.

Gate Biasing

Under normal conditions, the gate-source junction of a JFET is reverse biased. In the circuit configuration shown in Figure 5–5, a negative V_{GG} supplies the voltage to the gate to ensure a reverse bias. Because there is no gate current, there is no voltage drop across R_G. Therefore, $V_{GS} = V_{GG}$.

FIGURE 5–5 JFET gate biasing.

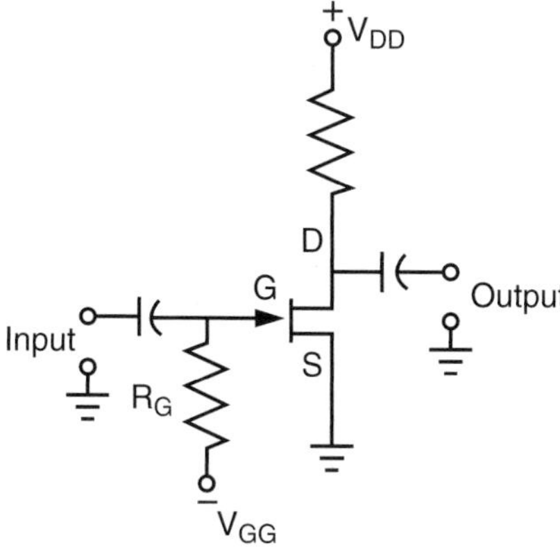

Self-biasing

A simpler biasing configuration is found in the self-bias circuit, as shown in Figure 5–6. In this circuit, the V_{GG} is replaced by a source resistor (R_S). The source resistor provides a voltage on the source that is always positive with respect to the gate. Note that $I_S = I_D$ and that $V_S = I_D \times R_S$. Because there is no current on the gate, the voltage across R_G is 0 and the corresponding voltage on the gate is 0. If the gate voltage is 0, the voltage drop on V_S will be greater. Therefore, the gate will remain reverse biased.

FIGURE 5–6 JFET self-biasing.

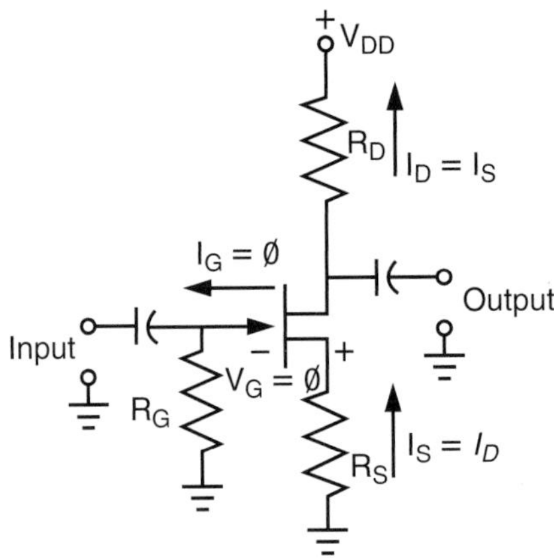

Voltage Divider Biasing

A very stable biasing configuration is the voltage divider. This method is also used in BJTs. The formula is almost the same for finding the gate voltage as for calculating the base voltage. Figure 5–7 shows a voltage divider biasing circuit. This type of biasing is the most commonly employed due to its relative simplicity and stability.

$$V_G = V_{DD}\left(\frac{R_2}{R_1 + R_2}\right)$$

FIGURE 5–7 JFET voltage divider biasing.

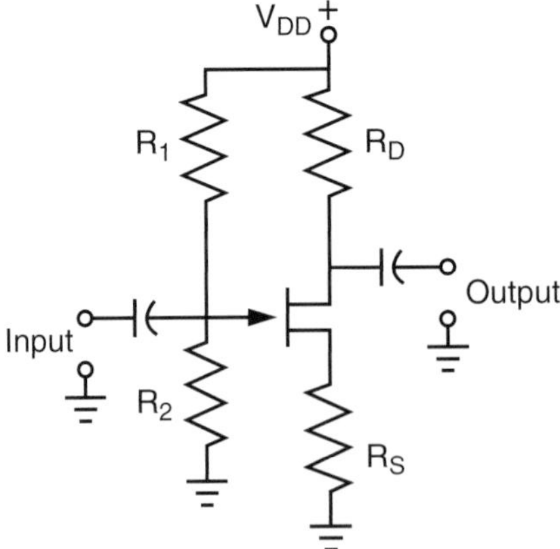

Current Source Biasing

The most stable of all of these biasing circuits is current source biasing (see Figure 5–8). This circuit provides the most Q-point stability. In this circuit, the value of $I_D = I_C$. I_C is independent of the JFET characteristics, as is the value of I_D. Even though this type of biasing is highly stable, the complexity and added components of the circuit limit its usefulness.

FIGURE 5–8 JFET current source biasing.

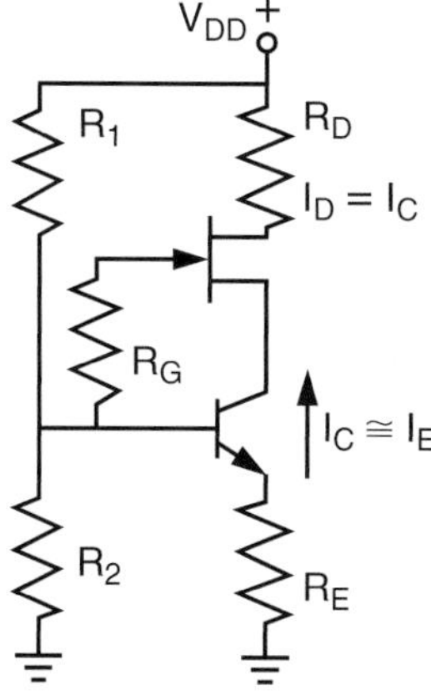

5.4 JFET Application

The high input impedance of the JFET makes it very useful where source loading is not desirable. For instance, the JFET shown in Figure 5–9 is acting as a "buffer amplifier" to the source amplifier. The JFET's high impedance (typically multiple megohms) presents no static load to the source amplifier and allows nearly all of the gain to be passed through to the load. Where exceptionally high input impedance is required—such as for thermocouple amplifiers or high-impedance microphone preamplifiers—JFETs have largely been replaced by low-power signal MOSFETs.

FIGURE 5–9 JFET amplifier buffer application.

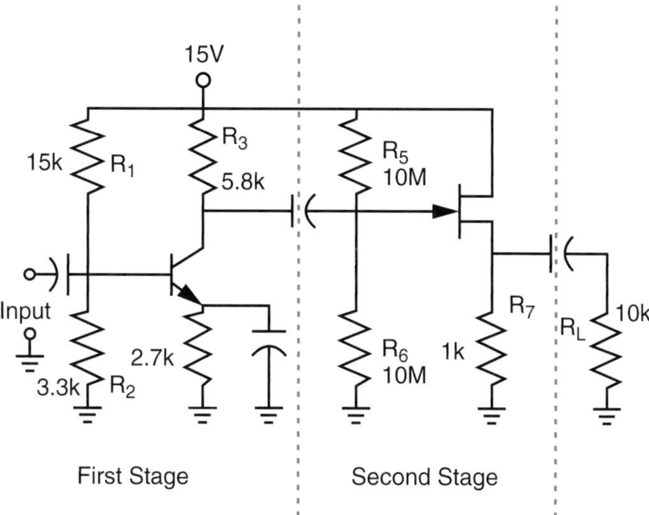

Unlike BJTs, which are current-controlled devices, the JFET is a voltage-controlled device. The BJT is normally off, but the JFET is normally on and draws practically no current from the driving source. Finally, it should be noted that just like BJTs, the JFET can be biased several ways: common-gate, common-drain, common-source configurations. JFETs are more vulnerable than BJTs to accidental damage from static discharge.

METAL-OXIDE SEMICONDUCTOR FIELD EFFECT TRANSISTOR

Metal-oxide semiconductor (MOS) technology is used extensively in most calculators, watches, desktop computers, and other appliances. The main advantage of MOS circuits is their low current requirement. With less current, there is less heat dissipation. Another advantage is that MOS components can be made much smaller than BJT components.

5.5 MOSFET Construction

There are two basic types of MOSFETs: the enhancement MOSFET (E-MOSFET) and the depletion MOSFET (D-MOSFET). The depletion types are more versatile because they can operate in enhancement or depletion mode. The enhancement type can operate only in enhancement mode.

The physical difference between the enhancement and depletion types is the channel between the source and drain. The depletion type has a channel; the enhancement type does not. Figure 5–10 shows how the two types of MOSFETs are constructed. The E-MOSFET is normally an "off" device, but the proper gate voltage will attract carriers to the gate region and form a conductive channel. Either a negative or positive voltage potential to the gate will allow current to flow between the source and the drain.

The D-MOSFET is normally on because of the physical channel, and only when a negative voltage is applied to the gate does the channel resistance increase and current reduce. When a D-MOSFET has a negative gate voltage, it is operating in the depletion mode. Note that FETs do not require any gate current to operate. Therefore, the gate region is not physically connected to the channel.

FIGURE 5–10 Depletion and enhancement MOSFET construction.

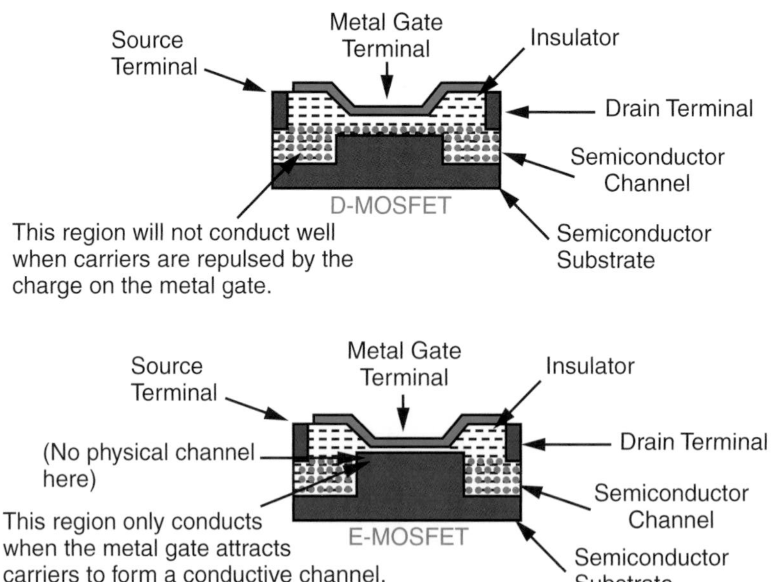

Both types have an insulation layer between the gate and the rest of the semiconductor. This layer is made of silicon dioxide (SiO_2), and because the layer is thin it can easily be damaged by static electricity. Many of the MOSFETs manufactured today have Zener diodes etched into their construction. The diodes are between the gate and the source and are designed to operate above the rated voltage of the V_{GS}. These diodes protect the gate from damage due to static electricity. Figure 5–11 shows the schematic symbols for the E- and D-type MOSFET, along with a schematic for the diode static electricity protection.

FIGURE 5–11 E- and D-type MOSFET schematic symbols.

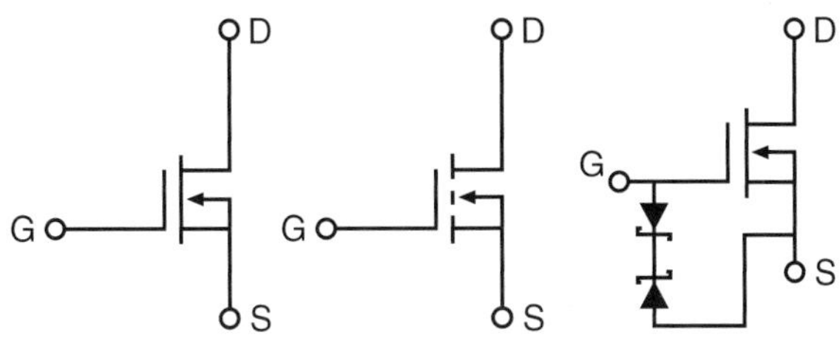

D-Type MOSFET E-Type MOSFET MOSFET Static Protection

(The arrow pointing "in" represents an N-channel device.)

Also note the dashed line in the E-Type MOSFET. This indicates the channel is not always present and it is normally off.

Like the JFET, the MOSFET has very high input impedance. For example, one MOSFET has a maximum gate current rating of 10 pA (picoamperes) and a V_{GS} of 35 V. A D-MOSFET has two other advantages over the JFET. It can operate in zero bias (Figure 5–12 shows an example circuit for zero biasing) and can operate in enhancement mode.

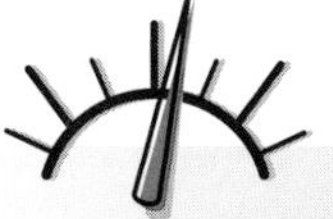

TechTip!

Testing JFETs and MOSFETs is not recommended without a curve tracer or transistor tester designed for the task. In a pinch, a DMM resistance check can be made from gate to source. Normal resistance would be in the megohm region. Source-to-drain measurement may not be reliable unless the transistor is "dead shorted" (0 ohms). Be forewarned: JFETs and MOSFETs are electrostatic discharge sensitive. Handling these devices without proper electrostatic discharge precautions, especially out of circuit, may cause their destruction.

FIGURE 5–12 MOSFET zero bias circuit.

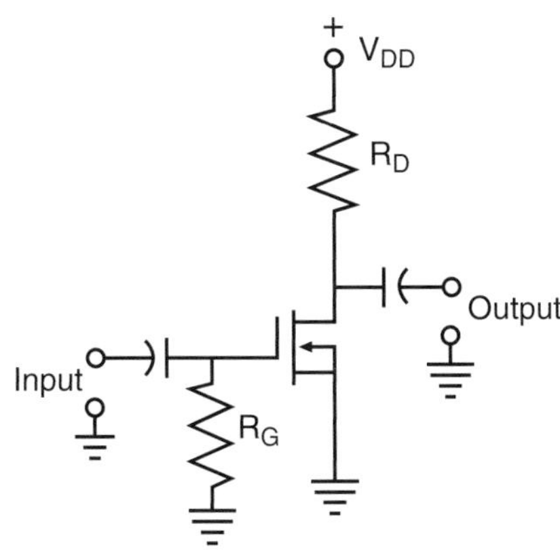

Analog
Referring to a class of circuits in which output varies as a continuous function of input.

Digital
Referring to a class of circuits in which output operates like a switch (on or off), with no intermediate values.

D-MOSFETs are usually used for analog applications such as oscillators and amplifiers. Although E-MOSFETs can also be used in **analog** circuits, they are most widely used in **digital** circuits. A simple and inexpensive portable AM/FM receiver utilizes analog circuits. A bedside digital clock utilizes digital circuits.

5.6 VMOS Power Applications

In modern communications, most signals are transmitted and received in digital form. The receiver converts the incoming signal and information back into its original form. This requires the use of a power amplifier that can produce high-speed, high-current digital outputs (i.e., respond to on/off signals very quickly with no distortion). This is the perfect application for power MOSFET amplifier drivers. VMOS transistors also function well as analog power amplifiers when appropriately biased.

PMOSFET (Figure 5–13) is also known as VMOS. The VMOS name comes from the way the semiconductor is constructed. The gate is V shaped and is surrounded by heavier doping of the N-type material. This heavier doping produces a wider channel and allows the VMOS to handle higher drain current levels. The VMOS operates only in enhancement mode.

FIGURE 5–13 VMOS construction.

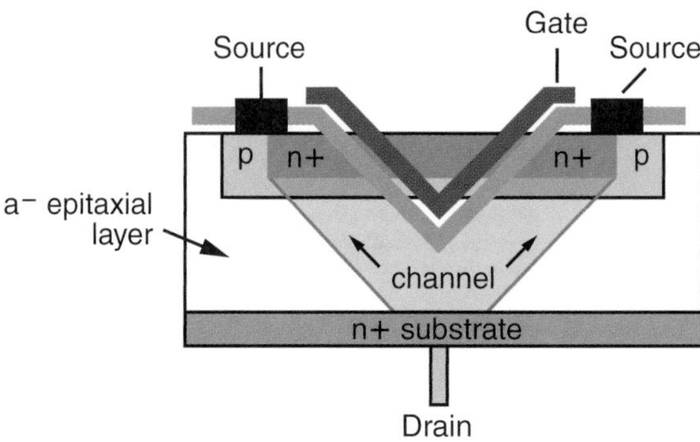

Note in Figure 5–13 that the VMOS current flow is vertical. This short and wide channel results in a higher current capacity, greater power dissipation, and improved frequency response. Because the VMOS is an enhancement mode device, it is normally off and has no physical channel. Current flows between the source and the drain when the gate is made positive with respect to the source.

VMOS transistors look like any other power transistor. Figure 5–14 shows an example of a typical output characteristic curve for a VMOS device. Note the large V_{gs} swing permissible. There is, however, a built-in Zener diode between the VMOS gate and source. This Zener limits positive spikes, reducing the likelihood of damage from excessively high input voltage spikes. Nevertheless, care should be taken to avoid applying excessive gate-source voltage. Figure 5–15 shows a simple VMOS application that takes advantage of its very high input impedance and high output current capability.

FIGURE 5–14 VMOS output characteristic curve.

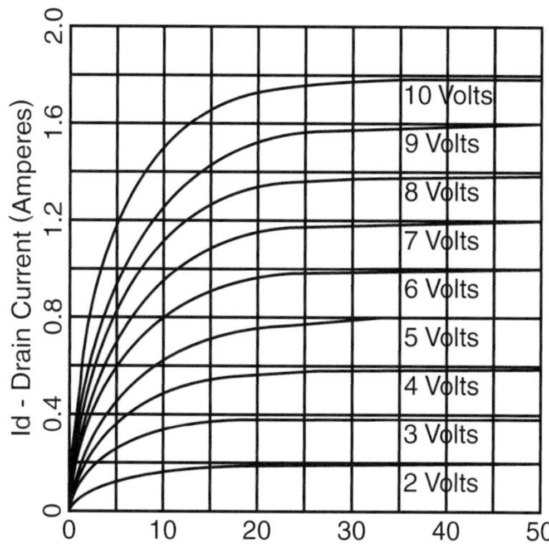

FIGURE 5–15 VMOS touch-activated alarm bell.

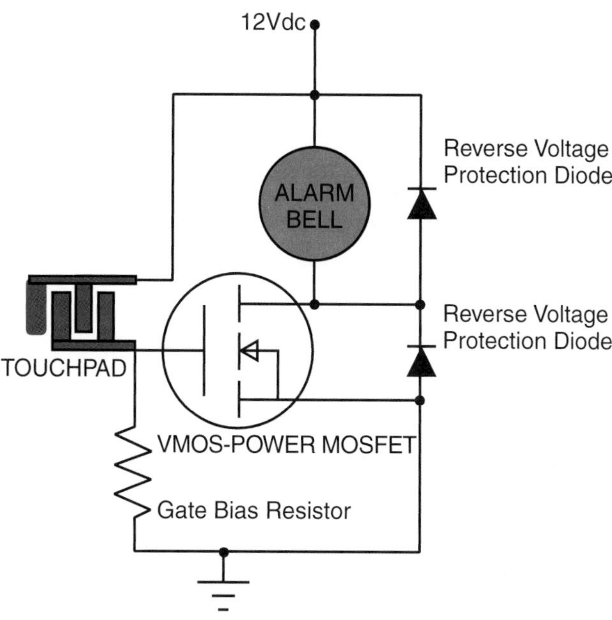

5.7 Other Types of MOSFET Semiconductors

Two other types of MOSFET semiconductors are the dual gate and the LDMOS (lateral double-diffused MOSFET). One limiting factor in the use of MOSFETs is the inherent capacitive coupling between gate and channel. A dual-gate MOSFET employs two smaller gates rather than one large gate. Therefore input capacitance is smaller for each gate.

One gate is typically employed for input, whereas the other is usually grounded. The reduced input capacitance allows the dual-gate MOSFET to produce respectable gain well into the ultra-high frequency range (UHF, 300 to 3,000 MHz).

The dual-gate MOSFET may also be employed as a signal "mixer" or "mixer-oscillator" by applying different signals at each of the two gates.

Although transistors came into the marketplace in the 1950s and 1960s, it was not until recent years that solid-state power amplifiers could perform well above 100 MHz. LDMOS is an enhancement-type MOSFET. Its main advantage is low channel resistance and the ability to handle very high drain currents without generating a lot of heat dissipation. A very small channel and a heavily doped N-type substrate make this possible.

LDMOS performs well into the UHF range. Its linearity and wideband frequency response make the LDMOS a good choice for cellular communication high-power amplifiers. LDMOS power dissipation ($P_{DISSIPATION} = I_{DRAIN} \times V_{DRAIN-SOURCE}$) ranges from 5 to 250 watts. The schematic symbol is that of a conventional enhancement mode transistor.

UNIJUNCTION TRANSISTOR

5.8 UJT Construction

The UJT has three leads but only two doped regions. The terminals are emitter, base 1 (B1) and base 2 (B2). Figure 5–16 shows the schematic symbol for a UJT. The emitter is heavily doped P-type material, and the base is only slightly doped N-type material. Note that the construction of the UJT is very similar to that of the JFET. The only difference is that the P-type gate material of the JFET surrounds the n-channel, whereas the P-type material of the emitter does not surround the base in the UJT.

FIGURE 5–16 UJT schematic symbol.

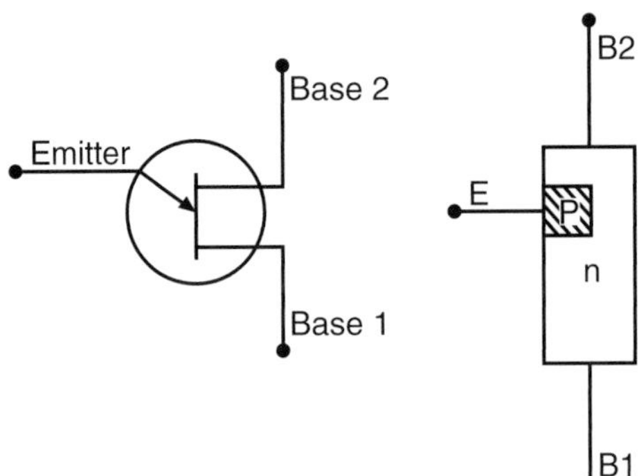

5.9 UJT Operating Characteristics

The main function of the UJT is switching. Oscillators make frequent use of UJTs. UJTs are not used as amplifiers. One application of a UJT is elevator floor-arrival gongs (one up, two down).

Refer to Figure 5–17 for the following discussion. When a base biasing voltage (V_{BB}) is applied between B1 and B2, the emitter-B1 (E-B1) junction is seen as open until the voltage across V_{EB1} is increased to a specific value.

This value is called peak voltage (V_P). When V_P is reached, the E-B1 junction triggers or fires. After firing, the UJT will remain on until the emitter current falls below a specific value. This value is called peak current (I_P). When the emitter current falls below I_P, the UJT turns off.

FIGURE 5–17 UJT biasing.

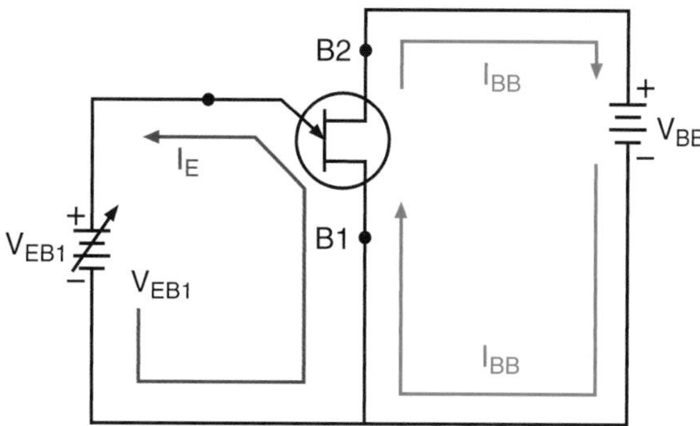

Figure 5–18 shows the UJT characteristic curve. When the emitter current (I_E) increases above I_P, the value of the internal resistance R_{B1} drops dramatically. This results in a reduction of V_{EB1} as I_E increases. The reduction in V_{EB1} continues until I_E reaches the valley current (I_V).

Increasing I_E beyond I_V puts the UJT into saturation. Note that the region where the internal resistance is reduced (between I_P and I_V) is called the **negative resistance region**. Negative resistance is when any device has current and voltage values that are inversely related to each other.

Negative resistance region With reference to a UJT, the range of emitter current depicted on the VEB1/IE characteristic curve, where VEB declines as emitter current increases.

FIGURE 5–18 UJT characteristic curve with negative resistance region.

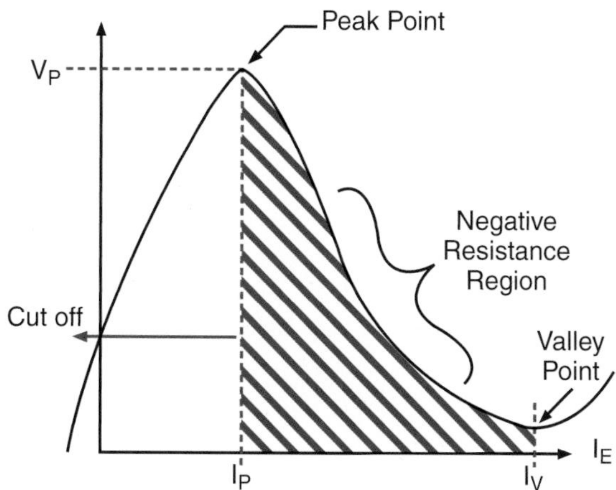

The relationship between the resistance of B1 and B2 in the UJT can be expressed as a voltage divider using an equivalent circuit. Figure 5–19 shows an example of this equivalent circuit. The actual voltage drops and operating principle are straightforward. The E-B1 diode must be forward biased. This means that a voltage drop of 0.7 V is required from the anode to the cathode. This forward bias potential (V_K) is calculated using the voltage divider rule.

$$V_K = V_{BB}\left(\frac{R_{B1}}{R_{B1} + R_{B2}}\right)$$

The resistance ratio $\left(\frac{R_{B1}}{R_{B1} + R_{B2}}\right)$ is called the intrinsic standoff ratio (η). This ratio is listed on the specification sheet for any UJT.

$$V_K = \eta\ V_{BB}$$

$$V_P = \eta\ V_{BB} + 0.7\text{ V}$$

FIGURE 5–19 UJT equivalent circuit.

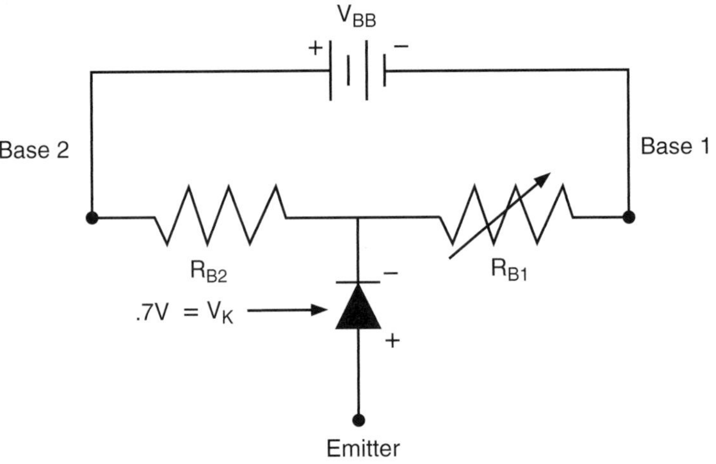

5.10 UJT Example Circuit

Example

Calculate the V_P required to trigger the UJT in the following circuit (see Figure 5–20). Assume the following: η max from the specification sheet for this UJT is 0.75.

$$V_P = \eta\ V_{BB} + 0.7\text{ V}$$

$$V_P = [0.75(15\,V)] + 0.7\text{ V}$$

$$V_P = 11.95\text{ V}$$

FIGURE 5–20 Example UJT circuit.

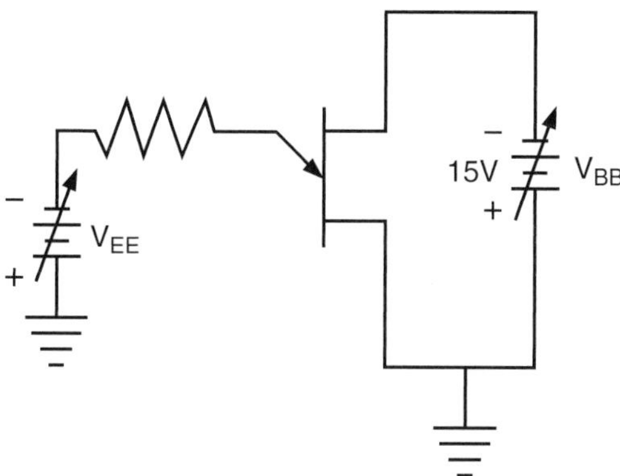

Because one of the primary functions of the UJT is to trigger triacs or SCRs into conduction, an RC time constant could be added to the circuit of Figure 5–20 to provide alternating charge and discharge pulses. An example of a UJT circuit with added RC components is shown in Figure 5–21. Note that the resistance is variable. This allows the frequency of the oscillator to be varied.

Once the capacitor charges to 11.95 V, the UJT turns on. When the capacitor discharges below V_P, the UJT will turn off and the capacitor C_T will begin to recharge to V_p (repeating the cycle).

FIGURE 5–21 UJT circuit with RC time constant.

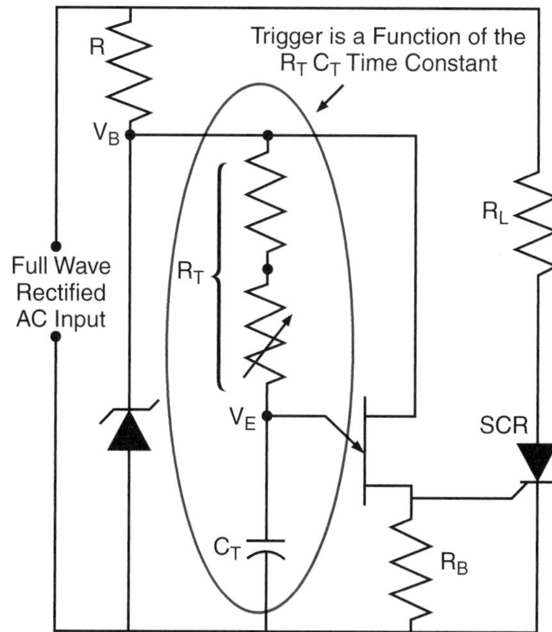

INSULATED GATE BIPOLAR TRANSISTOR

The insulated (or isolated) gate bipolar transistor (IGBT) is not a newcomer to solid-state power switching, but the latest generation of these devices represents much improved versions. It is a composite of bipolar and insulated-gate technology, producing a device with characteristics uniquely suited to fast high-power switching with low current drive requirements. Due to its rugged construction, the IGBT is usually found in switching power supplies and solid-state motor controllers. As with most digital switching devices, the IGBT is an enhancement-mode transistor. See Figure 5–22.

FIGURE 5–22 Schematic symbol of the IGBT.

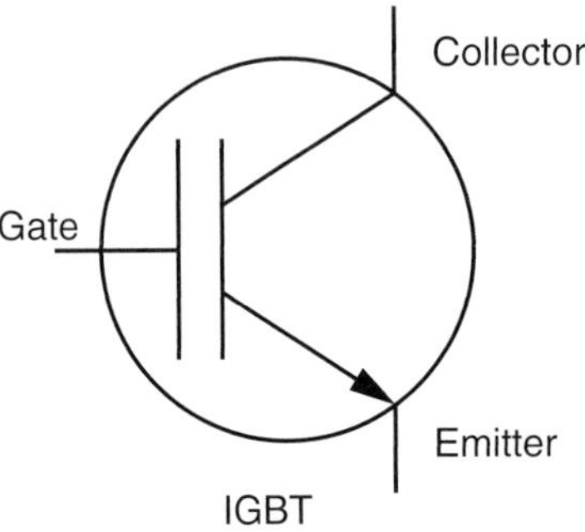

Digital multimeters may be used to check IGBTs only for short circuits. Because the gate is isolated, gate to collector and gate to emitter should measure in excess of 1 megohm. Because collector and emitter are not adjacent layers, resistance between them (with open gate) should be in excess of 100,000 ohms.

SUMMARY

In this chapter, you have learned about three major types of transistors: the JFET, MOSFET, and UJT. Each has its own advantages and disadvantages. Figure 5–23 summarizes the main characteristics of each.

The JFET has a relatively high input impedance and is used as a buffer against loading down the source amplifier and reducing overall amplifier gain. The JFET comes in two basic types, n and p. JFETs are voltage-controlled rather than current-controlled devices.

The MOSFET also comes in two basic types: depletion and enhancement. The depletion type can operate in either depletion or enhancement mode. The enhancement type also has a high-power version (VMOS). It is used when low heat dissipation and high currents are required. The MOSFET has a higher input impedance than the JFET and is used extensively in integrated circuitry. MOSFETs also require special handling to protect them from possible static electricity damage.

The UJT is not used as an amplifier, but as a switching device. Frequently, the UJT is used to trigger other devices—such as triacs and SCRs. UJTs are very similar in construction to JFETs but do not have the same lead configuration. Unlike FETs (which have a gate, a source, and a drain), UJTs have an emitter, a base 1, and a base 2.

Insulated Gate Bipolar Transistors (IGBTs), hybrids of BJT and insulated gate technology, are tough, high input impedance, enhancement power transistors well-suited for use in switching power supplies.

FIGURE 5–23 Summary of JFETs, MOSFETs, and UJTs.

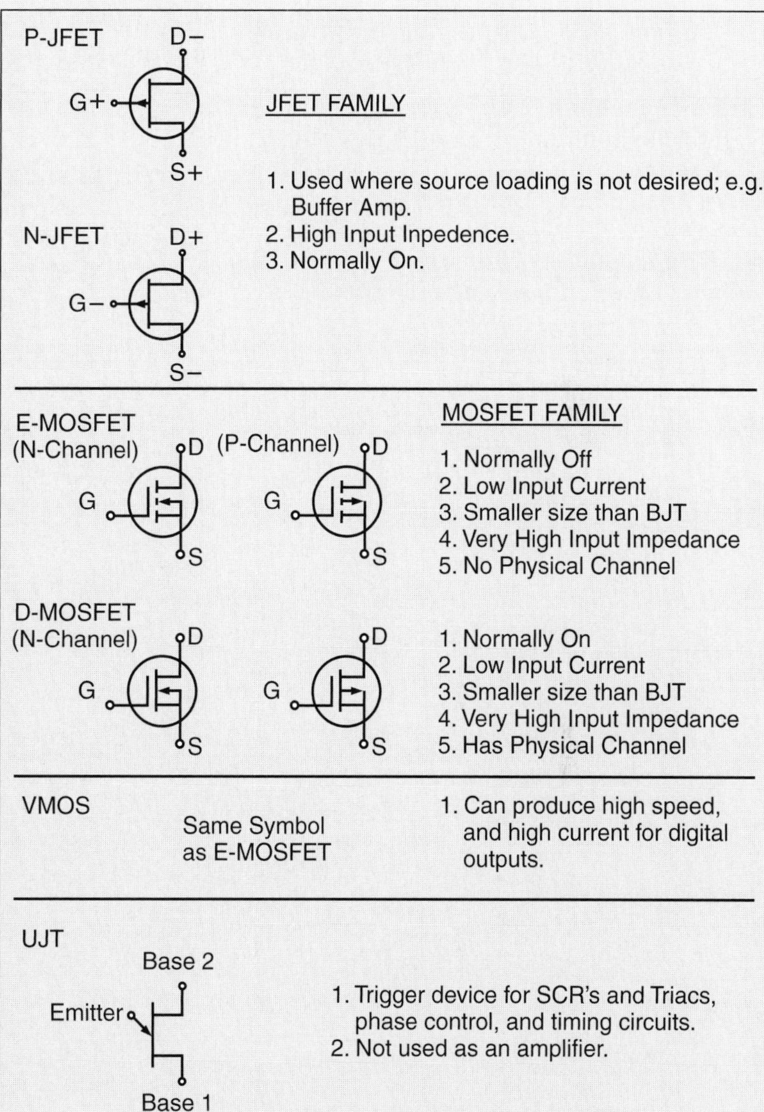

P-JFET
D–
G+
S+
JFET FAMILY
1. Used where source loading is not desired; e.g. Buffer Amp.
2. High Input Inpedence.
3. Normally On.
N-JFET
D+
G–
S–
E-MOSFET (N-Channel)
(P-Channel)
G
D
S
MOSFET FAMILY
1. Normally Off
2. Low Input Current
3. Smaller size than BJT
4. Very High Input Impedance
5. No Physical Channel
D-MOSFET (N-Channel)
1. Normally On
2. Low Input Current
3. Smaller size than BJT
4. Very High Input Impedance
5. Has Physical Channel
VMOS
Same Symbol as E-MOSFET
1. Can produce high speed, and high current for digital outputs.
UJT
Base 2
Emitter
Base 1
1. Trigger device for SCR's and Triacs, phase control, and timing circuits.
2. Not used as an amplifier.

REVIEW QUESTIONS

1. JFETs and BJTs are 3-pin semiconductor devices. The BJT pins are designated Emitter, Base, and Collector. What are the related pin designations of the JFET?
2. Compare and Contrast the input impedances of JFETs, MOSFETs, and BJTs.
3. Study Figure 5–24. Explain how the JFET may be reverse-biased, when the junction of R_1 and R_2, (Vgate) is above Ground potential.

FIGURE 5–24 JFET voltage divider biasing.

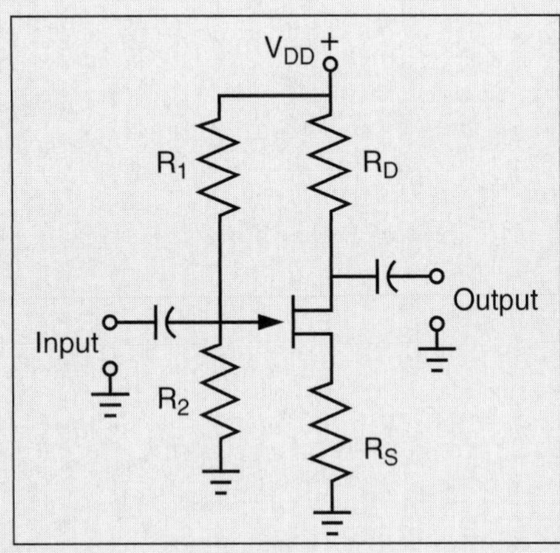

4. Describe the fundamental difference between Enhancement-mode MOSFETs, and Depletion-mode MOSFETs.
5. Zener diodes are often built-into MOSFET transistors. How do these diodes improve MOSFET reliability?
6. The Lateral Double-diffused MOSFET (LDMOS) overcomes one of the MOSFET's principle shortcomings, poor high frequency response. Name one or more applications made possible by the LDMOS structure.
7. What are the advantages of the Dual-gate MOSFET?
8. The negative resistance exhibited by the UJT makes it an ideal driver for what switching devices you have previously studied?
9. The Insulated-Gate Bipolar Transistor (IGBT) incorporates what advantages of both the MOSFET, and the BJT?
10. Discuss Digital Multimeter testing procedures for the IGFET.

6 Amplifiers

OUTLINE

OVERVIEW

In this chapter, you will learn about the basic types of amplifiers and how they operate. Amplifiers are used in most electronic circuits. In audio equipment, the normal input signal is too small to power the speaker system and must therefore first be amplified. The same is true with video equipment. The input signal must be amplified before it can be presented on a cathode-ray tube or liquid crystal display (a TV screen or computer monitor).

The electronic circuits that perform this job of increasing the voltage, current, or power level of electrical signals are called amplifiers. In the strict definition, power gain must be involved for the circuit to be an amplifier. A transformer, for example, will increase the voltage level of a signal but does not increase the power level. Consequently, the transformer is not an amplifier.

In electronics, you normally do not worry about such issues. Amplifier circuits may be designed for voltage gain, current gain, or both voltage and current gain. Because $P = IE$, we can safely say that all amplifiers make available more power at their outputs than delivered to their inputs. Although there are exceptions, amplifiers are usually designed to deliver much the same wave shape at their outputs as found at their inputs.

OBJECTIVES

After completing this chapter, the student should be able to:

1. Define the various terms associated with amplifier circuits.
2. Draw basic circuit configurations for the three types of amplifiers—common emitter, common collector, and common base.
3. Describe the unique characteristics of each type of amplifier—common emitter, common collector, and common base.

AMPLIFIER GAIN

An **amplifier** provides **gain**—the ratio of the output to the input. An amplifier achieves gain by converting one thing to another. For example, a lever converts force to movement or movement to force. One pound of force on the end of a beam may lift 3 pounds on the other end, but it will not lift it far. An electronic amplifier uses circuit DC power and divides that power between the output terminal and a load resistor based on the strength of the input signal. Figure 6–1 shows a block diagram of an amplifier circuit.

FIGURE 6–1 Amplifier block diagram.

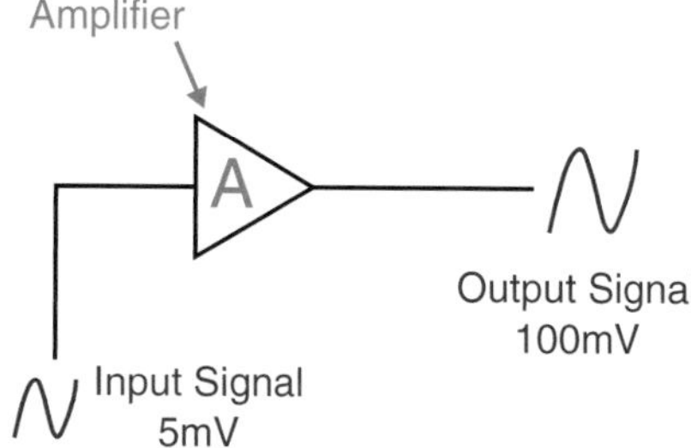

6.1 Types of Gain

If the objective for an amplifier is to produce gain, a measure of the gain of the amplifier is a measure of the amplifier's success. Gain is expressed as the ratio of output to input. Gain is shown in formulas using the symbol A with a subscript of $_P$, $_I$, or $_V$ to indicate power, current, or voltage, respectively.

Take care to write the result correctly; A_P, A_I, and A_V are unitless. This is because A is calculated by dividing power by power, current by current, or voltage by voltage. By not tacking units on the ends of **amplification** factors, statements such as "We have six times as many watts at the output than at the input" can be avoided. Normally, gain is expressed as a dimensionless number—"The amplifier has a voltage gain of 10, 20, 50, 110, . . . "

6.2 Measuring Gain

To calculate voltage gain, divide the output voltage by the input voltage levels. Refer to Figure 6–2.

$$\text{Gain} = \frac{\text{Output signal}}{\text{Input signal}}$$

$$A_V = \frac{V_{OUT}}{V_{IN}}$$

$$A_V = \frac{10\text{ V}}{0.5\text{ V}}$$

$$A_V = 20.0$$

Amplifier
A device that provides gain without much change in the original signal waveform.

Gain
The ratio of the output signal to the input signal of an **active component**.

Active component
Components of an electronic circuit that use a power source to process a signal. The processing usually involves amplification or some other change in the signal that requires additional power. BJTs, FETs, and UJTs are examples of active components.

A_P
Power gain.

A_I
Current gain.

A_V
Voltage gain.

Amplification
The process of increasing the voltage, current, or power of a signal.

FIGURE 6–2

Voltage gain.

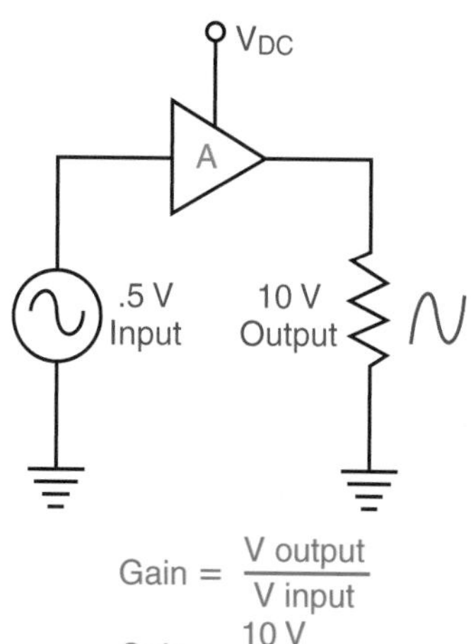

Note that the volts symbols cancel each other and the final result is unitless. As another example, assume an amplifier has a 1 mW input and an output level of 10 mW, as shown in Figure 6–3. What is the amplifier's power gain?

$$A_P = \frac{P_{OUT}}{P_{IN}}$$

$$A_P = \frac{10 \text{ mW}}{1 \text{ mW}}$$

$$A_P = 10$$

FIGURE 6–3 Power gain.

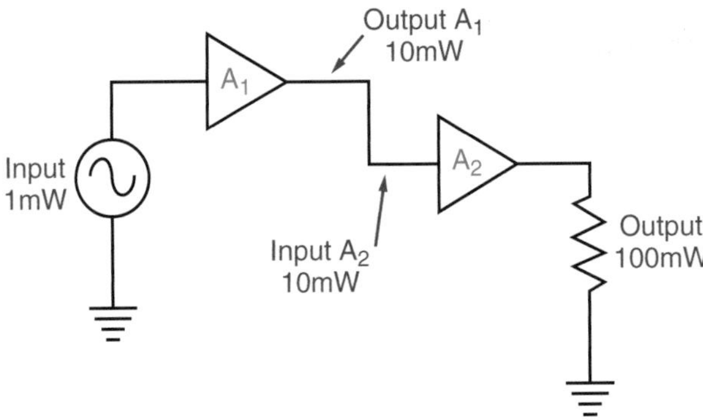

If another amplifier takes the 10 mW input and yields a 100 mW output level, what is this amplifier's power gain?

$$A_P = \frac{P_{OUT}}{P_{IN}}$$

$$A_P = \frac{100 \text{ mW}}{10 \text{ mW}}$$

$$A_P = 10$$

What is the gain of both stages from input to output?

$$A_P = \frac{P_{OUT}}{P_{IN}}$$

$$A_P = \frac{100 \text{ mW}}{1 \text{ mW}}$$

$$A_P = 100$$

Note that the total gain of both stages is the product of the individual gains.

6.3 Gain in Decibels

Power Versus Audio Response

If we listen to the sound of a tone that starts at 1 mW and increases to 100 mW in two steps (as in the previous two examples), we might expect to be deafened by the volume of the second step.

The gain can be calculated as 100. In fact, we would perceive the sound to be only four times as loud as the original. Furthermore, the 10 mW tone would sound twice as loud as the 1 mW tone and the 100 mW tone would sound twice as loud as the 10 mW tone.

The human ear has a logarithmic response to sound. This means that we perceive the relative volume of two sounds instead of the absolute volume. We perceive other things in the same way. A poor person will look at the gift of a dollar in a very different way than a millionaire. This is to say that the effect of the sound on our ears is relative instead of absolute (in math calculation terms).

Decibels

Because of the logarithmic nature of human hearing, the decibel system of expressing power levels was developed. It is used to describe the performance of audio systems, but it is also useful for describing any system of electrical amplifiers or relative amplitude levels. The power gain in Bels is calculated as:

$$A_{G(B)} = \log_{10}\left(\frac{P_{OUT}}{P_{IN}}\right)$$

For most "real-world" applications, the Bel is simply too large a unit to work with. A more convenient unit is the decibel (dB), which is defined as:

$$A_{G(dB)} = 10 \times A_{G(B)}$$

$$A_{G(dB)} = 10 \times \log_{10}\left(\frac{P_{OUT}}{P_{IN}}\right)$$

A decibel is, as its name implies, a DECI-BEL, or, $\frac{1}{10}$ of a Bell. (The *Bel* in the term honors Alexander Graham Bell.) In the previous example, the difference between 1 and 10 mW is 10 decibels or 10 dB. This is calculated by using the following formula.

$$A_{P(dB)} = 10 \times \log_{10}\left(\frac{10\text{ mW}}{1\text{ mW}}\right)$$

$$A_{P(dB)} = 10 \times \log_{10} 10$$

$$A_{P(dB)} = 10 \times 1$$

$$A_{P(dB)} = 10\text{ dB}$$

Note that the gain of the second stage is also equal to 10 dB. The gain of both stages together is calculated as:

$$A_{P(dB)} = 10 \times \log_{10}\left(\frac{100\text{ mW}}{1\text{ mW}}\right)$$

$$A_{P(dB)} = 10 \times \log_{10} 100$$

$$A_{P(dB)} = 10 \times 2$$

$$A_{P(dB)} = 20\text{ dB}$$

The conversion is easy with a scientific calculator. With an algebraic entry calculator, the following steps will do the job.

1. Press the left parenthesis [(].
2. Key in the output power.
3. Press the divide key [÷].
4. Key in the input power.
5. Press the right parenthesis [)].
6. Press the log button. [Be sure to use log to the base 10, not the *ln* button, which calculates logs to the natural (Naperian) base *e*.]
7. Press the multiply key [×].
8. Key in the number 10.
9. Press the equals key [=].

The result is the power gain expressed in decibels. For an RPN-type calculator (such as many Hewlett-Packard models), the procedure is slightly different.

1. Key in the output power.
2. Press the [Enter] key.
3. Key in the input power.
4. Press the divide key [÷].
5. Press the log key. (As before, be sure you use the common logarithm, not the natural logarithm.)
6. Key in the number 10.
7. Press the multiply key [×].

The result is the power gain expressed in decibels.

6.4 Multistage Gain

It is unusual to find more than three amplifiers of the same type cascaded (connected one after another). One reason for this is that all electric and electronic devices generate undesirable signals (thermal noise) on their own. This noise adds to noise passed on by previous devices. Using too many amplifier stages can result in noise that may reach unacceptable levels when compared with signals the amplifiers were intended to enhance.

Figure 6–4 shows the gains of each stage of a multistage amplifier. They are 14 dB, −4 dB, and 18 dB. Note that a gain of −4 dB is actually a reduction in signal level, as you will see. To calculate gain using decibels, simply add the decibel gains as follows.

$$A_{P(dB)} = 14 \text{ dB} + (-4 \text{ dB}) + 18 \text{ dB} = 28 \text{ dB}$$

System gain can also be calculated using ratios, as illustrated in Figure 6–5. The gain ratio for each stage is found by dividing output power by input power. Those gains are then multiplied, to obtain the total system gain. This gain measurement has no unit of measure.

FIGURE 6–4 Multistage gain in decibels.

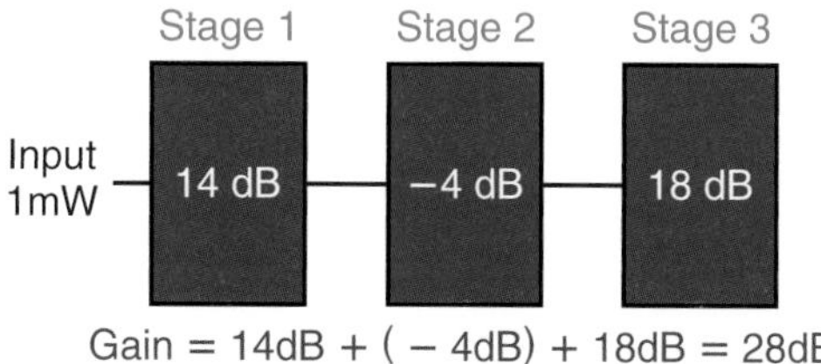

FIGURE 6–5 Multistage gain in linear amplification factors.

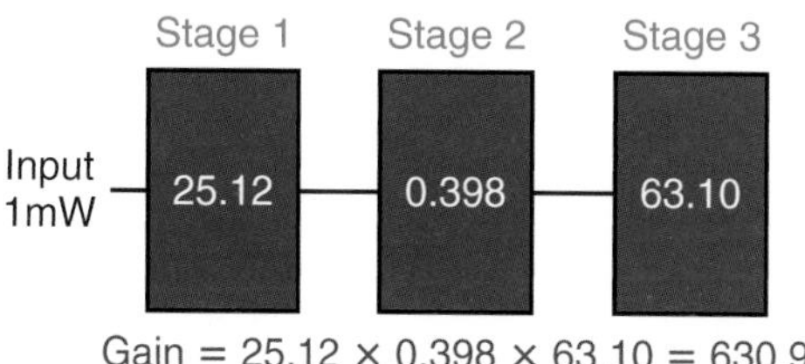

Start by considering the definition of the decibel.

$$A_{P(dB)} = 10\log_{10}\left(\frac{P_{OUT}}{P_{IN}}\right)$$

Solving this formula for the ratio $\frac{P_{OUT}}{P_{IN}}$ involves a little algebra.

$$\frac{A_{P(dB)}}{10} = \log_{10}\left(\frac{P_{OUT}}{P_{IN}}\right) \Rightarrow \left(\frac{P_{OUT}}{P_{IN}}\right) = 10^{\left(\frac{A_{P(dB)}}{10}\right)}$$

Calculating the ratio gain of the three stages (see Figure 6–5) gives:

$$A_{P(Stage\ 1)} = 10^{\left(\frac{14}{10}\right)}$$

$$A_{P(Stage\ 1)} = 10^{1.4}$$

$$A_{P(Stage\ 1)} = 25.12$$

$$A_{P(Stage\ 2)} = 10^{\left(\frac{-4}{10}\right)}$$

$$A_{P(Stage\ 2)} = 10^{-0.4}$$

$$A_{P(Stage\ 2)} = 0.398$$

$$A_{P(Stage\ 3)} = 10^{\left(\frac{18}{10}\right)}$$

$$A_{P(Stage\ 3)} = 10^{1.8}$$

$$A_{P(Stage\ 3)} = 63.10$$

$$A_{P(Total)} = 25.12 \times 0.398 \times 63.10$$

$$A_{P(Total)} = 630.9$$

To check your work, you can put this value into the original formula.

$$A_{P(dB)} = 10 \times \log_{10}(630.9) = 28.00\ \text{dB}$$

The slight difference in result is caused by rounding error. If you perform the previous steps using a scientific calculator and do not round, you will get a much closer answer. Perhaps you can see that it is easier to work with decibels than linear amplification factors. Using decibels generally causes fewer mistakes.

6.5 Voltage Gain in Decibels

The voltage gain in decibels uses a slightly different formula than the power gain in decibels. The reason for the difference is the logarithmic nature of decibels and the relationship between voltage and power. Recall that $P = \frac{V^2}{R}$. Therefore,

$$A_{P(dB)} = 10 \times \log_{10}\left[\frac{\frac{V^2}{R_{OUT}}}{\frac{V^2}{R_{IN}}}\right]$$

If R_{OUT} equals R_{IN}, the resistances will divide out of the equation leaving only:

$$A_{P(dB)} = 10 \times \log_{10}\left(\frac{V_{OUT}^{\;2}}{V_{IN}^{\;2}}\right)$$

Note that you can divide the resistances out only if they are equal. If they are not, you must include them and use the slightly more complicated formula. From logarithms, recall the following:

$$\log X^2 = 2 \times \log X$$

This means that:

$$A_{P(dB)} = A_{V(dB)}$$

$$A_{V(dB)} = (2 \times 10)\log_{10}\left(\frac{V_{OUT}}{V_{IN}}\right)$$

$$A_{V(dB)} = 20 \times \log_{10}\left(\frac{V_{OUT}}{V_{IN}}\right)$$

Note the change from power gain to voltage gain. This occurs because the ratio of the logarithm $\left(\frac{V_{OUT}}{V_{IN}}\right)$ is a voltage ratio. Remember that this formula applies *only* if the resistances across which the two voltages are dropped are equal.

6.6 Practical Uses of dB Ratios

The advantage of dB ratios is the logarithmic ability to express large numbers with small numbers. The decibel (1/10th of a Bel) is normally used in sound systems because the number scaling is more natural. By applying the dB power formula (gain in decibels), you will find that if a 400 watt audio amplifier were replaced with an 800 watt amplifier there would be only an actual 3 dB gain. The power difference between the two amplifier gains is twofold. Therefore, the application of the power formula is:

$$dB = 10 \times \log_{10}\left(\frac{P_{OUT}}{P_{IN}}\right)$$

$$dB = 10 \times \log_{10}\frac{800\text{ W}}{400\text{ W}}$$

$$dB = 10 \times \log_{10} 2$$

$$dB = 3.01\text{ dB}$$

TABLE 6–1 Power Values (Watts) Versus Equivalent dB Levels

Power Values (Watts)	Levels in dB (Compared to 1 Watt)
1.0	0
1.25	1
1.6	2
2.0	3
2.5	4
3.15	5
4.0	6
5.0	7
6.3	8
8.0	9
10.0	10
100	20
200	23
400	26
800	29
1,000	30
2,000	33
4,000	36
8,000	39
10,000	40
20,000	43
40,000	46
80,000	49
100,000	50

Table 6–1 outlines the relationships between power values and dB levels. For example, an 800 watt amplifier output is equal to 29 dB ($dB = 10 \times \log_{10}(800) = 29$). In this case, we are comparing the 800 watts to 1 watt, and the gain is simply selected from the 800 watt row.

In comparison, if a 400 watt amplifier were replaced with an 800 watt amplifier the results would be a 3 dB gain. Here, we are comparing the new 800 watt amplifier to the replaced 400 watt amplifier. By subtracting the 26 dB in the 400 watt row from the 29 dB in the 800 watt row, the result is 3 dB.

Therefore, Table 6–1 represents the values in dB compared to 1 watt or by subtracting compared to other wattage ratings within the table. If you turned your stereo up from 10 watts to 100 watts and someone commented "It's too loud," you could say "But I only turned it up ?? dB."

By subtracting the dB in the 10 watt row from the dB in the 100 watt row, the answer is found to be 10 dB. The decibel expresses the "logarithmic" ratio of two acoustic quantities (sound levels) and is widely used in audio and communication electronics.

An electrical power level is expressed as dBm, where the *m* stands for milliwatts and is the reference. The formula for dBm is:

$$dBm = 10 \times \log_{10}\left(\frac{P_{mW}}{1\text{mW}}\right)$$

The dBm has no direct relationship to impedance or voltage. Originally, the dBm was devised from a 600 ohm telephone load, and in most cases today 0 dBm is referenced to 0.775 V (see following equation), but can be referenced to other voltage levels.

$$P = \frac{V^2}{R}$$

$$P = \frac{0.775^2\ \text{V}}{600\ \Omega}$$

$$P = 1\ \text{mW}$$

A voltage level is expressed as dBu and is referenced to 0.775 V. It is not dependent on the resistance load. The voltage expressed by dBu and dBm are equal only if the dBm voltage is derived with a 600 ohm load. A sound system console specification could list its maximum output level as +20 dBu into an 8k ohm (or higher) impedance load.

The dBm and dBu mainly express small wattage and voltage levels. The dBW was derived to express large wattage outputs of large amplifier systems. Zero dBW is equal to 1 watt. Therefore, an 800 watt amplifier is a 29 dBW amplifier. This is found by applying the following power formula.

$$\text{dBW} = 10 \times \log\left(\frac{800}{1}\right) = 29\ \text{dBW}$$

AMPLIFIER CLASSIFICATION

6.7 Amplifier Operating Characteristics

There are two major operating characteristics that affect the classification of an amplifier: function and frequency response.

1. Function: an amplifier is usually designed to amplify either voltage or power.
2. Frequency response: an amplifier may be designed as:
   - Audio amplifier: designed to amplify frequencies from 5 Hz to 20 kHz
   - Radio frequency amplifier: designed to amplify frequencies between 10 kHz and 100,000 MHz
   - Video amplifier: designed to amplify wide bands of the frequencies between 10 Hz and 6 MHz
   - DC amplifiers: designed to amplify DC or near DC signals, such as from temperature sensors

6.8 Amplifier Classes

The class of operation of an amplifier is determined by the amount of time the current flows in the output circuit compared to the input signal. There are four classes of operation: A, AB, B, and C. Each of these classes has certain uses and characteristics, but no one class of operation is considered better than another class. The selection of the class to use is determined by how the amplifier is used. The best class of operation for a servo-feedback loop would not be the same as for a radio transmitter.

Class A Amplifier

Figure 6–6 illustrates the basic operation of a class A amplifier. In a class A amplifier, current will flow in the output during the entire input signal period. The output is an exact copy of the input except for increased **amplitude.** A class A amplifier is called a fidelity amplifier because of its ability to recreate a larger-scale image of the input signal with little or no distortion.

Amplitude
The size of a signal. Most commonly, the amplitude is expressed in terms of the signal voltage.

FIGURE 6–6 Class A amplifier.

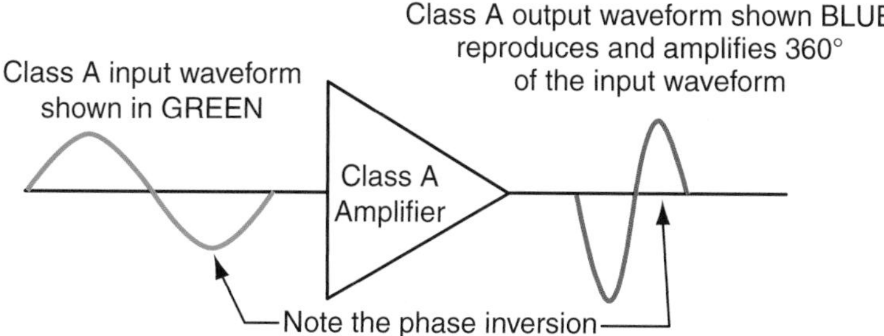
Class A output waveform shown BLUE
reproduces and amplifies 360°
of the input waveform
Class A input waveform
shown in GREEN
Class A
Amplifier
Note the phase inversion

A class A amplifier may change the phase of a signal, but not the shape. Because the class A amplifier operates during the entire time the input signal is present, it is less efficient than an amplifier that operates only during half the input. However, it produces better fidelity. Class A amplifiers are commonly utilized in audio and video circuits, as well as instrumentation amplifiers.

Class AB Amplifier

The class AB amplifier operates during 51% to 99% of the input signal. Because it does not operate for the complete input cycle, the output signal cannot be the same shape as the input signal. In this case, the output signal will be distorted. Distortion is any undesired change in signal from the input to the output. Figure 6–7 shows the clipping that occurs during the positive part of the output.

FIGURE 6–7 Class AB amplifier.

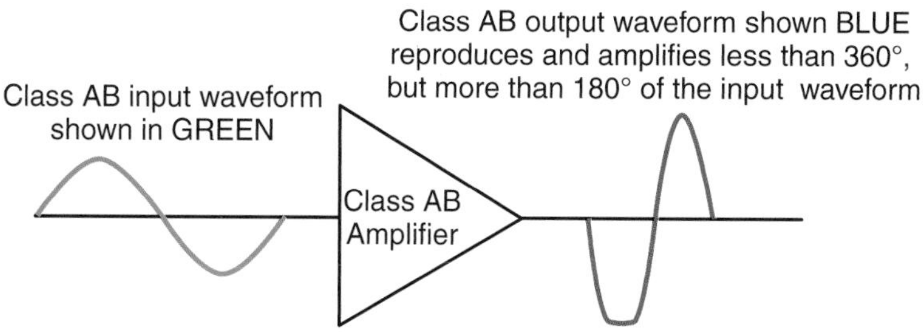
Class AB output waveform shown BLUE
reproduces and amplifies less than 360°,
but more than 180° of the input waveform
Class AB input waveform
shown in GREEN
Class AB
Amplifier

Note that not all of the positive part is clipped. In a class AB amplifier, the transistor going into cutoff causes the clipping. Thus, it is more efficient than the class A amplifier because it does not always amplify. Nor does it produce a fidelity signal because the signal is distorted due to clipping. Class AB amplifiers are often used in audio power amplifiers and other applications requiring linear operation.

Class B Amplifier

The class B amplifier operates for 50% of the input signal. The amplifier will amplify one half of the signal very well, but the other half is completely lost. This amplifier is twice as efficient as the class A amplifier and is used when only half the signal is needed. Figure 6–8 shows an example of a class B amplifier.

FIGURE 6–8 Class B amplifier.

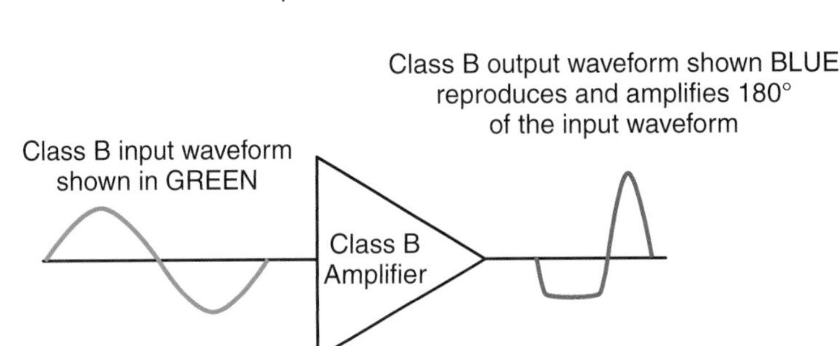
Class B output waveform shown BLUE
reproduces and amplifies 180°
of the input waveform
Class B input waveform
shown in GREEN
Class B
Amplifier

No current flows in the class B amplifier when no signal is applied to its input. This fact allows the circuit to operate more efficiently, run cooler, and demand less from the power supply. Class B amplifiers are generally used in radio frequency amplifiers where efficiency is more important than precise waveform reproduction.

Class C Amplifier

The class C amplifier is the most efficient of the four classes because it operates only during a small portion of the input. It also produces the most distortion of all the amplifiers. It is useful when only a very small part of a signal must be amplified. Figure 6–9 shows an example of a class C amplifier. The class C amplifier, just as the class B, draws no current when no signal is applied to its input. Class C amplifiers are most commonly used as radio frequency power amplifiers.

FIGURE 6–9 Class C amplifier.

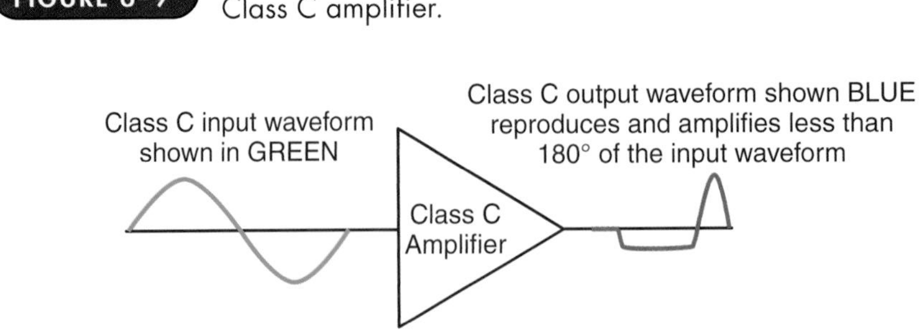
Class C output waveform shown BLUE
reproduces and amplifies less than
180° of the input waveform
Class C input waveform
shown in GREEN
Class C
Amplifier

6.9 Amplifier Configurations

Figures 6–10, 6–11, and 6–12 show the three types of BJT amplifiers discussed in an earlier chapter. They are the common emitter, common collector, and common base. Each of these configurations can function as an amplifier.

FIGURE 6–10 Common-emitter amplifier.

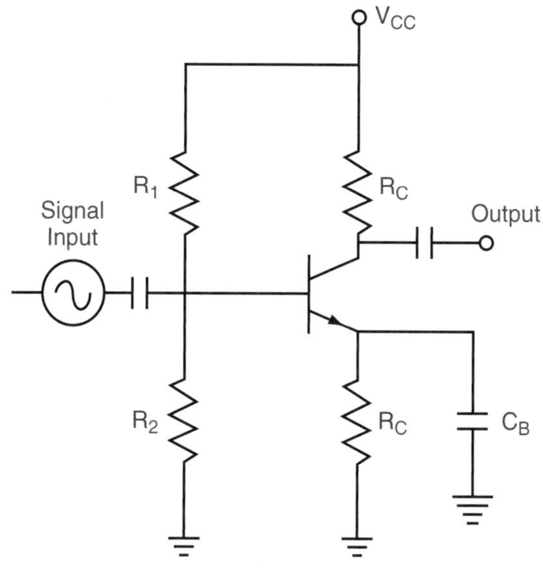

FIGURE 6–11 Common-collector amplifier.

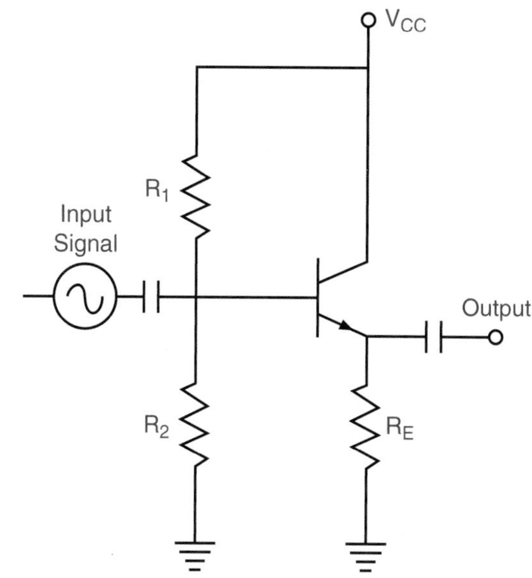

FIGURE 6–12 Common-base amplifier.

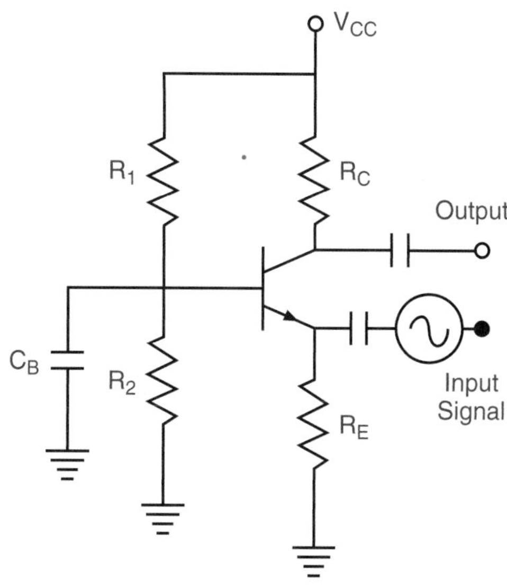

The characteristics of a common-base amplifier's gain are very different from those of a common emitter or common collector. Table 6–2 outlines the relationship of the gain characteristics among the three configuration types.

TABLE 6–2 Amplifier Gain Characteristics

Configuration Types	Power Gain	Voltage Gain	Current Gain
Common Emitter	High	Moderate	Moderate
Common Collector	Moderate	Very low, less than 1	Moderate
Common Base	Moderate	Moderate	Very low, less than 1

Gains: Low = less than 100, Moderate = 100 to 1,000, and High = greater than 1,000.

6.10 Load Line and the Operating Point

Load Line Development

Transistor amplifiers are current-controlled devices. The amount of current in the base of a transistor controls the amount of current in the collector. The secret to understanding amplifiers is to remember the fact that current controls the gain. If the current is controlled in the amplifier, simply decreasing or increasing the current that flows through the collector can control the output voltage.

If only 5% of the total current flows through the base, this small portion of the current will control the larger collector current flow. The capability to control the larger collector current flow with the smaller base flow is the basic concept of a transistor amplifier. This is why the load line characteristics of a transistor amplifier are so important. The load line is developed from two assumptions:

1. The transistor in Saturation $I_{C(Sat)}$
2. The transistor in Cutoff $V_{CE(Cutoff)}$

$$I_{C(Sat)} = \frac{V_{CC}}{R_C + R_E}$$

$$V_{CE(Cutoff)} = V_{CC}$$

Use the values of Figure 6–13 to calculate $I_{C(Sat)}$:

$$I_{C(Sat)} = \frac{V_{CC}}{R_C + R_E}$$

$$I_{C(Sat)} = \frac{15\text{ V}}{(5{,}000\ \Omega + 0\ \Omega)}$$

$$I_{C(Sat)} = \frac{15\text{ V}}{5{,}000\ \Omega}$$

$$I_{C(Sat)} = 3\text{ mA}$$

$$V_{CE(Cutoff)} = V_{CC}$$

$$V_{CE(Cutoff)} = 15\text{ V}$$

FIGURE 6–13 Load line for amplifier.

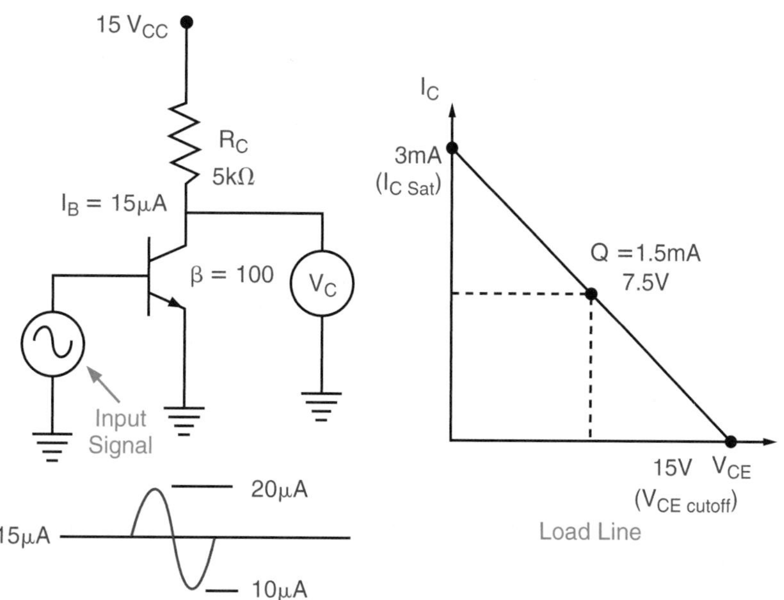

On the right-hand side of Figure 6–13, these two values have been plotted and connected with a straight line. The straight line is the load line and can be used to determine the various operating values for the amplifier.

Q-Point (Quiescent Point)

The Q-point is the point on the load line that indicates the values of V_{CE} and I_C when there is no active (AC) input signal. Figure 6–14 shows an example of how the input AC signal causes the amplifier output of Figure 6–13 to vary above and below the load line Q-point. Note that the Q-point of the amplifier in Figure 6–13 is $V_{CE} = 7.5$ and $I_C = 1.5$ mA. This shows a class A amplifier because the entire signal is amplified and because the biasing is at midpoint on the load line.

If the input signal becomes too large, the amplifier is overdriven and the collector current will reach saturation ($I_{C\ Sat}$). This is what causes the clipping in audio signals, and the result is poor-quality sound. An example of this can be seen in Figure 6–13. If the input signal rises to 45 μA, the resulting I_C would be:

$$I_C = I_B \times \beta$$

$$I_C = 45\ \mu\text{A} \times 100$$

$$I_C = 4.5\ \text{mA}$$

Thus, 4.5 mA would produce an output signal across R_C of (4.5 mA × 5,000) = 22.5 V. This is greater than V_{CC}. Therefore, the amplifier would be driven into saturation and the last 7.5 V of signal would be clipped or distorted.

FIGURE 6–14 AC signal effect on operating (Q) point.

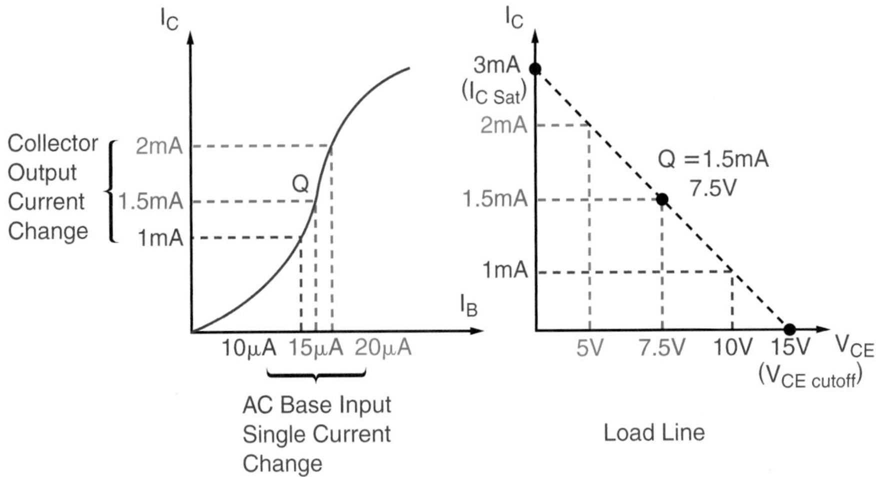

FIGURE 6–15 Common-emitter example circuit.

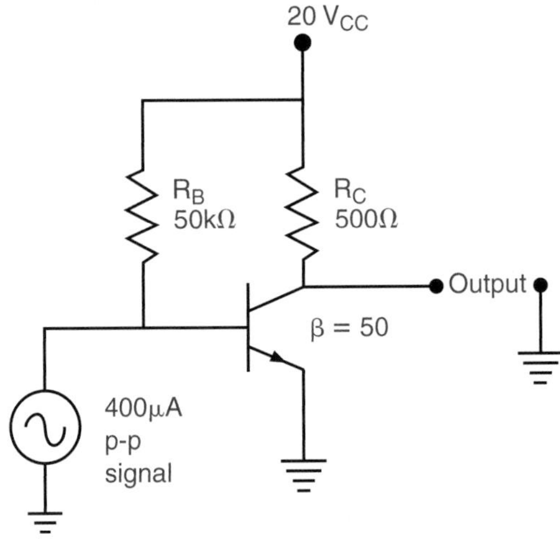

Example

Use the common-emitter circuit in Figure 6–15 for the following example. Assume a gain (β) of 50 and solve for the following:

- $I_{C(Sat)}$
- $V_{CE(Cutoff)}$
- I_B
- I_{CQ} (operating current with static AC signal)
- V_{CQ} (operating voltage with static AC signal)

(continued)

- Minimum and maximum base current due to the dynamic swing of the AC signal input
- Minimum and maximum collector current due to the dynamic swing of the AC signal input
- Minimum and maximum voltage output due to the dynamic swing of the AC signal input

Step 1: Determine $I_{C(Sat)}$.

$$I_{C(Sat)} = \frac{V_{CC}}{R_C + R_E}$$

$$I_{C(Sat)} = \frac{20\text{ V}}{500\ \Omega + 0\ \Omega}$$

$$I_{C(Sat)} = \frac{20\text{ V}}{500\ \Omega}$$

$$I_{C(Sat)} = 40\text{ mA}$$

Step 2: Determine $V_{\text{CE cutoff}}$.

$$V_{CE(Cutoff)} = V_{CC}$$

$$V_{CE(Cutoff)} = 20\text{ V}$$

Step 3: Determine the quiescent base current (I_B). (Quiescent means with no AC signal.)

$$I_B = \frac{V_{CC} - V_{BE}}{R_B}$$

$$I_B = \frac{20\text{ V} - 0.7\text{ V}}{50{,}000\ \Omega}$$

$$I_B = \frac{19.3\text{ V}}{50{,}000\ \Omega}$$

$$I_B = 386\ \mu\text{A}$$

Step 4: Determine the quiescent collector current (I_{CQ}).

$$I_{CQ} = \beta \times I_B$$

$$I_{CQ} = 50 \times 386\ \mu\text{A}$$

$$I_{CQ} = 19.3\text{ mA}$$

Step 5: Determine the quiescent output voltage.

$$V_{CQ} = V_{CC} - V_C$$

$$V_{CQ} = 20\text{ V} - (500\ \Omega \times 19.3\text{ mA})$$

$$V_{CQ} = 20\text{ V} - 9.65\text{ V}$$

$$V_{CQ} = 10.35\text{ V}$$

(continued)

Step 6: Determine the minimum and maximum base current due to the dynamic swing of the AC signal input.

$$I_{B(Min)} = I_{BQ} - \Delta I_B$$

$$I_{B(Min)} = 386\ \mu\text{A} - 200\ \mu\text{A}$$

$$I_{B(Min)} = 186\ \mu\text{A}$$

$$I_{B(Max)} = I_{BQ} + \Delta I_B$$

$$I_{B(Max)} = 386\ \mu\text{A} + 200\ \mu\text{A}$$

$$I_{B(Max)} = 586\ \mu\text{A}$$

Step 7: Determine the minimum and maximum collector current due to the dynamic swing of the AC signal input.

$$I_{C(Min)} = \beta \times I_{B(Min)}$$

$$I_{C(Min)} = 50 \times 186\ \mu\text{A}$$

$$I_{C(Min)} = 9.3\ \text{mA}$$

$$I_{C(Max)} = \beta \times I_{B(Max)}$$

$$I_{C(Max)} = 50 \times 586\ \mu\text{A}$$

$$I_{B(Max)} = 29.3\ \text{mA}$$

Step 8: Determine the minimum and maximum voltage output due to the dynamic swing of the AC signal input.

$$V_{CE} = V_{CC} - V_C$$

$$V_{CE(Min)} = 20\ \text{V} - (29.3\ \text{mA} \times 500\ \Omega)$$

$$V_{CE(Min)} = 20\ \text{V} - 14.65\ \text{V}$$

$$V_{CE(Min)} = 5.35\ \text{V}$$

$$V_{CE(Max)} = 20\ \text{V} - (9.3\ \text{mA} \times 500\ \Omega)$$

$$V_{CE(Max)} = 20\ \text{V} - 4.65\ \text{V}$$

$$V_{CE(Max)} = 15.35\ \text{V}$$

SUMMARY

In this chapter, you learned about the different classifications of amplifiers: classes A, AB, B, and C. The class A amplifier has the most fidelity in reproducing the input signal, but it is the least efficient. The class AB operates over a range of 51 to 99% of the input signal. The class B operates over exactly 50% of the input signal, and the class C operates for less than 50% of the input signal. The class C is the most efficient of the classes.

Amplifier configurations were also reviewed. There are three basic types: common emitter, common collector, and common base. As presented earlier, the input and output to the amplifier determine which part of the transistor is considered "in common." For example, if the signal input is at the base and the output is taken from the collector the emitter is common to both.

Amplifiers produce gain. The gain can be in current, voltage, or power. The different configurations are designed for one or two of these gain types. Decibels are used as a measurement of audio signal strength and gain. Multistage amplifiers' gain (in decibels) can be simply added to obtain the total gain for the circuit configuration. Decibels are not linear but logarithmic.

A load line was developed for an example amplifier. The two key points that determine the load line are $I_{C(Sat)}$ and $V_{CE(Cutoff)}$. The Q-point (or operating point) on the load line is approximately centered for a class A amplifier. The operating limits of the circuit are at $I_{C(Sat)}$ and $V_{CE(Cutoff)}$. Attempted operation beyond these points will cause the amplifier to have a clipped and distorted signal.

REVIEW QUESTIONS

1. Define and discuss gain. What is it? How is it calculated? Does a transformer have gain?
2. An amplifier has an input voltage of 0.25 V and an output voltage of 15 V. What is its voltage gain?
3. What is the relationship between gain expressed as a ratio and decibels?
4. How do you calculate the gain of a two-stage amplifier if you know the gain of each stage expressed in dB?
5. A certain amplifier has a gain of −3 dB. What does this imply about the amplifier?
6. Why does a decibel value always have to have a reference?
7. Discuss the advantages and disadvantages of the various classes of amplifier.
8. Which of the classes of amplifier has the best fidelity? Which has the worst?
9. What happens to the output of an amplifier that is driven so hard its collector voltage reaches the cutoff value?
10. Name some applications for each of the four classes of amplifier.

PRACTICE PROBLEMS

1. An amplifier has a power gain of 15. What is its gain in decibels?
2. An amplifier has a power gain of 0 dB. What is its gain expressed as a ratio?
3. An amplifier has a voltage (V_{IN}) of 5 V and an input resistance (R_{IN}) of 1,000 Ω. The output voltage (V_{OUT}) is 10 V, and the output resistance (R_{OUT}) is 100 Ω. What is the amplifier's power gain in decibels?
4. Use the common-emitter circuit in Figure 6–16 for the following example. Assume a gain (β) of 110 and solve for the following:
   a. $I_{C\ Sat}$
   b. $V_{CE\ cutoff}$
   c. I_B
   d. I_{CQ} (operating current with static AC signal)
   e. V_{CQ} (operating voltage with static AC signal)
   f. Minimum and maximum base current due to the dynamic swing of the AC signal input
   g. Minimum and maximum collector current due to the dynamic swing of the AC signal input
   h. Minimum and maximum voltage output due to the dynamic swing of the AC signal input
5. Looking at your results from question 4, answer the following.
   a. Does this amplifier have a high fidelity?
   b. What class of amplifier is this?

FIGURE 6–16 Common-emitter example circuit.

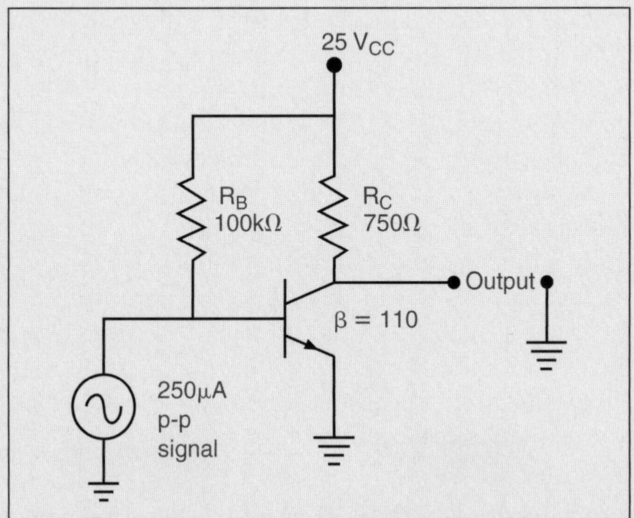

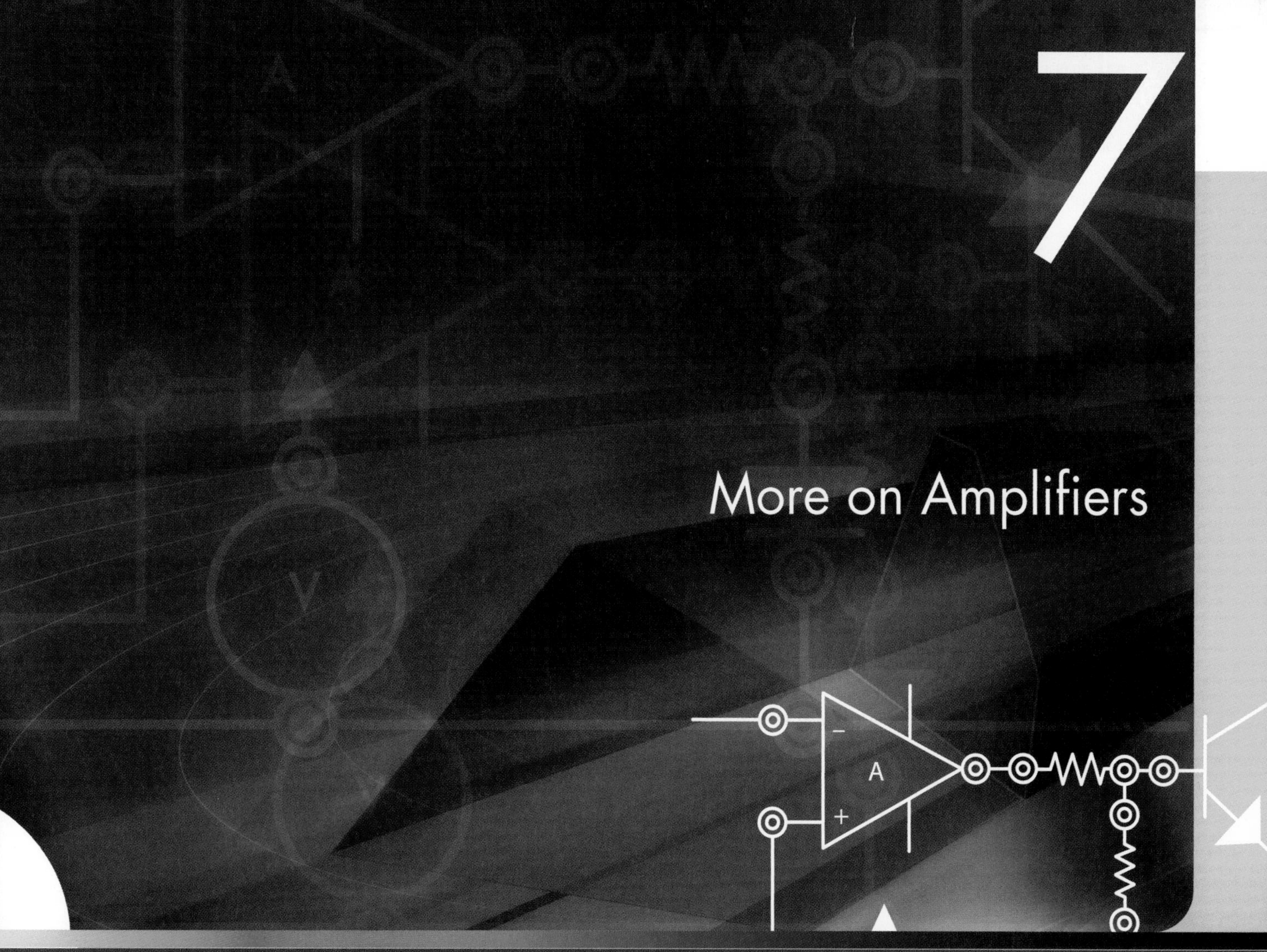

7 More on Amplifiers

OUTLINE

OVERVIEW

This chapter builds on the last chapter's subject of amplifiers. The common-emitter circuit configuration is used to show the analysis of an amplifier circuit and how a DC-biased circuit is used to further demonstrate the usefulness of coupling transistors to achieve a specific response. This lesson expects you to "add to" previous knowledge to further your abilities to interpret and predict the operation of transistors.

Coupling amplifiers is used in practically every modern solid-state device. Variable-speed drives, measuring instrumentation, photosensitive lighting, process control, security systems, and voice, video, and data systems are just a few of the components you may be required to troubleshoot over the years of your electrical career. Understanding the operation of amplifiers used in these devices will provide you with the tools you will need.

OBJECTIVES

After completing this chapter, the student should be able to:

1. Understand the voltage gain of cascaded amplifiers.
2. Draw the load line for a common-emitter circuit.
3. Describe the unique characteristics of direct coupling, capacitative coupling, and transformer coupling.

COMMON-EMITTER ANALYSIS

Solve for the missing values for transistor Q_1 in Table 7–1. Use Figure 7–1.

TABLE 7–1 Common-Emitter Analysis Data

	R_C (kΩ)	I_C (mA)	V_{RC}	V_{CE}	I_B (μA)	R_B (kΩ)
Q_1 in Cutoff	.8	0	0	20	0	∞
Q_1 in Normal Operation	.8	15				
Q_1 in Saturation	.8	25	20	0	250	77.2

FIGURE 7–1 Common-emitter analysis.

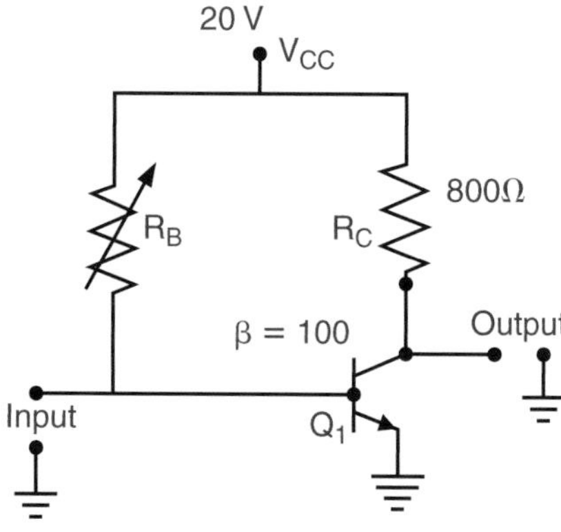

7.1 Row 1 (Cutoff)

When Q_1 is in cutoff, $V_{CE} = V_{CC}$ and $I_C = 0$ mA. This means that there is 0 V across R_C, that there is 0 μA for I_B, and that R_B is seen as ∞.

7.2 Row 2 (Normal Operation)

Given the value of I_C as 15 mA, the rest of the table values can be solved using the following steps.
Step 1:

$$I_B = \frac{I_C}{\beta}$$

$$I_B = \frac{15 \text{ mA}}{100}$$

$$I_B = 150\ \mu\text{A}$$

Step 2:

$$R_B = \frac{V_{CC} - V_{BE}}{I_B}$$

$$R_B = \frac{20\ \text{V} - 0.7\ \text{V}}{150\ \mu\text{A}}$$

$$R_B = \frac{19.3\ \text{V}}{150\ \mu\text{A}}$$

$$R_B = 128.67\ \text{k}\Omega$$

Step 3:

$$V_{RC} = R_C \times I_C$$

$$V_{RC} = 800\ \Omega \times 15\ \text{mA}$$

$$V_{RC} = 12\ \text{V}$$

Step 4:

$$V_{CE} = V_{CC} - V_{RC}$$

$$V_{CE} = 20\ \text{V} - 12\ \text{V}$$

$$V_{CE} = 8\ \text{V}$$

7.3 Row 3 (Saturation)

When Q_1 is in saturation, $V_{CE} = 0$ and $I_C = 25$ mA. Recall that this can be calculated by using the following formula.

$$I_{C(Sat)} = \frac{V_{CC}}{R_{CC}}$$

$$I_{C(Sat)} = \frac{20\ \text{V}}{800\ \Omega}$$

$$I_{C(Sat)} = 25\ \text{mA}$$

This means that

$$I_B = \frac{I_{C(Sat)}}{\beta}$$

$$I_B = \frac{25\ \text{mA}}{100}$$

$$I_B = 250\ \mu\text{A}$$

With an I_B of 250 μA, then

$$R_B = \frac{(V_{CC} - V_{BE})}{I_B}$$

$$R_B = \frac{20\ \text{V} - 0.7\ \text{V}}{250\ \mu\text{A}}$$

$$R_B = \frac{19.3\ \text{V}}{250\ \mu\text{A}}$$

$$R_B = 77.2\ \text{k}\Omega$$

Now that you have reviewed the basic analysis for a common-emitter amplifier, the next step is to look at amplifier coupling.

AMPLIFIER COUPLING

7.4 Multiple Stages

Many times, it is either impractical or impossible to amplify a small signal to a usable level with a simple one-transistor amplifier. This could be compared to trying to enlarge a slide of a single-celled organism so that an entire audience could see it straight from the microscope (Figure 7–2). To project the image so that the class could see it would require a lot of light. If we tried to illuminate the specimen with enough light that we could project the image straight through the microscope to the wall, we would incinerate the little bug. The only reasonable answer to this problem is to amplify in stages.

FIGURE 7–2 From microscopic image to wall projection in one step.

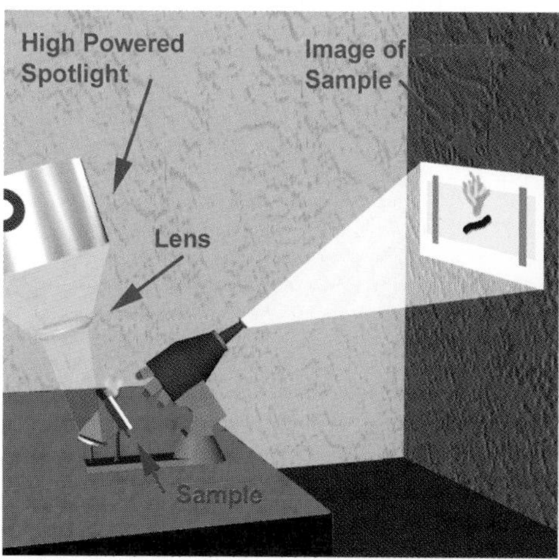

Amplifying in stages presents a problem. We have already talked about measuring the gain of multiple stages of amplification and discovered that the easiest way to deal with it is to use the decibel system of adding and subtracting gain factors. We have discussed how to produce voltage, current, and power gain with single-transistor amplifiers. Each of these amplifiers may be used as one stage of a multistage amplifier system. Now we tackle the problem of coupling the amplifier stages.

Coupling refers to the method of transferring a signal from one stage to the next. There are three methods of coupling amplifiers.

- *Direct coupling:* Connecting the output of one stage directly to the input of the next stage. The only allowed element between the two stages is a resistor.
- *Capacitive coupling:* Using a capacitor to couple two stages.
- *Transformer coupling:* Using a transformer to couple two stages.

7.5 Impedance Matching

Regardless of the coupling type used, a characteristic of the amplifier that must be taken into consideration is its impedance. The input impedance of the amplifier will directly affect how it loads the source. For example, an input signal from cable or an antenna could have an input impedance of 75 Ω, or the input could be from a microphone with more than 150 kΩ. Either way, to get the best power transfer (from input to amplifier) the impedances must be matched (or as closely matched as possible). The three block circuits shown in Figures 7–3a through c illustrate this principle. Table 7–2 summarizes the results of the three different loads.

FIGURE 7–3 Impedance matching. (a) $Z_L > Z_S$, (b) $Z_L = Z_S$, (c) $Z_L < Z_S$.

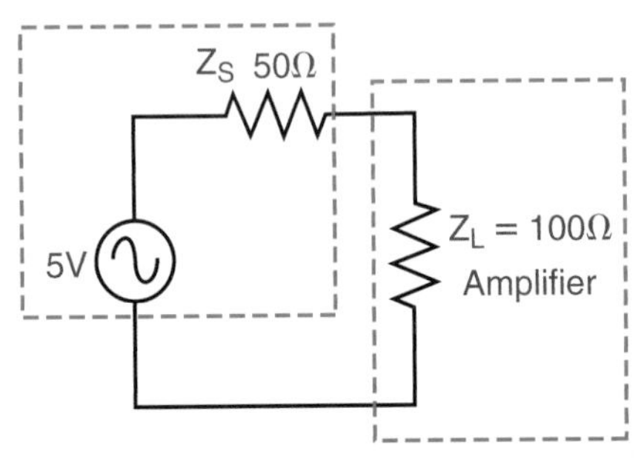

$$I = \frac{V}{Z} = .033A$$
$$P = I^2 \times R$$
$$P_{ZL} = (.033)^2 \times 100 = .1111W$$
$$P_{ZS} = (.033)^2 \times 50 = .0555W$$

(a)

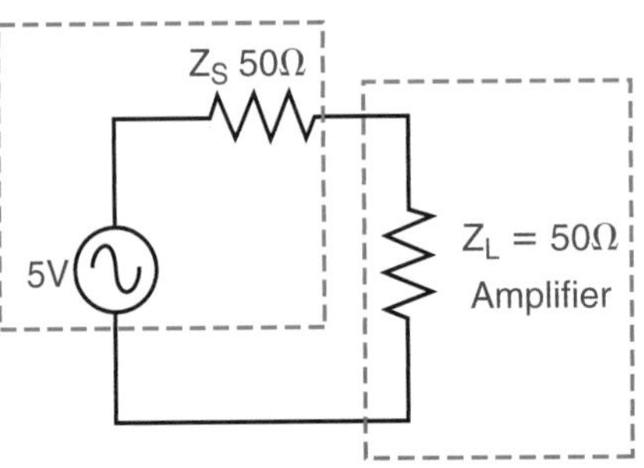

$$I = \frac{V}{Z} = .05A$$
$$P = I^2 \times R$$
$$P_{ZL} = (.05)^2 \times 50 = .125W$$
$$P_{ZS} = (.05)^2 \times 50 = .125W$$

(b)

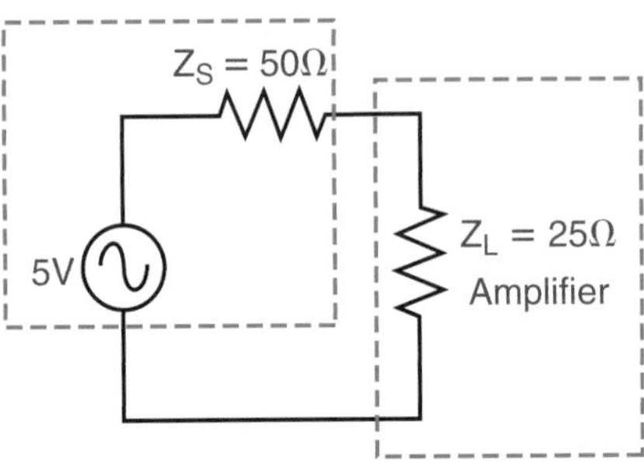

$$I = \frac{V}{Z} = .066A$$
$$P = I^2 \times R$$
$$P_{ZL} = (.066)^2 \times 25 = .1111W$$
$$P_{ZS} = (.066)^2 \times 50 = .2222W$$

(c)

TABLE 7–2 Summary of Power Transfers for Figure 7–3

Figure	R_S (Ω)	R_L (Ω)	P_S (Watts)	P_L (Watts)
7–3a	50	100	0.055	0.111
7–3b	50	50	0.125	0.125
7–3c	50	25	0.222	0.111

Note that the power delivered to the load is greatest when the impedance of the signal and the load are matched. Consequently, for maximum power transfer the impedance of the load must be equal to the impedance of the amplifier. In other words, the output impedance of one stage of an amplifier must be equal to the input impedance of the next stage.

Direct coupling
The transfer of current from the output of one stage directly to another by way of a conductor; capacitance and inductance are not employed.

7.6 Direct Coupling

The simplest method of coupling is called **direct coupling**. As the name implies, it is passing current straight from one stage to another over a wire or some other conductor. Both AC and DC signals are passed with direct coupling.

Because neither capacitors nor transformers are capable of passing a DC signal, they cannot be used in direct coupling. Direct coupling finds wide use in medical equipment, voltmeters, and other equipment that must amplify DC or very low-frequency inputs. Figure 7–4 shows a pair of common-emitter amplifier stages directly connected.

FIGURE 7–4 Directly coupled common-emitter amplifier stages.

An emitter resistor or a series base resistor may be used to reduce the high base input voltage
V_{CC}
Input
Output
Q_1
Q_2
Base Input Voltage could be quite high

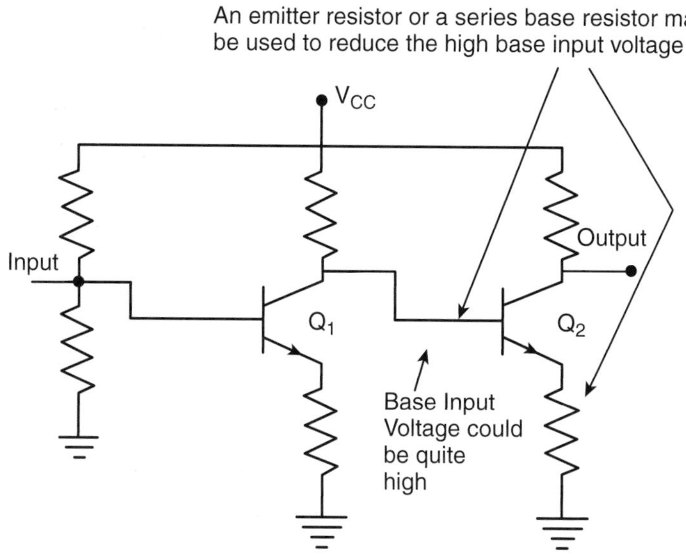

The output voltage of the first stage becomes the input voltage of the second stage. The arrangement itself is simple enough, but there are limitations. In most applications, the base voltage of Q_2 would become excessive and cause Q_2 to burn out. A resistor tied to the emitter of Q_2 can reduce the danger of Q_2 burnout, but there is a gain sacrifice through Q_2 as a result. A resistor can also be placed in series with the base of Q_2, but this also causes a loss of overall gain.

Another problem with direct-coupled amplifiers is temperature sensitivity. A slight rise in temperature in one of the early stages of amplification can result in a big error by the time the little bit of noise is amplified several times. Figure 7–5 shows the effects of amplifying temperature-induced noise.

FIGURE 7–5 Effect of temperature sensitivity.

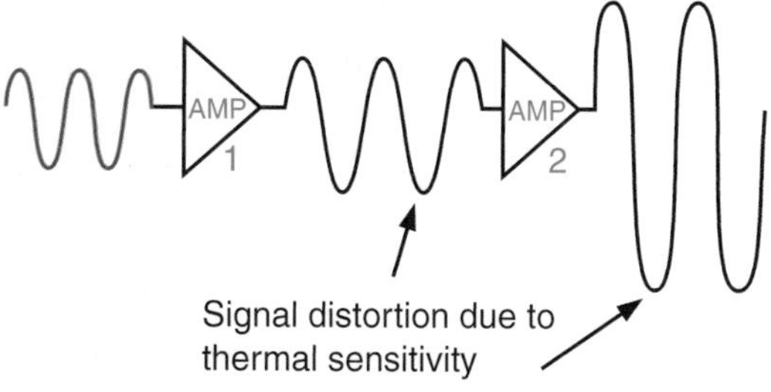
AMP
1
AMP
2
Signal distortion due to thermal sensitivity

Temperature also induces thermal drift. This is the reduction in internal resistance of transistors, such as BJTs, with the rise in temperature. The drift shows up as both voltage and current changes in various parts of the amplifier. Depending on the configuration, a little drift early on can result in clipping or saturation in later stages of amplification (as shown in Figure 7–6).

FIGURE 7–6 (a) Sample circuit subject to thermal drift. (b) Waveforms taken at points A, B, and C.

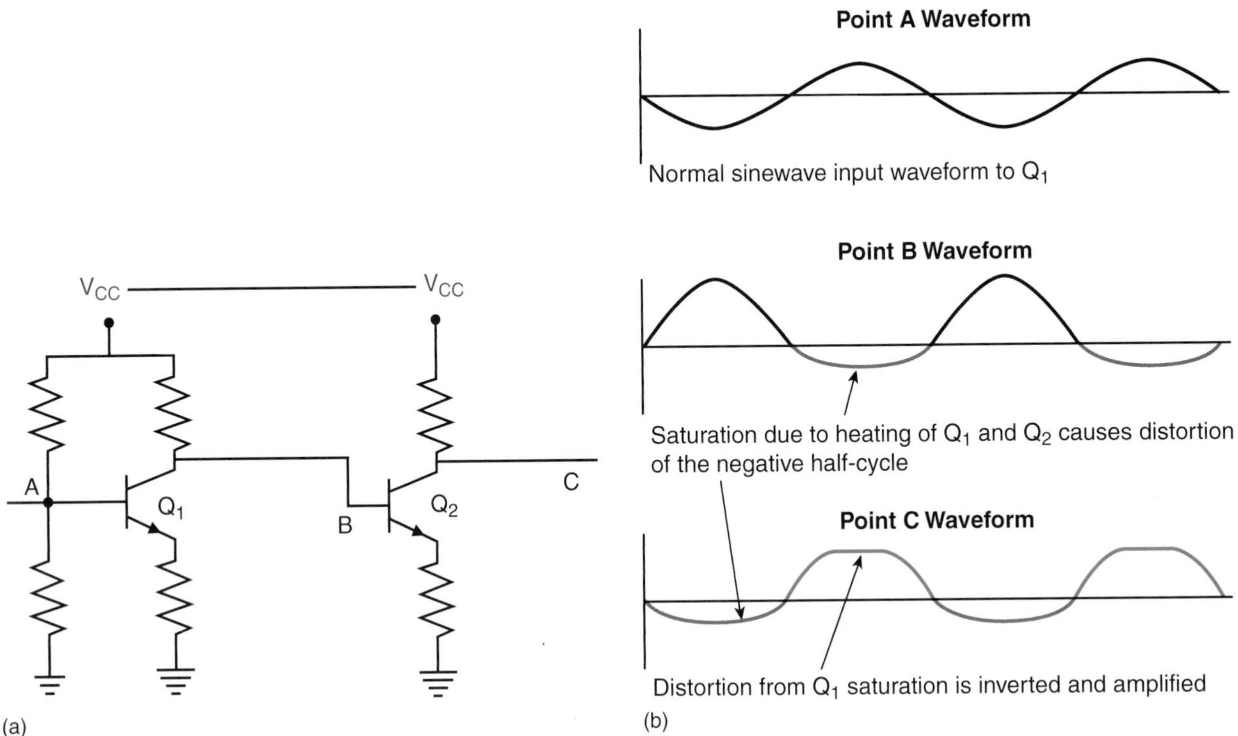
Vcc
Vcc
A
Q_1
B
Q_2
C
(a)
Point A Waveform
Normal sinewave input waveform to Q_1
Point B Waveform
Saturation due to heating of Q_1 and Q_2 causes distortion of the negative half-cycle
Point C Waveform
Distortion from Q_1 saturation is inverted and amplified
(b)

Thermal runaway
A circuit condition where an increase in transistor temperature results in higher current, and higher temperature, continuously, until saturation or a circuit breakdown occurs.

If uncontrolled, thermal drift can result in **thermal runaway**—where an increase in transistor temperature increases power dissipation in the transistor, resulting in ever higher temperatures. The outcome of thermal runaway is often destruction of the transistor. Proper feedback biasing is the antidote to thermal runaway.

One of the most common direct-coupled devices is the Darlington pair, shown in Figure 7–7. The advantages of this direct-coupled amplifier are simple circuitry, high input impedance, and an overall current gain that is slightly more than the mathematical product ($T_1 \times T_2$) of the current gain of the two transistors ($I_{total} > I_{Q1} \times I_{Q2}$). The Darlington pair is often sold packaged as a single transistor and used for small signal inputs for low-noise applications.

Direct coupling is great for DC or very low-frequency signals that do not respond well to transformers and capacitors, but its strength with these signals is its weakness with higher-frequency signals. It does not isolate them and may not pass them. If you were to use direct coupling of a speaker or microphone, the signal would be shunted to ground. Consequently, some other type of coupling should be used for such applications (see Figure 7–8).

FIGURE 7–7 The Darlington pair.

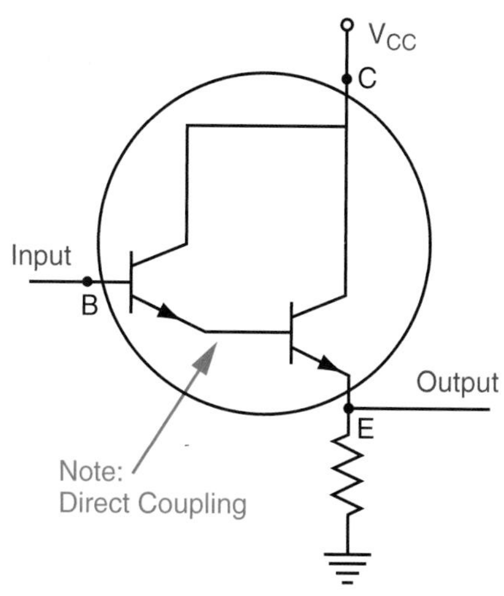

FIGURE 7–8 When direct coupling does not work.

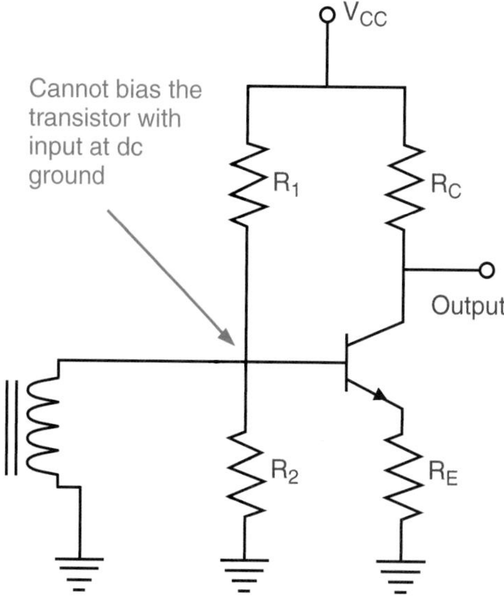

Figure 7–9 shows a simple example of direct coupling applied to the real world. A thermistor is a two-wire temperature-sensitive device. Most thermistors have negative temperature coefficients. That means that as temperature rises the resistance of the thermistor goes down. The thermistor and R_g form a voltage divider, supplying the MOSFET gate. When ambient temperature rises, thermistor resistance decreases—causing gate voltage to rise. In response to rising gate voltage, drain current increases.

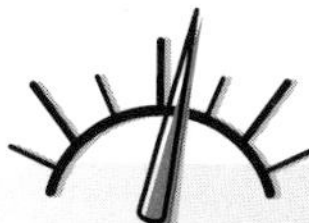

TechTip!

Take a look at the schematic symbol of a Darlington transistor. Note that there are two PN junctions between the base pin and the emitter pin. When testing this transistor with the diode function of a multimeter, you can see that the forward voltage meter reading should be closer to 1.4 than to 0.7 V.

FIGURE 7–9 Temperature sensor.

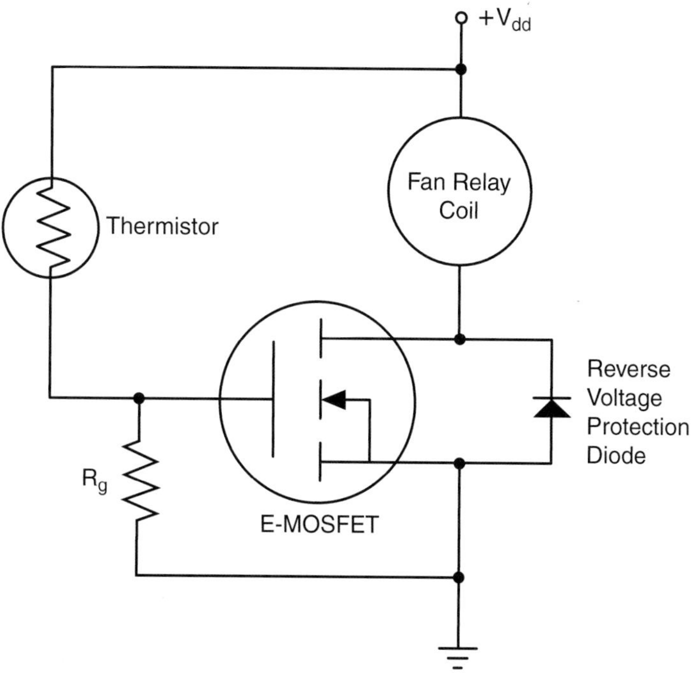

When drain current rises sufficiently, the fan relay coil will pull in its contacts. The fan, turned on by the fan relay, will cool the area surrounding the thermistor. The thermistor resistance will increase, the fan relay coil will deenergize, and the fan will turn off. The diode across the relay prevents CEMF (counter-electromotive force) from damaging the transistor when power is turned off.

7.7 Capacitive Coupling

Overview

FIGURE 7–10

RC-coupled amplifier.

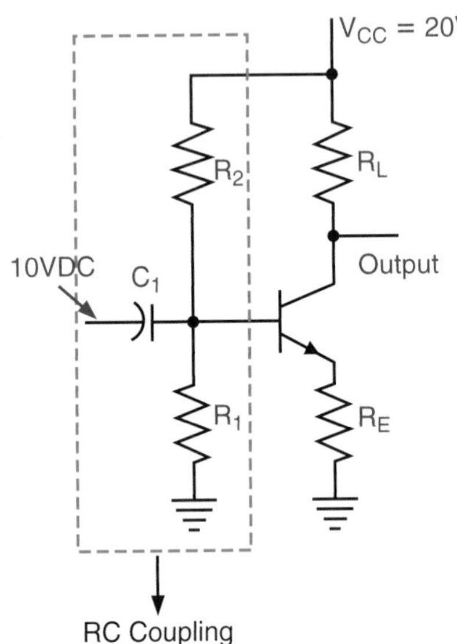

The most common form of coupling for analog signals is capacitive coupling, as shown in Figure 7–10. It is also known as RC (resistive-capacitive) coupling. RC coupling offers the advantages of inexpensive and simple design, with adequate frequency response over a relatively wide range of frequencies. Capacitive coupling also enables voltage divider bias by blocking DC between stages. Its principal drawbacks are:

1. Inability to match source impedance over a wide range of input signal frequencies. This is because capacitive reactance decreases as signal frequency increases.
2. The transistor input circuit voltage divider resistors consume (waste) a portion of signal power from the input signal source.

Coupling capacitors are employed to pass alternating current from an input source such as a microphone, to an amplifier, while blocking direct current.

Notice, again, the RC coupling network of Figure 7–10. The dotted box includes the base biasing resistors as part of the RC coupling network. This is because the biasing resistors, in addition to providing the correct base bias voltage, presents a significant portion of the AC load viewed from the signal source.

Also note that the coupling capacitor is often electrolytic. This is common in RC coupling, and care must be taken to install replacement capacitors with the correct polarity. Installing an electrolytic capacitor backward may destroy the capacitor dielectric, resulting in a loud pop followed by a burning smell when power is applied.

Refer to Figure 7–10 for the following discussion. The capacitive reactance of C_1 must be low in the frequency range of the desired signal, and the resistance of R_1 must be at least 10 times the capacitive reactance of C_1 to prevent signal loss across the coupling. Assume Q_1 is in the static state (Q-point quiescent—no signal applied) and that the biasing of Q_1 is set at 2 V by the voltage divider and R_E. C_1 will develop a charge equal to the static voltage of the input signal minus the bias (2 V) of Q_1.

Assume that the previous stage of amplification had an output level of 10 V. The resulting level of charge on C_1 is 10 V – 2 V = 8 V. As the input signal swings higher (a 1 V increase), C_1 charges through the base of Q_1 and R_1 (11 V – 2 V = 9 V). Note that the base current and the voltage across R_1 both increase during the charging of C_1 (see Figure 7–11).

When the input signal swings lower (see Figure 7–12), C_1 discharges through R_2. As the current through R_2 increases, the voltage drop across it increases—lowering the current through Q_1 and decreasing the voltage drop across R_1.

FIGURE 7–11 R_C coupling during positive half cycle.

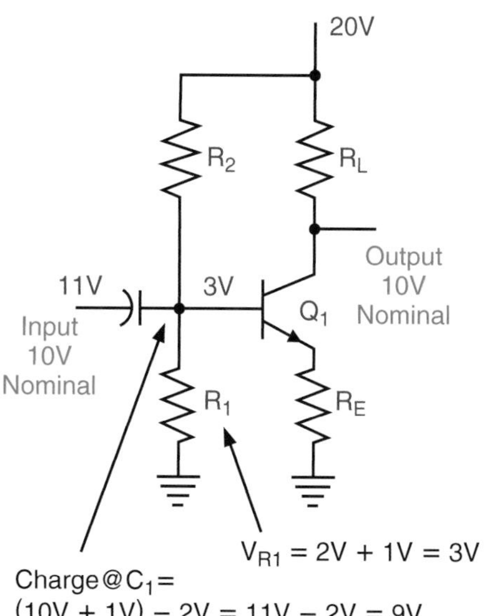

FIGURE 7–12 R_C coupling during negative half cycle.

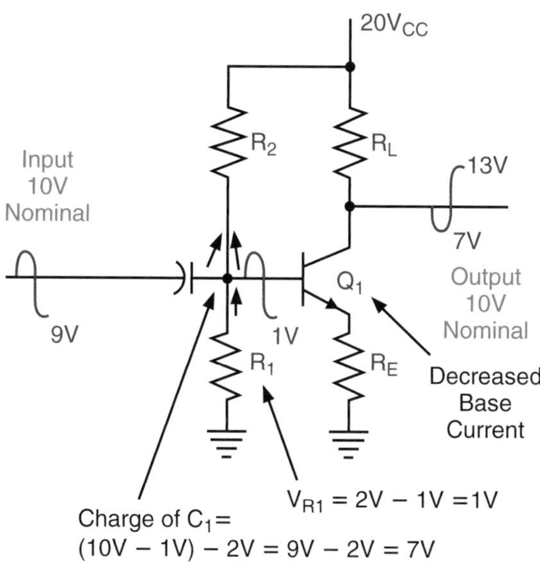

AC and DC Equivalent Circuits

The circuit effect of capacitors in amplifier stages can be seen in an AC-equivalent circuit. Figure 7–13a shows the full amplifier circuit. To construct the AC-equivalent circuit of this amplifier, perform the following.

- Short-circuit all capacitors by adding a jumper wire across them. This step assumes that the impedance of the capacitor is low in the frequency range of the amplifier. If it is not, you must substitute an impedance of the capacitor that is suitable for the range of frequencies. This, of course, will make the analysis more complex.

FieldNote!

In regard to electrolytic capacitors, a critical piece of electronic hardware at a geographically remote location had been inoperative for quite some time. The onsite technician directly responsible had diagnosed the problem as follows:

- The device had been turned in for repair, with a repair ticket indicating that the signal from the device was "complete garbage."
- The tech applied 208 VAC/60 Hz to the device, and confirmed the "complete garbage" analysis.
- Using an oscilloscope, he viewed the power amplifier output and found a strong 120 Hz ripple. **What would *you* conclude at this point?**

The tech concluded that there was a problem related to the full-wave center-tapped power supply because it should produce 120 Hz ripple at the rectifier diode just prior to the filter section. Severe 120 Hz will be present at the filter output given:

1. A low-resistance path from V+ to **GND**,
2. An open filter capacitor, or
3. A shorted V+ filter choke.

Because shorted chokes are rare, the tech shelved that possibility. Because possibilities 1 and 2 are very common occurrences, he decided to test the easiest possibility first. **Which choice would *you* make? How would you test your hypothesis?**

(continued)

GND (or Gnd)
A common abbreviation for the term "Ground".

FieldNote! *(cont'd)*

The tech chose possibility 2 (an open filter capacitor). He connected a known-good junk-box capacitor across the V+ filter output. If there had been an open filter cap, the parallel capacitor should have significantly smoothed the ripple seen on the scope. In this case, it did *not* smooth the observed ripple. The problem must be possibility 1 or 3. **Which possibility would *you* check next, and how would you conduct the test?**

The tech chose to eliminate possibility 3, the choke. He compared the winding resistance of the choke with that shown on the schematic diagram. The values compared, exactly, as they must. He had now eliminated two of the three possibilities. There must be a short (or very low resistance) between the filter output (V+) and GND (V−). There are numerous parallel paths along which the culprit may have been lurking. **How would *you* find the fault path?**

The technician measured (power off, naturally) resistance from the power supply filter output (V+) to GND and recorded a very much lower than normal resistance of 60 ohms. He disconnected 50% of the parallel branches from V+, and checked R from V+ to GND again. He measured 65 ohms. **What would *you* decide based on this clue?**

(continued)

- Replace all DC sources with a ground symbol. This is effectively the same thing as placing a short circuit across all DC sources.

See Figure 7–13b for the following discussion. The reasoning behind replacing all capacitors with a wire (straight line) is that the capacitor is built to offer little or no resistance to the AC signal frequency (recall that X_C is low in the frequency range of the signal). In practice, it acts as a short circuit. Replacing DC sources with a ground is based on similar logic. Recall from your study on batteries that the internal resistance of a DC source is extremely low. Ideally, it is 0 Ω. This causes the wire (straight line) DC source replacement to be at ground potential.

Note that C_2 is considered an open for the DC calculations of the circuit (see Figure 7–13c). If C_2 did not exist, the impact of R_{C1} on the base of Q_2 would be considerable. R_{C1} would be in parallel with R_3 and the resulting R_{EQ} would impact the voltage divider value used for Q_2. This is shown in the equation for R_{EQ}.

V_B without the coupling capacitor C_2:

$$R_{EQ} = \frac{(R_{C1} \times R_3)}{(R_{C1} + R_3)}$$

$$R_{EQ} = \frac{(7.2\text{ k}\Omega \times 20\text{ k}\Omega)}{(7.2\text{ k}\Omega + 20\text{ k}\Omega)}$$

$$R_{EQ} = \frac{144\text{ M}\Omega}{27.2\text{ k}\Omega}$$

$$R_{EQ} = 5.3\text{ k}\Omega$$

Using R_{EQ} to solve for V_B:

$$V_B = V_{CC}\left(\frac{R_4}{R_{EQ} + R_4}\right)$$

$$V_B = 20\text{ V} \times \left(\frac{4.4\text{ k}\Omega}{5.3\text{ k}\Omega + 4.4\text{ k}\Omega}\right)$$

$$V_B = 20\text{ V} \times \frac{4.4\text{ k}\Omega}{9.7\text{ k}\Omega}$$

$$V_B = 20\text{ V} \times .4536$$

$$V_B = 9.1\text{ V}$$

V_B with the coupling capacitor C_2:

$$V_B = \text{V}_{CC}\left(\frac{R_4}{R_3 + R_4}\right)$$

$$V_B = 20\text{ V} \times \left(\frac{4.4\text{ k}\Omega}{20\text{ k}\Omega + 4.4\text{ k}\Omega}\right)$$

$$V_B = 20\text{ V} \times \left(\frac{4.4\text{ k}\Omega}{24.4\text{ k}\Omega}\right)$$

$$V_B = 20\text{ V} \times .1803$$

$$V_B = 3.6\text{ V}$$

FIGURE 7–13 R_C coupled amplifier. (a) Full circuit. (b) AC equivalent. (c) DC equivalent.

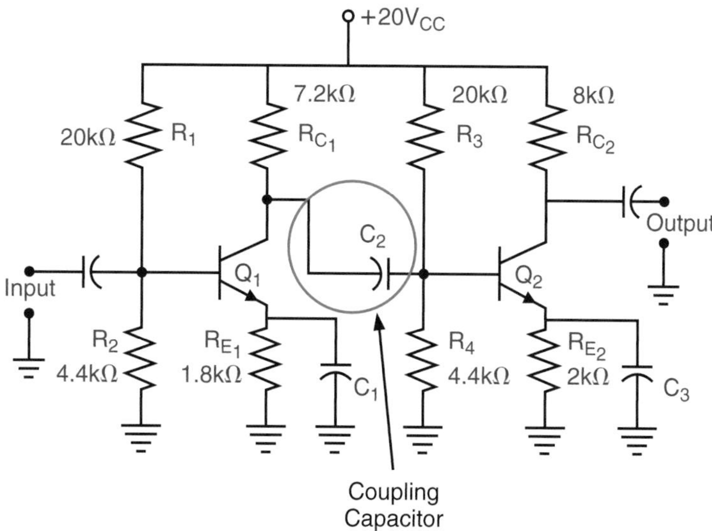

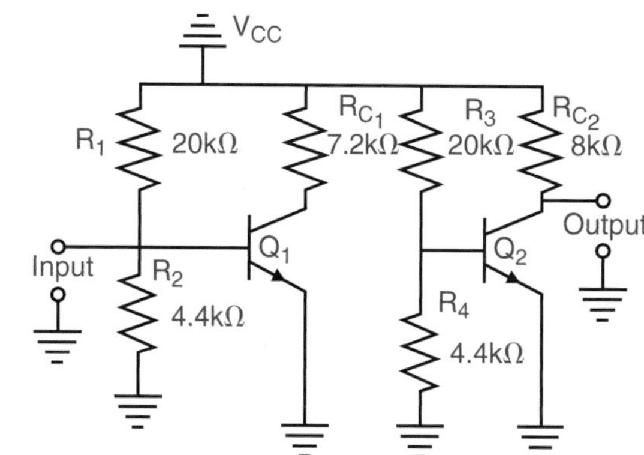

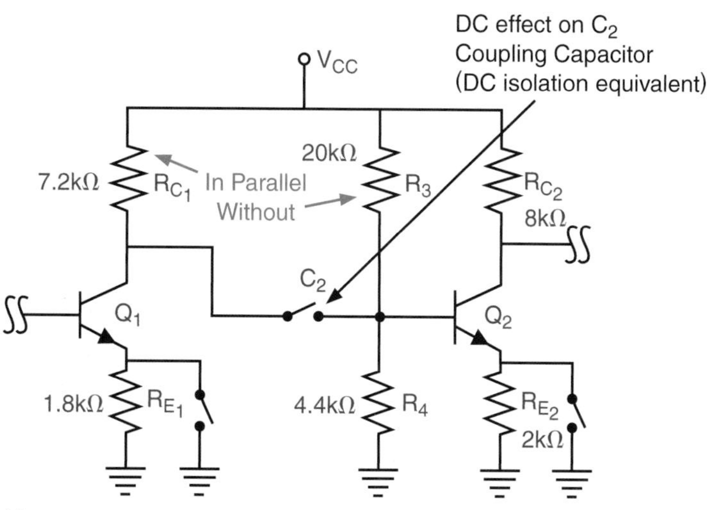

FieldNote! *(cont'd)*

The tech figured the problem was not associated with the disconnected branches because they did not significantly affect the leakage resistance. He removed the remaining branches from V+. The 65 ohms became 66 ohms between V+ and GND. The faulty component must be the filter capacitor at the power supply filter! The technician replaced the defective electrolytic filter capacitor in the power supply filter section. The device started blowing fuses. **What would *you* do at this point?**

The tech ordered yet another hard-to-get replacement cap. The senior technician flew into the remote site to review the problem. Upon making a visual inspection of the circuit, the senior tech noted that the electrolytic capacitor had been installed backward. Reversal of the two electrolytic capacitor wires cleared up the ripple problem. Resistance between V+ and GND jumped up to 875 ohms. The device was returned to service.

Polarized capacitors such as electrolytics will often act like short circuits, or self-destruct (sometimes explosively) when reverse connected. Double check the polarity when reinstalling polarized capacitors.

Figure 7–14 shows a schematic diagram of an audio preamplifier section. This circuit illustrates the use of capacitive (or R_C) coupling. Note the coupling capacitor between the microphone (signal source) and the D-MOSFET gate. Also observe a second and third at the input and output of the BJT. X_C (capacitive reactance) of these coupling capacitors is higher for lower signal frequencies $\left(Xc = \frac{1}{2\pi fC}\right)$. Therefore, less signal passes to the load at lower frequencies.

Note the false-bass tone control (FBTC) circuit between the two transistors. FBTC is a simple approach to reducing the effect of poor low-frequency response inherent in R_C coupling. When the variable resistance of the FBTC is lowered, more higher-frequency components of the signal are passed through the FBTC rather than the B-E junction of the BJT. This is the way simple tone controls operate on low-end AM and FM broadcast receivers.

FIGURE 7–14 AF preamplifier.

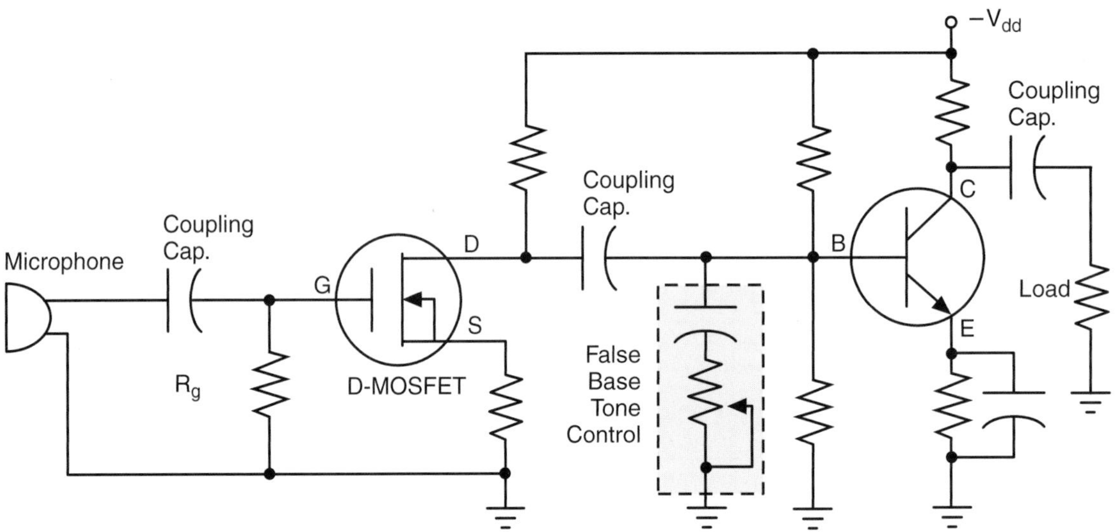

7.8 Transformer Coupling

As the name implies, a transformer-coupled amplifier (Figure 7–15) uses a transformer between the collector of the transistor and the load. The load can be the final output, or it could be another stage of amplification. Transformer coupling is mainly used for its ability to match impedance and provide maximum power transfer.

Consider Figure 7–16. As you learned previously, the maximum power transfer will occur between the two amplifiers if the output impedance of A_1 is equal to the input impedance of A_2. Transformer coupling provides DC isolation from the primary to secondary windings, and it can provide the necessary impedance matching.

Recall from earlier lessons that the relationships in the transformer are:

$$\frac{N_p}{N_s} = \frac{V_p}{V_s} = \frac{I_s}{I_p}$$

and

$$\left(\frac{N_p}{N_s}\right)^2 = \frac{Z_p}{Z_s}$$

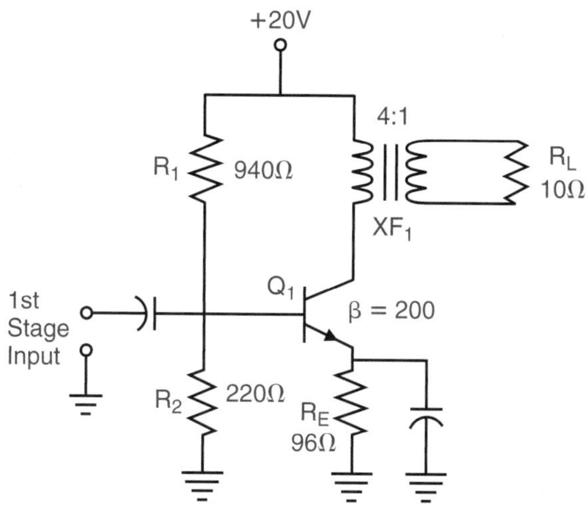

FIGURE 7–15 Transformer-coupled amplifier.

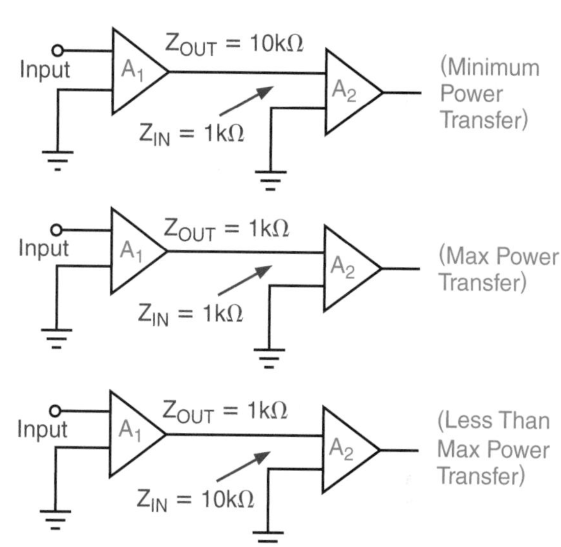

FIGURE 7–16 Importance of amplifier impedance matching.

The DC biasing of a transformer-coupled amplifier and the analysis on the collector curves (Figure 7–17) are slightly different than a direct-coupled amplifier. The analysis must take into account that the transistor sees two different values for R_{CE}. The DC current does not see the load resistance. This is because the transformer does not pass DC current, and thus the load resistance is effectively disconnected from the transistor for DC values. The AC current, on the other hand, does see the load resistance reflected back by the transformer. The analysis starts with the DC circuit.

DC Analysis

Refer to Figure 7–15. To promote linear operation, the quiescent voltage is usually selected as one-half ($\frac{1}{2}$) V_{CC}. Assume a primary winding resistance of 4 Ω.

$$V_{CE\,(QUIESCENT)} = V_{CC} - V_{XF} - V_{RE}$$

Because we have decided on the Q-point of one-half-V_{CC}, we can state that:

$$V_{XF} + V_{RE} = 20\text{ V} - 10\text{ V} = 10\text{ V}$$

We know the transformer primary resistance and R_E, and thus quiescent current can be found by Ohm's law.

$$I_{C(Quiescent)} = \frac{(V_{XF} + V_{RE})}{(R_{XF} + R_E)}$$

$$I_{C(Quiescent)} = \frac{(10\text{ V})}{(4\ \Omega + 96\ \Omega)}$$

$$I_{C(Quiescent)} = \frac{10\text{ V}}{100\ \Omega}$$

$$I_{C(Quiescent)} = 100\text{ mA}$$

Note, however, that the Q-point on the curves (Figure 7–17) falls at $V_{CE} = 10$ V and $I_C = 105$ mA. This is because there is a slight resistance between the collector and the emitter. Therefore, it takes a little more current than we had previously supposed. If the collector curves were perfectly horizontal, this problem would not exist. However, that would also imply that the transistor is perfect—which it cannot be.

In any event, the cutoff voltage for the transistor is 20 V and 0 mA. This makes up another point on the load line. You can now draw the load line by connecting and extending the Q-point and 20 V.

FIGURE 7–17 Load lines of transformer-coupled amplifier.

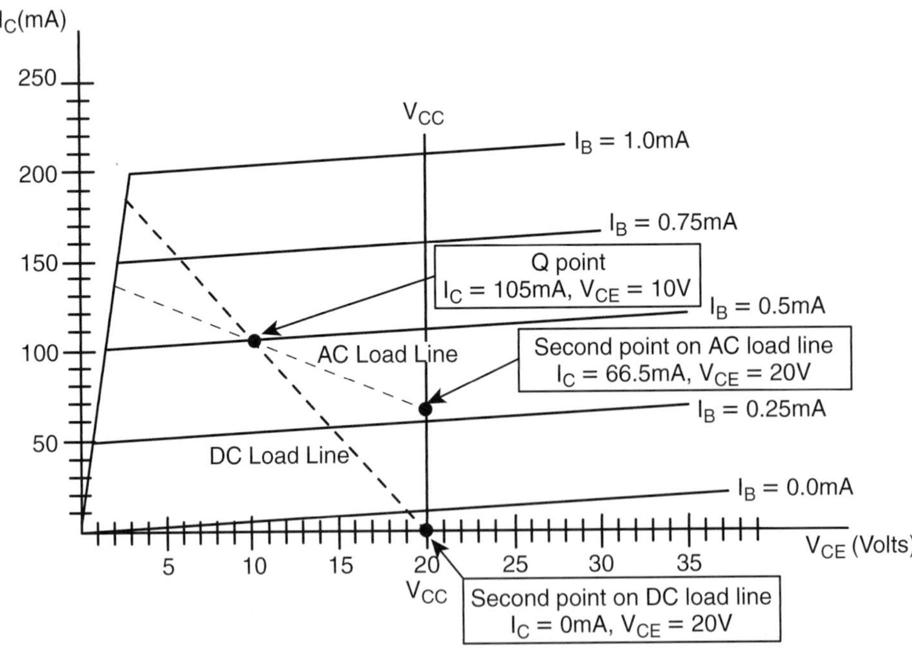

AC Analysis

The AC load line is a little more difficult to understand. Recall that any AC applied will pass through the reflected load impedance.

Step 1: Calculate the AC collector impedance (Z_{CE}).

$$Z_{CE} = \left(\frac{N_P}{N_S}\right)^2 \times Z_S = 4^2 \times 10\ \Omega = 160\ \Omega$$

Step 2: Determine the Q-point for AC operation.

The Q-point is, by definition, the "at rest" point for the amplifier. That is, it is the point at which the AC input is equal to zero. Clearly, this means that the Q-point is the same for both AC and DC operation.

Step 3: Calculate a second point on the AC load line.

Note that the total resistance seen by the AC signal is 160 Ω + 96 Ω + 4 Ω = 260 Ω. The 4 Ω is included because the transformer winding is still in the circuit. Because we are dealing with an AC value, we must work with the changes in voltage and current. Consider the following:

$$I_C\,(R_E + R_L + 4\ \Omega) + V_{CE} = V_{CC}$$

This is just Kirchhoff's voltage law for the collector circuit. Note that we have to include the 4 Ω resistance of the transformer primary as well as the reflected load resistance (R_L). Now consider if we look at the change in current versus the change in voltage for an AC signal.

$$\Delta I_C\,(R_E + R_L + 4\ \Omega) + \Delta V_{CE} = V_{CC} - V_{CC} = 0$$

Note the 0 on the right-hand side. This is because V_{CC} is the same at both points. Therefore, any AC quantity will not be reflected across the voltage supply. This can be further rearranged as

$$(R_E + R_L + 4\ \Omega) = -\left(\frac{\Delta V_{CE}}{\Delta I_C}\right),$$

which can be further reduced to

$$260\ \Omega = -\frac{V_{CE1} - V_{CE2}}{I_{C1} - I_{C2}}$$

Here, the values V_{CE1} and I_{C1} are one point on the load line and V_{CE2} and I_{C2} are another point on the load line. Take point 1 to be the Q-point, so that V_{CE1} = 10 V and I_{C1} = 105 mA. For the second point, we can take the cutoff value. Thus, V_{CE2} = 20 V, while I_{C2} is an unknown value. Substituting these values gives

$$260\ \Omega = -\frac{V_{CE1} - V_{CE2}}{I_{C1} - I_{C2}} = -\frac{10\text{ V} - 20\text{ V}}{105\text{ mA} - I_{C2}}$$

Rearranging terms derives

$$260\ \Omega \times (105\text{ mA} - I_{C2}) = -\ (10\text{ V} - 20\text{ V}),$$

which simplifies to

$$27.3\text{ V} - (260\ \Omega \times I_{C2}) = 10\text{ V} \Rightarrow I_{C2} = \frac{10\text{ V} - 27.3\text{ V}}{-260\ \Omega} = 66.5\text{ mA}$$

The AC load line can now be drawn by connecting the Q-point with the second point just calculated. Recalling geometry, you may note that the slope $\left(\frac{rise}{run}\right)$ of the DC load line is $-\frac{1}{100}$ and the slope of the AC load line is $-\frac{1}{260}$.

7.9 Tuned Transformers

A tuned transformer-coupled amplifier uses an LC (inductive-capacitive) parallel "tank" circuit. By varying the *L* and *C* values in the tank, you can control the total output impedance to the load. Figure 7–18 shows an example of a tuned transformer circuit.

FIGURE 7–18 Tuned transformer-coupled amplifier.

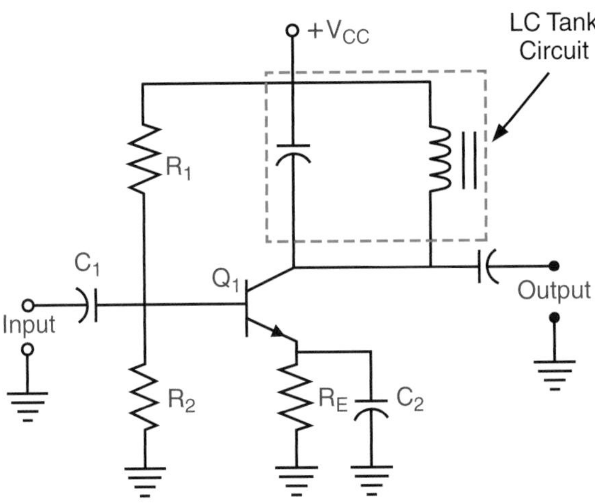

Figure 7–19 shows how the impedance (AC resistance) of a parallel LC circuit varies with frequency. Note that for frequencies well below the resonant frequency $\left(f_R = \frac{1}{2\pi\sqrt{(LC)}}\right)$ the impedance is very low. As the frequency increases, the impedance rises until at f_R the impedance is extremely high.

FIGURE 7–19 Impedance versus frequency characteristics for parallel LC circuit.

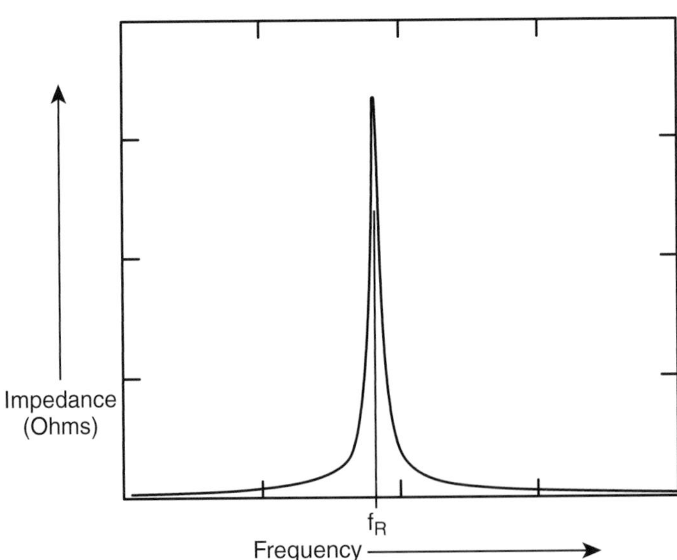

Theoretically, the impedance is infinite at f_R. The impedance starts to decrease above f_R in a mirror image of its behavior below f_R.

This behavior can be explained for the amplifier of Figure 7–18 by referring to Figure 7–20. There are three AC frequency conditions that can be described for the amplifier. Table 7–3 outlines the behavior of each.

FIGURE 7–20 AC equivalent tuned amplifier.

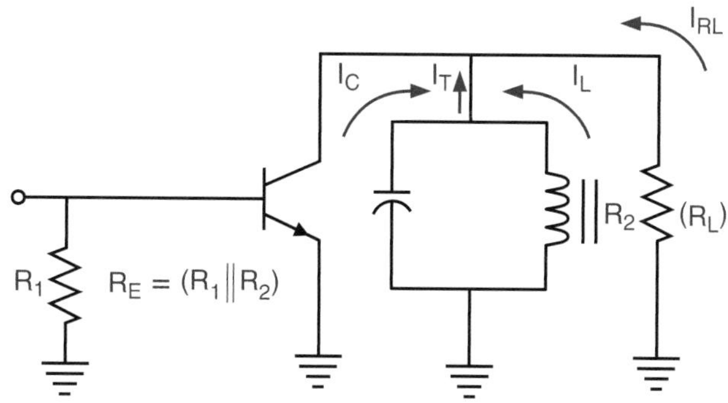

TABLE 7–3 Tuned Circuit Amplifier Response

Frequency In	Amplifier Response
$F_{IN} = f_R$	The tank circuit acts as an open circuit (infinite impedance). R_L becomes the only path to ground in the collector circuit. This results in all current flowing through R_L, which in turn causes the amplifier efficiency to be at maximum.
$F_{IN} < f_R$	I_L increases and I_C decreases as F_{IN} decreases. This causes the tank current (I_T) to increase. As I_T increases, the load voltage decreases—causing a loss in amplifier output efficiency.
$F_{IN} > f_R$	I_C increases and I_L decreases as F_{IN} increases. This causes an increase in the tank current (I_T). As the input frequency increases, the tank current increases. As I_T increases, the load voltage decreases—causing a loss in amplifier output efficiency.

Figure 7–21 illustrates a two-stage radio frequency amplifier. The amplifier input is supplied by an antenna-ground system. The antenna impedance is shown to be 75 ohms. Q_1 gate has roughly 1 megohm of resistance. Therefore, as the schematic indicates, the transformer primary winding should match the antenna impedance (75 ohms) and the tuned-secondary winding will present an exceptionally high impedance to roughly match Q_1 input impedance.

T_2 primary winding is matched to Q_1 drain. This is accomplished by tapping the T_2 primary winding. The load on the collector of Q_2 is the X_L of L_1. Coupling to the RF amplifier load is furnished by C_6. This is usually called impedance coupling. Due to nonlinearity, **impedance coupling** is rarely found in audio applications.

Impedance coupling
A coupling technique similar to capacitive (or RC) coupling wherein an inductor is used in lieu of a resistor.

FIGURE 7–21 Transformer-coupled RF amplifier.

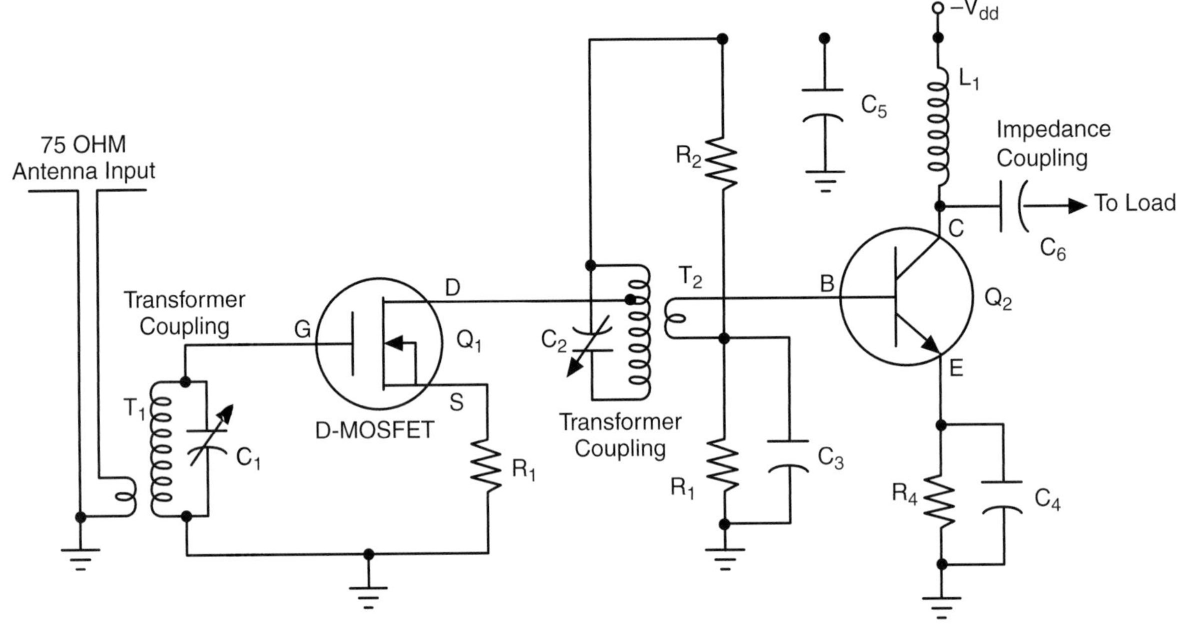

NEGATIVE FEEDBACK

Negative feedback
Any method used, whether intentional or unintentional, to couple a portion of amplifier output back to its input, 180° out-of-phase.

A technique widely used to reduce distortion in amplifiers is called **negative feedback**. It is also used in transistor circuits to

- Stabilize
- Increase bandwidth
- Improve linearity
- Improve noise performance
- Maintain the amplifier's frequency response

Negative feedback is defined as feeding a portion of an amplifier output signal back to the input. The signal is fed back in such a way that it is 180° out of phase with the input signal. This feedback reduces the gain of the amplifier, but it also reduces the amount of distortion in the output signal.

One technique of negative feedback has already been studied: the emitter-to-ground resistor. This resistor is used to stabilize the DC bias of an amplifier. When a capacitor is placed in parallel with R_E, the AC signal is bypassed to ground so that the effect of the feedback is not felt on the signal. Thus, the AC gain can be significantly higher.

Collector feedback
Generally, resistance coupling from a transistor collector to its base for the primary purpose of stabilizing collector current.

A second technique is called **collector feedback**. Collector feedback attenuates the signal, as well as stabilizes transistor bias. Figure 7–22 shows a schematic illustrating the simplicity of the collector feedback circuit. In common-emitter configuration, when base voltage goes up collector voltage goes down. In collector feedback, R_B is connected to the collector rather than to V_{CC}. Therefore, when higher collector current drives collector voltage down R_B feeds back a lower voltage to the base.

This lower base voltage, in turn, lowers collector current, and so on. Collector feedback is thus a self-regulating gain and bias circuit configuration. Collector feedback is often used as the first preamplifier in an audio amplifier. Simplicity is its advantage, and low gain compared to emitter feedback is its disadvantage.

FIGURE 7–22 Collector feedback.

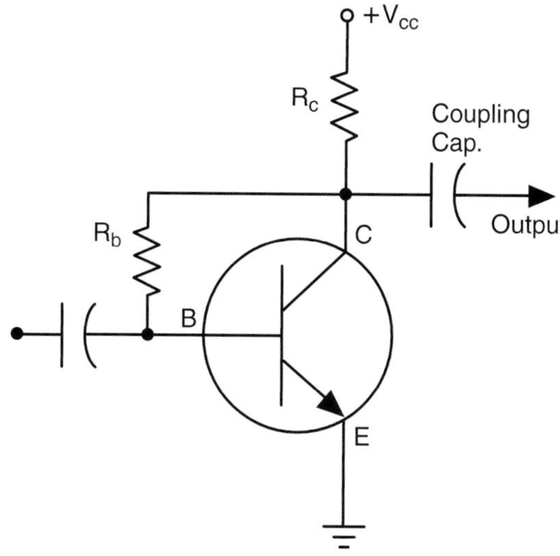

SUMMARY

In this chapter, we reviewed simple amplifier analysis using the common-emitter class A amplifier. We also learned about the three basic types of amplifier coupling: direct, capacitor, and transformer. Important amplifier characteristics reviewed included impedance matching, AC and DC equivalent circuits, and transformer tuning. Table 7–4 summarizes the three methods of coupling. Emitter feedback and the less common collector feedback methods were discussed.

TABLE 7–4 Summary of Coupling Methods

	Direct Coupling	Capacitor Coupling	Transformer Coupling
Response to direct current	Yes	No	No
Provides impedance match	No	No	Yes
Pros	Simple design when only a few stages are needed; Darlington pair very popular	Ensures fixed bias by blocking DC; easy to use, inexpensive	High efficiency; ability to tune a selective amplifier design (tank circuit); provide maximum power transfer
Cons	Difficult to design for many stages; temperature sensitive	Low-frequency applications may require high values of capacitance	Cost; transformer weight and size
Common usage	Medical equipment, voltmeters, and DC amplification	Common for coupling of analog signals	Applications where DC needs isolation; where total output impedance to the load needs to be controlled

REVIEW QUESTIONS

1. What is amplifier coupling?
2. Discuss the importance of impedance matching.
3. Which of the coupling methods will couple both AC and DC signals between stages?
4. What are the disadvantages of direct-coupled amplifiers?
5. In the ideal *RC* coupled amplifier, the capacitor behaves as a short circuit at signal frequencies and as an open circuit at nonsignal frequencies. Discuss how the real *RC* coupled amplifier behaves.
6. The AC and DC quiescent points (Q-points) are the same for a transformer-coupled transistor. Why?
7. A certain intermediate-frequency amplifier in a radio couples to the next stage using a tuned transformer circuit. Why?
8. The collector curves for a real BJT are not horizontal. Why?
9. Discuss one example of the importance of fidelity in an amplifier. Why is it important in your example?
10. What is negative feedback? What are its uses?

PRACTICE PROBLEMS

1. In Figure 7–23, assume the capacitor is 3 μF and the inductor is 2 *mH*. What is the resonant frequency (f_R) of the combination?

FIGURE 7–23 AC equivalent tuned amplifier.

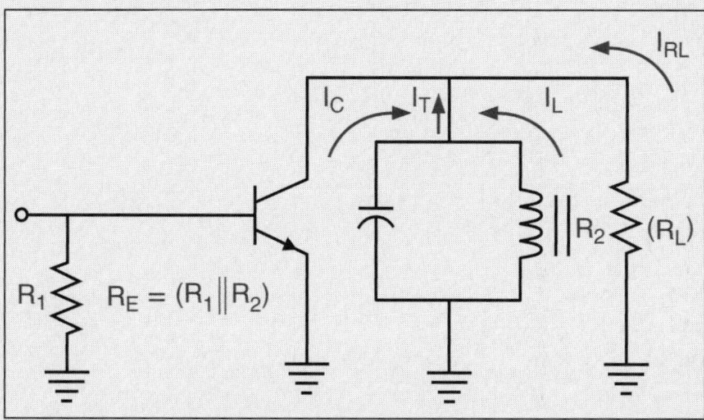

2. In Figure 7–24, what will the AC instantaneous collector current be for this amplifier with $I_B = 0.625$ mA?

FIGURE 7–24 Load lines of transformer-coupled amplifier.

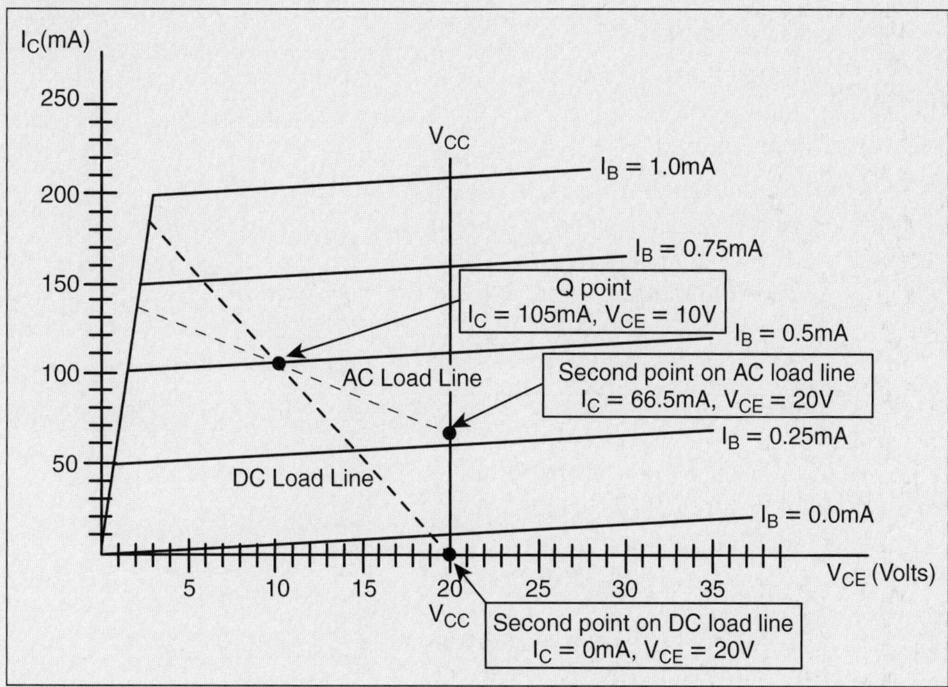

3. In Figure 7–25, what is the DC quiescent point (I_C and V_{CE}) assuming the following information:
   a. The transistor collector curves are given by Figure 7–24.
   b. $R_E = 100\ \Omega$.
   c. The collector inductor has a DC resistance (R_l) of 4 Ω.
   d. $V_{CC} = 30$ V.
   e. $V_{CE}Q = 15$ V.
   f. $\beta = 75$.

FIGURE 7–25 Tuned transformer-coupled amplifier.

+VCC
LC Tank
Circuit
R1
C1
Q1
Output
Input
R2
RE
C2

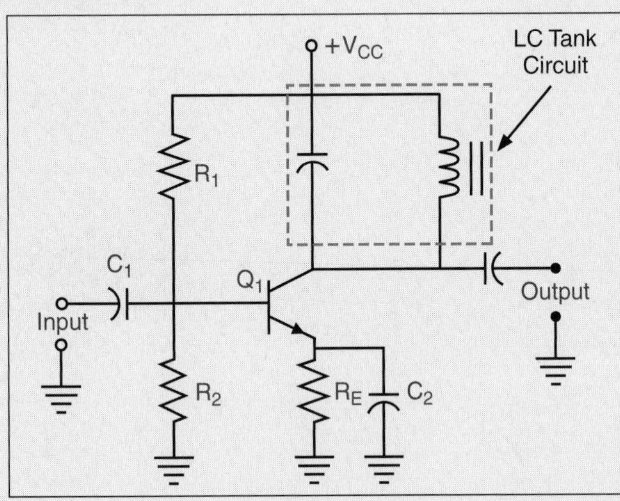

4. In Figure 7–25, what is the AC quiescent point assuming all of the same information as question 3 plus the following:
   a. The tank circuit inductance (L_T) = 6 μH.
   b. The tank circuit capacitance (C_T) = 0.4 μF.

5. In Figure 7–26, assume the tank circuit inductance is the primary of a transformer with a 10-to-1 turns ratio, and assume that the amplifier is operating at the resonant frequency of the tank circuit. The load resistance on the transformer secondary is 5 Ω. Draw the AC and DC load lines. Use the same circuit values as those given in problems 3 and 4.

FIGURE 7–26 Importance of amplifier impedance matching.

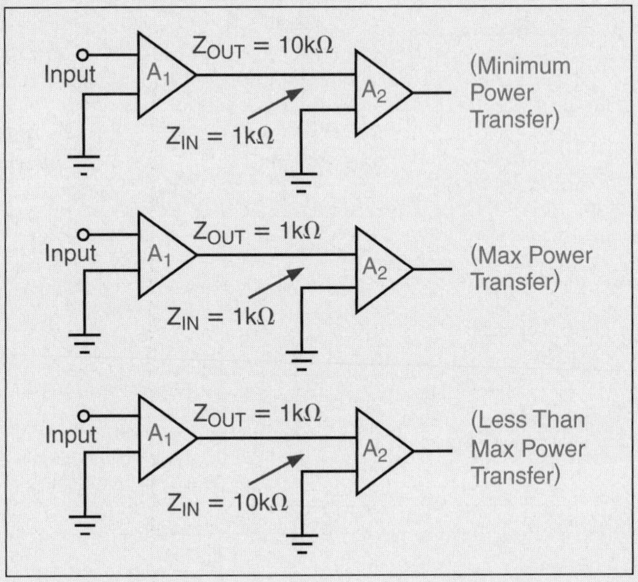

Input
A1
ZOUT = 10kΩ
A2
ZIN = 1kΩ
(Minimum
Power
Transfer)
Input
A1
ZOUT = 1kΩ
A2
ZIN = 1kΩ
(Max Power
Transfer)
Input
A1
ZOUT = 1kΩ
A2
ZIN = 10kΩ
(Less Than
Max Power
Transfer)

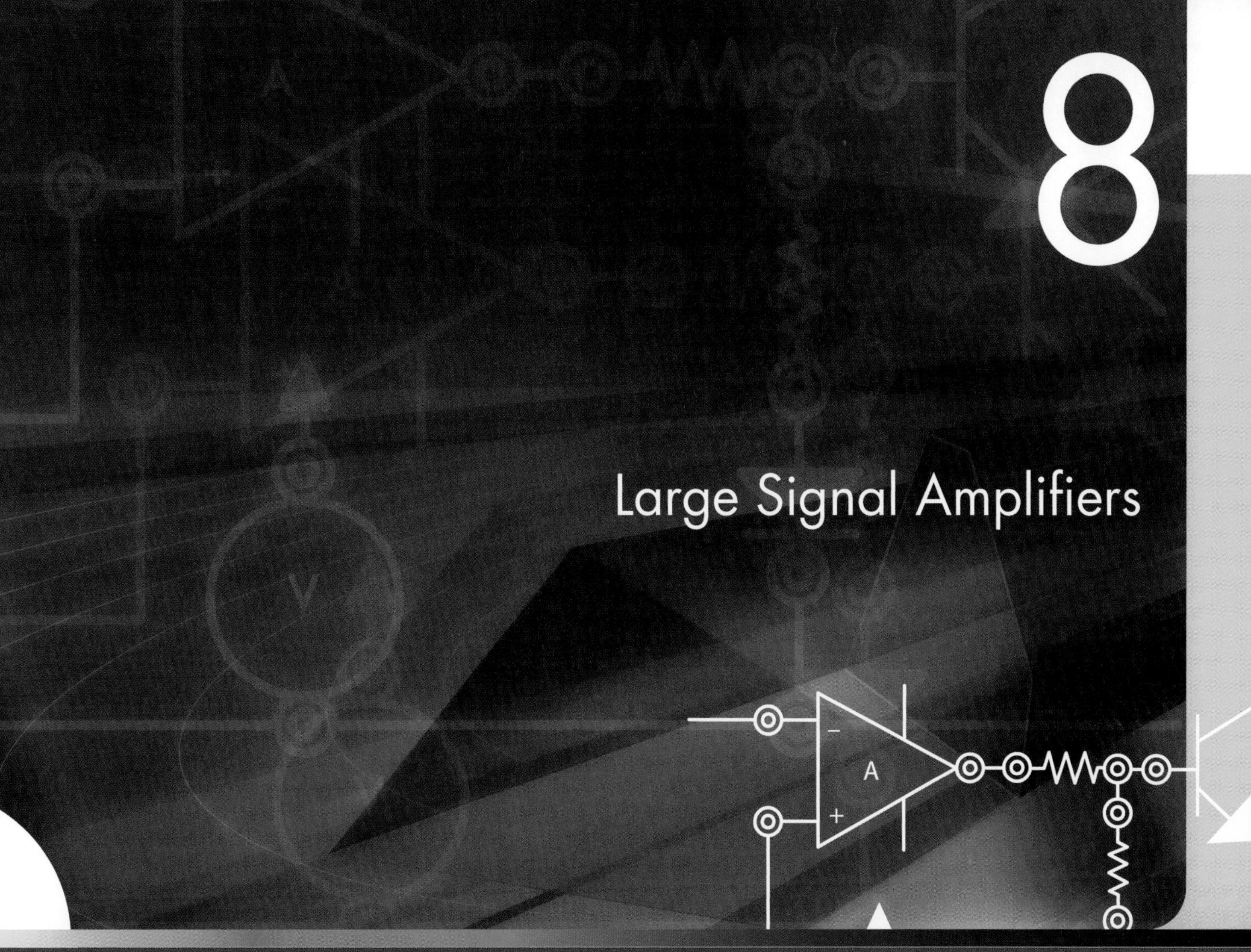

8 Large Signal Amplifiers

OUTLINE

OVERVIEW

In the last chapter, you learned about coupling amplifier stages to increase a small input signal to a larger signal for specific uses. Examples include amplifying a microphone signal for input to later stages or amplifying the very weak antenna signals for processing in a radio receiver. Voltage amplification is usually the critical task of these early stages.

There are also requirements to amplify larger signals. Consider Figure 8–1, for example. This might be the output stage of a small CD player or radio receiver. The input signal is an audio signal that has been generated by the earlier stages of the device. The final step is amplifying the signal to a power level that will allow it to drive the loudspeaker to a usable level. Such a stage is usually called a power amplifier.

Note that the amplifier shown in Figure 8–1 is being operated in class A. This means that the output is faithful to the input; that is, there is no distortion present. Class A is the least efficient of the four classes of amplifiers even though it has the greatest fidelity. In this chapter, you will learn more about the various classes of amplifiers and how they can be used to create larger signals for power applications.

OBJECTIVES

After completing this chapter, the student should be able to:

1. Calculate amplifier efficiency.
2. Identify a push-pull type of amplifier.
3. Determine whether an amplifier is operating in class A, AB, B, or C.

FIGURE 8–1 Simple class A amplifier.

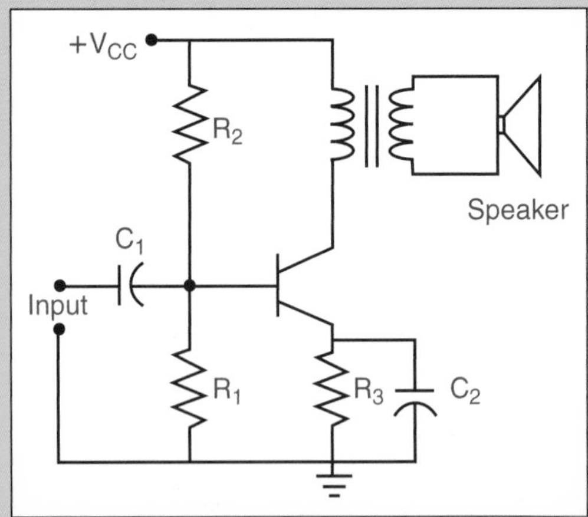

INTRODUCTION AND REVIEW

8.1 The Ideal Amplifier

Amplifiers work by converting some of the DC power supply energy to signal energy. In other words, the amplifier takes power from the DC power supply and "gives" it to the signal it is amplifying. The ideal power amplifier will deliver all (100%) of the power it uses from the power supply to the load. In reality, this never happens. The reason is that the circuit components dissipate some of the power as heat energy. The following formula is an expression of this loss in terms that can be measured. The Greek letter eta (η) is used as the symbol for efficiency.

$$\text{Amplifier Efficiency} = \eta\,(\%) = \frac{\text{AC output power}}{\text{DC input power}} \times 100$$

8.2 Power Relationships

The power efficiency equation shows that by using less DC input power for the same amount of AC output power the amplifier can become more efficient. DC input power is a function of DC biasing for the amplifier. The lower the Q-point on the DC load line, the lower the DC input power. Figure 8–2a shows a typical DC load line and the corresponding Q-points for the four basic classes of amplifiers. Note that the class A amplifier has a Q-point that is centered on the load line. Figure 8–2b shows that classes AB, B, and C do not conduct for the full 360° of signal. This is why the class A amplifier has the least signal distortion. The DC power that an amplifier uses from its power supply is calculated by:

$$P_{DC} = V_{CC} \times I_{CC}$$

FIGURE 8–2 Amplifier DC load line: (a) V_{ce} versus I_c, (b) operation characteristics.

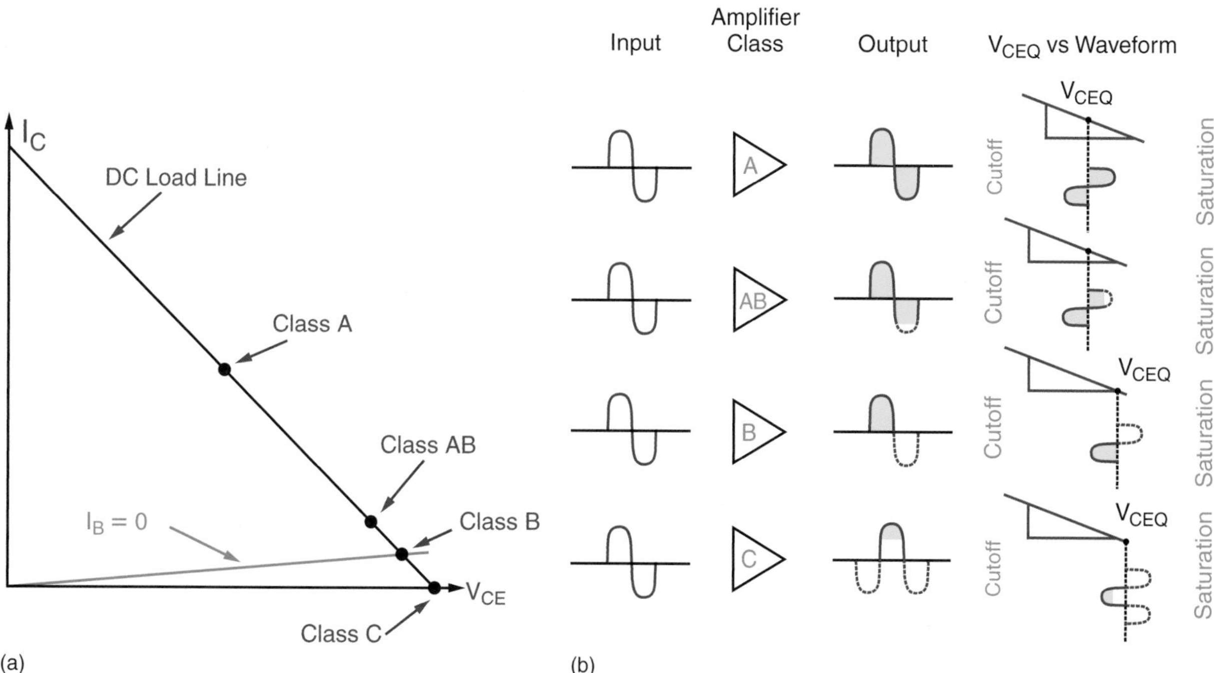

Figure 8–3 shows an example circuit.

FIGURE 8–3 Calculating I_{cc} for DC power input.

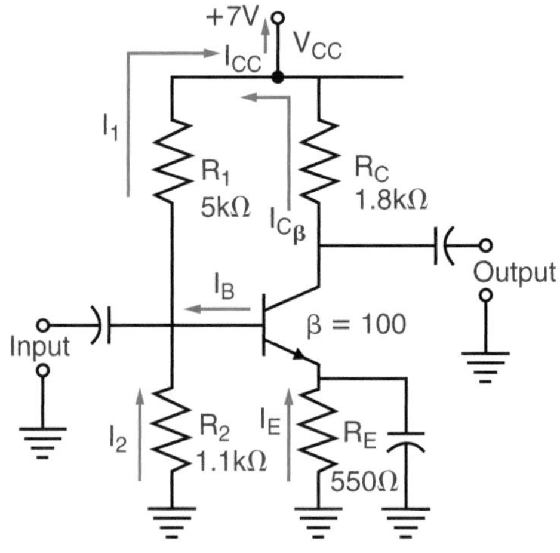

Example

Analyze this circuit and determine the DC input power. Solution: You can start by developing an equivalent circuit for Figure 8–3.

Step 1: Determine the value of I_1.

$$I_1 = \frac{V_{CC}}{R_1 + R_2}$$

$$I_1 = \frac{7\text{ V}}{5\text{ k}\Omega + 1.1\text{ k}\Omega}$$

$$I_1 = \frac{7\text{ V}}{6.1\text{ k}\Omega} = 1.15\text{ mA}$$

Step 2: Develop a Thevenin equivalent circuit for the base biasing circuit R_1 and R_2.

$$V_T = V_{CC} \times \frac{R_2}{R_1 + R_2}$$

$$V_T = 7\text{ V} \times \frac{1.1\text{ k}\Omega}{5\text{ k}\Omega + 1.1\text{ k}\Omega}$$

$$V_T = 7\text{ V} \times \frac{1.1\text{ k}\Omega}{6.1\text{ k}\Omega}$$

$$V_T = 7\text{ V} \times .180\ \Omega$$

$$V_T = 1.26\text{ V}$$

$$R_T = \frac{R_1 \times R_2}{R_1 + R_2}$$

$$R_T = \frac{5\text{ k}\Omega \times 1.1\text{ k}\Omega}{5\text{ k}\Omega + 1.1\text{ k}\Omega}$$

$$R_T = \frac{5.5\text{ M}\Omega}{6.1\text{ k}\Omega} = 901.6\ \Omega$$

(continued)

Step 3: Redraw Figure 8–3 using V_T and R_T. This is shown in Figure 8–4. Note that the collector circuit is shown as a current source of magnitude βI_B, where I_B is the base current. Note also that the emitter current is equal to $(1 + \beta)I_B$. Writing a voltage equation around the base-emitter circuit of Figure 8–4 yields:

$$[-(1 + \beta)I_B R_E] - V_{BE} + V_T = 0$$

Rearranging, we derive:

$$V_T = I_B R_E + V_{BE} + [(1 + \beta)I_B R_E],$$

which simplifies to:

$$1.26 \text{ V} = 550 \ \Omega \times I_B + 0.7 \text{ V} + [(1 + 100)(550 \ \Omega)I_B]$$

$$1.26 \text{ V} - 0.7 \text{ V} = [550 \ \Omega \times I_B + (101 \times 550 \ \Omega)I_B]$$

$$0.56 \text{ V} = I_B[550 \ \Omega + (101 \times 550 \ \Omega)]$$

$$I_B = \frac{0.56 \text{ V}}{[550 \ \Omega + (101 \times 550 \ \Omega)]}$$

$$I_B = \frac{0.56 \text{ V}}{[550 \ \Omega + 55{,}550 \ \Omega]}$$

$$I_B = \frac{0.56 \text{ V}}{56{,}100 \ \Omega}$$

$$I_B = 9.98 \ \mu\text{A}$$

Step 4: Calculate I_C.

$$I_C = \beta I_B$$

$$I_C = 100 \times 9.98 \ \mu\text{A}$$

$$I_C = 0.998 \text{ mA}$$

Step 5: Calculate the total DC current supplied by V_{CC}.

$$I_{CC} = I_C + I_1$$

$$I_{CC} = 0.998 \text{ mA} + 1.15 \text{ mA}$$

$$I_{CC} = 2.15 \text{ mA}$$

Step 6: Calculate P_S.

$$P_S = I_{CC} \times V_{CC}$$

$$P_S = 2.15 \text{ mA} \times 7 \text{ V}$$

$$P_S = 15 \text{ mW}$$

FIGURE 8–4 Equivalent circuit for Figure 8–3.

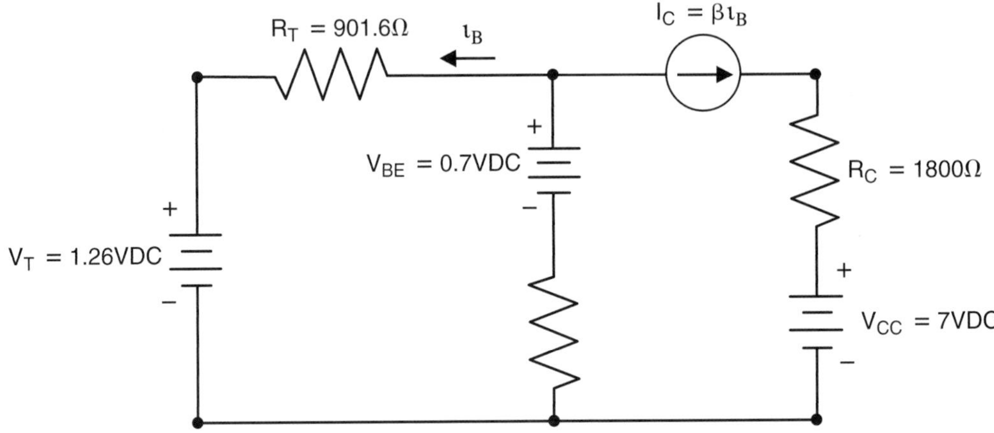

8.3 AC Load Power

Load power can be calculated using the familiar equation

$$P_L = \frac{V_L^2}{R_L},$$

Where:
V_L = load voltage
R_L = load resistance

Figure 8–5 shows the circuit you worked with earlier, except that now a load resistance and a coupling capacitor have been added. For the sake of this example, assume that the output voltage read by the meter is equal to 0.8 V_{RMS}. Use the previous formula.

$$P_L = \frac{0.8^2}{100} = 6.4 \text{ mW}$$

FIGURE 8–5 AC load power calculations.

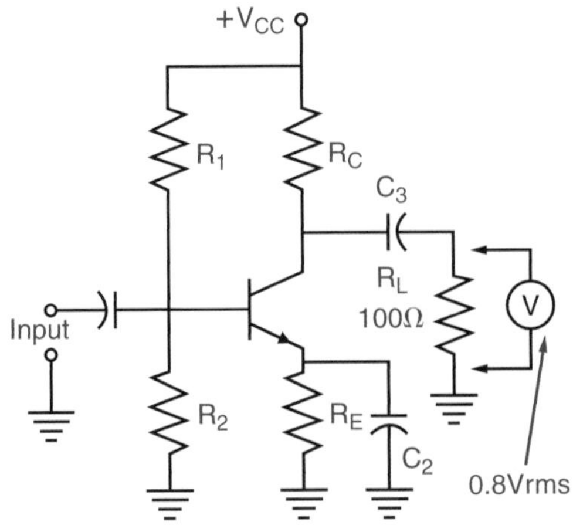

8.4 Amplifier Efficiency

Amplifier efficiency can now be calculated by the equation developed earlier.

$$\eta(\%) = \frac{P_L}{P_{DC}} \times 100$$

$$\eta(\%) = \frac{6.4 \text{ mW}}{15 \text{ mW} \times 100}$$

$$\eta(\%) = .4267 \text{ mW} \times 100$$

$$\eta(\%) = 42.67\%$$

CLASS B AMPLIFIERS

8.5 Class B Complementary Symmetry Operation

Basic Operation

The class B amplifier is biased exactly at cutoff. This means that a single-transistor class B amplifier will not faithfully reproduce its input. Full fidelity can be achieved in several ways. One of the best ways is the use of two transistors connected in what is called a complementary symmetry connection. Figure 8–6 shows an example of such a circuit.

Note that the AC input signal is applied to the base of each transistor. Because each transistor is biased to cutoff, and because Q_1 is PNP and Q_2 is NPN, each transistor will be forward biased on opposite half cycles of the input waveform. The output waveform is shown in Figure 8–7a.

FIGURE 8–6 A class B transistor amplifier using complementary symmetry.

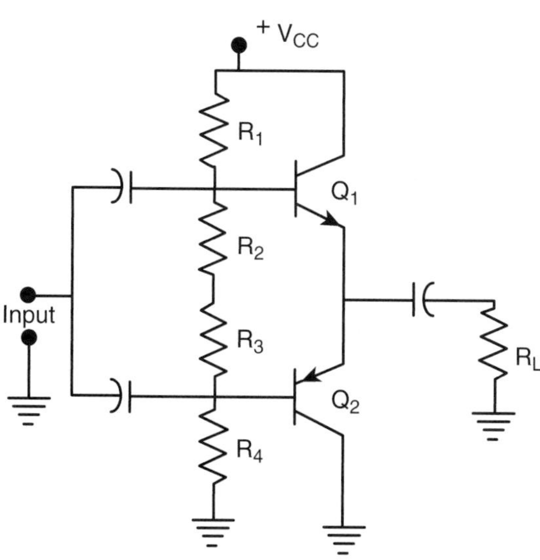

FIGURE 8–7 Output of Figure 8–6: (a) ideal, (b) with crossover distortion.

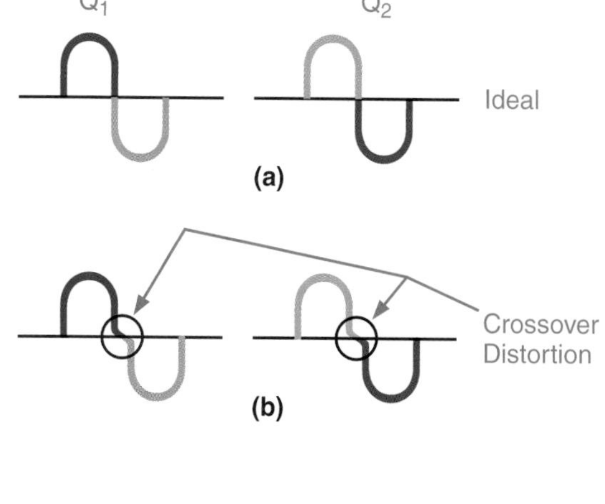

Because Q_1 and Q_2 are at cutoff with no AC input, the circuit will use no DC power at quiescence. This is the major efficiency advantage class B amplifiers have. They do not use any DC power unless they are actually amplifying. The circuit of Figure 8–6 is an emitter-follower type of operation because the load resistor is connected to the emitters of the two transistors.

Crossover Distortion

Figure 8–6 is a basic schematic diagram of a Complementary Symmetry amplifier. These amplifiers are commonly biased as class B, or class AB, in order to achieve a high degree of efficiency, with minimum distortion. The most obvious example of this High Fidelity requirement is the audio power amplifier.

Crossover distortion
The difference in a push-pull (or complementary symmetry) amplifier output compared to its input occurring at the crossover point.

Crossover
The point(s) in a signal waveform where an amplifier input transitions from positive-to-negative or negative-to-positive polarity.

Crossover distortion is a major problem unique to class B amplifiers. **Crossover** refers to the point(s) in a signal waveform where amplifier input transitions from positive-to-negative, or negative-to-positive polarity (Figure 8–7b). If both push-pull transistors conduct a small amount of current with zero signal input, they are actually operating as class AB—not B. If both transistors are just below cutoff, they are actually operating as class C—not B. Class B amplifiers are not often used because they are difficult to bias precisely at cutoff. Amplifier aging and thermal drift also contribute to the problem.

Study Figure 8–6. Output of the amplifier is taken from the emitters of both Q_1 and Q_2. Quiescent emitter current in Q_1 and Q_2 will be equal Kirchhoff's Current Law (KCL). No current flows in the load. We expect 50% of V_{CC} to be dropped across each transistor. If the transistors are precisely biased at cutoff, the slightest positive-going input signal will cause Q_1 to conduct and Q_2 will remain cutoff. Output voltage will move toward V_{CC}. The slightest negative-going input signal will cause Q_2 to conduct and Q_1 will remain cutoff. Output voltage will move toward ground.

Suppose the transistors are biased somewhat *below* their forward-conduction voltages: Quiescent emitter current in Q_1 and Q_2 will still be equal (KCL). No current flows in the load. Due to transistor leakage resistance, we still expect 50% of V_{CC} to be dropped across each transistor. However, to initiate transistor current input voltage must rise sufficiently to overcome the cutoff bias of the transistors. Therefore, the waveform of Figure 8–7b will result. If this circuit were to drive a speaker, the sound the listener would hear would not be a good representation of Figure 8–7a.

To eliminate this type of distortion, the class B amplifier may be biased as a class AB. This biasing operates the transistors slightly *above* cutoff. This means that Q_1 will not stop conducting until after Q_2 has started, and vice versa. Therefore, even the slightest change in input voltage will produce a change in output voltage. Figure 8–8 shows a circuit biased using diodes so that it operates as a class AB.

Diode Biasing

Figure 8–8 shows a schematic diagram of a complementary symmetry power amplifier. For this circuit to operate properly, Q_1 (an NPN transistor) and Q_2 (a PNP transistor) must be "matched." In other words, their beta, junction voltages, and other factors must be the same. In fact, complementary symmetry transistors usually state the part number of their matching complement.

D_1 and D_2, likewise, must be matched. Figure 8–8 shows Q_1 and Q_2 series connected, with their emitters joined.

When B-E junction voltages of both Q_1, and Q_2 are equal, R_{CE} of both transistors will also be equal. Therefore, voltage at the emitters will be $^1/_2\ V_{CC}$. Using Figure 8–8 as an example, $V_E = \frac{V_{CC}}{2} = \frac{10\ V}{2} = 5$ V. It is the function of the bias network R_1, D_1, D_2, and R_2 to set V_{BE} for Q_1 and Q_2.

As the applied signal voltage becomes more positive, Q_1 will conduct harder and output voltage will rise above the 5 V quiescent. As the applied signal voltage becomes more negative, Q_2 will conduct harder and output voltage will be driven below the 5 V quiescent. Because a small base current is present, even with no input signal, crossover distortion is eliminated or greatly reduced.

FIGURE 8–8 Diode biasing of a complementary-symmetry amplifier results in class AB operation.

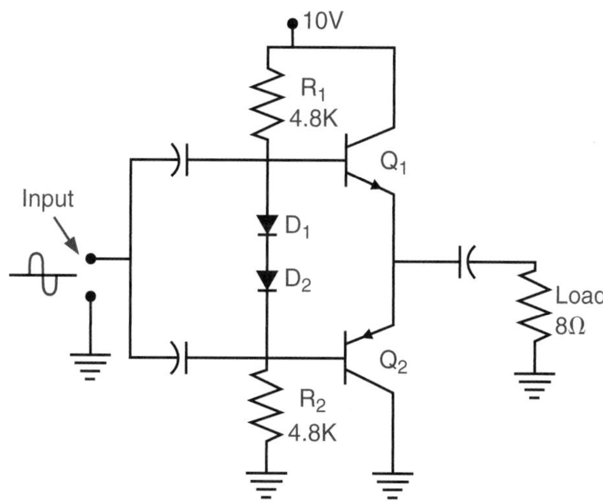

AC Operation

Assume that the peak output voltage in Figure 8–6 is at saturation. Then,

$$I_{CSat} = \frac{V_{CC}}{2 \times R_L}$$

This is because with both transistors biased to cutoff, the drop across each is $V_{CC}/2$ at the quiescent point. Consequently, the peak voltage across the load resistor is one-half V_{CC} (see Figure 8–9).

FIGURE 8–9 Output of amplifier shown in Figure 8–6.

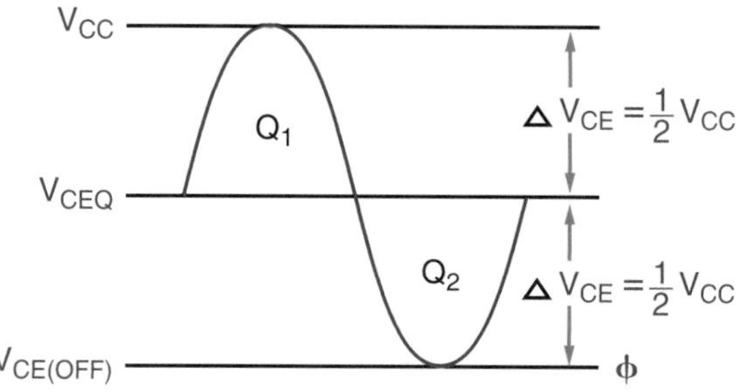

8.6 Class B Push-Pull

The class B push-pull amplifier uses two identical (matched) transistors and a center-tapped transformer. Figure 8–10 shows a typical class AB (or B push-pull) amplifier. The disadvantages of this circuit over the complementary pair are that this amplifier needs a center-tapped transformer to operate, it is more expensive, and it requires much more physical space and weight due to input and output transformer(s) in the circuit.

FIGURE 8–10 Push-pull amplifier.

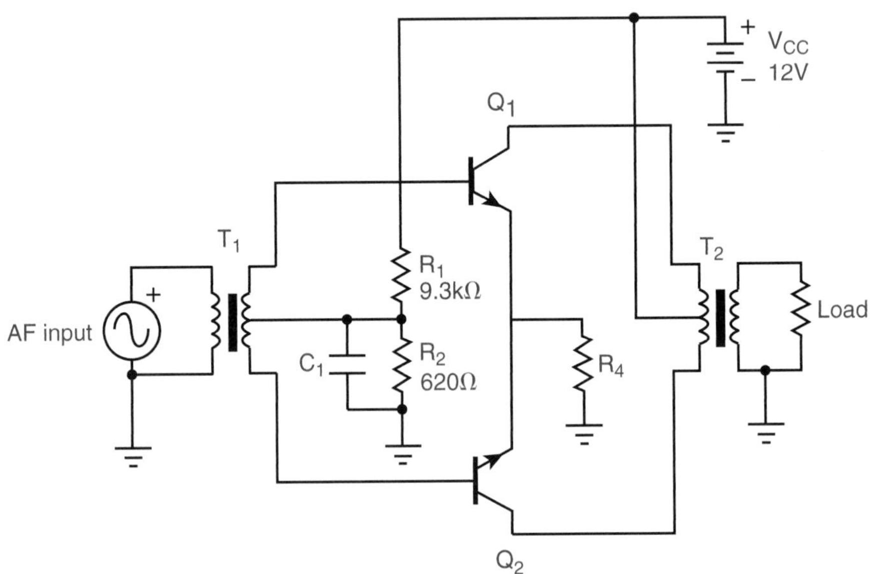

Just as the class B complementary pair amplifier of Figure 8–6, when operated as class B the push-pull amplifier of Figure 8–10 must be biased at the cutoff point. By using a center-tapped input transformer, the biasing circuit need only furnish a single bias voltage. Note that the bias voltage is connected to the transistor bases through the minimal resistance of the input transformer secondary winding. AC input signal for Q_1 is delivered between T_1 secondary center tap (CT) and Q_1 base. Input signal for Q_2 is delivered between T_1 secondary CT and Q_2 base. Because the base bias voltage is applied at T_1 secondary CT, it must not be directly grounded.

However, to ground the CT for signal purposes a capacitor (C_1) is used.

As a crude rule of thumb, the filter capacitor can be calculated to have X_C of 10% of the voltage divider resistance to ground. If the lowest frequency to be amplified is 300 Hz, then:

$$X_C = \frac{1}{2\pi fC} \Rightarrow C = \frac{1}{2\pi fX_C}$$

$$C = \frac{1}{6.28 \times 300\text{ Hz} \times \left(\frac{620\ \Omega}{10}\right)}$$

$$C = \frac{1}{116.8\text{ kHz}}$$

$$C = 8.6\ \mu\text{F}$$

8.7 Class B Power Calculations

The total input power formula for a class B amplifier is the same as for a class A amplifier: $P_{IN} = V_{CC} = I_{CC}$. The main difference is in how I_{CC} is calculated. The analysis that follows uses Figure 8–11 and assumes that the amplifier is delivering the maximum sinusoidal output.

$$I_{CC} = I_1 + I_{C(AVE)}$$

FIGURE 8–11 Calculating I_{cc} for class B amplifier of Figure 8–6.

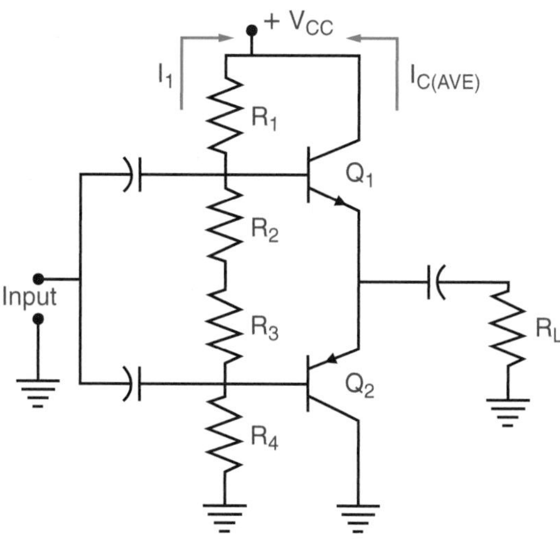

Here:

$$I_{C(AVE)} = \frac{I_{pk}}{\pi}$$

This is true because each transistor is producing a one-half sine-wave output. Because you are assuming that the amplifier is producing full output, $I_{pk} = I_{C\,Sat}$ and

$$I_{PK} = I_{C\,Sat} = \frac{V_{CC}}{2R_L}$$

Rearranging the terms, we find that:

$$I_{C(AVE)} = \frac{V_{CC}}{2\pi R_L}$$

The following example shows how to calculate the maximum power load and efficiency for a class B amplifier using Figure 8–12.

FIGURE 8–12 Class B amplifier total load power.

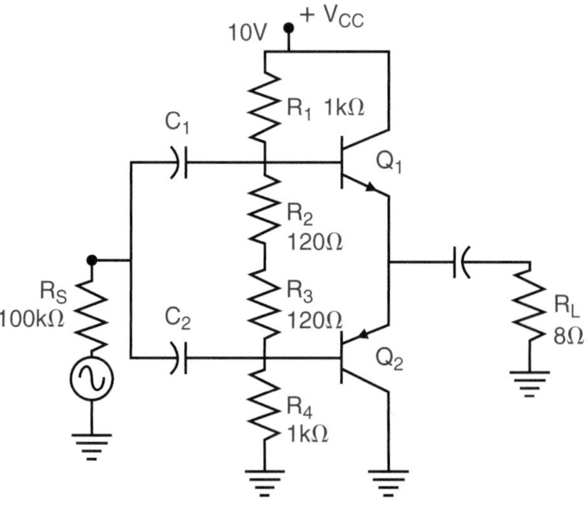

Example

Calculate the maximum power load.

Solution:

Step 1: Determine the maximum $V_{P\text{-}P}$ voltage for the amplifier.

This is called the amplifier's compliance. For a class B amplifier, where V_{CEQ} is the collector-to-emitter voltage at the quiescent (undriven) state,

$$V_{P\text{-}P} = 2V_{CEQ}.\ \text{And because } V_{CEQ} = \frac{V_{CC}}{2},\ \text{then } V_{P\text{-}P} \cong V_{CC}.$$

Step 2: Find the compliance ($V_{P\text{-}P}$) of the class B amplifier.

$$V_{P\text{-}P} \cong V_{CC} \cong 10\text{ V}$$

Step 3: Calculate $P_{L(MAX)}$.

The peak AC current supplied was calculated earlier as:

$$I_{PK} = I_{C\,Sat} = \frac{V_{CC}}{2R_L}$$

This means that the maximum AC power (a more detailed development is given in the summary section of this chapter) is:

$$P_{MAX} = \frac{1}{2} I_{PK}^2 \times R_L$$

$$P_{MAX} = \frac{1}{2} \times \left(\frac{V_{P\text{-}P}}{2R_L}\right)^2 \times \text{R}_\text{L}$$

$$P_{MAX} = \frac{V_{P\text{-}P}^2}{8R_L}$$

Substituting values gives:

$$P_{L\,MAX} = \frac{10^2}{8(8)}$$

$$P_{L\,MAX} = 1.56\text{ W}$$

Example

Calculate the efficiency (η) of the amplifier in Figure 8–12.

Solution:

Step 1: Determine the total power the amplifier draws from its DC power supply.

$$P_S = V_{CC} \times I_{CC}$$

Here:

$$I_{CC} = I_{C(AVE)} = I_1$$

and

$$I_1 = \frac{V_{CC}}{R_1 + R_2 + R_3 + R_4}$$

$$I_1 = \frac{10\text{ V}}{1\text{ k}\Omega + 120\ \Omega + 120\ \Omega + 1\text{ k}\Omega}$$

$$I_1 = \frac{10\text{ V}}{2240\ \Omega}$$

$$I_1 = 4.46\text{ mA}$$

$$I_{C(AVE)} = \frac{V_{CC}}{2\pi R_L}$$

$$I_{C(AVE)} = \frac{V_{CC}}{2\pi \times 8\ \Omega}$$

$$I_{C(AVE)} = \frac{10\text{ V}}{6.28 \times 8\ \Omega}$$

$$I_{C(AVE)} = \frac{10\text{ V}}{50.24\ \Omega}$$

$$I_{C(AVE)} = 199\text{ mA}$$

$$I_{CC} = 199\text{ mA} + 4.46\text{ mA}$$

$$I_{CC} = 203.46\text{ mA}$$

Substituting the values, we can now calculate P_S.

$$P_S = V_{CC} \times I_{CC}$$

$$P_S = 10\text{ V} \times 203.46\text{ mA}$$

$$P_S = 2.03\text{ W}$$

Step 2: Calculate amplifier efficiency.

$$\eta(\%) = \frac{P_L}{P_S} \times 100$$

$$\eta(\%) = \frac{1.56\text{ W}}{2.03\text{ W}} \times 100$$

$$\eta(\%) = .768\text{ W} \times 100$$

$$\eta(\%) = 76.8\%$$

CLASS C AMPLIFIERS

The class C amplifier in Figure 8–13 has one transistor that conducts for less than 180° of the AC input cycle. To accomplish this, the transistor is biased well into cutoff. The AC input signal causes Q_1 to turn on only while T_1 and T_2 are above 0 V. This is only for a short period of the entire input cycle (much less than 180°). This short conducting period gives the class C amplifier two distinguishing characteristics.

1. It is extremely efficient. Theoretically, it can have an efficiency rating of up to 99%. This is because it uses DC input power only during the short conducting time of the AC input signal.
2. The short conduction time caused by the negative base-emitter bias (refer to Figure 8–13) causes a lot of input signal distortion. For this reason, the class C amplifier is not suited as an audio amplifier. Because of the LC "tank" circuit's resonance frequency, it is used extensively in the radio frequency range.

Class C, and only class C, amplifiers may be used in frequency multiplier circuits in which an input signal of f_1 pulses the amplifier for a very small portion of the input cycle. The transistor collector (or vacuum tube plate) LC circuit will be tuned to a resonant frequency of 2 times f, 3 times f, 4 times f, or 5 times f. Energy injected in the LC circuit will "**flywheel**" until a new pulse is received from the input.

Flywheel effect
The maintenance of oscillation in an LC circuit resulting from the alternate charging of the capacitor, while the inductor field collapses, then increasing inductor field, as the capacitor discharges. Flywheeling dissipates as resistive factors, radiation, and coupling remove energy from the circuit. Mixer and multiplier circuits will be covered in a later chapter.

This circuit is common in communications receivers and transmitters. Communications receivers and transmitters also utilize class C circuits as mixers. Mixer and multiplier circuits are covered in a later chapter.

The output signal is generated by the cycling action of the LC "tank" collector circuit. In a parallel LC circuit, the inductor and capacitor alternate storing (charging) and discharging their energy as the AC signal rises and falls. This cycling action produces a sine wave (see the output signal of Figure 8–13).

FIGURE 8–13 Class C amplifier.

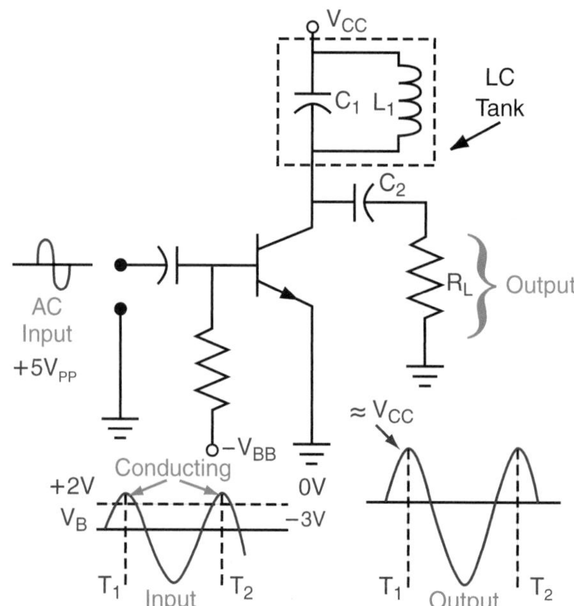

At the frequency where the capacitive reactance (X_C) equals the inductive reactance (X_L), a noninductive circuit condition called resonance exists. The total reactance of a "tank" (a parallel LC) circuit at resonance can be calculated as:

$$X_T = \frac{X_L \times X_C}{X_L - X_C}$$

When $X_L = X_C$, the results become:

$$X_T = \frac{X_L \times X_C}{0\ \Omega} = \infty\ \Omega$$

As a result, I_C will be minimum at the resonant frequency. If only reactances were present in the tank circuit, the impedance of the tank circuit would be infinite—producing zero signal current.

No capacitor or inductor is free of resistive losses. If we assume an ideal LC circuit (one having zero resistance), it is the resistive load (see R_L in Figure 8–13) that reduces the impedance of the collector load. Remember that impedance of the power supply must be considered 0 ohms. Therefore, in Figure 8–13 the tank and R_L are electrically in parallel. The collector signal current is calculated as:

Where I_{TANK} is 0 mA (because $Z_{TANK} = \infty$)

$$I_{SIGNAL} = \sqrt{I_{TANK}{}^2 + I_{RL}{}^2} = \sqrt{0\ \text{mA}^2 + I_{RL}{}^2}$$

$$I_{SIGNAL} = I_{RL}$$

The formula for calculating the resonant frequency of an LC "tank" circuit is:

$$f_R = \frac{1}{2\pi\sqrt{LC}}$$

The class C amplifier shown in Figure 8–13 is a tuned collector amplifier. The LC circuit is tuned to a specific input signal frequency to provide maximum output power transfer at that frequency, and reduced power transfer at unwanted frequencies both above and below the resonant frequency. The parallel tuned circuit at the collector results in *minimum line current* and *maximum circulating current at resonance.*

This maximized circulating current in a parallel tank at resonance lends itself to transformer coupling of the desired signal frequency. Transformer coupling can easily be employed to match standard transmission line impedances such as 50, 75, 300, and 600 ohms. In addition, transformer coupling limits amplifier tank circuit de-tuning resulting from sometimes dramatic variations of inductive and capacitive load factors.

THE POWER SWITCH (CLASS D)

A transistor can be used in a digital switch mode. This on/off configuration is used extensively with inductive motor loads and with voltage regulators. Figure 8–14 shows a typical switching regulator operation. Note that there are four basic circuit areas: the power switch, the switch driver, the filter circuit, and the control circuit.

FIGURE 8–14 Typical power switching circuit.

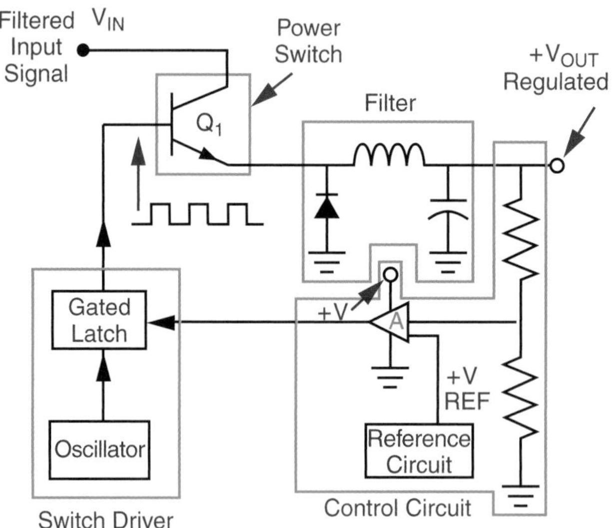

In Figure 8–14, Q_1 is rapidly switched between cutoff and saturation. When in saturation, the power switch provides a current path between the regulator's input and output. When in cutoff, the path is opened. This results in higher regulator efficiency and higher regulator power-handling capability.

This ability to be on or off for short periods of time increases the overall efficiency and reduces the heat buildup on the transistor at high power loads. This reduction in heat buildup is important because excess heat due to high power can damage the transistor. Many transistors that routinely handle high power or currents are mounted on heat sinks. A heat sink allows the transfer of the excess heat (dissipates the heat) from the transistor to the sink and the environment.

The circuit shown in Figure 8–14 operates as follows. When the control circuit senses a change in the regulator's output voltage, it sends a signal to the switch driver. If the change is negative (a lower output voltage), the oscillator changes the gated latch to increase the signal to the base and increase the conduction of the power switch.

Because the oscillator signal is a fixed magnitude and frequency, the gated latch combines the output of the control circuit with the oscillator to change the magnitude (voltage) of the digital pulse to the base. The magnitude of the pulse determines how much the power switch will conduct (determines the forward bias). This will cause the output voltage to rise. The opposite occurs when the control circuit senses an increase in output voltage.

Two other methods are also used to control the power switch. The first is pulse-width modulation. Here, the signal from the gated latch varies in pulse width without affecting the cycle time (see Figure 8–15). The second method uses a stable pulse width but changes the cycle time. This is called variable off-time modulation (see Figure 8–16).

FIGURE 8–15 Pulse-width modulation.

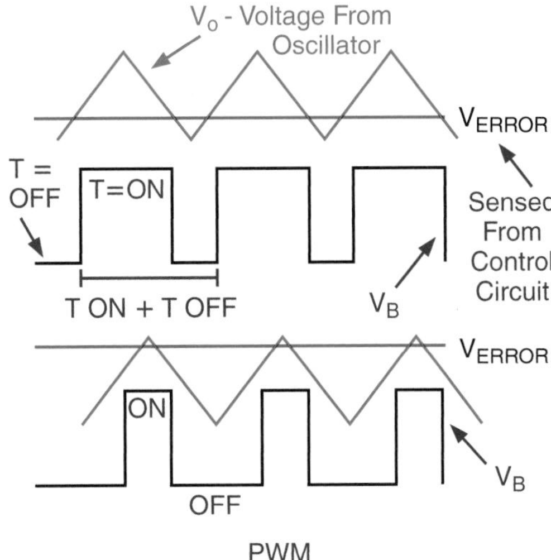

FIGURE 8–16 Variable off-time modulation.

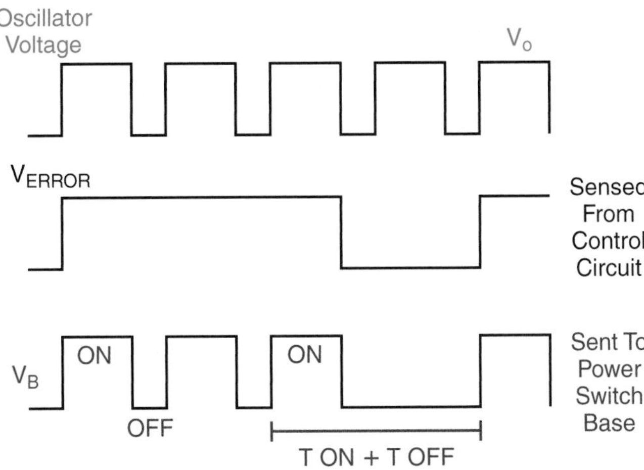

FIGURE 8–17 TO-204 transistor case style as typically shown in manufacturer specifications.

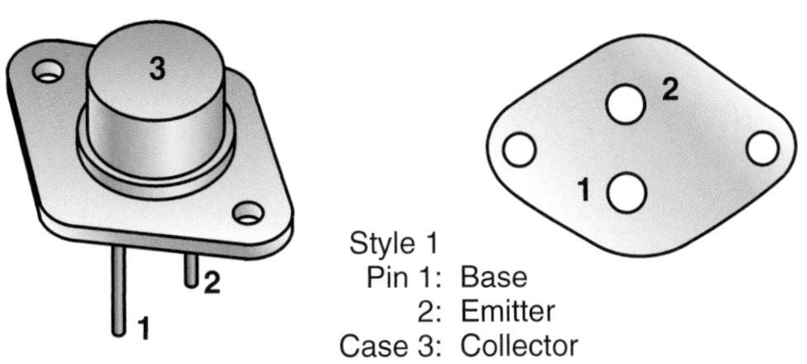

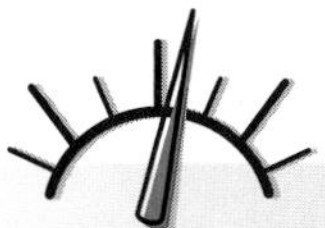

TechTip!

Sometimes we are stumped by the simple things. When a power transistor circuit fails, indicating very high collector current, the technician may be inclined to check the transistor junctions. Not a bad idea. However, it is often a quick and easy check of collector-to-heat sink, chassis, backplate, or backplane that results in a very quick fix to the problem. See Figure 8–17.

Note that this is a TO-204 (formerly TO-3). The collector is the case. There is generally a vinyl or mica insulator placed under the TO-204 to electrically isolate it from ground. Sometimes a metallic "spur" on the plate works through the insulator, shorting to the collector. Polish it off, replace the heat sink compound if required, remount the transistor (if it still checks out), and turn it on.

SUMMARY

Amplifiers are often needed in large signal applications. Such applications include audio power amplifiers, radio frequency amplifiers, controllers, variable-frequency drives, regulating power supplies, and a host of other such equipment.

In this chapter, you practiced techniques you have already learned and applied them to new circuitry—in particular, the class B complementary symmetry amplifier and the class B push-pull amplifier. Table 8–1 summarizes the various amplifier classes.

Earlier in the chapter, a simple explanation of the formula $P_L = \frac{V_{P\text{-}P}^2}{8R_L}$ was given. The following information provides a more detailed, and perhaps more interesting, development.

$$P_L = \frac{V_L^2}{R_L}$$

Here:

V_L is the RMS load voltage. With this case, the $V_{P\text{-}P}$ must be converted to RMS.

$$V_{RMS} = \frac{V_{P\text{-}P}}{2} \times \frac{1}{\sqrt{2}}$$

Substituting this into the $P_L = \frac{V_L^2}{R_L}$ formula,

$$P_L = \frac{\left(\frac{V_{P\text{-}P}}{2} \times \frac{1}{\sqrt{2}}\right)^2}{R_L}$$

$$P_L = \frac{\left(\frac{V_{P\text{-}P}}{2\sqrt{2}}\right)^2}{R_L}$$

$$P_L = \frac{\left(\frac{V_{P\text{-}P}^2}{2^2 \times \sqrt{2^2}}\right)}{R_L}$$

$$P_L = \frac{V_{P\text{-}P}^2}{8\ R_L}$$

TABLE 8–1 Summary of Amplifier Classes

Class	Description
A	It has the lowest possible efficiency of all the classes, a maximum of 50%. Uses a single transistor that conducts 360° of the input signal and provides the least signal distortion. The class A amplifier is biased near the center of the load line. The main use of class A amplifiers is in small signal applications.
AB	It has a greater efficiency rating than the class A amplifier and slightly less than the class B. It has two transistors in audio applications, and an improved signal distortion over the class B and is biased near cutoff. The class AB conducts between 181° and 359° of the AC signal input cycle. Diodes are commonly used in this class to provide consistent B-E function voltage. Therefore, it maintains a forward bias for both transistors throughout their respective portion of the input signal. The main use of class AB amplifiers is in high-power stages of coupled amplifiers and in audio and radio frequency applications.
B	The class B has two transistors in audio applications, and has a potential maximum efficiency rating of 78.5%. This amplifier is biased at cutoff and conducts during 180° of the AC signal input cycle. It has a high signal distortion problem due to crossover. The main use of class B amplifiers is in high-power stages of coupled amplifiers and in radio frequency applications. The push-pull amplifier is also a form of class B amplifier. It uses input and output transformers (center tapped) that increase the cost of circuit construction.
C	It has the highest efficiency potential of all the amplifiers, approximately 99% maximum. This amplifier is biased well below cutoff (reverse biased) and conducts during less than 180° of the AC input signal. Usually this is about 90°. The signal distortion is very high due to the limited conduction time, and is therefore not used for audio applications. The ability of the amplifier to be tuned removes much of the signal distortion thanks to tank circuit flywheel effect. The main use of class C amplifiers is in stages of coupled amplifiers for radio frequency applications, especially where signal mixing and frequency multiplication is desirable.
D	The ability to rapidly switch between saturation and cutoff (on/off) makes the class D an ideal digital switch that provides high regulator efficiency. Its excellent heat dissipation characteristics allow for higher regulator power-handling capability than the other amplifier types. The main uses of the class D amplifier are as power switches for inductive motor loads and as voltage regulators.

REVIEW QUESTIONS

1. Describe the concept of crossover distortion in a class B push-pull amplifier.
   a. What causes it?
   b. If it is severe in an audio amplifier, what would the effect be on the listener?
   c. How is it corrected?
2. How does the class C amplifier shown in Figure 8–18 create a pure sine wave when the transistor conducts for less than half of the input cycle.
3. Discuss efficiency in the various classes of amplifiers.
   a. Which has the highest efficiency?
   b. What are the tradeoffs for high efficiency?
4. Consider Figure 8–19.
   a. Why does the total DC power include I_1 and I_C?
   b. Why does the output capacitor have no effect on the DC input power?
5. The text mentions the use of the class D amplifier in inductive loads and regulators. What other loads might be fed by such amplifiers?

FIGURE 8–18 Class C amplifier.

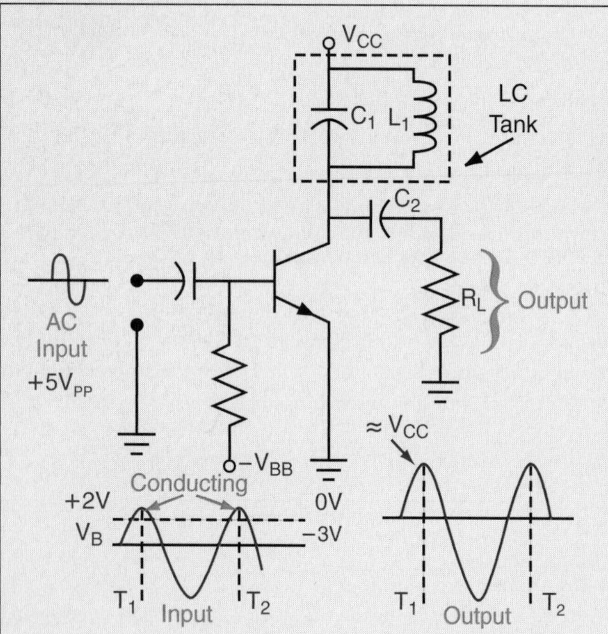

FIGURE 8–19 Calculating I_{cc} for DC power input.

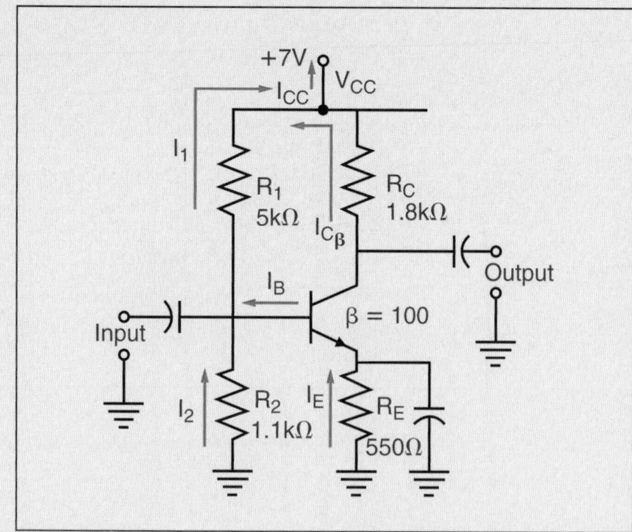

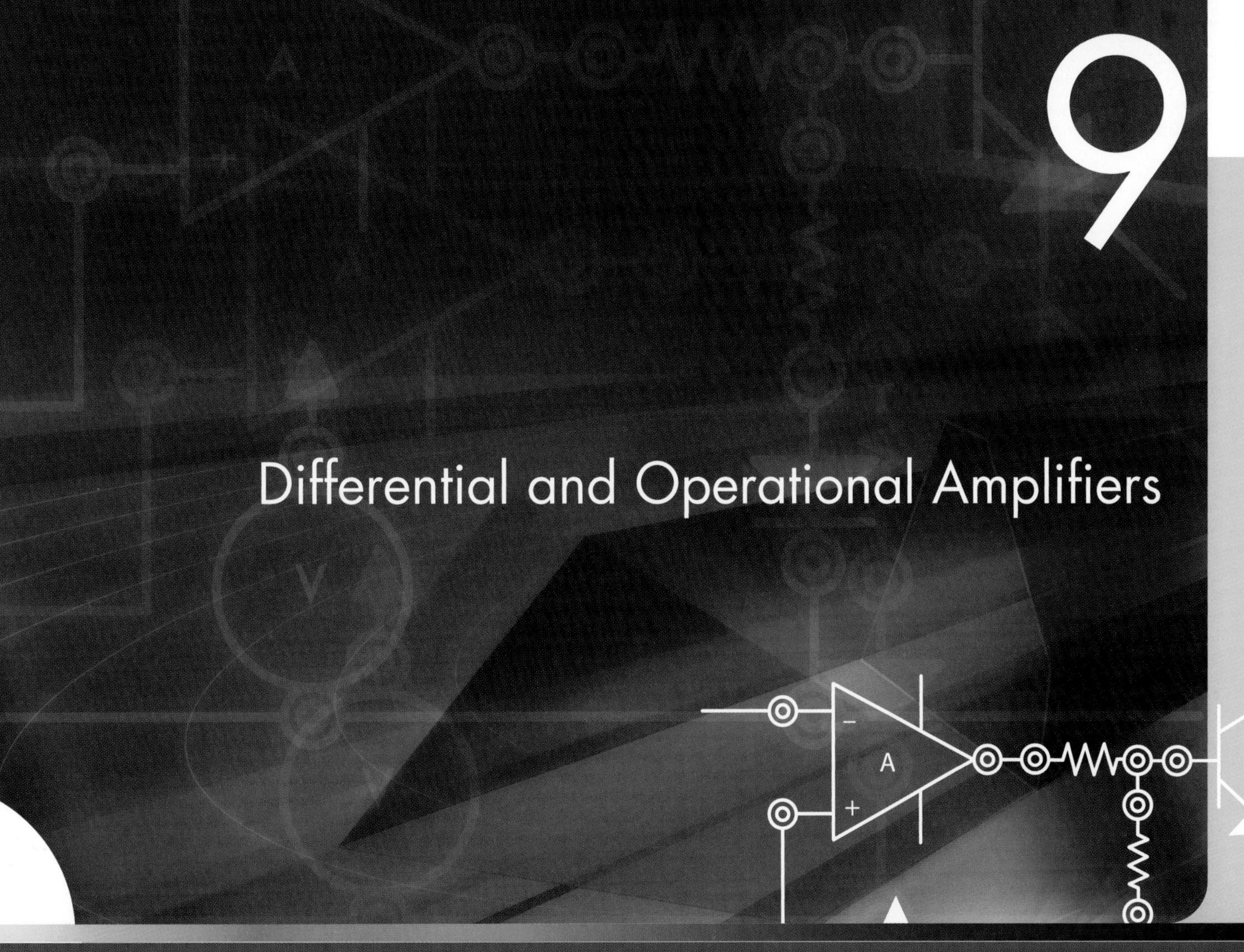

9 Differential and Operational Amplifiers

O U T L I N E

OVERVIEW

In this chapter, you will learn about two very commonly used amplifier circuits: differential amplifiers and operational amplifiers. Both of these amplifier circuits are very versatile and are widely used in linear applications. Their popularity is partially due to the fact that they are available as integrated circuits (ICs) at a very low cost. Both of these types of amplifiers are extremely easy to use and allow the electrician to build useful circuits without worrying about the complex internal circuitry.

OBJECTIVES

After completing this chapter, the student should be able to:

1. Draw the schematic symbols of differential and operational amplifiers.
2. Describe the operation of differential and operational amplifiers.
3. Describe the basic operations of circuits using differential and operational amplifiers.

DIFFERENTIAL AMPLIFIERS

9.1 Basic Circuit

The differential amplifier gets its name from the fact that it produces an output that is proportional to the difference between its two inputs. Figure 9–1 shows the basic arrangement of a differential amplifier. The transistors Q_1 and Q_2 must be as close to identical as possible. Ideally, they are exactly alike. R_{C1} and R_{C2} are also identical to keep the symmetry in the circuit. The input to transistor Q_1 is called the **inverting input** (I), and the input to transistor Q_2 is called the **noninverting input** (NI).

Inverting input
The differential op-amp input that drives the output 180° out-of-phase.

Noninverting input
The differential op-amp input that drives the output in-phase.

FIGURE 9–1 A differential amplifier.

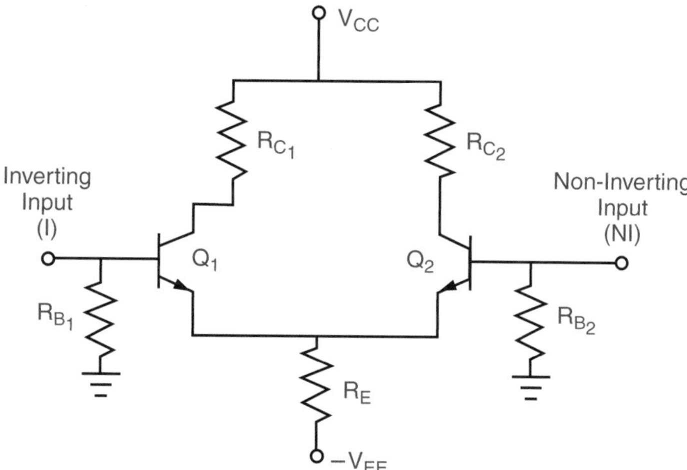

The circuit requires two power supplies: $+V_{CC}$ connected to the collector circuit and $-V_{EE}$ connected to the emitter circuit. Figure 9–2 shows a bipower supply that can be used for this purpose. When a BJT is used in an amplifier, the collector-base junction must be reverse biased and the emitter-base junction must be forward biased. The transistors used in Figure 9–1 are NPN transistors, and the voltages $+V_{CC}$ and $-V_{EE}$ establish the required biasing conditions in the amplifier.

FIGURE 9–2 A bipolar DC power supply.

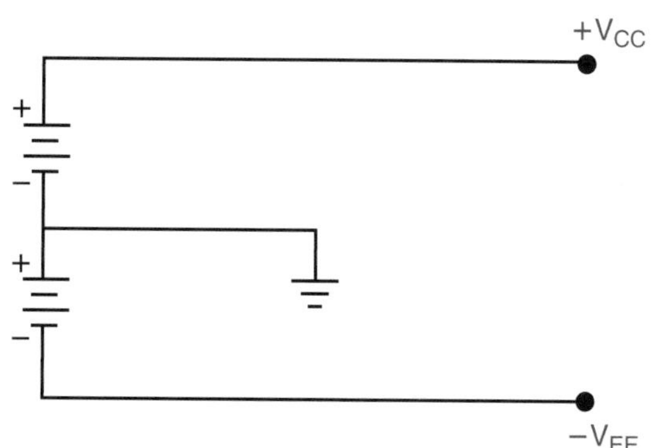

9.2 Modes of Operation

There are three basic modes of operation for a differential amplifier, each based on the way input is fed into the amplifier.

1. Single-ended mode
2. Differential mode
3. Common mode

Single-Ended Mode

A differential amplifier is operating in the single-ended mode when an active signal is connected to only one of its inputs. The inactive input is normally connected to ground (directly or through a resistor). The amplifier is classified as an inverting amplifier or a noninverting amplifier depending on which input is active.

Figure 9–3 shows a differential amplifier driven at the inverting input (Q_1). If the input increases in the positive direction, Q_1 (NPN transistor) will increase in conduction. This will cause the collector current (I_{C1}) to increase, which in turn causes the voltage drop across the collector resistor R_{C1} ($I_{C1} \times R_{C1}$) to increase. The voltage at the collector of Q_1 ($V_{CC} - I_{C1} \times R_{C1}$) decreases. Thus, an increase in the input to Q_1 (inverting input) causes a decrease in the collector voltage of Q_1. This means that the output of Q_1 is inverted, or 180° out of phase with its input.

FIGURE 9–3 Differential amplifier operating in the single-ended mode.

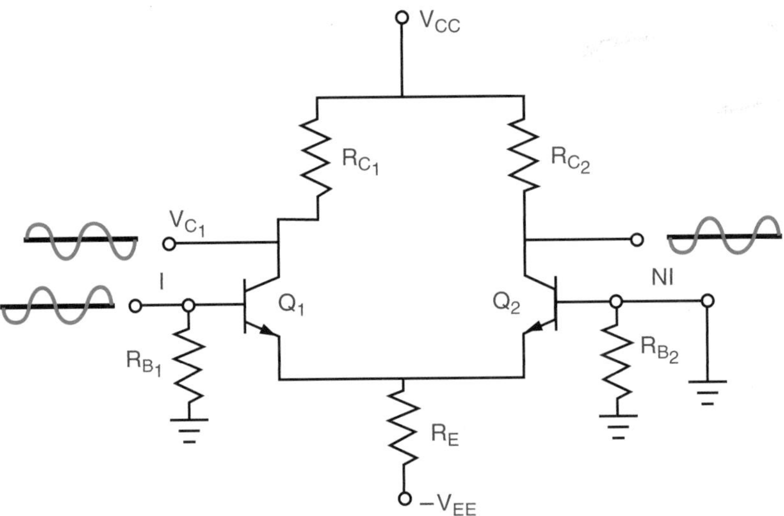

Now consider what is happening to the collector of Q_2. When Q_1 is driven to conduct by the positive-going input, its emitter current increases—thereby increasing the voltage drop across R_E. The emitter of Q_2 (which is also connected to R_E) also experiences a higher voltage. The base of Q_2 (noninverting input) is connected to ground, and thus its base-emitter PN junction experiences a smaller forward bias and conducts less current. In fact, the magnitude of this voltage is almost equal to the magnitude of the input signal. The only difference is the voltage drops of the PN junctions.

This causes the collector current (I_{C2}) to be small, which in turn results in a smaller voltage drop across resistor R_{C2}. The voltage at the collector of Q_2 ($V_{CC} - I_{C2} \times R_{C2}$) will increase in the positive direction. Consequently, the collector of Q_2 increases in the positive direction as the input increases, thus causing a noninverted output at the collector of Q_2.

This shows that the circuit in Figure 9–3 can be used to obtain an inverted (out-of-phase) and a noninverted (in-phase) output. The output is said to be single ended if it is measured between the collector of Q_1 or Q_2 and ground. The output is said to be differential if it is measured between the collector of Q_2 and the collector of Q_1. The differential output has twice the swing of the individual outputs.

As an example, consider an input signal of 1 $V_{p\text{-}p}$ applied to the inverting input while the noninverting terminal is connected to ground. This creates an amplified inverted signal being produced at the collector of Q_1, whereas an amplified noninverted output is produced in the collector of Q_2. If the transistors in the circuit amplify by a factor of 3, the output (single ended) at Q_1 will be 3 $V_{p\text{-}p}$ (inverted) and the output at Q_2 will be 3 $V_{p\text{-}p}$ (noninverted). This is shown in Figure 9–4.

FIGURE 9–4 Single-ended and differential outputs.

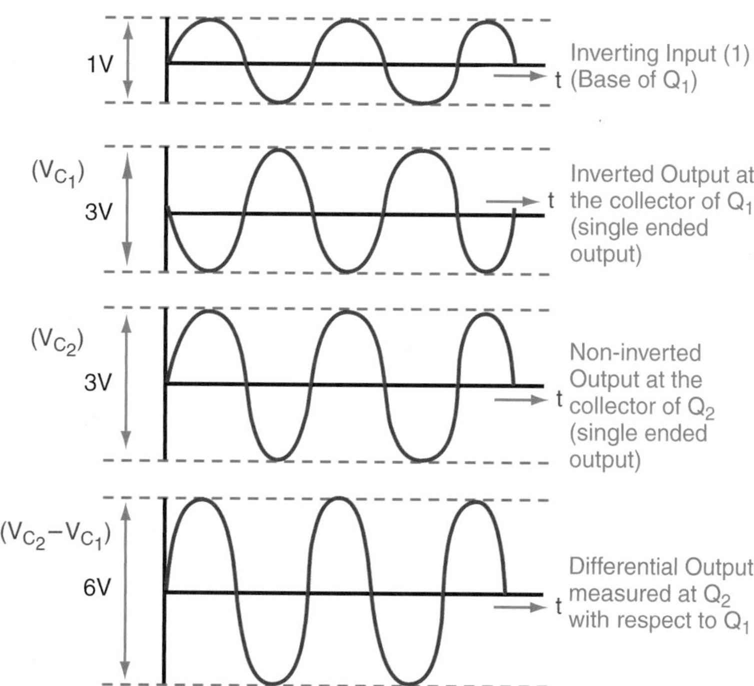

When the collector of Q_1 goes to its negative maximum (−1.5 V), the collector of Q_2 goes to its positive maximum (+1.5 V). The differential output measured between Q_2 and Q_1 is (+1.5 V) − (−1.5 V) = +3 V. When the collector of Q_1 goes to its positive maximum (+1.5 V), the collector of Q_2 goes to its negative maximum (−1.5 V).

The differential output measured between Q_2 and Q_1 is $(-1.5\text{ V}) - (+1.5\text{ V}) = -3$ V. Thus, we can see that the differential output in Figure 9–4 has a swing from +3 V to −3 V for a total peak-to-peak voltage of 6 V. This is twice the swing of the individual single-ended outputs.

It must be noted that in the previous example the difference between the two inputs is 1 $V_{p\text{-}p}$ − 0 V (GND) = 1 $V_{p\text{-}p}$. The single-ended output amplified this difference three times, whereas the differential output amplified the difference six times.

Differential Mode

In the differential mode of operation, two separate signals are applied to both amplifier inputs. The magnitude of the output voltage is the difference between the two individual inputs. Figure 9–5 shows a differential amplifier driven in the differential mode.

FIGURE 9–5 Differential amplifier operating in the differential mode.

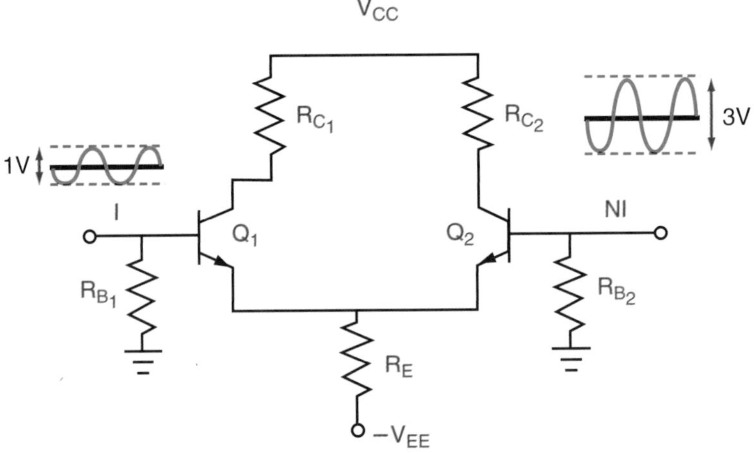

Assume the following conditions for Figure 9–5.

- The inverting input (Q_1) is fed with 1 $V_{p\text{-}p}$.
- The noninverting input (Q_2) is fed with 3 $V_{p\text{-}p}$.
- Both input voltages are in phase and are of the same frequency.
- Each of the two transistor circuits has a voltage gain of 3.

Both transistors, being identical, will conduct proportional to their inputs. This means the following:

- The collector of Q_1 (which has an input voltage of 1 $V_{p\text{-}p}$) will have an output voltage of 3 $V_{p\text{-}p}$.
- The collector of Q_2 (which has an input of 3 $V_{p\text{-}p}$) will have an output voltage of 9 $V_{p\text{-}p}$.
- Each transistor's output will be inverted from its input.

The outputs at the collectors of Q_1 and Q_2 are shown in Figure 9–6. Note that both outputs with respect to ground are out of phase with their corresponding inputs with respect to ground.

The differential output measured at Q_2 with respect to Q_1 can be calculated as the difference between the two outputs:

$$V_O = V_{Q2} - V_{Q1}$$

$$V_O = 9\ V_{P\text{-}P} - 3\ V_{P\text{-}P}$$

$$V_O = 6\ V_{p\text{-}p}$$

or the product of the gain and the difference of the two inputs:

$$V_O = A_V \times (V_{Q2(In)} - V_{Q1(In)})$$

$$V_O = 3 \times (3\ V_{p\text{-}p} - 1\ V_{p\text{-}p})$$

$$V_O = 6\ V_{p\text{-}p}$$

FIGURE 9–6 Input and output waveforms of differential amplifier operating in the differential mode.

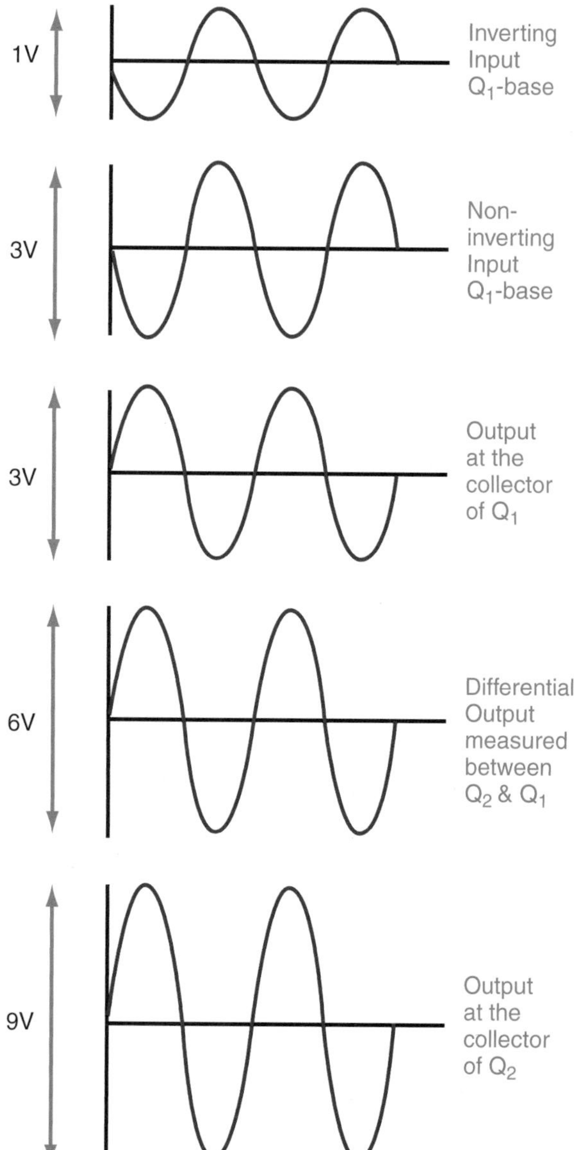

Note that when the inputs are made equal the differential output becomes zero. Also note that if either of the inputs is phase shifted by 180° the inputs effectively add.

$$V_O = 9\ V_{p\text{-}p} - (-3\ V_{p\text{-}p})$$
$$V_O = 12\ \text{V}_{\text{p-p}}$$

Common Mode

In the common mode of operation, identical signals are applied simultaneously to both the inverting and the noninverting inputs. Ideally, identical transistors will produce identical outputs with identical inputs. The collector voltages being identical throughout the cycle produces a differential output of 0 V. The advantage of this mode of operation is that any noise or undesired signal that appears simultaneously at the two inputs does not generate an output from the circuit.

Power-line hum is a commonly occurring noise signal that interferes with high-gain amplifiers. The 60 Hz frequency of the power line radiates signals that may be picked up by electronic circuits. Such hum is usually sensed in common mode. That is, it affects both wires of a signal circuit equally.

In Figure 9–7, the power-line noise is being impressed on lines a and b equally with respect to ground. The desired signal, however, is being applied across lines a and b. If the input signal line is connected to a differential amplifier, as shown in Figure 9–8, the 60 Hz noise is being connected equally to both the inverting and noninverting inputs. As you saw earlier, this means that the hum will be canceled at the output terminals.

The desired signal, on the other hand, is changing positive at one input and negative at the other. This means that it will be amplified at the output. In other words, hum and noise constitute the common-mode signal and are canceled.

FIGURE 9–7 Common mode.

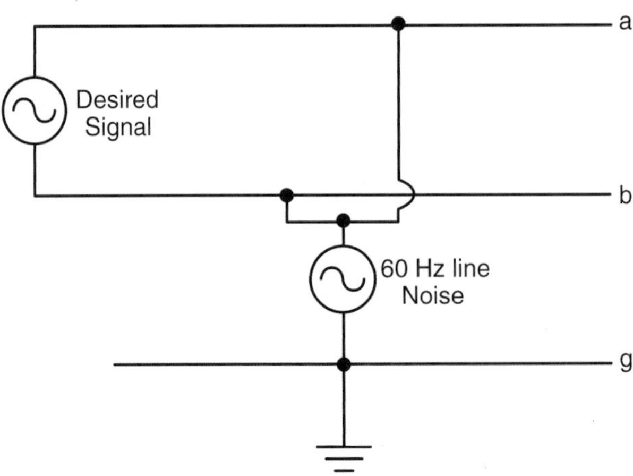

The inputs to the inverting and the noninverting inputs after the hum or noise is picked up are shown in Figure 9–9. The hum is of the same phase in both the inputs, and thus the differential output is unaffected by it. Identical transistors and resistors produce identical collector voltages, and hence the differential output measured between Q_1 and Q_2 is zero for the noise. In other words, the differential amplifier has rejected the common-mode signals (noise). Conversely, the input signal is fed in differential mode and is amplified.

FIGURE 9–8 Differential amplifier with hum fed in common mode and desired signal in differential mode.

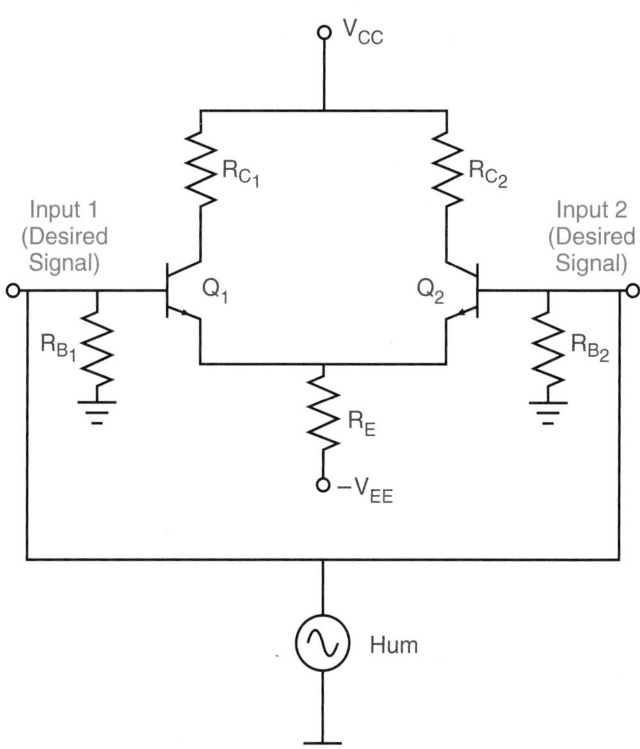

FIGURE 9–9 Input and output waveforms demonstrating noise rejection.

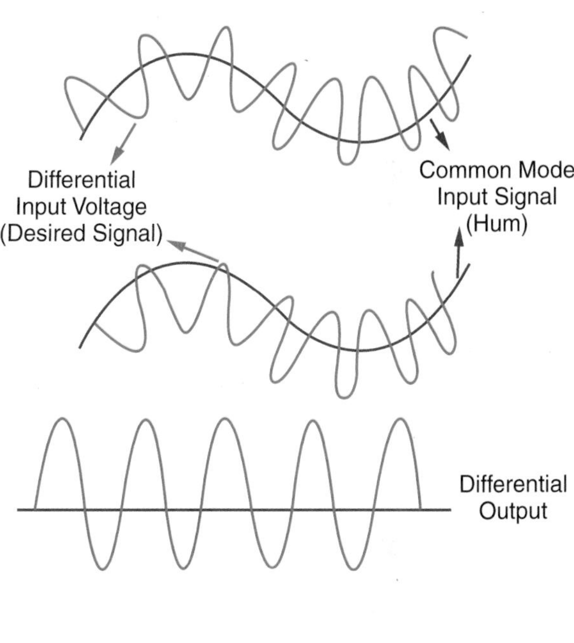

9.3 Common-Mode Rejection Ratio

Common-mode signal rejection occurs only if the two transistors and their associated resistors are perfectly matched. In reality, the differential amplifier will not have perfect balance and one transistor may have a higher gain than the other. This causes some common-mode signal to appear at the output. The ability of a circuit to reject common-mode signals is called the common-mode rejection ratio (CMRR).

$$\text{CMRR} = \frac{A_{V(DIFF)}}{A_{V(COMM)}}$$

Here:

$A_{V(DIFF)}$ = voltage gain of the amplifier for differential signals
$A_{V(COMM)}$ = voltage gain of the amplifier for common-mode signals
If $A_{V(DIFF)}$ and $A_{V(COMM)}$ are expressed in dB, then
$CMRR_{dB} = A_{V(DIFF)\,dB} - A_{V(COMM)\,dB}$

Example

A differential amplifier has the following input and output signals:

- Common-mode input = 1 V
- Common-mode output = 0.02 V
- Differential input = 0.1 V
- Differential output = 10 V

Solve for $CMRR_{dB}$

Solution:

Step 1:

$$A_{V(DIFF)} = \frac{Differential\ Output}{Differential\ Input}$$

$$A_{V(DIFF)} = \frac{10\text{ V}}{0.1\text{ V}}$$

$$A_{V(DIFF)} = 100$$

Step 2:

$$A_{V(DIFF)\ dB} = 20 \times \log_{10} A_{V(DIFF)}$$

$$A_{V(DIFF)\ dB} = 20 \times \log_{10} 100$$

$$A_{V(DIFF)\ dB} = 40\text{ dB}$$

Step 3:

$$A_{V(COMM)} = \frac{Common\ Mode\ Output}{Common\ Mode\ Input}$$

$$A_{V(COMM)} = \frac{0.02\text{ V}}{1\text{ V}}$$

$$A_{V(COMM)} = 0.02$$

Step 4:

$$\text{A}_{V(COMM)\ dB} = 20 \times \log_{10} A_{V(COMM)}$$

$$\text{A}_{V(COMM)\ dB} = 20\ \log_{10} 0.02$$

$$\text{A}_{V(COMM)\ dB} = -33.98\ dB$$

Step 5:

$$CMRR = \frac{A_{V(DIFF)}}{A_{V(COMM)}}$$

$$CMRR = \frac{100}{0.02}$$

$$CMRR = 5{,}000$$

(continued)

Step 6:

$$CMRR_{dB} = A_{V(DIFF)\,dB} - A_{V(COMM)\,dB}$$

$$CMRR_{dB} = 40 \text{ dB} - (-33.98\ dB)$$

$$CMRR_{dB} = 73.98 \text{ dB}$$

or

$$CMRR_{dB} = 20 \times \log_{10}(CMRR)$$

$$CMRR_{dB} = 20 \times \log_{10} 5{,}000$$

$$CMRR_{dB} = 73.98 \text{ dB}$$

9.4 Analysis

The characteristics and properties of differential amplifiers can be demonstrated by performing DC and AC analysis on the circuit.

DC Analysis

Figure 9–10 shows a differential amplifier operation with +10 V($+V_{CC}$) and −10 V(V_{EE}). In this circuit, do not consider any AC inputs applied because you are going to strictly analyze DC currents and voltages.

FIGURE 9–10 A differential amplifier circuit.

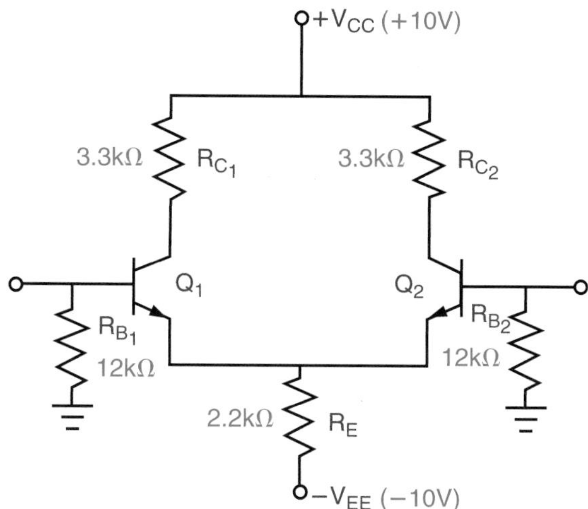

To ease the analysis, redraw the circuit as shown in Figure 9–11. Here, the emitter resistor (R_E) has been changed into two parallel resistors of 4.4 kΩ each. Because the differential amplifier is balanced, you can now separate the two circuits as shown in Figure 9–12. Any calculations done on one half of the circuit can be applied symmetrically to the other half.

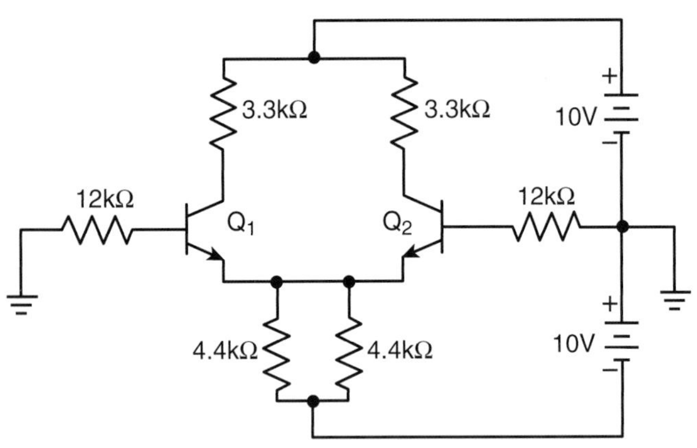

FIGURE 9–11 The circuit of Figure 9–10 with R_E split into two equal resistors in parallel.

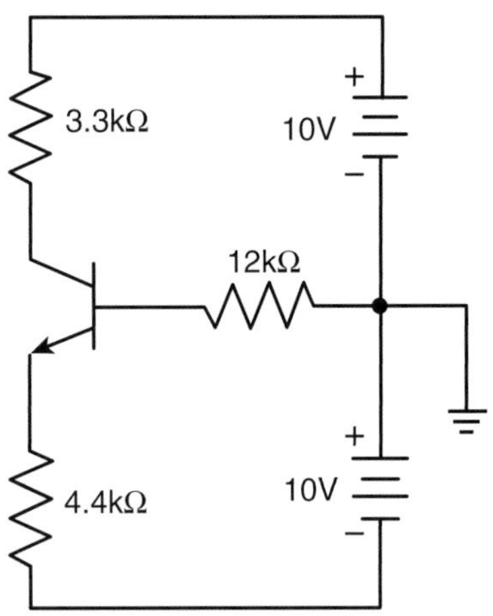

FIGURE 9–12 The right half of Figure 9–11 used to calculate the DC quiescent values.

Example

The following steps use the circuit of Figure 9–12.

Solution:

Step 1: Write a KVL around the base-emitter circuit.

$$-V_{(Source)} + V_{RB} + V_{BE} + V_{RE} = 0$$

Where:

$$V_{RB} = R_B \times I_B$$

$$V_{BE} = 0.7 \text{ V}$$

$$V_{RE} = R_E \times I_E$$

$$-10 \text{ V} + (12 \text{ k}\Omega \times I_B) + V_{BE} + [4.4 \text{ k}\Omega \times (1 + \beta) \times I_B] = 0$$

Substituting values and assuming that $\beta = 200$ gives:

$$-10 \text{ V} + (12 \text{ k}\Omega \times I_B) + 0.7 \text{ V} + [4.4 \text{ k}\Omega \times (1 + 200) \times I_B] = 0$$

Simplifying and collecting terms gives:

$$-10 \text{ V} + (12 \text{ k}\Omega \times I_B) + 0.7 \text{ V} + [4.4 \text{ k}\Omega \times (1 + 200) \times I_B] = 0$$

$$(12 \text{ k}\Omega \times I_B) + 0.7 \text{ V} + [4.4 \text{ k}\Omega \times 201 \times I_B] = 10 \text{ V}$$

$$10 \text{ V} - 0.7 \text{ V} = (12 \text{ k}\Omega \times I_B) + (4.4 \text{ k}\Omega \times 201 \times I_B]$$

$$10 \text{ V} - 0.7 \text{ V} = 9.3 \text{ V}$$

$$(12{,}000 \ \Omega \times I_B) = (4{,}400 \ \Omega \times 201 \times I_B) = 9.3 \text{ V}$$

$$(12{,}000 \ \Omega + 884{,}400 \ \Omega)I_B = 9.3 \text{ V}$$

$$896{,}400 \ \Omega \times I_B = 9.3 \text{ V}$$

$$I_B = \frac{9.3 \text{ V}}{896{,}400 \ \Omega}$$

$$I_B = 10.4 \ \mu\text{A}$$

(continued)

Step 2: Current through the emitter resistor is (Figure 9–12):

$$I_{4.4\,k\Omega} = (1 + \beta) \times I_B$$

$$I_{4.4\,k\Omega} = (1 + 200) \times 10.4\ \mu A$$

$$I_{4.4\,k\Omega} = 201 \times 10.4\ \mu A$$

$$I_{4.4\,k\Omega} = 2.09\ mA$$

Step 3: The collector current is:

$$I_C = \beta \times I_B$$

$$I_C = 200 \times 10.4\ \mu A$$

$$I_C = 2.08\ mA$$

Step 4: Write a KVL around the collector-emitter circuit:

$$-V_{(Source)} + V_{RC} + V_{CE} + V_{RB} = 0$$

Where:

$$V_{RC} = R_C \times I_C$$

$$V_{RE} = R_E \times I_E$$

$$-20\ V + (3.3\ k\Omega \times 2.08\ mA) + V_{CE} + (4.4\ k\Omega \times 2.09\ mA) = 0$$

$$-20\ V + (6.86\ V) + V_{CE} + (9.2\ V) = 0$$

Simplifying and collecting terms gives:

$$V_{CE} = 20\ V - (6.86\ V + 9.2\ V)$$

$$V_{CE} = 20\ V - 16.06\ V$$

$$V_{CE} = 3.94\ V$$

AC Analysis

When calculating values for transistor circuits, AC values are normally expressed in lowercase (e.g., r_E) and DC values are expressed in uppercase (e.g., R_E). AC analysis is used to determine the values of $A_{V(DIFF)}$ and $A_{V(COM)}$ using the values of resistors used in the circuit. To determine the gains, first determine the value of the transistor's AC emitter resistance r_E. Note that the emitter voltage drop for BJTs is usually considered relatively constant for small-signal work. In this example, use 50 mV as the value typical of the types of BJTs used for this application.

$$r_E = \frac{50\ mV}{2.09\ mA}$$

$$r_E = 23.92\ \Omega$$

The AC base resistance is calculated using

$$r_B = \frac{R_B}{\beta}$$

$$r_B = \frac{12\ k\Omega}{200}$$

$$r_B = 60\ \Omega$$

The value of $A_{V(DIFF)}$ is calculated with the formula

$$A_{V(DIFF)} = \frac{R_C}{(2 \times r_E) + r_B}$$

$$A_{V(DIFF)} = \frac{3.3\ \text{k}\Omega}{(2 \times 23.92\ \Omega) + 60\ \Omega}$$

$$A_{V(DIFF)} = \frac{3.3\ \text{k}\Omega}{107.84\ \Omega}$$

$$A_{V(DIFF)} = 30.6$$

The value of $A_{V(COM)}$ is calculated with the formula

$$A_{V(COM)} = \frac{R_C}{(2 \times R_E)}$$

$$A_{V(COM)} = \frac{3.3\ \text{k}\Omega}{(2 \times 2.2\ \text{k}\Omega)}$$

$$A_{V(COM)} = \frac{3{,}300\ \Omega}{4{,}400\ \Omega}$$

$$A_{V(COM)} = 0.75$$

OPERATIONAL AMPLIFIERS

9.5 Basic Characteristics

Operational amplifiers are circuits that use differential amplifiers and exhibit characteristics that make them very easy to use in electronic circuits. Some of the characteristics are:

- High gain
- High input impedance
- Low output impedance
- Common-mode rejection
- Ease of design by using external circuit values

The term *operational* in operational amplifiers (op amps) is derived from the fact that they were originally designed to perform mathematical operations such as addition, subtraction, division, multiplication, and even differentiation and integration. Although modern-day digital computers with their speed and accuracy have taken the place of op amps when it comes to mathematical operations or computations, op amps are still used in a multitude of applications today. Some of the present-day uses include signal conditioning, process control, and communications for which they are used as amplifiers, oscillators, filters, and comparators.

9.6 Stages of an Op Amp

Characteristics of an op amp such as high gain, high input impedance, and low output impedance cannot be attained by a single transistor. An op amp is an integrated combination of several stages of amplifiers. Figure 9–13 shows the different stages of an op amp.

- Stage 1 (input stage) is a differential amplifier that provides a high input impedance and high CMRR.

- Stage 2 is another differential amplifier whose inputs are derived from the outputs of the first stage. This stage serves to improve on the differential voltage gain and common-mode rejection provided by the first stage.
- Stage 3 (output stage) is a common-collector (emitter-follower) stage. This stage has very low output impedance. The output is a single terminal and is referred to as a single-ended output.

Note in Figure 9–13 that the op amp does not provide a differential output. The single-ended output shows only one phase with respect to ground. When the input is applied to the noninverting terminal, the output and input are in phase. When the input is applied to the inverting terminal, the output is 180° out of phase with the input. Operational amplifiers are normally represented by a triangle with a plus sign (+) to indicate a noninverting input and a minus sign (−) to indicate an inverting input. Figure 9–14 shows a simplified block diagram of an op amp.

FIGURE 9–13 A simplified representation of the stages of an op-amp.

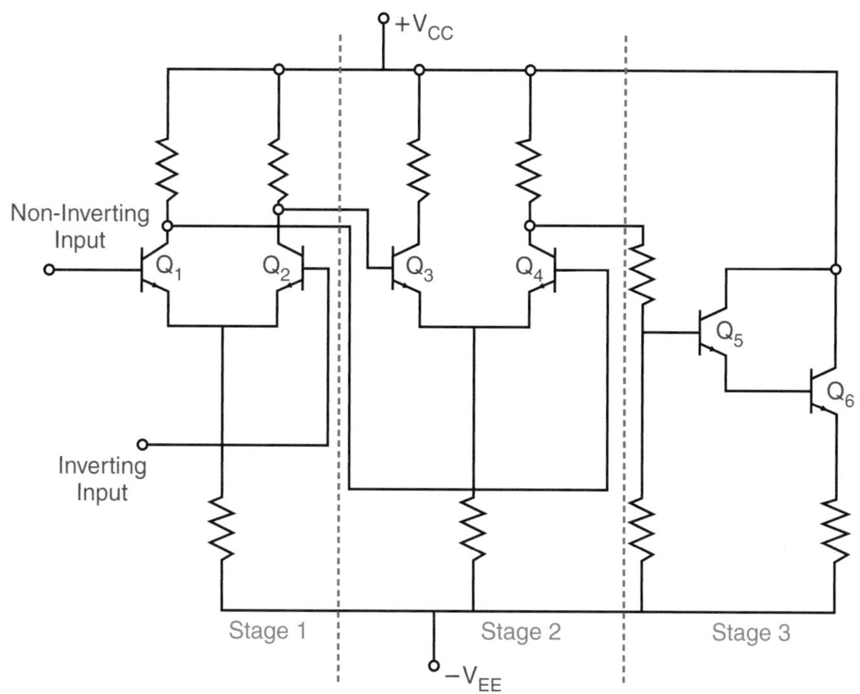

FIGURE 9–14 Simplified block diagram of an op-amp.

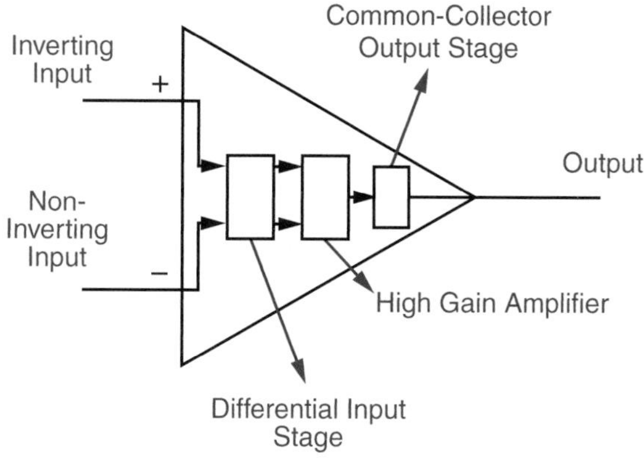

9.7 A Typical Op Amp

Circuitry and Schematic Symbols

Op amp stages are packaged in the form of single or multiple ICs. Figure 9–15a shows a typical small-signal differential op amp installed on a circuit board. Figure 9–15b shows the schematic diagram of an IC op amp. The offset null terminals are used to take care of DC-offset error. As has been already discussed, it is not possible to perfectly match the transistors and resistors of the op amp. This mismatch creates an error in the output and can be corrected using the nulling terminals.

FIGURE 9–15 (a) Typical 8-pin surface mounted op-amp. (b) Schematic diagram of an integrated circuit op-amp.

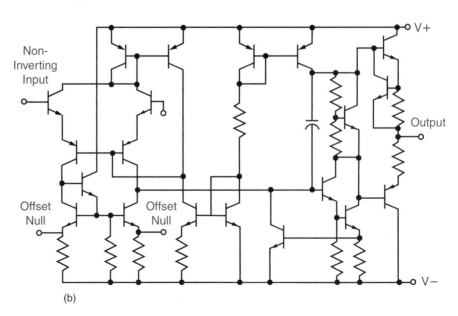

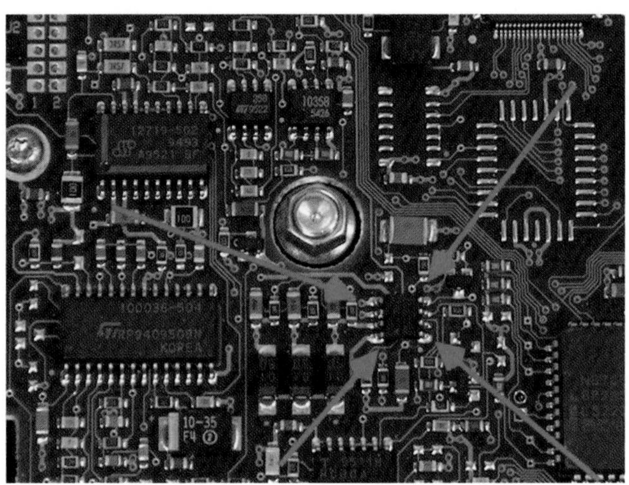

(a)

V+
Non-Inverting Input
Output
Offset Null
Offset Null
V−

(b)

A typical op amp is represented by the symbols shown in Figures 9–16a and b. Figure 9–16a shows the inputs, outputs, and biasing supply connections. Figure 9–16b shows only the inputs and outputs and represents the manner in which the op amp is usually drawn in a circuit.

FIGURE 9–16 Schematic symbols of an op-amp; (a) showing power connections, (b) as usually seen in circuit diagrams.

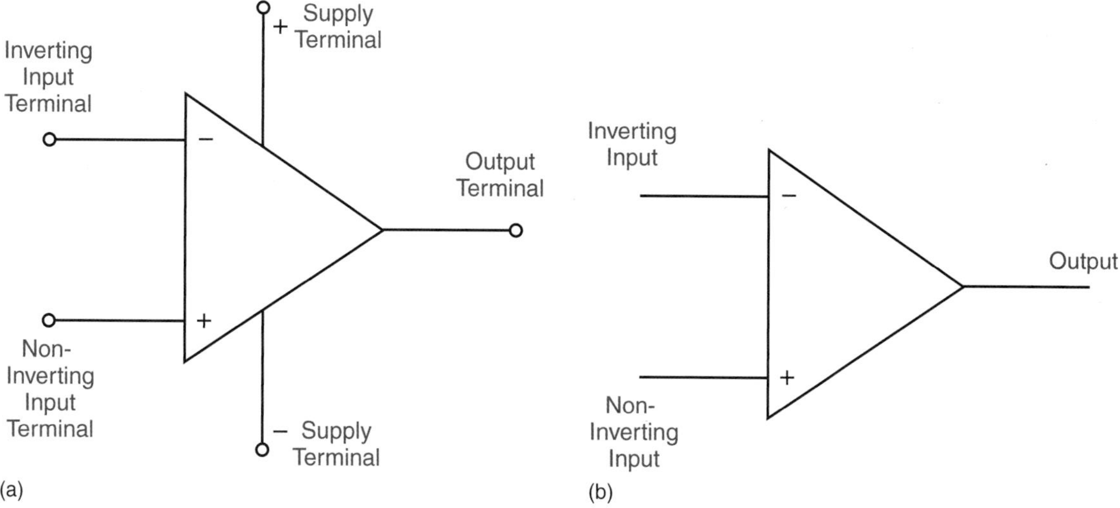

External Nulling Circuitry

Figure 9–17 shows an op-amp nulling circuit. When no input voltage is applied, the output voltage of the op amp should be zero with respect to the ground. However, if an offset exists the output voltage is likely to be different from zero. This can be easily corrected by adjusting the potentiometer setting in the nulling circuit until the output voltage becomes zero.

FIGURE 9–17 An op-amp nulling circuit.

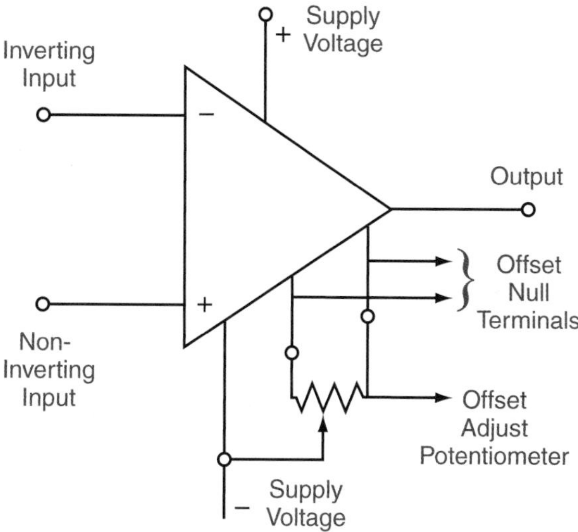

Slew Rate

Slew rate is defined as the maximum rate of change of the output voltage of an op-amp. The output voltage cannot make an instantaneous change. If the slew rate is 0.04 V/mS, it means that the output voltage changes over a range of 40 mV in 1 mS. A high value of slew rate means that the output of the op amp is not able to slew (change) fast enough with the input. This produces slew-rate distortion. Slew-rate distortion is an important factor to consider at high frequencies. Figure 9–18 shows an op amp's response to an instantaneous change in the op amp's input.

FIGURE 9–18 Op-amp's response to an instantaneous change in input.

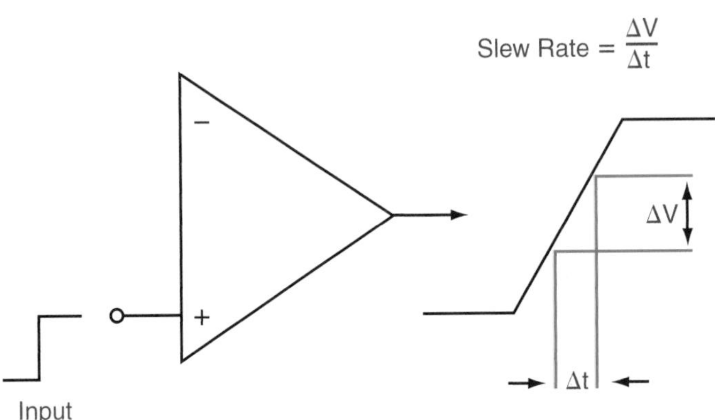

Figure 9–19 shows slew-rate distortion. Note that when a square-wave input signal is applied the output signal becomes triangular because of the op amp's inability to follow the rate of change of input. This deviation of the output waveform from the input constitutes slew-rate distortion.

FIGURE 9–19 Slew rate distortion.

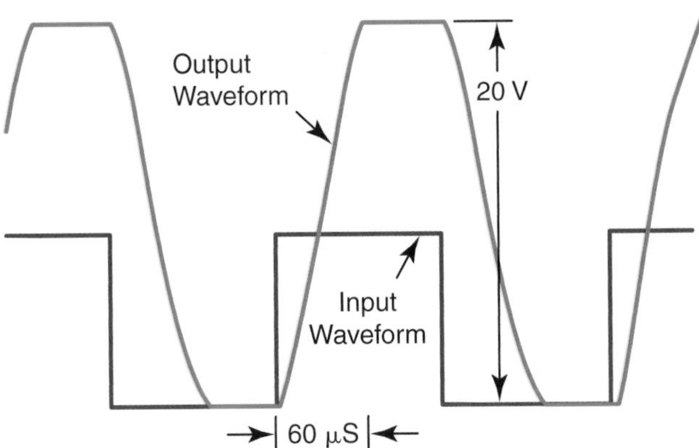

Slew rate also limits the output of the op-amp from producing its full output swing. We can use the following equation to predict the value of the maximum signal frequency that can be applied to an op amp without distortion being produced at the output.

$$f_{max} = \frac{\text{slew rate}}{2\pi V_P}$$

Here:
f_{max} = the maximum frequency that can be applied
V_P = peak output voltage

If an op amp can produce a maximum output swing of 13 V when using a 15 V power supply and the slew rate is $\frac{0.4\text{ V}}{\mu S}$, the maximum operating frequency of the op amp can be determined as follows.

$$f_{max} = \frac{Slew\ Rate}{2\pi \times V_P}$$

$$f_{max} = \frac{\frac{0.4\text{ V}}{\mu S}}{(2 \times 3.14) \times 13\text{ V}}$$

$$f_{max} = \frac{\frac{0.4\text{ V}}{.000001\ S}}{6.28 \times 13\text{ V}}$$

$$f_{max} = \frac{400{,}000}{81.64} = 4.90\text{ kHz}$$

Typical operating values for op amps such as LM 741C include the following:

- Open-loop voltage gain (A_{OL}), which is the value of the voltage gain without any feedback, is very large and typically about 200,000 (106 dB).
- Output impedance is typically small and has a value of 75 Ω.
- Input impedance is typically large and has a value of 2 MΩ.
- CMRR is very large and is 90 dB.
- Offset adjustment range is ±15 mV.
- Small-signal bandwidth is 1 MHz.
- Output voltage swing is ±13 V.
- Slew rate is $\frac{0.5\text{ V}}{\mu S}$.

9.8 Open-Loop Operation of the Op Amp

The open-loop operation of the op amp involves circuits that have no feedback. That is, there is no connection between the output and the input. The op amp's input terminals are called differential input terminals because the output voltage (V_O) depends on the difference in voltage (E_d) between them and the open-loop gain of the amplifier (A_{OL}).

$$V_O = E_d \times A_{OL}$$

Where:
E_d = [voltage at the (+) input] − [voltage at the (−) input]

Figures 9–20a and b show op-amp circuits that operate in the open-loop mode. In this mode, they essentially function as comparators comparing the inputs to the inverting (−) and noninverting (+) terminal and amplify the difference. Keep in mind that A_{OL} is extremely large, on the order of 200,000 or more. Typical operating voltages of the op-amp are ±15 V, which limits the output voltages of the op-amp to $\pm V_{Sat}$ (±13 or ±14 V). Even with an E_d as low as ±65 μV, the output will go to $\pm V_{Sat}$.

$$V_O = \pm 65\ \mu\text{V} \times 200{,}000 = \pm 13\ \text{V} = \pm V_{Sat}$$

FIGURE 9–20 Op-amp input and output characteristics; (a) noninverting input more positive, (b) inverting input made more positive.

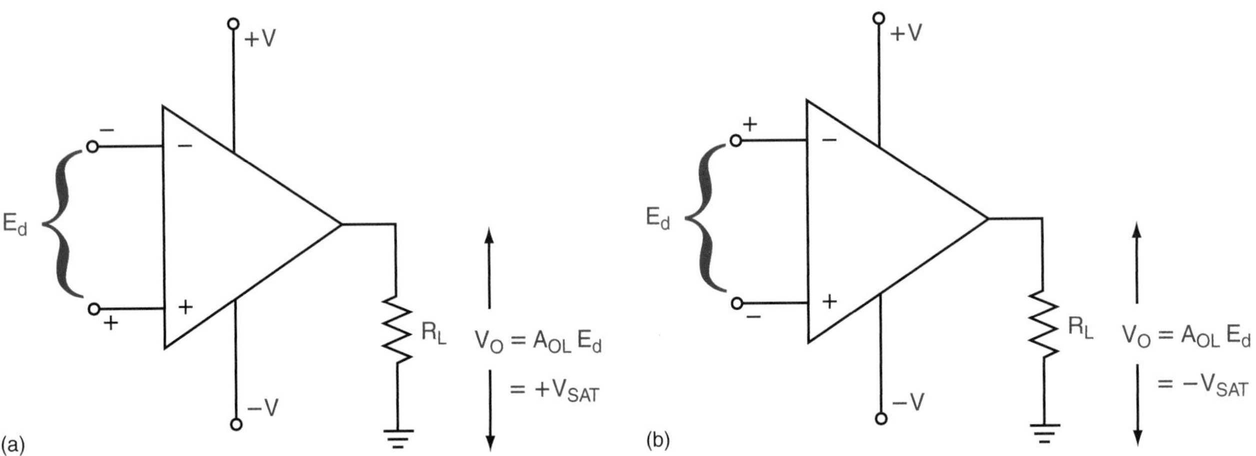

Clearly, this high-gain amplifier forces the output V_O to stay between $+V_{Sat}$ and $-V_{Sat}$ depending on whether E_d is positive or negative. In Figure 9–20a, the positive (+) terminal is made more positive with respect to the negative (−) terminal, and this causes the output V_O to be positive ($+V_{Sat}$). In Figure 9–20b, the positive (+) input is made negative with respect to the negative (−) input, and this causes the output V_O to be negative ($-V_{Sat}$). Table 9–1 outlines data for three examples of open-loop operation.

TABLE 9–1 Example Problem

Voltage at positive (+) input	Voltage at negative (−) input	$E_d = (+V) - (-V)$	$V_O = A_{OL} \times E_d$	Actual V_O
2 V	0.5 V	+1.5 V	300,000 V	$+V_{Sat} = +13$ V
1 V	2 V	−1 V	−200,000 V	$-V_{Sat} = -13$ V
30 mV	0 V	30 mV	6,000 V	$+V_{Sat} = +13$ V

9.9 Closed-Loop Operation of the Op Amp

When a portion of the output signal is fed back to the input, the circuit is called a closed-loop circuit. Op amps are generally operated with feedback in the closed-loop mode.

There are two types of feedback used in op amp circuits.

1. Negative feedback
2. Positive feedback

Negative Feedback Circuits

In negative feedback circuits, a portion of the output is fed back to the inverting negative (−) input of the op-amp. Negative feedback is used to reduce the gain and increase the bandwidth of an amplifier. Amplifier circuits constructed using op amps use negative feedback. The explanations of operation that follow depend on two important points.

1. The voltage between the positive (+) terminal and negative (−) terminal is zero. This is easily proven by looking at Figure 9–13.
   a. Note that the emitters of Q_1 and Q_2 are connected and are therefore at the same potential.
   b. Both Q_1 and Q_2 are forward biased. Consequently, the base voltages for both transistors are given by the formula $V_B = V_E + V_{BE}$ where V_{BE} is the 0.7 V for the base-emitter junction.
2. The input impedance of the op amp is extremely high. (The LM 741C has a 2 MΩ input impedance.) This means that there is negligible current drawn into the op amp's input terminals.

   Keep these two points in mind as you read the following sections.

Inverting Amplifiers

Figure 9–21 shows an inverting amplifier circuit. The resistor R_f connected between the output and negative (−) or inverting terminal provides negative feedback. The input V_i to the amplifier is connected to the inverting terminal via the resistor R_1, and the noninverting terminal is grounded. Because of the first point, that the two terminals are at the same potential, the inverting input (−) is also at ground potential and is known as virtual ground. The output of the amplifier is 180° out of phase with the input. This means that it is inverted with respect to the input.

Derivation of the voltage gain (A_V) is as follows:

The existence of **virtual ground** at the negative (−) input causes the right end of R_1 to be grounded. Therefore, $I_1(I_{R1})$ can be calculated from Ohm's law as:

Virtual ground
A point in a circuit which, although not hardwired to a ground point, maintains a zero-volt potential.

$$I_1 = \frac{V_1}{R_1}$$

The output voltage V_O causes a current I_2 to flow through R_f.

$$I_2 = \frac{V_O}{R_f}$$

There is almost no current flow through the input terminals of the op amp (from the second point previously discussed). Therefore, $I_1 = I_2$, and

$$\frac{V_O}{R_f} = \frac{V_i}{R_1}$$

Rearranging terms yields the closed-loop voltage gain $A_{V(CL)}$:

$$A_{V(CL)} = \frac{V_O}{V_i} = -\frac{R_f}{R_1}$$

Note the simple elegance of the op-amp as an amplifier. The gain of the amplifier is dependent only on the input and feedback resistor values. The negative sign in the gain indicates that there is a phase inversion present. For the circuit shown in Figure 9–21, the voltage gain is calculated as:

$$A_{V(CL)} = -\frac{R_f}{R_1}$$

$$A_{V(CL)} = \frac{-100\ \text{k}\Omega}{10\ \text{k}\Omega}$$

$$A_{V(CL)} = -10$$

A voltage gain of -10 suggests that if an input of 1 V is applied to the circuit shown in Figure 9–21 the output voltage is -10 V. On the other hand, if an input voltage of -1 V is applied the output voltage is $+10$ V.

FIGURE 9–21 Inverting amplifier.

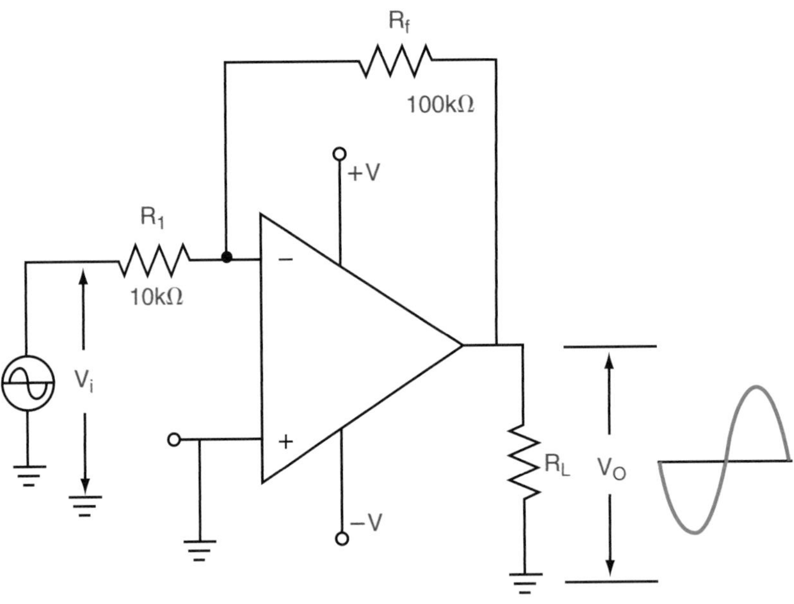

Noninverting Amplifiers

Figure 9–22 illustrates a noninverting amplifier. In this amplifier, the input (V_i) is applied to the positive (+) input. Once again, note the presence of negative feedback in the form of the resistor R_f connected between the output and input. In this amplifier, the output voltage is in phase with the input.

The voltage gain is slightly different in this connection. Based on the first point, that the two input terminals are at the same potential, the inverting terminal is at a potential V_i. It can be seen that V_O is divided between the resistors R_f and R_L. The input voltage is calculated using the voltage divider rule and given by:

$$V_i = V_O\left(\frac{R_1}{R_1 + R_f}\right)$$

Rearranging terms gives the voltage gain $A_{V(CL)}$

$$\frac{V_O}{V_i} = A_{V(CL)} = \frac{R_1 + R_f}{R_1}$$

The gain of the circuit in Figure 9–22 is calculated as

$$A_{V(CL)} = \frac{R_1 + R_f}{R_1}$$

$$A_{V(CL)} = \frac{10\ \text{k}\Omega + 100\ \text{k}\Omega}{10\ \text{k}\Omega}$$

$$A_{V(CL)} = 11$$

This formula can be reduced to

$$A_{V(CL)} = (R_f \div R_1) + 1$$

$$A_{V(CL)} = \frac{100\ \text{k}\Omega}{10\ \text{k}\Omega} + 1$$

$$A_{V(CL)} = 10 + 1$$

$$A_{V(CL)} = 11$$

A voltage gain of +11 suggests that when an input voltage of 1 V is applied the output voltage is +11 V and that when the input voltage applied is −1 V the output voltage is −11 V. If we know an op-amp's inverting gain, its noninverting voltage gain is simply

$$A_{V(NON\text{-}INVERTING)} = \frac{R_f}{R_1} + 1$$

FIGURE 9–22 Noninverting amplifier.

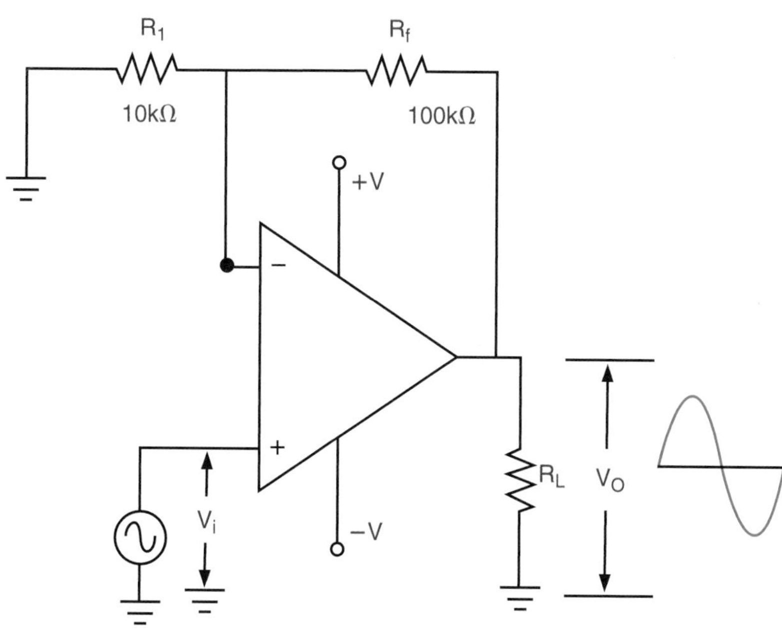

Voltage Followers

This is a special case of a noninverting amplifier, with $A_{V(CL)} = 1$. Figure 9–23 illustrates a voltage follower. Negative feedback is still present in the circuit, but here the resistors R_f and R_1 are absent. Because the two input terminals are at the same potential, the voltage V_i applied to the noninverting (+) terminal also appears at the inverting (−) terminal. The output is connected to the inverting (−) input by a wire, and hence the output voltage V_O follows the input voltage V_i.

FIGURE 9–23 Voltage follower.

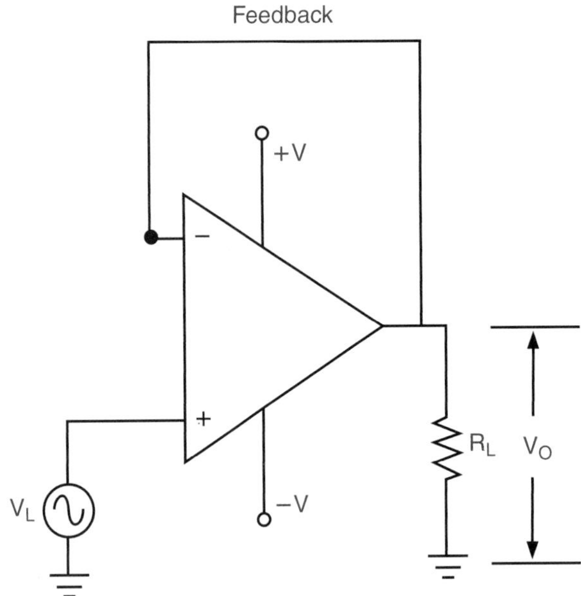

The voltage gain of this amplifier is 1. You may wonder what the use of this amplifier with a gain of 1 is. Although the amplifier has no voltage gain, it serves as a buffer that offers high input impedance and low output impedance and can be used for the purposes of impedance matching. The 741 op-amp has a noninverting input impedance of 2 megohms and an output impedance of 75 ohms. Therefore, for a voltage gain of 1 the power gain at low frequencies would be:

$$A_{(Power)} = \frac{P_{(Out)}}{P_{(In)}}$$

$$A_{(Power)} = \frac{\dfrac{V^2}{75\ \Omega}}{\dfrac{V^2}{2{,}000{,}000\ \Omega}}$$

$$A_{(Power)} = \frac{V^2}{75\ \Omega} \times \frac{2{,}000{,}000\ \Omega}{V^2}$$

$$A_{(Power)} = \frac{2{,}000{,}000\ \Omega}{75\ \Omega}$$

$$A_{(Power)} = 26{,}667$$

Be aware, however, that maximum amplifier gain declines linearly as frequency increases. In the case of the 741 op-amp, maximum power gain declines to 1 at 1 MHz.

Summing Amplifiers

An inverting amplifier connected with two or more inputs is a summing amplifier or mixer. Figure 9–24 shows a summing amplifier with two inputs, V_1 and V_2, applied to the inverting input. The output voltage is the inverted sum of the input voltages. These summing amplifiers are used to add or mix AC or DC voltages. The output voltage V_O for the summing amplifier of Figure 9–24 is given by the following equation.

$$V_O = -\left(\frac{R_f}{R_1} V_1\right) - \left(\frac{R_f}{R_2} V_2\right)$$

If V_1 is 3 V; V_2 is 5 V; and the resistors R_1, R_2, and R_f are each of 10 kΩ, the output voltage for the summing amplifier shown in Figure 9–24 is given by

$$V_O = -\left(\frac{10\ \text{k}\Omega}{10\ \text{k}\Omega} \times 3\ \text{V}\right) - \left(\frac{10\ \text{k}\Omega}{10\ \text{k}\Omega} \times 5\ \text{V}\right)$$

$$V_O = -(1 \times 3\ \text{V}) - (1 \times 5\ \text{V})$$

$$V_O = -3\ \text{V} - 5\ \text{V}$$

$$V_O = -8\ \text{V}$$

The inputs V_1 and V_2 can be scaled by any factor by changing the values of R_1 and R_2. If R_1 is made 5 kΩ, while R_2 and R_f remain the same (10 kΩ), the output voltage V_O is found by

$$V_O = -\left(\frac{R_f}{R_1} V_1\right) - \left(\frac{R_f}{R_2} V_2\right)$$

$$V_O = -\left(\frac{10\ \text{k}\Omega}{5\ \text{k}\Omega} \times 3\ \text{V}\right) - \left(\frac{10\ \text{k}\Omega}{10\ \text{k}\Omega} \times 5\ \text{V}\right)$$

$$V_O = -(2 \times 3\ \text{V}) - (1 \times 5\ \text{V})$$

$$V_O = -6\ \text{V} - 5\ \text{V}$$

$$V_O = -11\ \text{V}$$

The number of inputs of the summing amplifier/mixer can be extended. By scaling the gains of the different inputs, the inputs can be mixed in different proportions. A summing amplifier can thus be used in an audio mixer to mix the inputs from different microphones. The virtual ground on the inverting input provides isolation between the individual microphones, and there is no interaction between inputs. By scaling the gains of different inputs, a very low voice can be made more audible than a loud guitar. Figure 9–25 shows an audio mixer.

FIGURE 9–24 Summing amplifier.

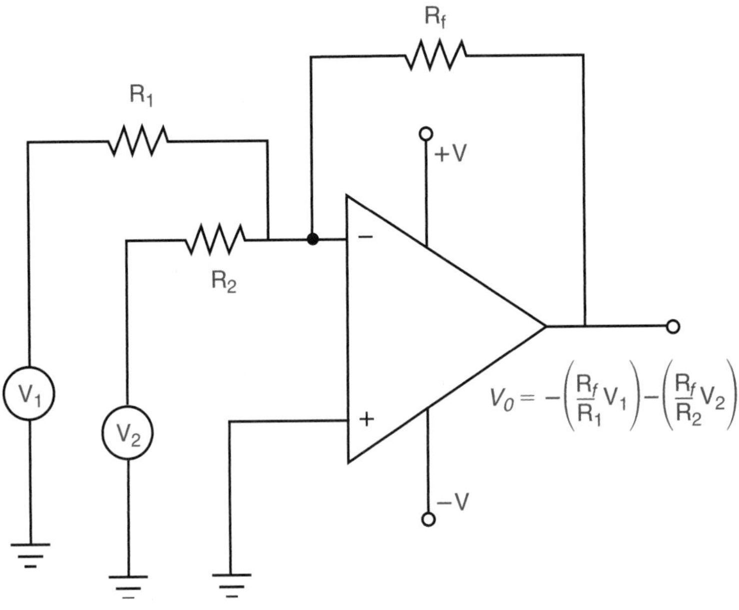

FIGURE 9–25 An audio mixer.

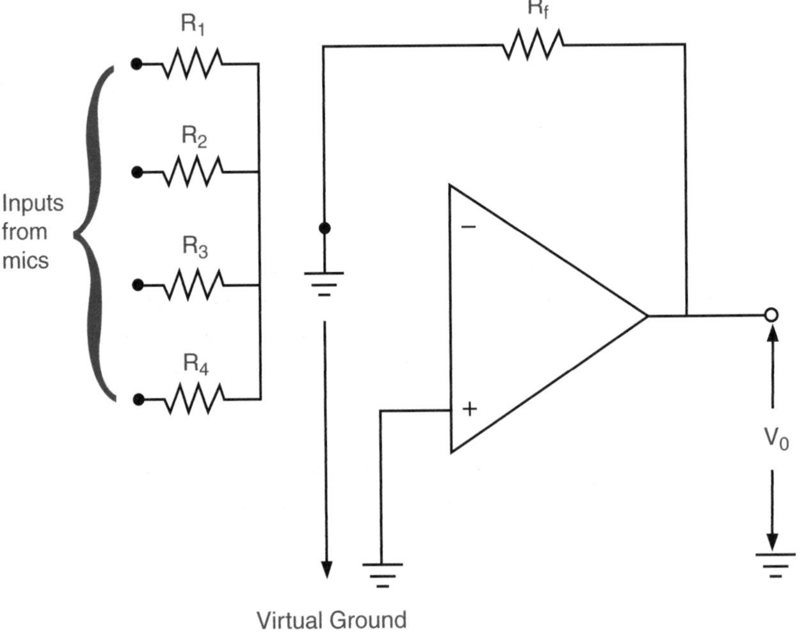

Positive Feedback Circuits

In positive feedback circuits, signal from the output of the op-amp is applied to the noninverting (+) input of the op-amp. Figure 9–26 shows an op-amp circuit with positive feedback. Positive feedback increases the gain, and thus the output of the op-amp circuit changes between $+V_{Sat}$ and $-V_{Sat}$. Positive feedback serves to reinforce comparator action, and makes the circuit immune to noise voltages.

FIGURE 9–26 A Schmitt trigger.

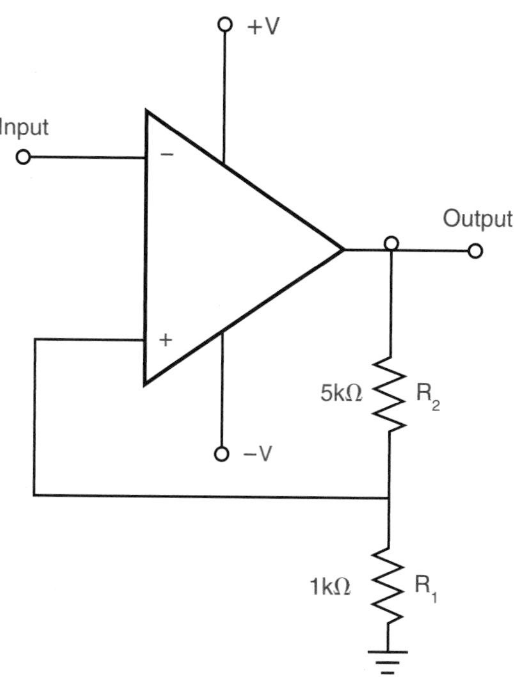

The op-amp circuit of Figure 9–26 is called a Schmitt trigger. The output of this circuit changes between $+V_{Sat}$ and $-V_{Sat}$. Resistors R_1 and R_2 divide the output voltage and apply a portion to the noninverting input. When the output is at its positive maximum ($+V_{Sat}$), the voltage fed back to the negative (−) input is V_{UTP} (or upper threshold point) and is given by applying the voltage divider rule:

$$V_{UTP} = V_{Sat} \times \frac{R_1}{R_1 + R_2}$$

When the output is at its negative maximum ($-V_{Sat}$), the voltage fed back to the positive (+) input is called V_{LTP} (or lower threshold point).

$$V_{LTP} = -V_{Sat} \times \frac{R_1}{R_1 + R_2}$$

Example

Suppose that the Schmitt trigger of Figure 9–26 operates at $\pm V_{Sat} = \pm 15$ V. R_1 and R_2 are respectively 1 kΩ and 5 kΩ. The values of V_{UTP} and V_{LTP} can be calculated as:

$$V_{UTP} = V_{SAT} \times \frac{R_1}{R_1 + R_2}$$

$$V_{UTP} = +15 \text{ V} \times \frac{1 \text{ k}\Omega}{1 \text{ k}\Omega + 5 \text{ k}\Omega}$$

$$V_{UTP} = +15 \text{ V} \times \frac{1 \text{ k}\Omega}{6 \text{ k}\Omega}$$

$$V_{UTP} = +15 \text{ V} \times .167$$

$$V_{UTP} = +2.5 \text{ V}$$

$$V_{LTP} = -V_{SAT} \times \frac{R_1}{R_1 + R_2}$$

$$V_{LTP} = -15 \text{ V} \times \frac{1 \text{ k}\Omega}{1 \text{ k}\Omega + 5 \text{ k}\Omega}$$

$$V_{LTP} = -15 \text{ V} \times \frac{1 \text{ k}\Omega}{6 \text{ k}\Omega}$$

$$V_{LTP} = -15 \text{ V} \times .167$$

$$V_{LTP} = -2.5 \text{ V}$$

Figure 9–27 shows the Schmitt trigger in operation. The input to the Schmitt trigger is a sinusoidal signal. First suppose that the output is at $+V_{Sat}$. The voltage applied to the noninverting terminal is V_{UTP}. As the input goes positive, it crosses the upper threshold point at 2.5 V. This causes the inverting input to be more positive than the noninverting input, and thus the output switches to $-V_{Sat}$. The voltage fed to the noninverting terminal is V_{LTP}. As the input voltage crosses the lower threshold (−2.5 V), the output switches to $+V_{Sat}$.

The advantage of positive feedback is that it makes the circuit immune to noise signals. In normal comparators, there is a chance of false triggering due to noise. However, when the noisy signal passes through the Schmitt trigger's upper threshold point and lower threshold point false triggering does not occur. Figure 9–28 compares the operation of the Schmitt trigger to that of regular comparators.

FIGURE 9–27 Operation of Schmitt trigger.

FIGURE 9–28 Elimination of false triggering using Schmitt trigger.

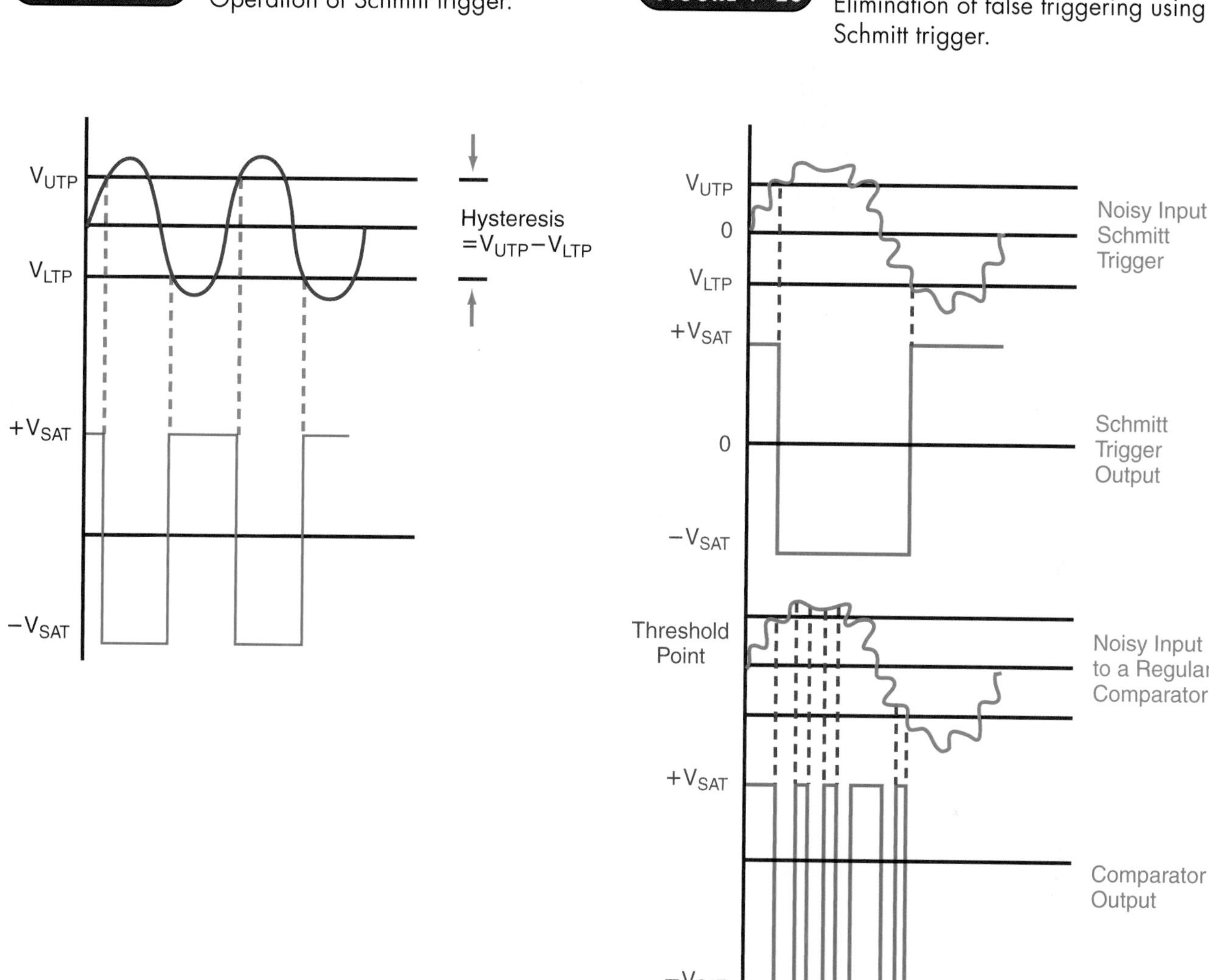

Practical Op-Amp Circuits

Figure 9–29 shows a working circuit utilizing a small op-amp such as the 741, an N-channel E-mode MOSFET, a thermistor, and six resistors. Both inverting and noninverting inputs are utilized in this circuit. The purpose of the circuit is to operate a small temperature-controlled oven for small-circuit testing. The sensor is a negative-temperature coefficient thermistor placed inside the oven along with the heating element (R_6). When R_6 is cold, its resistance is high. The low voltage applied to the inverting input drives the op amp output voltage higher.

FIGURE 9–29 Small temperature-controlled oven.

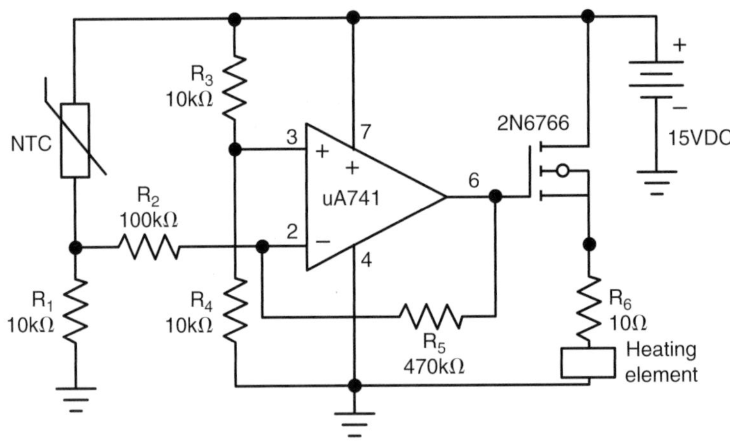

The MOSFET, which is operating as a source follower current, will rise heating R_6 (a 10 ohm 25 watt resistor inside the oven). As oven temperature rises, the NTC thermistor resistance decreases, op-amp output voltage goes down, MOSFET current decreases, R_6 cools, and oven temperature begins to decline. R_5, the feedback resistor, can be adjusted to set the op-amp gain so that the oven heats fast enough. However, oven temperature does not overshoot. The temperature set point can be adjusted by varying the value of R_4.

Figure 9–30 shows a schematic diagram of a 0.5 watt AF power amplifier using a 386 power op-amp. This simple single-ended amplifier can be used as drawn, as a portable guitar amplifier, as a baby monitor, or as anything your imagination can conjure. V_{CC} as low as 6 V can be used, but will reduce the maximum power output. When constructing op-amp circuits, be sure to keep ground wires short because low-frequency oscillation "motorboating" may occur due to high currents and high gain.

FIGURE 9–30 General purpose op-amp power amplifier.

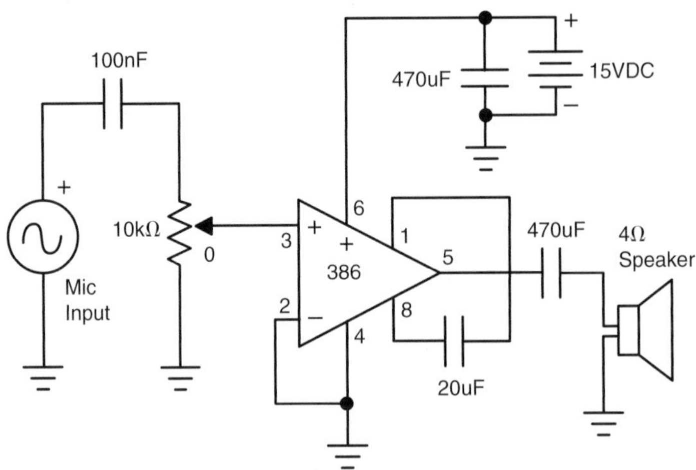

9.10 Active Filters

Electronic filters are frequency-selective circuits. Those that utilize amplifiers are called **active filters.**

Active filter
A frequency-selective electronic filter utilizing amplification, as opposed to passive frequency-selective filters.

Filters constructed using only resistors, inductors, and capacitors are called passive filters. Active filters may be easily constructed from discrete transistors (or vacuum tubes). However, today's technology enables us to use IC operational amplifiers at the heart of active filters to great advantage. Table 9–2 outlines active and passive filter advantages.

TABLE 9–2 Active Versus Passive Filters

Filter Type	Advantage
Passive	No power supply required: Passive filters use no active devices. Power dissipation within the passive filter is limited to losses of the filtered signal.
	Low component count: Passive filters utilize as few as two components. Although there is no maximum component count, accumulated signal losses place a practical limit on the number of R, L, and C components.
Active	Excellent isolation: Input impedance can be in excess of 1 megohm, whereas output impedance can easily be less than 100 ohms.
	Low cost: A sophisticated active filter consists of only resistors, capacitors, and one or more op-amps. Therefore, component cost can be as little as a few cents.
	Adjustable gain: Op-amp gain can be easily controlled without affecting frequency response.
	Small footprint: Op-amps and small resistors/caps can replace very large passive filters.
	Tunability: Programmable or manually adjustable frequency response over a range of 1000:1 is possible, far outstripping the capability of passive filters.
	Little shielding required: Small overall size and absence of inductors make active filters largely immune to induced noise.
	Section cascadeability: A variety of filters may be cascaded without the prohibitive losses of passive devices.
	Light weight: Absence of inductors and large capacitors dramatically reduces weight of filters.

Active filters constructed using op-amps employ both positive and negative feedback. These filters are constructed using resistors and capacitors in the input and feedback paths. The capacitor in the feedback path has a lagging effect on the current and simulates an inductor. Active filters can be classified as follows:

- Active low-pass filter
- Active high-pass filter
- Active band-pass filter
- Active band-stop (or band-reject) filter

Cutoff frequency is defined as the point at which the voltage gain drops to 70.7% ($\frac{1}{2}\sqrt{2}$) of the maximum value. This is generally referred to as the **half-power point, critical point,** or the **filter cutoff frequency.**

Cutoff frequency
The frequency at which output power is half of the peak output power. This point is also called: Half-power point, Critical point, and Filter cutoff frequency.

FIGURE 9–31 Low-pass active filters: (a) first-order filter, (b) second-order filter, (c) third-order filter.

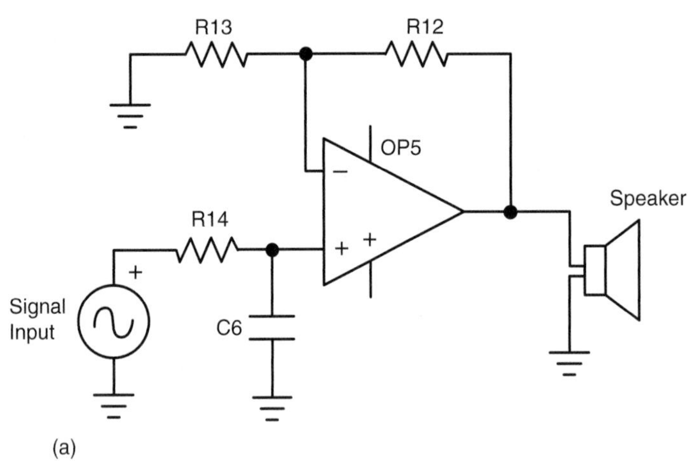

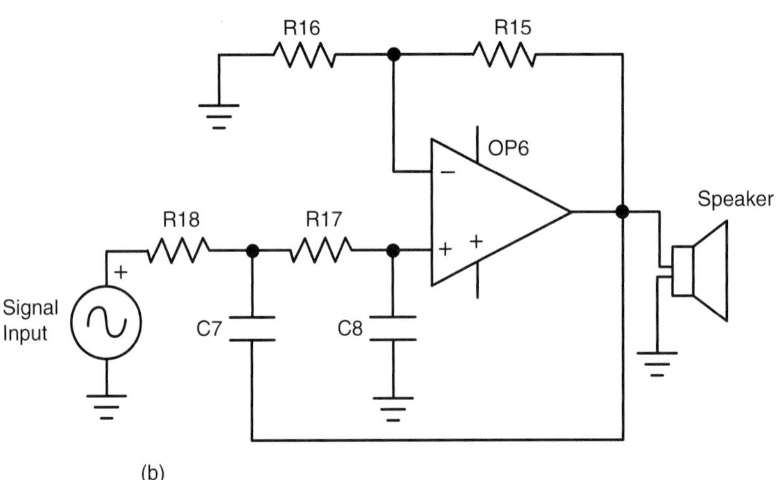

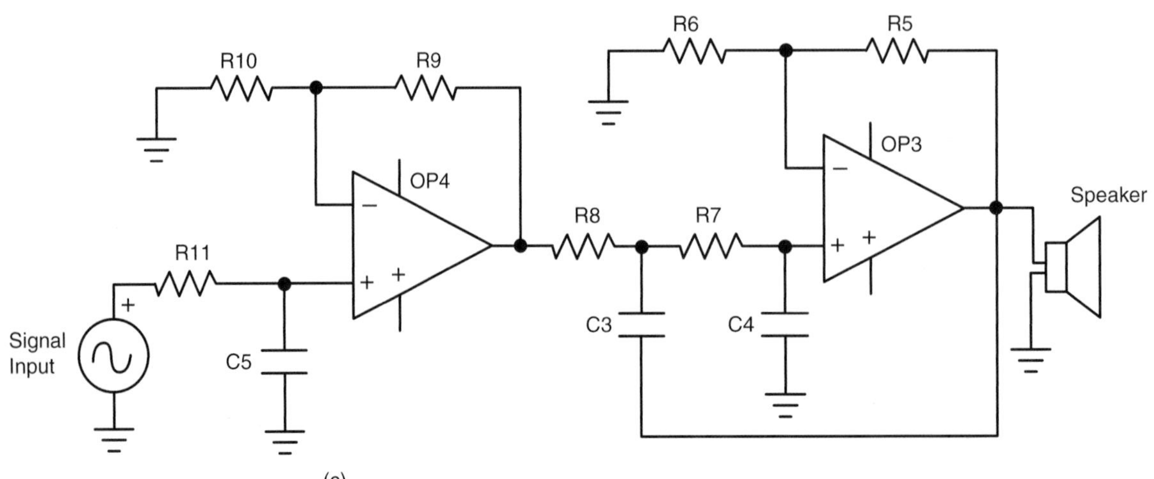

There are two fundamental forms of single-amplifier active filter: the first-order filter requiring one RC network (Figure 9–31a), and the second-order filter requiring two RC networks (Figure 9–31b). Figure 9–31c shows a first-order low-pass filter cascaded with a second-order low-pass filter, to produce a third-order low-pass filter. Any combination of first and second order filters may be cascaded. For example, a sixth order low-pass, or high-pass filter can be built using three second-order filters.

Filters may also be described by the number of poles employed in their construction. In the case of low or high-pass filters, a single-pole, and a first-order filter are the same thing. Only one capacitor is required in the full filter circuit. A fifth-order low-pass filter is the same thing as a five-pole filter. There are five capacitors in the full filter circuit. However, bandpass, and bandstop filters are different. These circuits require a minimum of two capacitors per opamp. Cascading two bandpass, or bandstop filter sections would require four capacitors. Therefore, these would be fourth-order, two-pole filters.

The 0.5 power point, as described previously, is −6 dB(V) below the peak filter output level (often called the critical point). A single-pole filter typically produces a −6 dB per **octave** rolloff. For the first-order (one-pole) low-pass filter of Figure 9–31a, if maximum output extends from 0 Hz to 1.5 kHz one could anticipate that the output will drop −6 dB (the 0.5 power point) at 3.0 kHz, −12 dB at 6 kHz, −18 dB at 12 kHz, and so on. A second-order (two-pole) low-pass filter produces a steeper ramp of −12 dB per octave. Theoretically, filter responses will correspond with Table 9–3.

Octave
An octave is a range of frequencies in which the highest frequency in the range is double the lowest frequency in the range. For example, 3,000 Hertz to 6,000 Hertz is one octave.

TABLE 9–3 Ideal Filter Response

Filter Order Number	Low-pass or High-pass (dB)	Band-pass or Band-stop (dB)
1	−6	n/a
2	−12	6
3	−18	n/a
4	−24	12
5	−30	n/a
6	−36	18

Active Low-Pass Filter

An active low-pass filter passes low frequencies and attenuates (blocks) high frequencies. Figure 9–32 shows the characteristics of an active low-pass filter. It indicates that gain is maximum at lower frequencies, and that as frequency increases, gain starts dropping. Active low-pass filters can be practical at frequencies as low as approximately 0.01 Hz. The physical size and precision of capacitors available today dictate this limit.

Figure 9–33 illustrates an op-amp circuit that functions as a low-pass filter. Note the presence of negative feedback in the form of resistor R_f connected between the output and the inverting input terminal. The function of the negative feedback is to limit the gain of the circuit, and thus simulate the Q of the filter. The output signal is fed back to the noninverting input through C_1, forming a low-pass L-shaped filter with R_2. R_3 and C_2 form the second L-shaped low-pass filter, making this a second-order filter circuit.

FIGURE 9–32

Active low-pass filter characteristics.

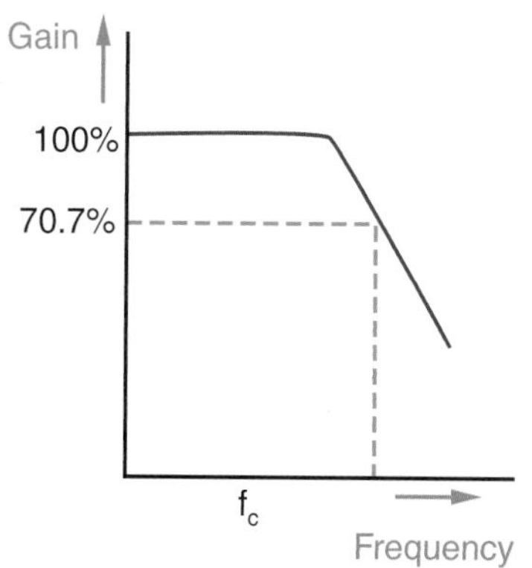

FIGURE 9–33 Active low-pass filter.

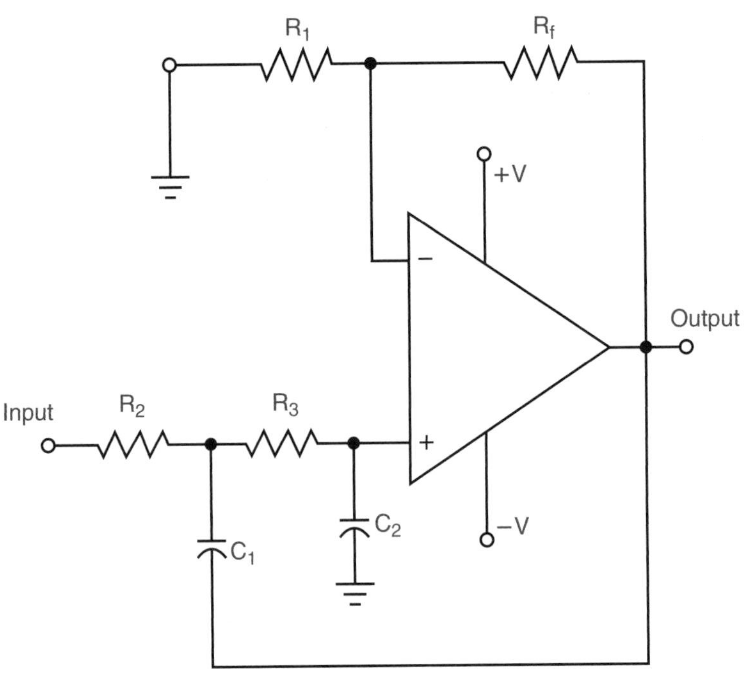

Figure 9–34 shows a practical circuit of a two-pole low-pass filter designed for input impedance of 1 kΩ, with 1 kHz cutoff (0.5 power point). To change input impedance to 10 kΩ without changing the frequency response, multiply all resistor values by 10. This will result in a tenfold *decrease* in cutoff frequency.

FIGURE 9–34 Practical low-pass active filter.

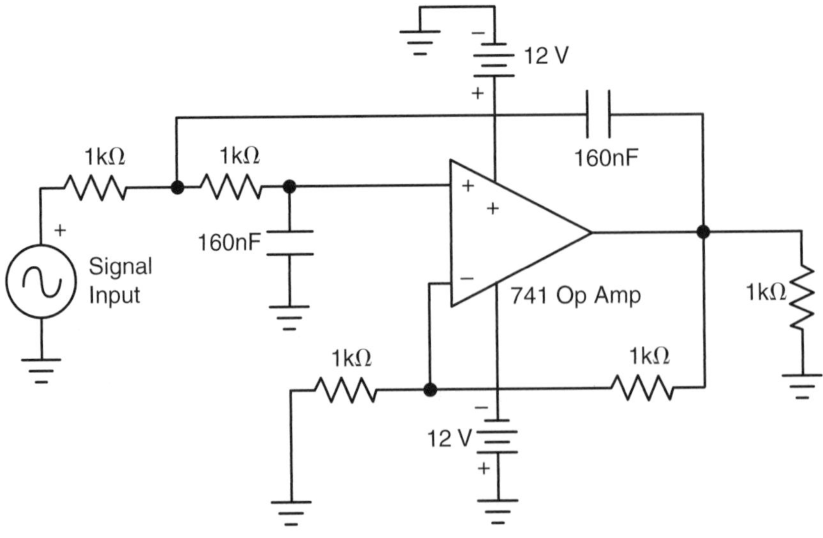

Because $f = \frac{1}{RC}$ a corresponding tenfold decrease in all capacitor values must be made to bring the cutoff back to 1 kHz. By using this linear relationship, you can design a filter for any frequency range within the useful bandwidth of the op-amp having any Q and any input impedance you wish.

One application of an active low-pass filter is as a "noise snooper." A sixth-order LP filter can be used to help locate vibration sources in a mechanical system. Connect a microphone to the filter input, and headphones to the LP filter output. Tune out frequencies above the vibration noise frequency, and probe with the microphone to find the loudest point.

Active High-Pass Filter

An active high-pass filter attenuates low frequencies and passes or amplifies high frequencies. Figure 9–35 shows the characteristics of an active high-pass filter. The graph indicates that the gain is low at lower frequencies and starts increasing as the frequency increases. The upper limit of an active high-pass filter is a function of the bandwidth of the op-amp and the precision of exceedingly small-value capacitors.

FIGURE 9–35 Active high-pass filter characteristics.

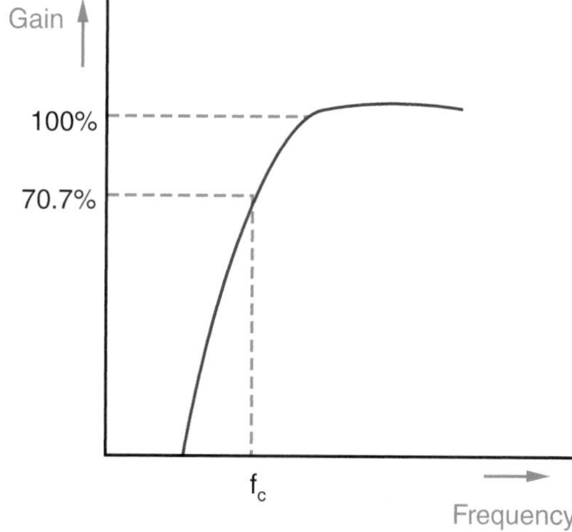

Figure 9–36 illustrates a practical two-pole high-pass filter. The same practical design rules apply to high-pass component value calculations as described for low-pass active filters. Active filters are extensively used in audio amplifier tone control circuits and equalizers. High-pass filters can easily be used to improve or customize frequency response of audio amplifier and speaker systems. An AF system with rolloff at the high end can employ a high-pass active filter to boost gain at high audio frequencies. This technique is especially useful for compensating for poor high-frequency speaker performance.

FIGURE 9–36 One kHz, 1 kΩ high-pass active filter.

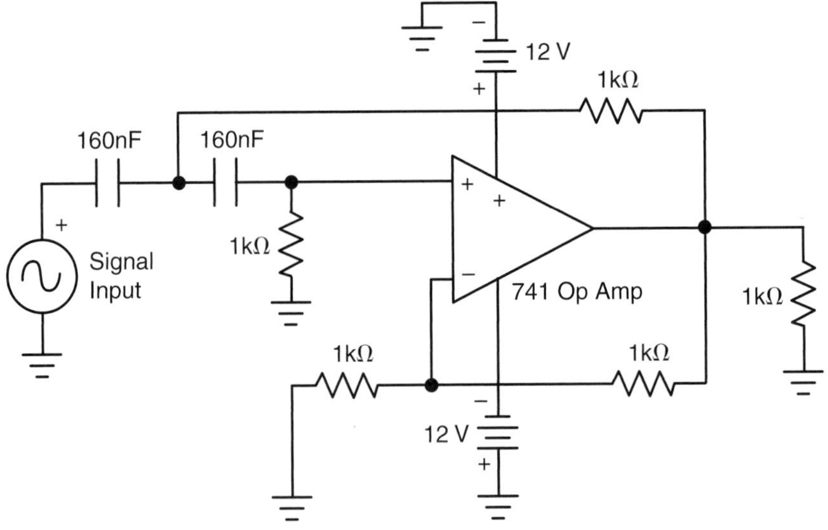

Active Band-Pass Filter

An active band-pass filter passes a band of frequencies while attenuating the amplitude of signals below and above that band. Figure 9–37 shows the characteristics of a band-pass filter. The graph indicates that the gain is high at the band of frequencies between f_{C2} and f_{C1}. At frequencies less than f_{C1} and greater than f_{C2}, the gain is significantly lower.

FIGURE 9–37 Active band-pass filter characteristics.

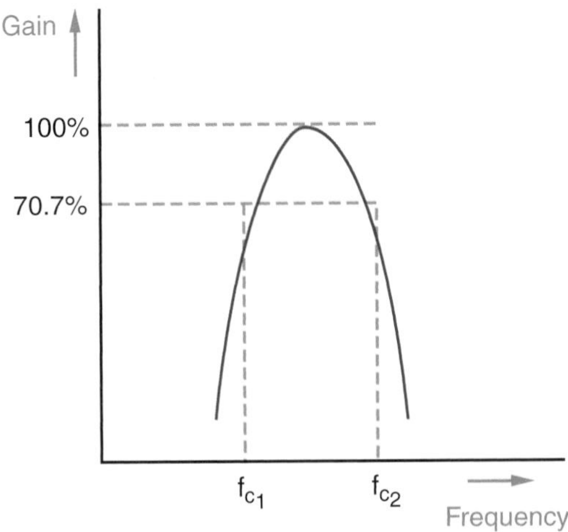

Figure 9–38 illustrates a practical op-amp circuit that functions as a band-pass filter. There are many applications for the active band-pass filter. Some low-cost broadcast band receivers use active band-pass amplifiers rather than tuned transformers to assist in selecting radio stations to avoid interference. These active filters operate with band-pass of approximately 100 kHz.

FIGURE 9–38 One kHz, 1 kΩ band-pass active filter.

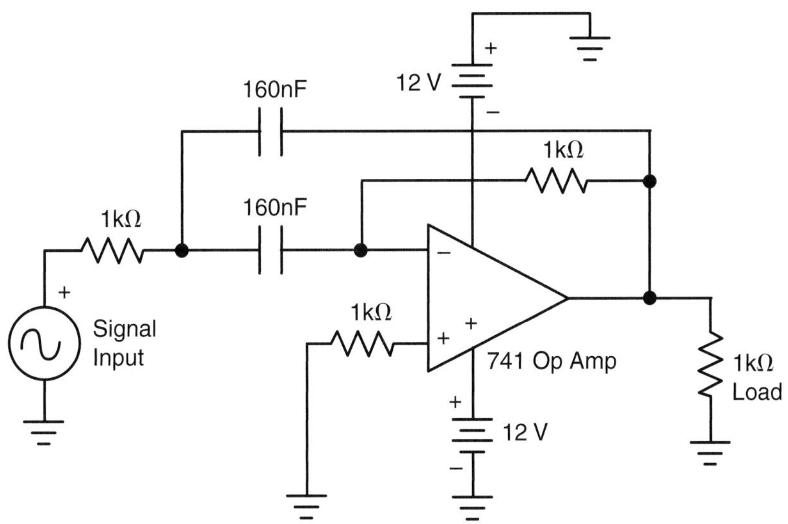

1 kHz, 1 kΩ Band-Pass Active Filter

FieldNote!

The chief engineer on a electrical construction project was in a hurry.

The 300 VDC pump motor was not running when commanded. 208V/3PH/60Hz was applied to the motor drive. The drive was supposed to rectify the 208 V and turn the pump motor via additional control circuitry and thyristers. The engineer unpacked his oscilloscope, opened the proper disconnect switch, and connected scope common to the rectifier output common on the motor.

He plugged the scope into a nearby 120 VAC outlet, and then threw on the disconnect. There was a flash/pop, and smoke curled from the oscilloscope. The engineer had been in too big a hurry to study the circuit and had not realized the scope probe common was tied to the scope chassis ground. The DC rectifier ground is not necessarily the same electrical potential as chassis ground!

Active Band-Stop Filter

An active band-stop filter stops or attenuates a range of frequencies without attenuating frequencies above or below that band. The band-stop filter is also called a notch (or reject) filter. Figure 9–39 shows the characteristics of an active band-stop filter. The graph shows that the gain is low at the band of frequencies between f_{C2} and f_{C1}. At frequencies less than f_{C1} and frequencies greater than f_{C2}, the gain is significantly higher.

FIGURE 9–39 Active band-stop filter characteristics.

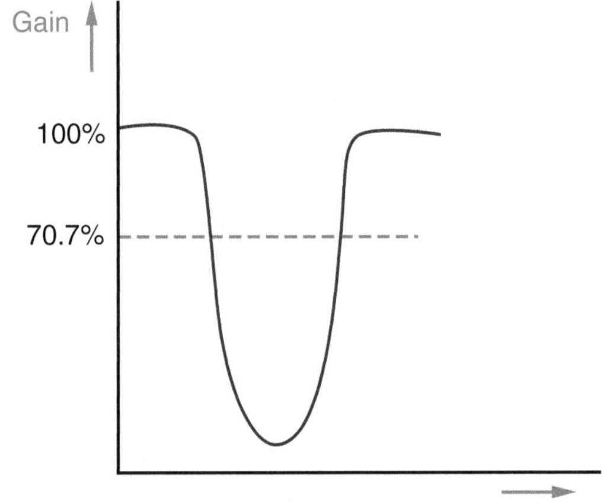

A band-stop filter is constructed as the sum of low-pass and high-pass filters. Like the band-pass filter, a band-stop filter requires a minimum second-order single-pole filter to achieve −6 dB per octave. Figure 9–40 shows a practical band-stop filter. In this case, a fourth-order two-pole filter is needed to achieve −12 dB of attenuation at the 60 Hz notch frequency. This circuit may be used in an audio amplifier to limit 60 Hz line frequency "hum" (perhaps induced in a PA amplifier microphone cable) from reaching the speaker output.

FIGURE 9–40 Sixty Hz two-pole notch active filter.

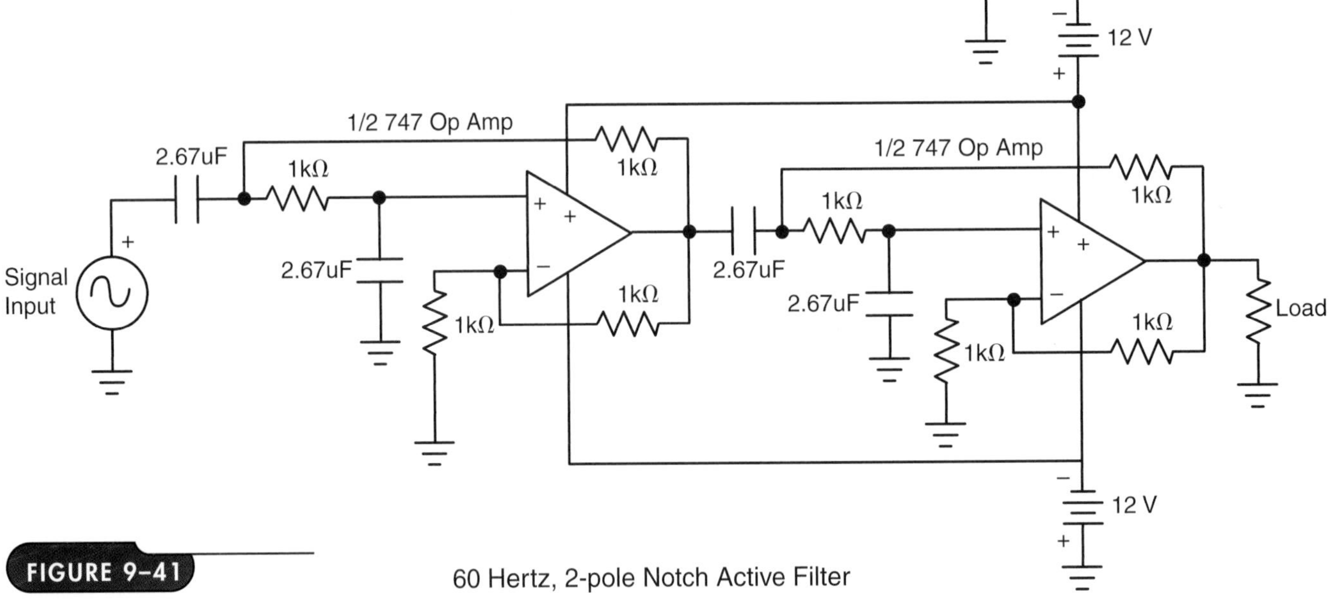

FIGURE 9–41 Grounds.

Common Ground Symbols

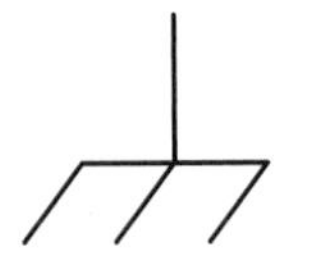

Chassis Ground

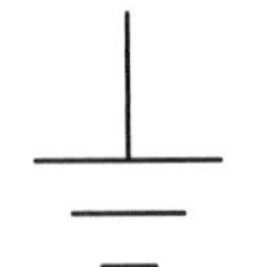

Circuit Ground

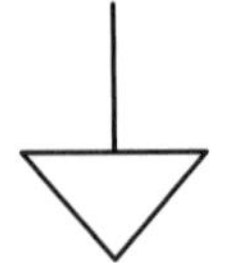

Alternative Circuit Ground

9.11 Grounds

Electricians, electronics technicians, instrumentation mechanics, and engineers absolutely must have a good grasp of grounding. Human life, safety, and protection of equipment rely on our understanding of grounds. Grounds are not necessarily all connected to the same reference point. It is common for very high resistance to exist between or among various circuits within a piece of equipment. As a result, voltages and the potential for damaging current exist when we make grounding errors.

Figure 9–41 shows several commonly used ground symbols. It is not unusual to find unfamiliar ground symbols on equipment schematic drawings. Usually, however, a description can be found somewhere on the drawing set. If a chassis contains digital logic, small-signal operational amplifiers, 24 VDC control circuits, 25,000 VDC HV, 24 VAC or DC relay circuits, and 120 VAC circuitry, there is a strong likelihood the commons for these circuits will be intentionally isolated from one another.

Tying unintended grounds together can result in unpredictable equipment behavior, induce unwanted noise into circuits, and risk application of excessive voltages to sensitive circuits. In addition, voltage and waveform measurements may make no sense at all to the technician who chooses the wrong ground point for his measurement.

SUMMARY

Differential amplifiers can be operated or fed in three different ways.

- *Single-ended mode:* The input signal is fed to only one of the transistor base circuits. In this mode, the differential amplifier works very much like a single-transistor amplifier. The amplifier may be either inverting or non-inverting, depending on which of the two transistors is driven.
- *Differential mode:* In this mode, the input is applied between the two bases so that when the signal sends one base positive it drives the other base negative. The gain in this condition is quite high and is a function of the difference between the two signals.
- *Common mode:* When the same signal is applied to both bases, it is canceled in the output. This is because one of the transistors creates a negative signal and the other creates a positive. Thus, the two cancel in the output circuit.

Differential amplifiers are used whenever high amplification, input isolation from ground, and removal of common-mode noise are desired. Differential amplifiers are also used as the first two stages of an operational amplifier. Operational amplifiers got their name from their early use in analog computers for performing mathematical operations such as addition, subtraction, and so on. In this application, they were often employed in analog computers.

With the advent and improvement of digital computers, analog computers have all but disappeared. Op-amps are now used in a wide variety of applications, including linear high-gain amplifiers, integrators for high-power walk-in circuits, trigger circuits, summing amplifiers, active filters, and many other such applications.

REVIEW QUESTIONS

1. The differential amplifier uses a positive and negative voltage supply, with the two connected in the middle at common. Discuss this structure and try to determine reasons for such a method.
2. Describe the three modes of operation for a differential amplifier. List some possible applications for each mode.
3. One input of a differential amplifier is called the inverting input. Why? What is the difference between this input and the other?
4. Look at the differential amplifier and its waveforms shown in Figures 9–42 and 9–43. What would the output be if the inputs were changed to 3 V and 3 V? 8 V and 5 V? 3 V and −3 V?

FIGURE 9–42 Differential amplifier operating in the differential mode.

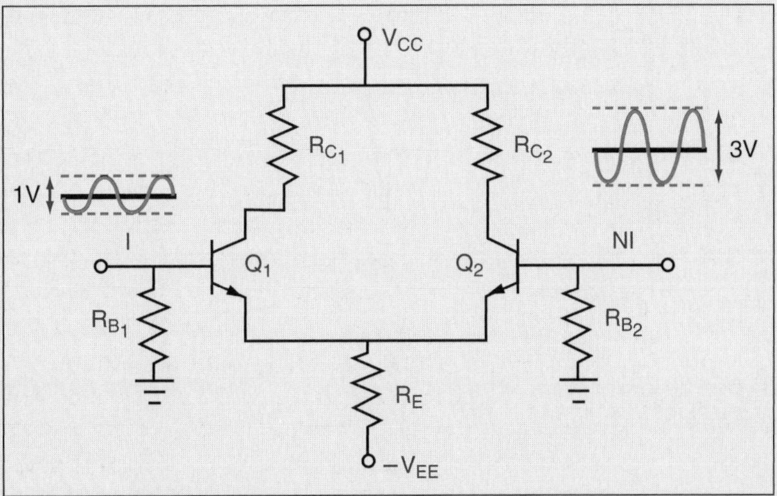

FIGURE 9–43 Input and output waveforms of differential amplifier operating in the differential mode.

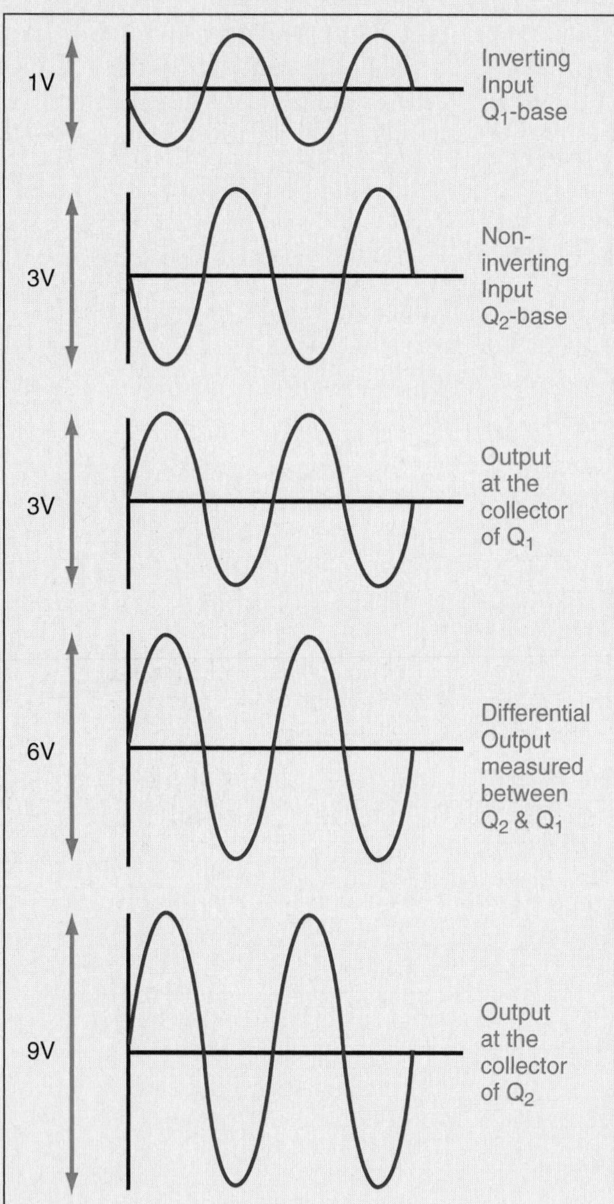

5. A certain differential amplifier has a differential mode voltage gain of 200 and a common-mode voltage gain of 0.06. What is the CMRR?
6. Look at the circuit of Figure 9–44. How would the quiescent point be changed if R_E were changed to 5.6 kΩ? What if R_E were left alone and both collector resistors were changed to 5 kΩ?
7. Discuss the concept of offset null. What is it? Why is it important?
8. What is slew rate and how does it affect the frequency response of an op amp?
9. Describe negative feedback and positive feedback. What are their uses in operational amplifier circuits?
10. You are designing an audio stage for a communications receiver. This particular receiver needs a low-pass filter that will allow voice signals but no extraneous noise above approximately 2.5 kHz. How would you provide this feature?

FIGURE 9–44 A differential amplifier circuit.

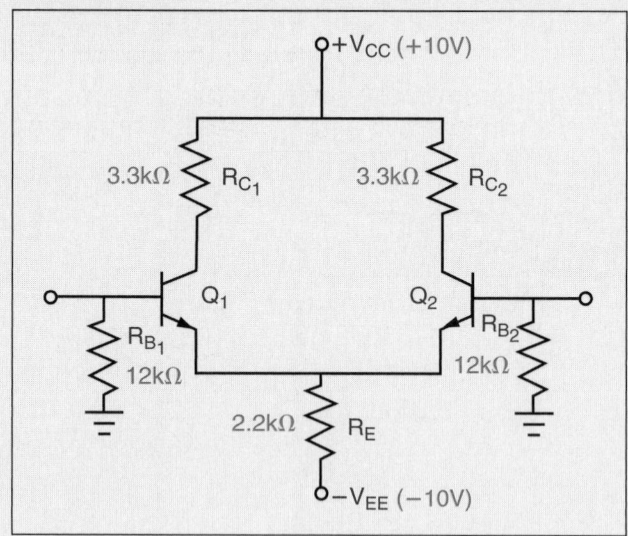

10

Oscillators

OUTLINE

OVERVIEW

An oscillator is an amplifier circuit that uses positive feedback to provide output waveforms without an input (external) signal. Because an oscillator has no input, it can be considered a signal generator. In other words, an oscillator converts the DC used to bias the circuit to an AC output signal.

Oscillators provide various waveforms for a variety of equipment. Through the use of solid-state components and circuitry, oscillators are used to create stable waveforms for use in control circuitry. Oscillators are also used in the control circuitry for variable-speed motor controls and other output devices. In this chapter, you will learn about a variety of commonly used oscillators and how they are used in modern electrical equipment.

OBJECTIVES

After completing this chapter, the student should be able to:

1. Use the concepts of gain and feedback to analyze the operation of oscillators.
2. Predict the frequency of operation for oscillators using the circuit diagrams and component values.
3. Determine the causes of undesired oscillations.
4. Identify the different types of oscillators.

OPERATING PRINCIPLES

10.1 Feedback Principles

Consider the wrecking ball of a construction crane, as in Figure 10–1a. Originally, the ball is hanging straight down. However, when the crane arm moves the ball begins to swing back and forth. Assume that the ball first swings to the right. We call this the positive direction. Eventually, the ball will expend all of its energy and will start to swing in the other direction. The ball will pass back through the starting point and swing out to the left. Again, it will reach the end of its swing and start back to the right. This type of motion is called **oscillation.**

Oscillation
The movement from one point to another in a smooth rhythmic manner. This term can refer to either electrical or mechanical motion.

This swinging action is shown graphically in Figure 10–1b. Note the angles shown in Figure 10–1b, as indicated by the degree symbol (°). Although the pendulum does not actually swing to plus or minus 90°, we can refer to one full oscillation as 360°. Therefore, one-fourth of the total swing is 90°. Two important points need to be understood about this swinging action.

1. The ball will swing at a definite frequency determined primarily by the length of the cable holding it.
2. Friction and air resistance will gradually cause the swinging action to reduce, until the ball finally comes to rest again.

FIGURE 10–1 Oscillations and sine waves: (a) the physical motion of a wrecking ball, (b) time versus distance graph of the wrecking ball.

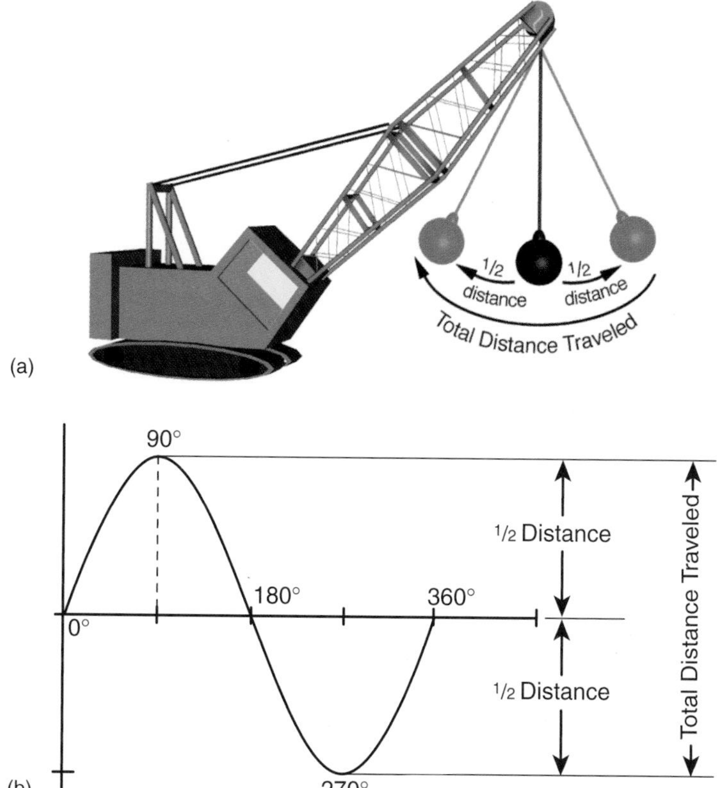

Damping
The gradual reduction of an oscillation, usually caused by resistance in an electric circuit.

Oscillator
An electronic circuit that creates an alternating (AC) output waveform. The waveform may be sinusoidal or any other regular shape, such as a square wave or triangular wave.

Feedback
The transfer of a portion of the output signal back to the input.

Positive feedback
Signal returned from circuit output to circuit input in such a way that the returned signal is in phase (additive) with other signal input.

Regenerative feedback
See *positive feedback*.

The gradual reduction of the length of the swing is called **damping.** Note that even though the distance traveled with each swing is reduced the time each swing takes remains the same for a simple pendulum. This is because the frequency of the swing depends primarily on the length of the cable.

To offset this damping effect, we need to add energy to the motion of the ball, thus replacing the energy taken out by friction and air resistance. This added energy in an **oscillator** circuit is called **positive feedback.** Positive feedback must be in phase with the original sine wave so that it adds the needed energy to overcome and maintain signal loss. Another name for this type of feedback is **regenerative feedback** (see Figure 10–2).

The needed positive feedback effect can be accomplished only if the feedback is greater than the signal loss and if the feedback is in phase.

FIGURE 10–2 Damped oscillations and positive feedback.

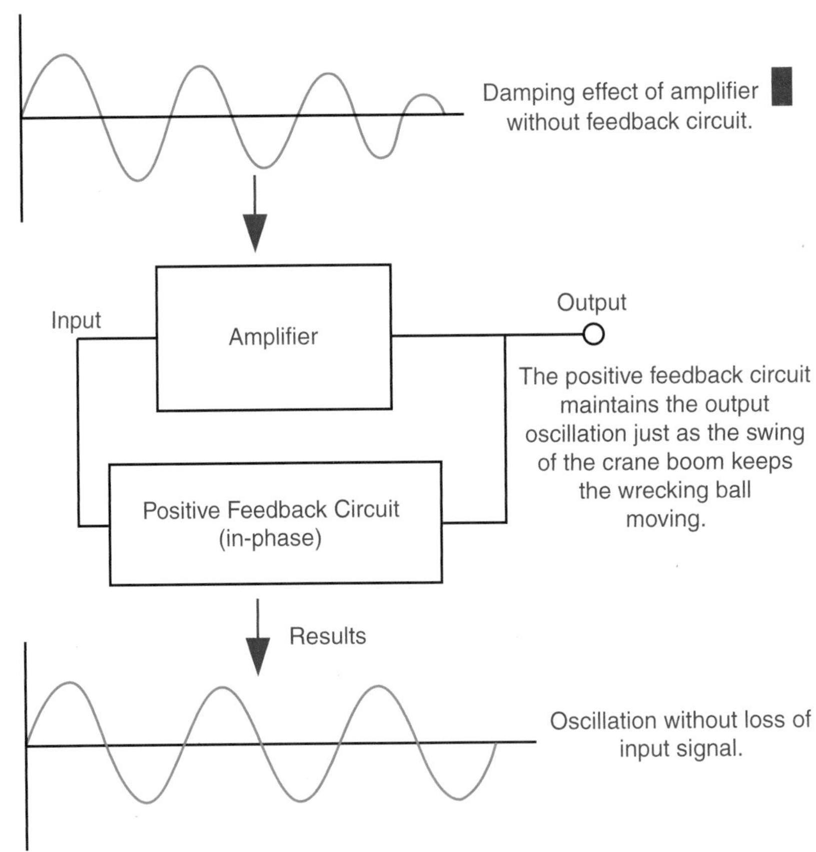
Damping effect of amplifier without feedback circuit.
Input
Amplifier
Output
The positive feedback circuit maintains the output oscillation just as the swing of the crane boom keeps the wrecking ball moving.
Positive Feedback Circuit (in-phase)
Results
Oscillation without loss of input signal.

10.2 The Barkhausen Criterion

This criterion (a standard on which judgment or decision may be made) expresses the relationship between the circuit feedback factor and the voltage gain of the circuit for proper oscillator operation (see Figure 10–2).

The Barkhausen criterion is expressed as

$$\alpha_v \times A_v = 1,$$

where:
α_v = fraction of the output signal fed back to the input
A_v = output voltage gain.

This says that the product of the feedback and the gain must be exactly equal to 1 in order to maintain the oscillation at a constant level. If $\alpha_v \times A_v < 1$, the oscillations will fade within cycles. If $\alpha_v \times A_v > 1$, the oscillations increase each cycle and will eventually drive the oscillator into clipping and saturation.

The effects of this criterion can be seen in Tables 10–1, 10–2, and 10–3. Each table is used to analyze an amplifier with a forward voltage gain of $A_v = 100$. In each case, we assume a starting point of 0.1 V peak. The inputs and outputs of subsequent cycles are given in the tables.

In Table 10–1, the fraction of the output fed back with each cycle is set to 0.0075 of the output. Thus, the Barkhausen criterion is 0.75 less than 1. The four rows are analyzed as follows.

1. The starting point is assumed to be 0.1 V peak. With a gain of 100, the output will be 10 V. The fraction fed back is equal to 0.0075, or in this case 0.075 V.
2. With 0.075 V now at the input, the amplifier produces an output of 7.5 V. The feedback is now 0.0075 × 7.5 V = 0.0563 V.
3. The 0.0563 V at the input produces 5.63 V at the input. The feedback is given by .0075 × 5.63 V = .042 V.
4. In the fourth oscillation, the output is 100 × .042 V = 4.2 V. The feedback is .0075 × 4.2 = .0315 V.

TABLE 10–1 Negative Relationship Between Feedback and Gain

Sine-Wave Cycle	V_{IN} (V-pk)	V_{OUT} (A_V) (V pk)	Feedback (α_V) (V pk)
First	0.1	10 Vpk	0.075
Second	0.075	7.5 Vpk	0.0563
Third	0.0563	5.63 Vpk	0.042
Fourth	.042	4.2 Vpk	0.0315

Obviously, the output is dying off and will eventually reach zero. Thus, you see that with a Barkhausen criterion of less than 1 the oscillator will not sustain its output. Table 10–2 outlines the same oscillator with a Barkhausen criterion of 5. Therefore, the feedback fraction is .05.

1. With the same starting point, the output will again be 10 V. The amount of feedback is 0.05 × 10 V = 0.5 V.
2. With 0.5 V at the input, the output is now 50 V and the feedback is 50 V × 0.05 = 2.5 V.
3. With 2.5 V at the input, the output is 250 V, and so on.

TABLE 10–2 Positive Relationship Between Feedback and Gain

Sine-Wave Cycle	V_{IN} (V pk)	V_{OUT} (A_V) (V pk)	Feedback (α_V) (V pk)
First	0.1	10	0.5
Second	0.5	50	2.5
Third	2.5	250	12.5
Fourth	12.5	1,250	62.5

Clearly, with this activity the oscillator will eventually reach saturation. Table 10–3 analyzes the circuit with a Barkhausen criterion of $100 \times .01 = 1$. Note that in each case the feedback is exactly what is needed to sustain the input and the output at their starting points. Such an oscillator will be stable and continue its output.

TABLE 10–3 Oscillator Feedback and Gain Equal 1

Sine-Wave Cycle	V_{IN} (V pk)	V_{OUT} (A_V) (V pk)	Feedback (α_V) (V pk)
First	0.1	10	0.1
Second	0.1	10	0.1
Third	0.1	10	0.1
Fourth	0.1	10	0.1

10.3 Oscillation Frequency

The basic oscillation-producing components of an oscillator are the capacitor and inductor. This LC circuit is the familiar "tank" circuit. Recall that these components, connected in parallel, provide alternating charge and discharge outputs. Refer to Figure 10–3. There are two formulas you need to recall. The first is for the resonant frequency of an RC circuit. The second is for the resonant frequency of an LC "tank" circuit. For an RC circuit:

$$f_r = \frac{1}{2\pi RC}$$

For an LC circuit:

$$f_r = \frac{1}{2\pi\sqrt{LC}}$$

The oscillator can have varying frequency ranges by varying the values of the RC or LC components. Oscillators that have this capability are called variable-frequency oscillators (VFOs). One method of frequency control is the lead-lag network. Figure 10–4 shows how a series-parallel RC network can be configured.

FIGURE 10–3 LC tank oscillations.

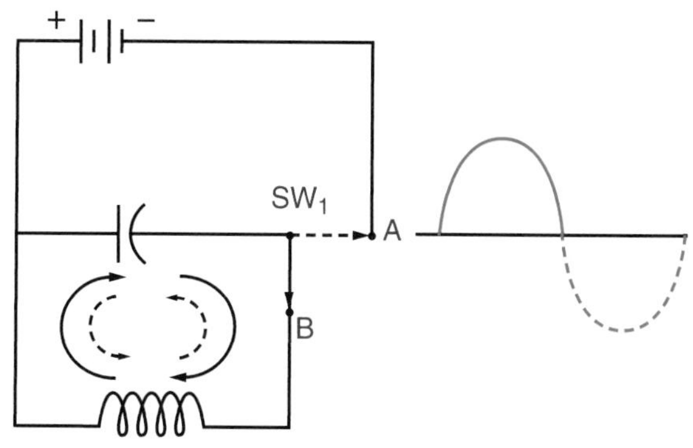

FIGURE 10–4 Lead-lag RC networks.

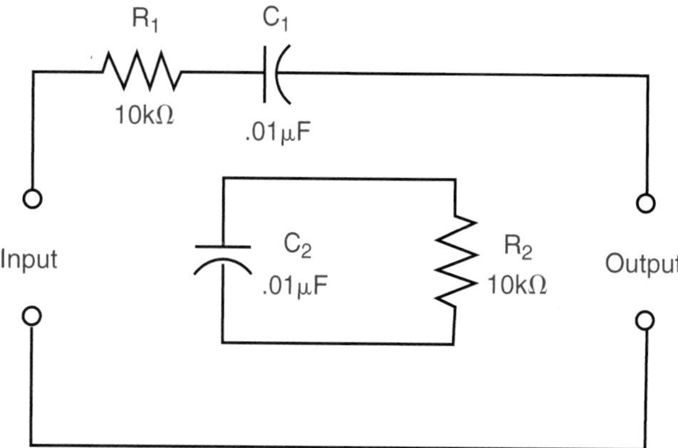

Example

Solve for the resonant frequency (f_r) of the circuit in Figure 10–4.

$$f_r = \frac{1}{2\pi RC}$$

$$f_r = \frac{1}{2\pi \times 10{,}000\ \Omega \times .00000001\ \text{F}}$$

$$f_r = \frac{1}{6.28 \times 0.0001\ \text{Sec}}$$

$$f_r = \frac{1}{0.000628}$$

$$f_r = 1{,}592\ \text{Hz}$$

By redrawing the circuit in Figure 10–4 and replacing the resistor with an inductor (refer to Figure 10–5), we can calculate the resonant frequency of the "tank" circuit. In this case, the resonant frequency is calculated as:

$$f_r = \frac{1}{2\pi \times \sqrt{.0005\ \text{H} \times 0.00000001\ \text{F}}}$$

$$f_r = \frac{1}{6.28 \times \left(\sqrt{5 \times 10^{-12}}\right)}$$

$$f_r = \frac{1}{6.28 \times (2.236 \times 10^{-6})}$$

$$f_r = \frac{1}{(1.404 \times 10^{-5})}$$

$$f_r = 71{,}225\ \text{Hz}$$

By reducing the LC component values, the oscillator can be made to create an output signal in the high megahertz range. For example, substitute the values of 10 pF and 5 μH for the capacitor and the inductor, respectively.

$$f_r = \frac{1}{2\pi\sqrt{LC}} \Rightarrow \frac{1}{2\pi \times \sqrt{5\ \mu\text{H} \times 10\ \text{pF}}}$$

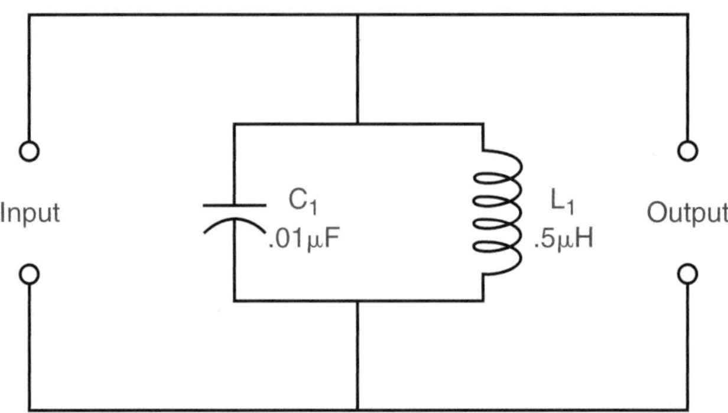

FIGURE 10–5 LC tank oscillator circuit.

VFOs are sometimes built using such circuits, with continuously variable inductors or capacitors to adjust the frequency as required. Varactors (studied in a previous chapter) are particularly suited to use in LC tuned circuits.

10.4 Oscillator Characteristics

Oscillators can be sensitive to stray or line capacitance. This is capacitance that comes from the skin effect of wires and inter-electrode capacitances of transistor junctions. These excess capacitance sources can cause false readings or increase lag time at very high frequencies.

Recall the ELI and ICE saying, "ELI the ICEman, from your basic electricity lessons. Voltage (E) leads Current (I) in Inductive (L) circuits, and Current Leads Voltage in Capacitive (C) circuits. This is still true when dealing with the needed phase shifts of an oscillator. Unknown feedback paths can be developed in complicated circuits because of the unwanted capacitance and inductance.

Another impact on oscillator stability is the power supply voltage. An increase or decrease in power supply voltage can cause the biasing of the circuit to become unbalanced. For example, cutoff points could be reached earlier or saturation levels exceeded. Other effects are lowering or increasing the need for feedback from the output. In a crystal oscillator, the increased voltage will vary the amplitude of the frequency.

OSCILLATOR TYPES

10.5 Phase-Shift Oscillators

The RC-Coupled Oscillators

RC oscillators are generally used only below a frequency of 1 MHz. RC circuits are used to produce the required phase shift for an oscillator to function. The RC oscillator is made of three basic sections: the amplifier, the phase-shift network, and the feedback loop. Figure 10–6 shows a typical common-emitter amplifier with the phase-shift network and the feedback loop.

FIGURE 10–6 RC phase-shift common-emitter oscillator.

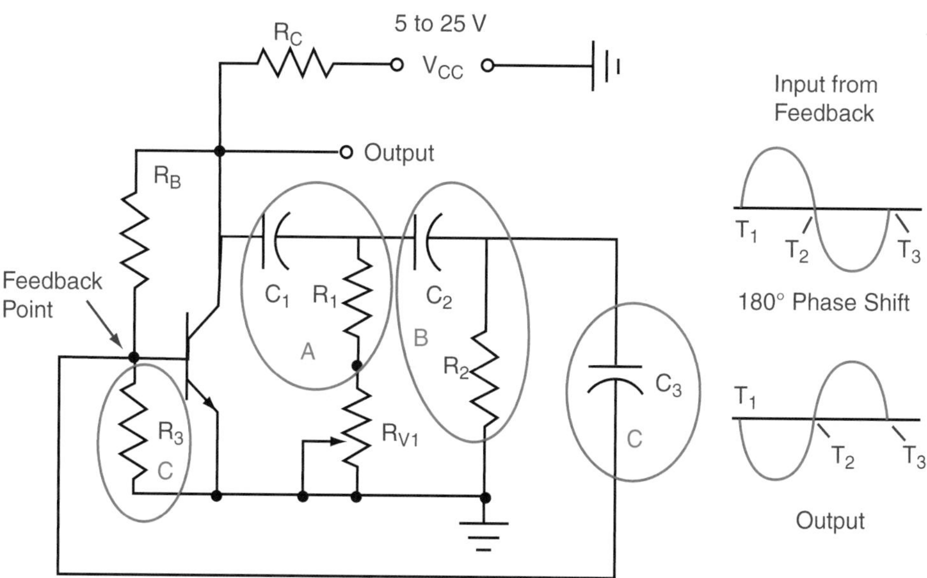

RC phase-shift oscillators are built from three RC circuit combinations. In Figure 10–6, note the circled areas labeled *A*, *B*, and *C*. Each RC combination provides a phase shift of approximately 60°. The variable resistor (R_{V1}) provides for fine-tuning the oscillator's phase shifting to ensure 180° feedback to the input of the amplifier. The 180° phase shift is necessary because the output signal of a common-emitter amplifier is 180° out of phase with the input. The 180° phase shift of the output becomes in phase with the input and thus provides signal stability, provided $A_V\alpha_V = 1$ at the resonant frequency of the RC network.

Example

In Figure 10–6, each capacitor is equal to 0.062 μF, and R_1, R_2, and R_3 are equal to 6.2 kΩ. At what frequency will a 180° phase shift occur for the proper operation of the oscillator? Note: A new formula for calculating the combined three RC networks for the 180° phase shift is

$$f = \frac{1}{15.39 \times RC}$$

Using this new formula, solve for the frequency.

$$f = \frac{1}{15.39 \times (0.062\ \mu\text{F} \times 6.2\ \text{k}\Omega)}$$

$$f = \frac{1}{0.00592\ \text{Sec}}$$

$$f = 169\ \text{Hz}$$

Although the phase-shift oscillator is one of the easiest to understand, it is rarely used because it is not stable and is difficult to tune.

Phase-Shift Bridged-T Oscillators

This is a variation on the phase-shift oscillator that allows for very good frequency stability. What is sacrificed in this oscillator circuit is the range of frequencies at which the oscillator operates. Figure 10–7 shows the circuit for this type of oscillator.

FIGURE 10–7 Phase-shift bridged-T oscillator.

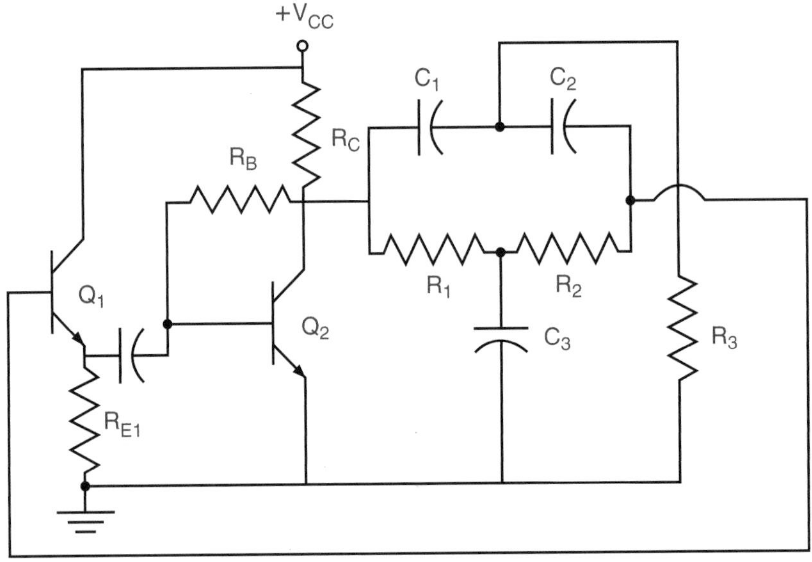

C_1, C_2, and R_3 provide a high-pass filter to the network. This is because capacitive reactance decreases as frequency increases. Note that these three components can be drawn in a T shape, thus the name. R_1 and R_2 connected as a *T* with C_3 create a low-pass filter to the network. This is because C_3 acts as a shunt to ground at high frequencies. The circuit values are selected so that the frequency of oscillation is between the low-pass cutoff and high-pass cutoff of the two bridge circuits.

Wien-Bridge Oscillators

The Wien-bridge oscillator is a common low-frequency RC oscillator. This oscillator provides no phase shift at resonant frequency. The result of this "non–phase-shifting" characteristic is that neither the amplifier nor the feedback network produces a phase shift.

This oscillator uses two feedback circuits, one positive and one negative. Figure 10–8 shows a schematic of the Wien-bridge oscillator. The positive feedback circuit is used to control the operating frequency of the oscillator. The **negative feedback** circuit is used to control the gain of the oscillator.

Negative feedback
Feedback that decreases input.

FIGURE 10-8 Wien-bridge oscillator.

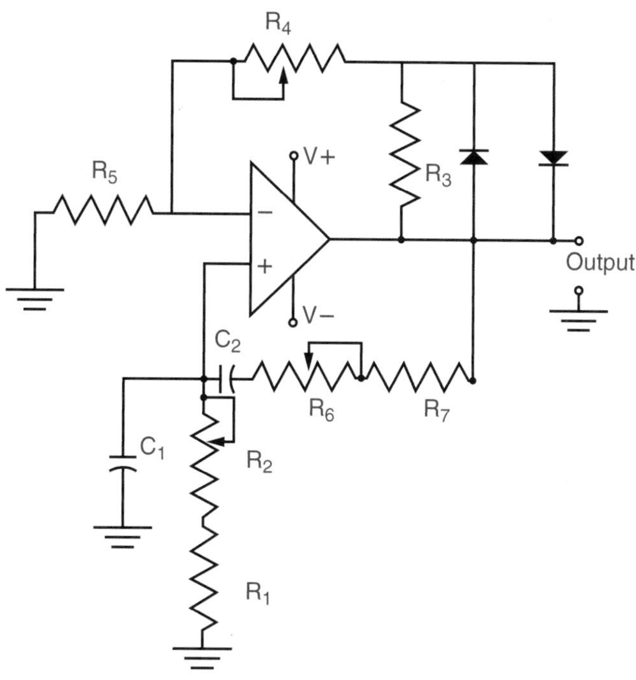

Positive feedback circuit

Note that $R_1 C_1 = R_7 C_2$. R_2 and R_6 are added trimming pots to ensure "fine-tuning" of the positive feedback circuit. Given these equivalent values, the resonant frequency is equal to

$$fr = \frac{1}{2\pi RC}$$

Negative Feedback Circuit

The negative feedback path is a closed loop consisting of R_4, R_3, and two diodes in parallel with R_3. R_4 is a potentiometer used to control the gain of the circuit. The closed-loop voltage gain of a noninverting amplifier is

$$Av = \frac{R_f}{R_i} + 1.$$

$R_i = R_5$ and $R_f = R_4 + R_3$ as long as the diodes are not conducting. If the oscillator output tries to go above the voltage drop of $R_4 + R_5$ by more than 0.7 V, one of the diodes will "turn on" and effectively short out R_3—reducing the voltage gain to

$$Av = \frac{R_4}{R_5} + 1$$

10.6 Hartley and Colpitts Oscillators

The Hartley and Colpitts oscillators are very similar in design. The major difference is how the feedback circuit is produced. Figures 10–9 and 10–10 show a comparison between the Hartley and Colpitts oscillators. Note that the feedback circuit from the Hartley oscillator is taken from the tapped coil of the inductor or auto-transformer (L_1) and that the feedback from the Colpitts oscillator is taken from the tap between C_1 and C_2.

In the Hartley oscillator (Figure 10–9), C_1 prevents the RF (radio frequency) choke (RFC) and L_1 from shunting V_{CC} to ground. The value of C_1 is made very high to prevent it from affecting the resonant frequency circuit calculations. Note that the output voltage is developed across L_1 and that the feedback voltage is developed across L_2. The ratio of L_1 to L_2 is very important because the amount of feedback percentage is determined by where the tap is placed.

In a similar manner, the output of the Colpitts oscillator is taken from C_1, and the feedback is developed across C_2. Table 10–4 lists the gain and the resonant frequency for each of the oscillators.

FIGURE 10–9 Hartley oscillator.

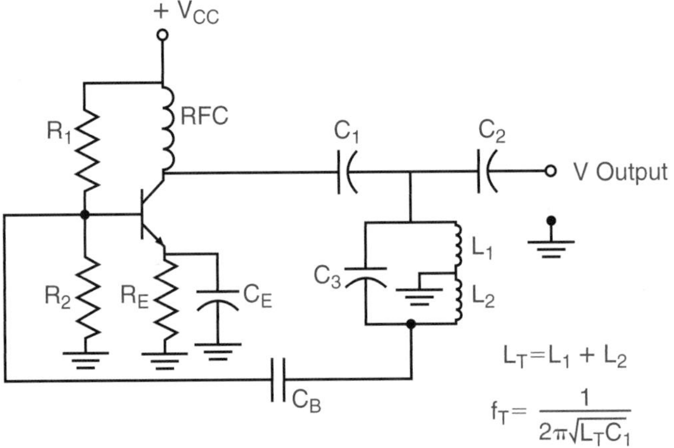

FIGURE 10–10 Colpitts oscillator.

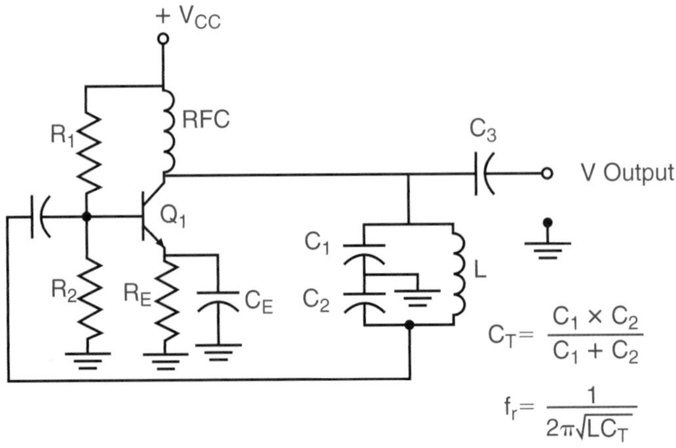

TABLE 10–4 Gain and Resonant Frequency of the Hartley and Colpitts Oscillators

Oscillator Type	Gain (A_V)	Resonant Frequency (f_r)
Hartley	$\frac{L_2}{L_1}$	$f_r = \frac{1}{2\pi\sqrt{(L_1 + L_2)C_1}}$
Colpitts	$\frac{C_2}{C_1}$	$f_r = \frac{1}{2\pi\sqrt{L\left(\frac{C_1 \times C_2}{C_1 + C_2}\right)}}$

Any parallel resonant tank circuit will lose efficiency when loaded. A transformer is used to couple the output of the oscillators to the load. This prevents false readings and loss of oscillator efficiency. Figures 10–11 and 10–12 show Hartley and Colpitts oscillators using transformers to couple their outputs to the load.

FIGURE 10–11 Transformer coupling for a Hartley oscillator.

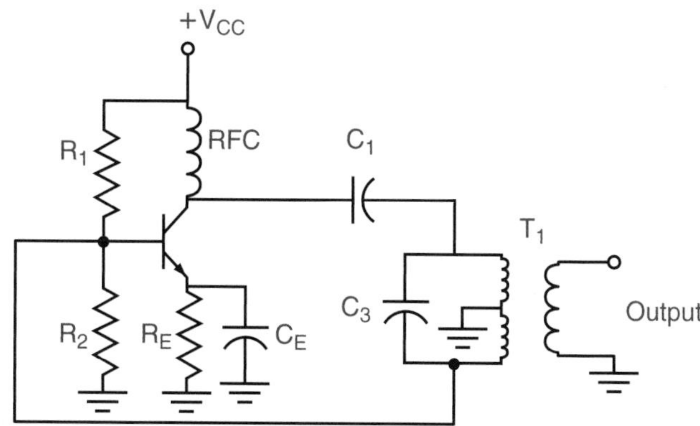

FIGURE 10–12 Transformer coupling for a Colpitts oscillator.

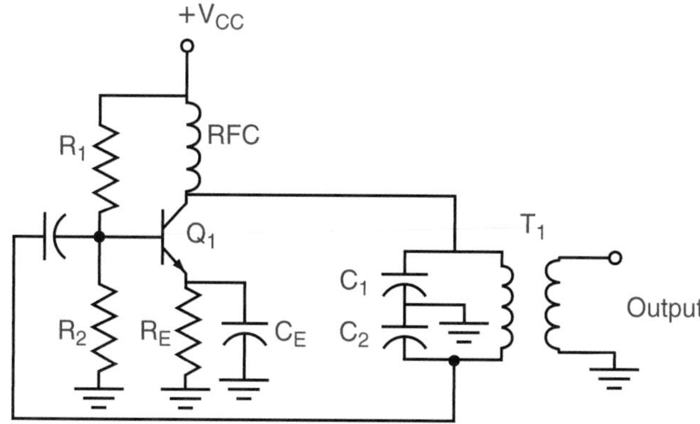

10.7 Crystal Oscillators

Piezoelectric effect
The effect exhibited by some crystals in which pressure causes electric charge, and electric charge causes a change in shape.

The most stable of the oscillators is the crystal-controlled oscillator. Crystal oscillators have a quartz crystal used to control the operational frequency. Certain types of crystals exhibit what is called the **piezoelectric effect.** This means that under compression the crystal produces an electric charge on its surface.

The reverse is also true. When an electric charge is put across a crystal, it expands and contracts. The frequency of this expansion and contraction depends on the physical dimensions of the crystal. You can produce different frequencies by cutting crystals to different dimensions.

There are three common crystals used in oscillators: quartz, tourmaline, and Rochelle salt. Rochelle salt is the best from a performance standpoint, but it breaks easiest. Tourmaline is the most rugged, but it is the least stable. Quartz is the best all-around crystal for oscillator work, and it is most used. Quartz crystals are made of silicon dioxide, SiO_2. This is the same material used for the insulating layer of the MOSFET gate. Figure 10–13 shows a typical quartz crystal. Figure 10–14 shows the equivalent circuit for a crystal. Note that the crystal has all four components of an oscillator circuit.

- C_C = crystal capacitance
- C_M = mounting capacitance
- L = crystal inductance
- R = crystal resistance

FIGURE 10–13 Quartz crystal.

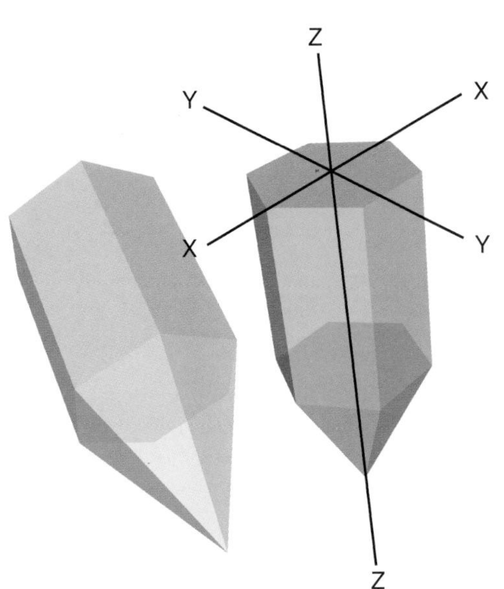

FIGURE 10–14 Crystal equivalent circuit.

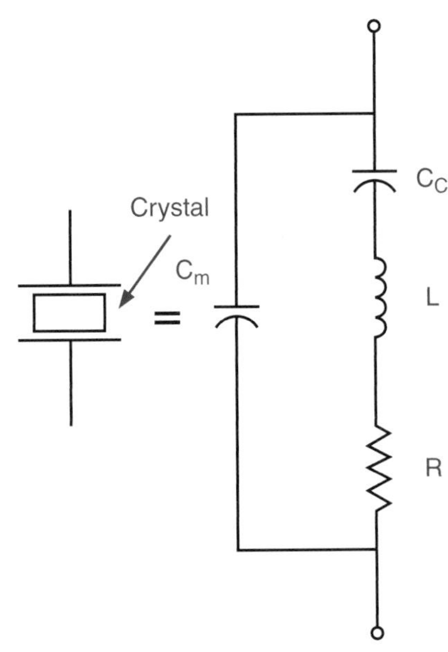

The "resonant" frequency of a crystal is determined by its thickness. The thinner the crystal the higher the frequency. Crystals also have a very high **Q** (quality) because of their natural frequency of vibration. This high Q is important because the higher the Q the higher the frequency stability. That is, with a high Q the oscillator will operate at resonance within a very narrow frequency band.

Q
The measure of the quality of an inductor, or capacitor $\left(\frac{X}{R}\right)$. The determining factor in selectivity of a resonant circuit.

Crystal oscillators can operate in parallel or in series. Figure 10–15 shows a crystal-controlled Colpitts oscillator. Note that the crystal (Y_1) is in series with the "tank" circuit. At resonance, the series crystal will have a minimum resistance and allow full feedback from the tank circuit. At frequencies other than resonance, the crystal will act as a band-stop circuit and prevent feedback from returning to the transistor's base.

FIGURE 10–15 Crystal-controlled Colpitts oscillator.

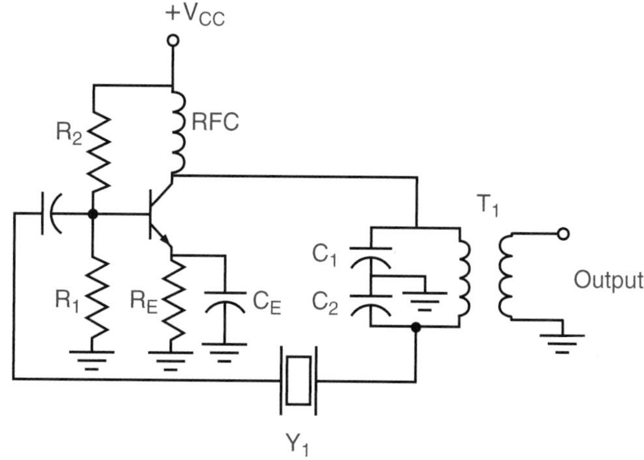

10.8 Relaxation Oscillators

The UJT, or unijunction transistor, can be used in an oscillator circuit called a relaxation oscillator. The relaxation oscillator is a circuit that uses the charge/discharge cycle of a capacitor or inductor to produce a pulse output. The output pulse is normally used to fire or trigger an SCR or a triac. The output pulse waveform is either sawtooth or rectangular. Figure 10–16 shows a simple relaxation oscillator circuit.

FIGURE 10–16 UJT relaxation oscillator.

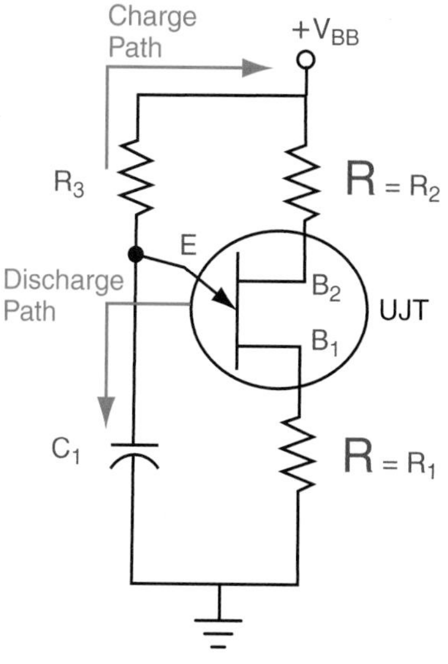

Figure 10–17 shows the equivalent circuit for the UJT and is used for the following discussion. Recall from an earlier lesson that the equivalent circuit of a UJT is a diode connected to a voltage divider. For the UJT to trigger or conduct, the emitter-base junction must be 0.7 V more positive than its cathode. The cathode potential is determined by a combination of V_{BB}, R_{B1}, and R_{B2}. The normal voltage divider equation $\left(\frac{R_{B1}}{R_{B1} + R_{B2}}\right)$ is called the intrinsic standoff ratio, η. Thus,

$$V_K = \eta \, V_{BB}$$

The triggering voltage is

$$V_P = \eta \, V_{BB} + 0.7 \text{ V}$$

FIGURE 10–17 UJT equivalent circuit.

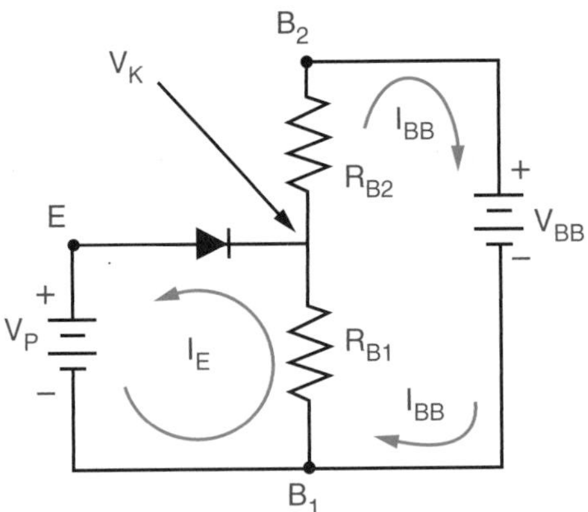

Figure 10–18 shows the waveforms for the operating relaxation oscillator. As the capacitor charges, the voltage across the emitter builds until it is equal to V_P. At V_P, the negative resistance of B_1 goes to almost 0 Ω and the capacitor rapidly discharges. This rapid discharge produces a waveform similar to that shown in Figure 10–18b.

FIGURE 10–18 Relaxation oscillator charge/discharge waveforms.

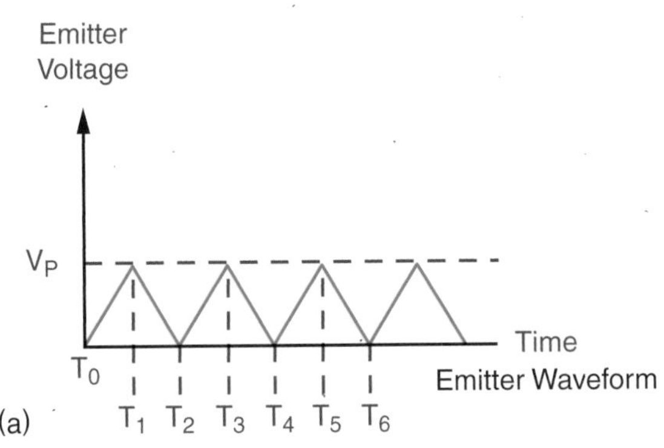

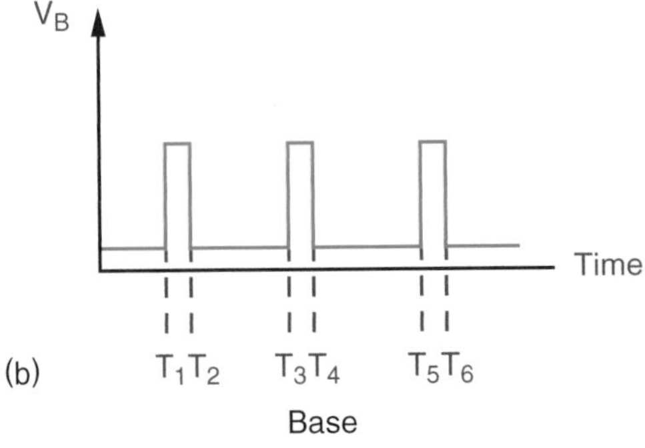

Both of the waveforms in Figure 10–18 can be used as an output, depending on where the output is taken. The frequency of the oscillator is controlled by the RC time constant. Referring to Figure 10–16, you see that R_3 and C_1 provide the time constant ($T = R_3C_1$) for the charge/discharge cycle. Recall that the frequency is equal to the inverse of the time constant. Thus, the frequency is equal to $\frac{1}{R_3C_1}$.

Note that this is slightly different from the previous calculations because we are dealing with pulses per second. Recall also that the formula to charge the capacitor to one time constant (63%) is $T = RC$. For example, assume that $R_3 = 20\ \text{k}\Omega$ and that $C_1 = 15$ mF. The frequency for the oscillator would be

$$f = \frac{1}{\text{RC}}$$

$$f = \frac{1}{20\ \text{k}\Omega \times 15\ \mu F}$$

$$f = 3.3\ \text{Hz}$$

Assume that the desired oscillator frequency is 200 Hz and the value of R_3 is 30 kΩ. What would be the required value of C_1? Start by rearranging the frequency equation and then insert the given values.

$$C = \frac{1}{f \times R}$$

$$C = \frac{1}{200\ \text{Hz} \times 30\ \text{k}\Omega}$$

$$C = \frac{1}{6{,}000{,}000}$$

$$C = .167\ \mu F$$

10.9 Direct Digital Synthesis

When many operational frequencies are needed, there are several options. Two are the phase-locked loop (PLL) frequency synthesizer and the direct digital synthesizer (DDS). The synthesizer of choice is the DDS. The advantage of a DDS system over a crystal is that the DDS can be programmed for a high number of narrow-band frequencies.

Keep in mind, however, that stability of the DDS and PLL are dependent on a fixed-frequency oscillator. This base-frequency oscillator is most likely crystal controlled. Another name for the DDS oscillator is the numerically controlled oscillator. The DDS system has the following components (see Figure 10–19):

- *Frequency tuning word:* The input number to the phase accumulator for the value of the next phase increment.
- *Phase accumulator:* Produces (with the tuning word and clock) the binary digit for the sine lookup table.
- *Sine lookup table:* A set of sine-wave voltage values that are sent one value at a time to the digital-to-analog (D/A) converter.

- *Digital-to-analog converter (DAC):* Produces an approximation of the sine wave voltage, at a specified phase value, for the input voltage from the sine lookup table.
- *Clock:* Provides a constant reference square pulse frequency for the phase accumulator and the DAC.
- *Low-pass filter:* Provides the filtering function for sine-wave cleanup generated from the DAC.

FIGURE 10–19 DDS oscillator.

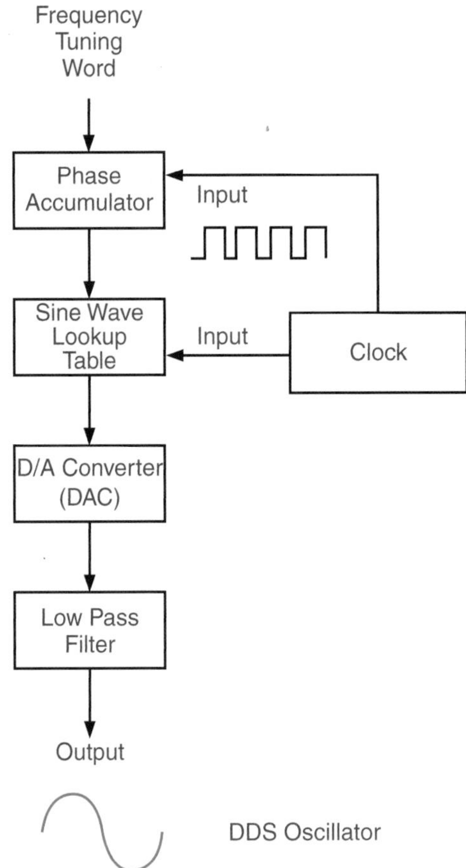

10.10 Unwanted Oscillation

It would seem from reading this chapter that specialty circuits (i.e., Colpitts, Hartley, phase-shift, and other oscillator types) are needed to generate AC signals electronically. Unfortunately, this is not the case. All too often, oscillation occurs when we do not want it. Amplifiers are susceptible to unwanted oscillation, especially if they employ op-amps or consist of two or more transistors. The following are issues to consider if you are experiencing unwanted circuit oscillation:

- *Decoupling:* When two or more amplifiers use the same power supply, they are likely to cause minute changes in supply voltage.

Those small changes are coupled to other stages, affecting their instantaneous bias voltages, possibly resulting in oscillation. The conventional solution is *decoupling* by an RC filter, as shown in Figure 10–20a. The time constant of the filter is usually about 10 times the lowest operating frequency. If circuit oscillation suddenly becomes a problem, check the decoupling capacitor(s) for an open.

- *Component leads and interconnecting wires have both inductance and capacitance (to nearby conductors):* Figure 10–20a shows symbolic capacitive coupling (in red) from output to input of the two-stage amplifier. To avoid this, try not to rearrange wiring and keep replacement component leads as short as practical. The higher the frequencies involved the more important this becomes. Figure 10–20b, with no input signal, illustrates that no oscillation appears at amplifier output when stray capacitance is at a minimum. Sloppy wiring practices often result in regenerative feedback. This undesirable feedback can result in totally unusable amplifier output (see Figure 10–20c). Note that the output does not look anything like the input waveform. Reducing the stray capacitance should produce a nice clean output (as in Figure 10–20d).
- Sometimes probing circuit test points produces oscillation. This inconvenience can often be eliminated by moving test leads away from adjacent components. If using an oscilloscope, isolate the scope from the circuit by using the X10 probe rather than X1 probe.

FIGURE 10–20 (a) Stray capacitance.

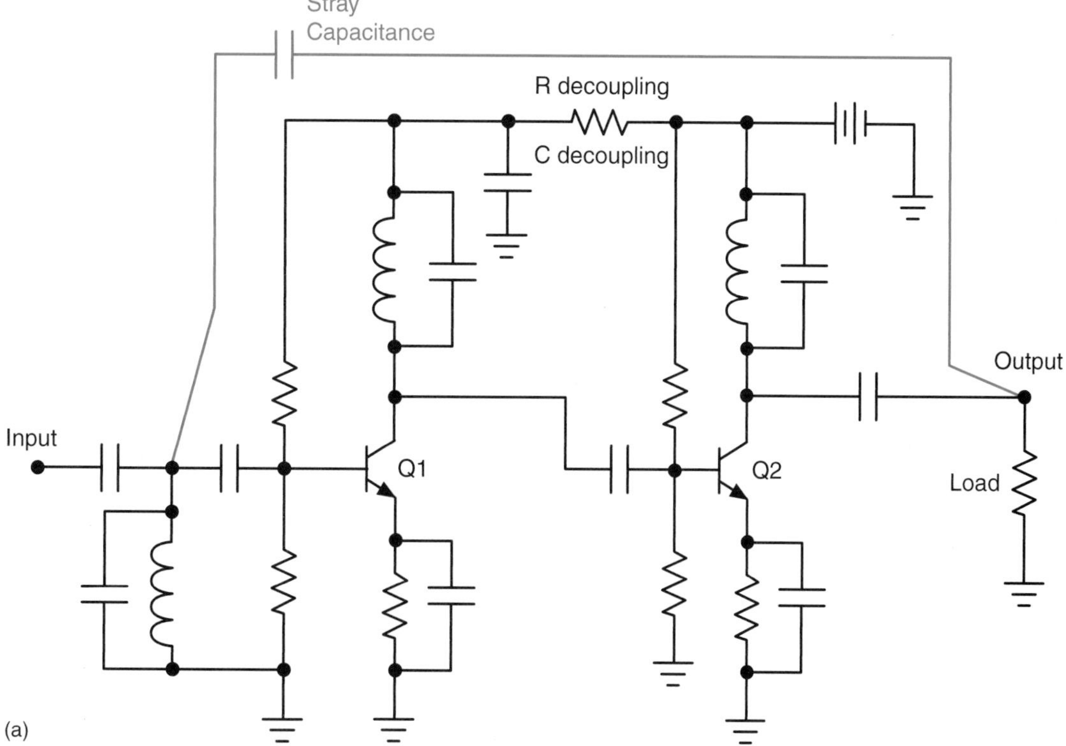

FIGURE 10–20 (b) Low regenerative feedback. (c) Normal and stray capacitance signal waveforms. (d) Normally operating amp.

No output signal is present

No input signal is present

(b)

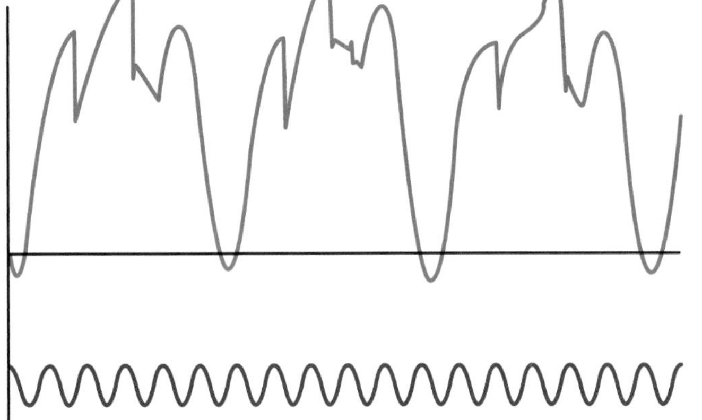
Stray capacitance causes amp oscillation to be evident in the output waveform
Normal amplifier input signal

(c)

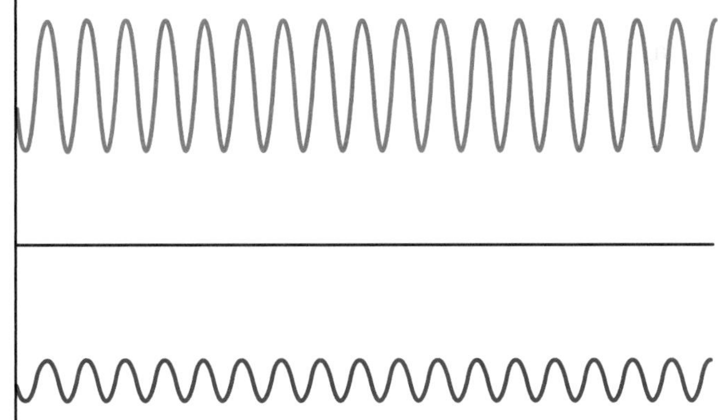
Normal output. Regenerative feedback is not sufficient to cause amp oscillation.
Normal amplifier input signal

(d)

10.11 Practical Oscillator Circuits

RC Oscillators

Phase-Shift Oscillators

Figure 10–21 shows a practical schematic of a phase-shift oscillator. Using the component values shown on the schematic diagram, the resonant frequency is

$$f_r = \frac{1}{15.39 \times RC}$$

$$f_r = \frac{1}{15.39 \times 10\ \text{k}\Omega \times 50\ \text{nF}}$$

$$f_r = 130\ \text{Hz}$$

FIGURE 10–21 Practical phase-shift oscillator circuit.

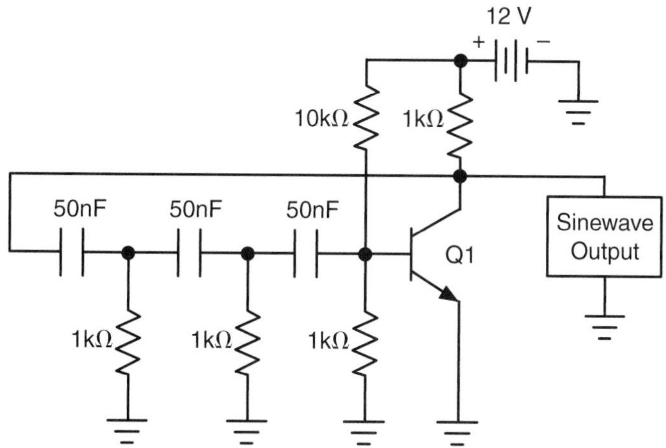

Although, like other RC oscillators, the phase-shift oscillator is not highly stable it can produce a very good sine-wave–shaped output. Therefore, as opposed to square-wave signals it can be used to simulate many naturally occurring sounds. This circuit, with an earphone, could be used to help tune a musical instrument. It could also be connected as a signal source to an audio amplifier input to assist in troubleshooting.

Multivibrators

Figure 10–22 shows a schematic of a very simple and common variation on the phase-shift oscillator known as the astable multivibrator. It requires two transistors, four resistors, and two capacitors. This easy-to-build circuit generates a square-wave output. Assuming that base resistors are equal and feedback capacitors are equal, output frequency is calculated by:

$$f_r = \frac{1}{\sqrt{2} \times \mathrm{RC}}$$

$$f_r = \frac{1}{\sqrt{2} \times 100\ \mathrm{k\Omega} \times 50\ \mathrm{nF}}$$

$$f_r = 141\ \mathrm{Hz}$$

FIGURE 10–22 Practical multivibrator oscillator circuit.

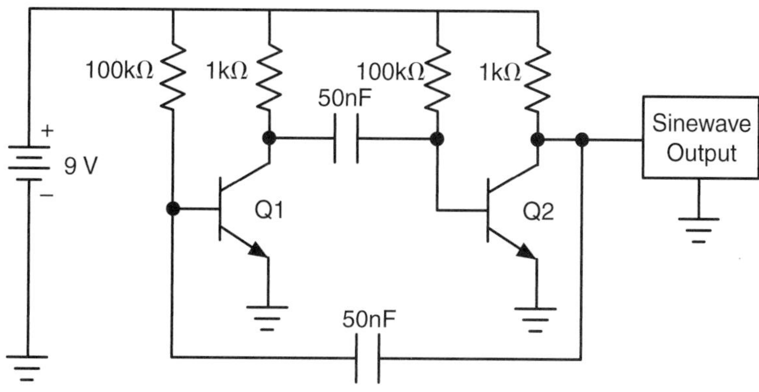

This is *not* a precision oscillator, and is generally only useful below 1 MHz. However, the astable multivibrator is widely used where precision and clean waveforms are not required, and where cost must be controlled. This circuit, with the component values shown and 9 V applied (as in Figure 10–22) will output approximately 8 $V_{P\text{-}P}$.

LC Oscillators

Hartley

The circuit of Figure 10–23 is a Hartley oscillator. Hartleys, like other LC oscillators, are used primarily in RF applications. There are several variations on the Hartley. If the inductor is tapped, the oscillator is a Hartley. The resonant frequency of the circuit shown in the schematic is found by

$$f_r = \frac{1}{2\pi\sqrt{LC}}$$

$$f_r = \frac{1}{\sqrt{2\pi\, 100\ \mu\text{H} \times 100\ \text{pF}}}$$

$$f_r = \frac{1}{2\pi\sqrt{1 \times 10^{-14}}}$$

$$f_r = 1.59\ \text{MHz}$$

There are few uses for a Hartley oscillator as a stand-alone circuit. However, Figure 10–23 provides for the addition of an AF oscillator (the previously cited multivibrator, for example) to the amplitude modulate the RF output of the Hartley oscillator. This circuit configuration would perform well as a model rocket or airplane locator beacon, or similar application.

FIGURE 10–23 Practical Hartley oscillator circuit.

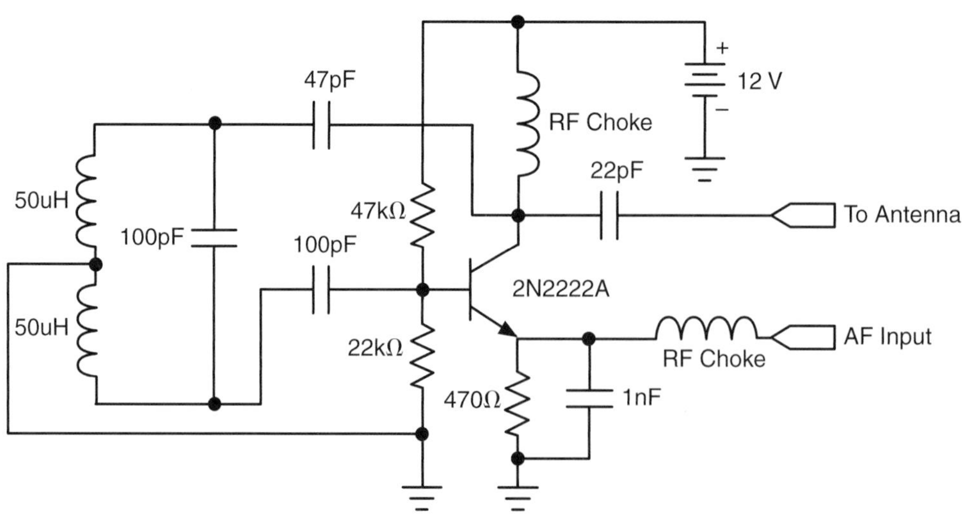

Crystal-Controlled Colpitts

Figure 10–24 shows a schematic diagram of a functional crystal-controlled Colpitts oscillator. The Colpitts oscillator is generally considered the most stable LC-controlled oscillator. The addition of crystal control makes the Colpitts the most stable oscillator available. Note that the only difference between the crystal-controlled of Figure 10–10 and LC-controlled Colpitts of Figure 10–12 is the inclusion of the crystal between the base circuit and the LC tank circuit.

As with the Hartley, there are variations to the Colpitts oscillator design. If the tank circuit includes dual capacitors, it is a Colpitts oscillator. Figure 10–24 substitutes the crystal for the inductor, and the feedback path is emitter to base rather than collector to base. The crystal-controlled Colpitts oscillator of Figure 10–24 is usually used as a local oscillator for superhetrodyne RF receivers or as a frequency-determining circuit in RF transmitters.

FIGURE 10–24 Practical crystal-controlled Colpitts oscillator circuit.

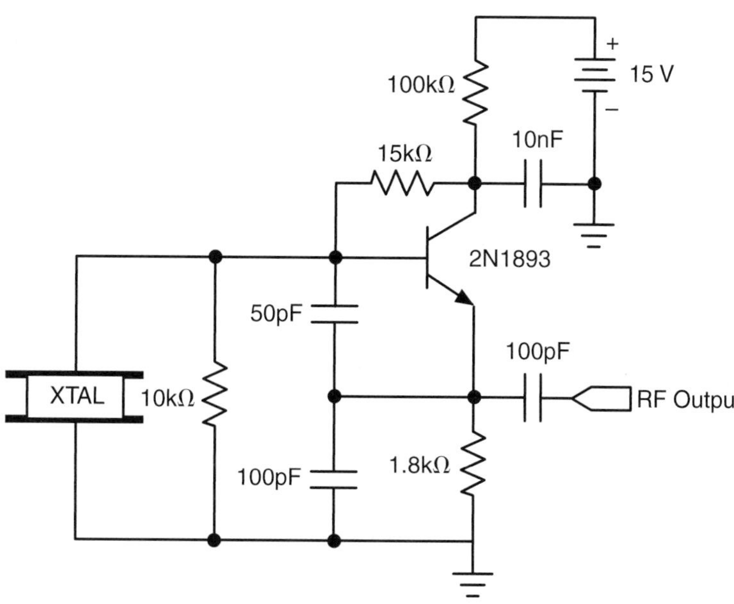

SUMMARY

The oscillator is an electronic circuit that uses positive feedback to sustain its output. A small signal, usually created by random noises in the input circuit, is amplified by the circuit. A small portion of the output signal is fed back to the input in such a way that the input is reinforced or strengthened. This is called positive feedback.

For the oscillator to operate, the gain of the amplifier multiplied by the percentage of the output that is fed back must be greater than 1. In this way, the input signal will be kept at its original level or higher. This product is called the Barkhausen criterion and is calculated using the following equation:

$$BC = \alpha_V A_V$$

Here:

α_V = the percentage of the output that is feedback to the input

A_V = the forward gain of the amplifier

Note that if the feedback is negative (i.e., 180° out of phase) the feedback will not occur. The frequency of an oscillation is determined by the passive elements in the circuit. Some oscillators use inductors and capacitors to determine their resonant frequency. Others, especially those that use integrated circuits or operational amplifiers, use resistors and capacitors for their frequency control.

The most stable of all standard oscillator types uses a quartz crystal for the frequency control element. This type of circuit takes advantage of the piezoelectric properties of a properly cut crystal. In junction with an RC or LC circuit, the crystal-controlled oscillator provides an extremely stable output.

The relaxation oscillator uses the charging of a capacitor through a resistor to produce a pulse type of output. The capacitor charges until the junction of a UJT is forward biased. When that happens, the capacitor discharges through the UJT. When the capacitor is completely discharged, the UJT is reset and the capacitor begins its charge again. The frequency of the relaxation oscillator is controlled by the RC time constant of the capacitor input circuit.

Modern digital circuits use a variety of circuits to produce stable, controllable output waveforms. The frequency of output is controlled by a clock that controls the input of a digital series of pulses to a sine-wave lookup table and a DAC converter. The output is smoothed to a sine wave by use of a low-pass filter. Unwanted oscillation can occur in amplifier circuits. Usually, the cause is attributable to loss of decoupling capacitance or mechanical rearrangement of nearby wires.

REVIEW QUESTIONS

1. Figure 10–25 shows half of a sine wave that is being generated by a rotating circle. Assume the circle has a radius of 1 unit. As the circle rotates, the distance from the point on the circumference to the horizontal diameter is measured. These distances are then plotted on a horizontal axis. What will the height of the sine wave be when the circle has rotated 45°, 135°, and 254°, respectively?
2. Examine Table 10–5. What would happen if the gain of the amplifier were reduced to 80? Increased to 120?
3. A certain oscillator has a gain of 50. What will happen if 2% of the output is fed back to the input? 1%? 3%?

FIGURE 10–25 Creating a sine wave with a rolling circuit.

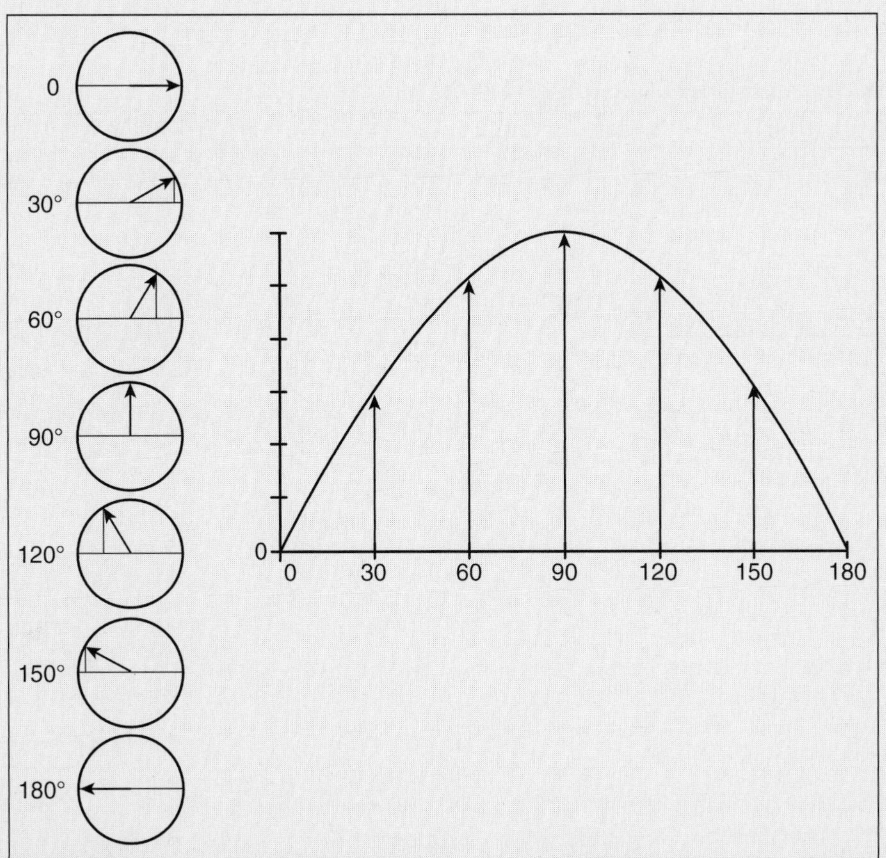

TABLE 10–5 Oscillator Feedback and Gain Equal 1

Sine-Wave Cycle	V_{IN} (V pk)	V_{OUT} (A_V) (V pk)	Feedback (α_V) (V pk)
First	0.1	10	0.1
Second	0.1	10	0.1
Third	0.1	10	0.1
Fourth	0.1	10	0.1

4. In Figure 10–26, which component provides the feedback path from output to input? Which components determine the frequency of the output?

FIGURE 10–26 RC phase-shift common-emitter oscillator.

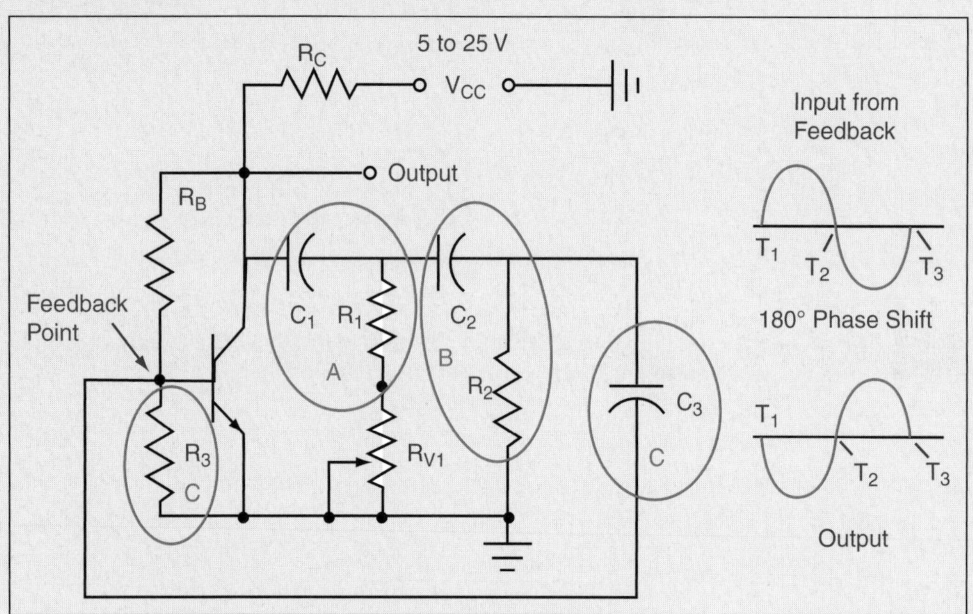

5. In Figure 10–27, what will happen if the size of resistor R_1 is increased? Decreased?

FIGURE 10–27

UJT relaxation oscillator.

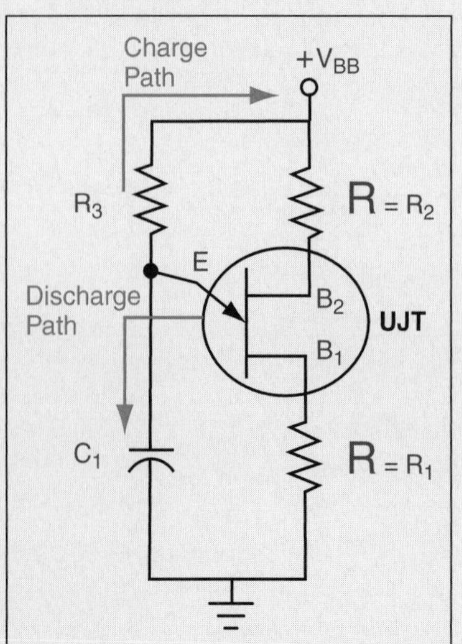

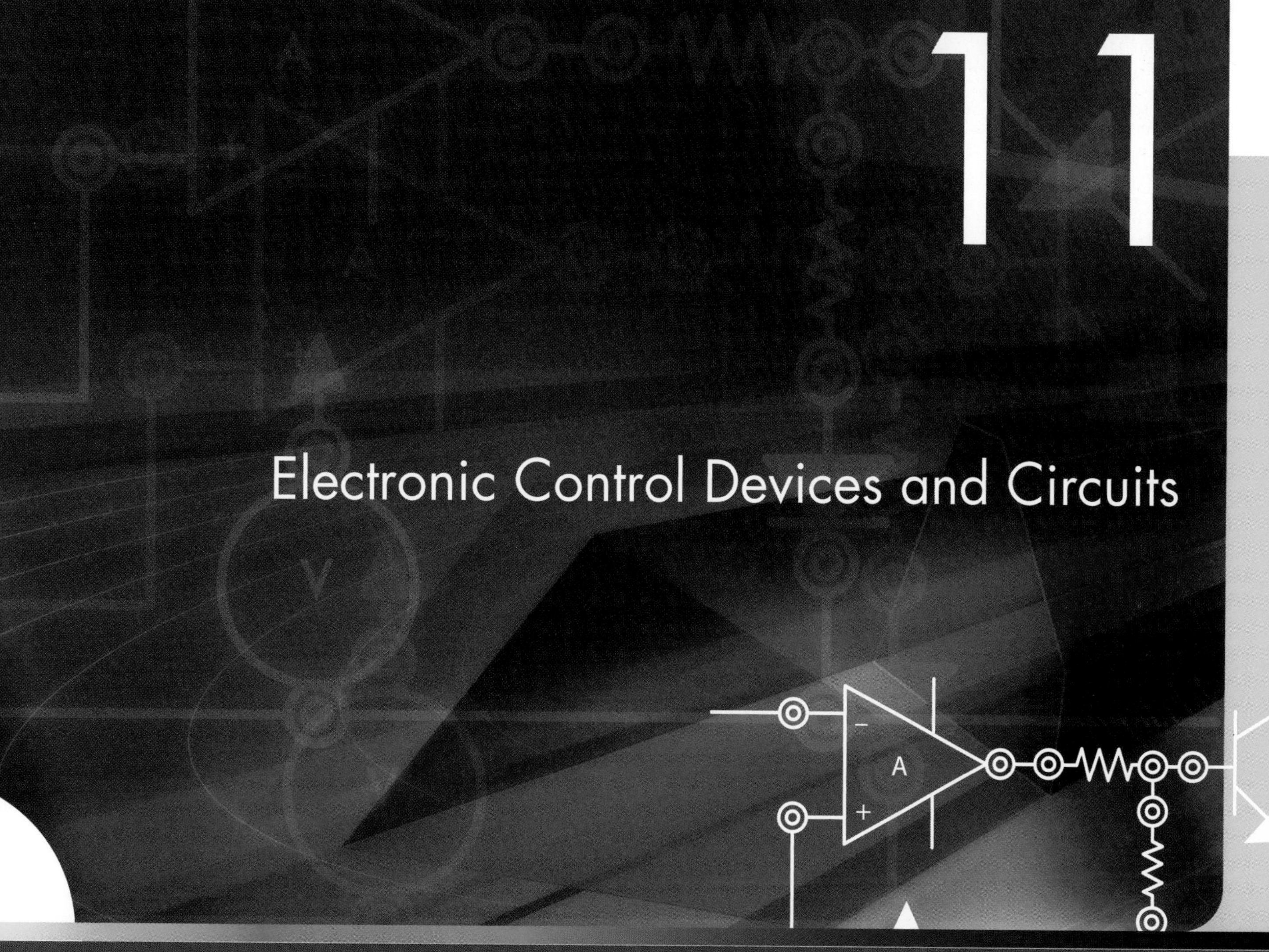

11 Electronic Control Devices and Circuits

OUTLINE

OVERVIEW

Control of current to the load is an area of circuit operation with a lot of significance. For instance, a motor, a lamp, or a heating element may serve as the load in a circuit. By regulating the current in each of these, the speed of the motor, brightness of the lamp, or temperature of the heating element can be controlled.

Control circuits can accurately set the currents and thereby control the loads. Although an adjustable resistor (rheostat) can vary or set current, it wastes a lot of energy. Solid-state devices can perform the same job much more efficiently. In this chapter, you will learn about many of the more important electronic control devices and how they are used in day-to-day applications.

OBJECTIVES

After completing this chapter, the student should be able to:

1. Identify the schematic symbols used for control devices.
2. Explain the AC and DC switching capabilities of the silicon-controlled rectifier (SCR).
3. Describe steps used to test an SCR, Diac, and Triac.
4. Describe the operation of thyristors.
5. Draw a schematic showing the use of a snubber network and varistor for noise and high-voltage suppression.
6. Understand the operation of diac and triac.

CONTROL CIRCUITS

11.1 Rheostat Control

A rheostat circuit used to control the brightness of a lamp is shown in Figure 11–1. The control device in the circuit is the rheostat. Unfortunately, although very simple, a rheostat control is extremely inefficient (as outlined in Table 11–1).

FIGURE 11–1 A rheostat control circuit.

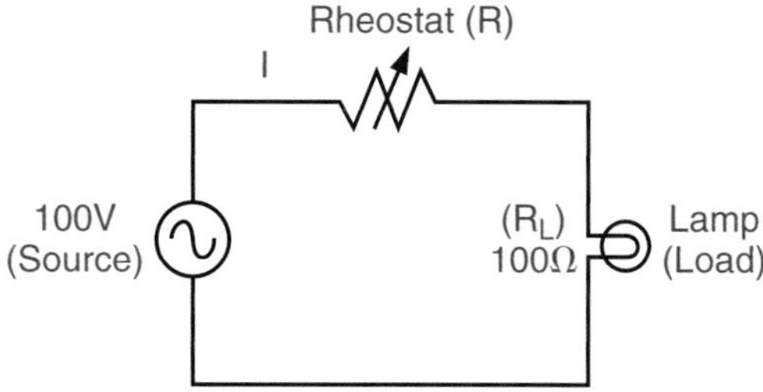

TABLE 11–1 Rheostat Control Circuit Analysis

Rheostat Value	Total Resistance $R_T = R + R_L$	Current $I = \frac{V}{R_T}$	Power Dissipated in Rheostat $P_R = I^2R$	Power Dissipated in Load $P_{RL} = I^2R_L$	Total Power $P_T = P_R + P_{RL}$
100 Ω	200 Ω	0.5 A	25 W	25 W	50 W
50 Ω	150 Ω	0.66 A	22 W	44 W	66 W
0 Ω	100 Ω	1 A	0 W	100 W	100 W

Table 11–1 indicates that as the resistance of the rheostat increases, the power dissipated in the rheostat increases and the power dissipated in the load decreases. Full power is delivered to the load only when the rheostat is set at 0 Ω. With increasing values of rheostat resistance, the power delivered to the load decreases because a big portion of the total power is dissipated in the control device itself. Efficiency (η) of the device is calculated by

$$\eta(\%) = \frac{P_{RL}}{P_T} \times 100$$

Using the values found in Table 11–1 and the formula for efficiency, you can determine that 100% efficiency exists only when the rheostat is set at 0 Ω. As the value of the rheostat increases, the efficiency drops. Poor efficiency contributes to high cost of operation due to power loss. In addition, the rheostat must be physically large enough to dissipate the large quantities of heat generated. In conclusion, it can be said that rheostat control is inefficient and generally a poor choice for a control application of this type.

11.2 Voltage Control

An alternative to the rheostat control circuit is voltage control, as shown in Figure 11–2. Here, the voltage applied to the load can be varied from 0 V to 120 V, and the power dissipated in the circuit will be 0 watts to 144 watts. Note that the only significant resistance in the circuit is the load, and the only place for power to be dissipated is in the load. The efficiency of this circuit will always be close to 100%. However, controlling the line voltage is a relatively expensive alternative because of the cost of the variable auto-transformer.

FIGURE 11–2 A voltage control circuit.

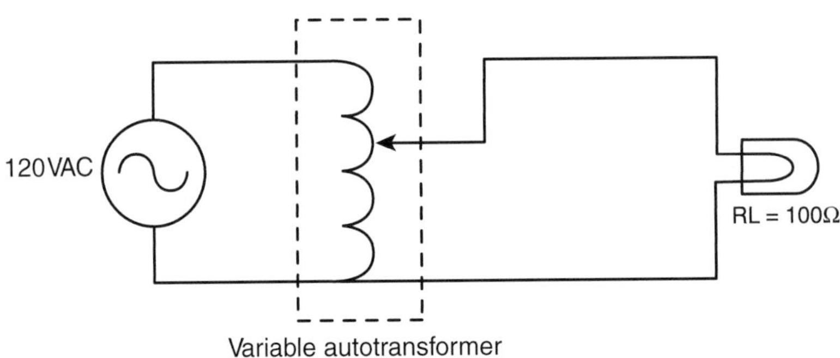

11.3 Switch Control

The switch control is a form of voltage control. However, it is less expensive and provides for very rapid automatic control. To be efficient, a control device must have very low resistance. A PN junction has very low resistance (ideally zero) when it is conducting (on) and has high resistance (ideally infinite) when it is nonconducting (off).

When on, the low resistance of the junction causes the power dissipation to be almost zero. When off, its very high resistance causes almost no current flow in the circuit. Hence, the power dissipated in it is almost zero. Thus, there is never any significant power dissipation in the control device. Because of this, a semiconductor device makes an ideal electronic switch.

Figure 11–3 shows a control circuit that employs an electronic switch. At this point in the discussion, it might seem as though the electronic switch just provides on/off control. However, imagine a fast switch that can open and close 60 times per second, with the switch being on for a portion of the entire AC cycle. Because the lamp is connected to the source only for a portion of the AC cycle, it will operate at a lower intensity. Because of the high speed of switching (60 times per second or more), the lamp will dim without a noticeable flicker.

A mechanical switch used in an application such as that shown in Figure 11–3 would not be nearly as efficient or rugged. The mechanical switch would generally be much slower, and the contacts would tend to wear out very quickly. Now that you have a basic understanding of electronic switching and control circuits, let's look at some electronic control devices. The control devices discussed in this chapter belong to a family of solid-state devices called **thyristors.**

Thyristor
A semiconductor device with three or more junctions that has the ability to turn on and/or off by application of an external signal.

FIGURE 11–3 A switch control circuit.

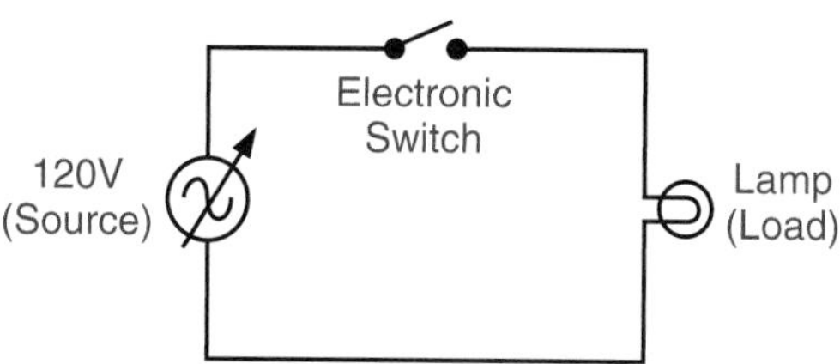

THE SILICON-CONTROLLED RECTIFIER

11.4 Basic Construction and Operation

The SCR is the most popular of the control switches. The SCR is a four-layered, three-terminal device also referred to as a four-layer diode. The three terminals are referred to as the anode (A), cathode (K), and gate (G). It is constructed with two P-type regions and two N-type regions in a p-n-p-n sequence, as shown in Figure 11–4a. It is called a diode because it conducts in one direction and blocks in the other. Figure 11–4b shows the schematic symbol for an SCR. The symbol is that of a solid-state diode with the addition of a gate lead.

FIGURE 11–4 The SCR: (a) construction, (b) schematic symbol.

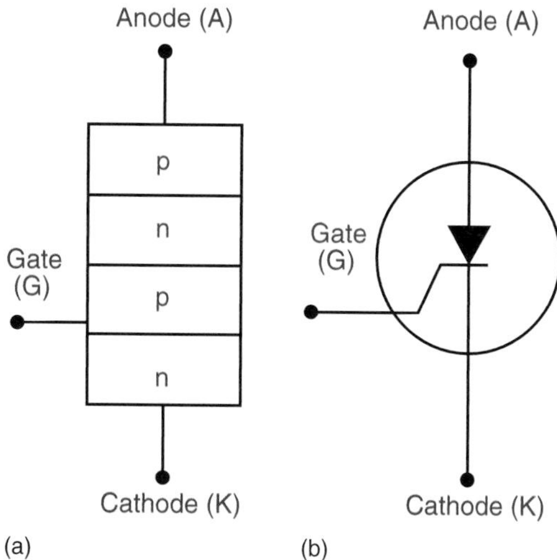

To understand the operation of an SCR, consider Figure 11–5a. Here you see an SCR with the middle PN junction split in half. This shows that the SCR can be modeled as two transistors, as shown in Figure 11–5b. The equivalent circuit shows two directly connected transistors, one PNP and the other NPN. Note that the collector of the PNP transistor feeds into the base of the NPN transistor, and vice versa.

FIGURE 11–5 SCR split construction and equivalent circuit.

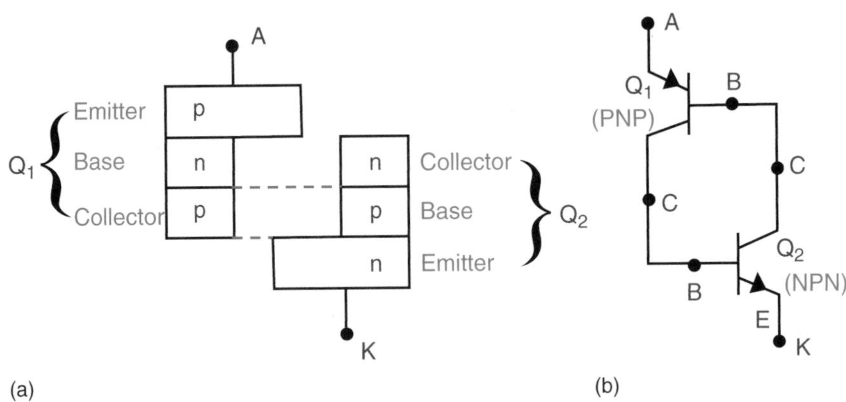

Now consider a switching circuit that uses the same type of two-transistor circuit (Figure 11–6). When the gate switch is closed, the positive side of the supply is applied to the base of the NPN transistor. This forward biases the base-emitter junction, and the NPN transistor turns on. This in turn causes current to be supplied to the base of the PNP transistor and hence the transistor comes on. With both transistors on, the circuit is now complete and current flows through the load.

FIGURE 11–6 A two-transistor switching circuit showing the operation of an SCR.

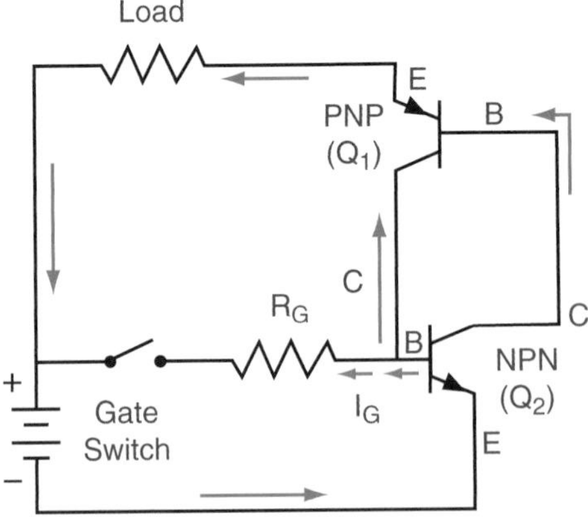

When both transistors are in conduction (on) and the gate switch is opened, the transistors will not go off because they are now supplying each other with base current. Therefore, you can see that once triggered on by the gating current the transistors will not turn off even if the gate switch is open and the gate current I_G is cut off. This type of circuit that stays on once triggered is called a "latch" circuit. Once latched on, the only way the transistors go off is if the source voltage is removed or the load circuit is opened.

The two-transistor equivalent circuit operation gives a basis for understanding the SCR operation. The SCR is a device that conducts when gate current is applied in addition to a positive anode-to-cathode voltage. The SCR then stays latched on until the current through it is interrupted.

11.5 Characteristics

Figure 11–7 shows the volt-ampere characteristics of an SCR for both forward (+V) and reverse bias (−V). When reverse biased, the SCR behaves like an ordinary diode in that it is off and very little current flows until the reverse breakover voltage is reached. Reverse breakover is avoided by using an SCR with ratings greater than the operating voltage of the circuit.

FIGURE 11–7 Volt-ampere characteristics of an SCR.

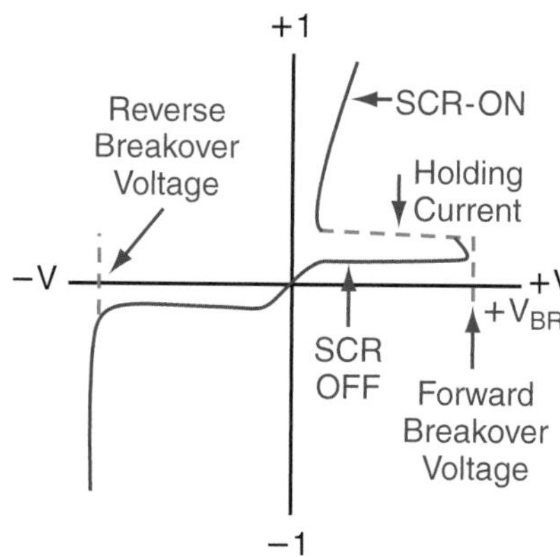

The SCR differs from an ordinary diode in its forward-biased operation. The SCR stays in its off state until forward breakover voltage is reached, at which point the SCR switches to the on state. The voltage drop across the SCR decreases rapidly, and the current increases. The holding current is the minimum current required to keep the SCR latched on.

Figure 11–7 does not show the effect of the gate current on the characteristics. The presence of the gate terminal distinguishes an SCR from an ordinary diode. The voltage at which forward breakover occurs is controlled by the gate current. Figure 11–8 shows how the gate current affects forward breakover. As the gate current increases, the forward voltage required for turning on the SCR decreases. This means that a voltage below maximum breakover can be applied and the SCR will not conduct until gate current is applied.

Normally, the SCR is not operated at high forward breakover voltages. A gate pulse large enough to turn the SCR on at relatively low forward voltages is chosen. Once triggered on, the SCR remains in conduction until the current flow through the SCR is reduced below the holding current. The SCR works like a switch, at least for turning it on.

FIGURE 11–8 The effect of gate current on breakover voltage.

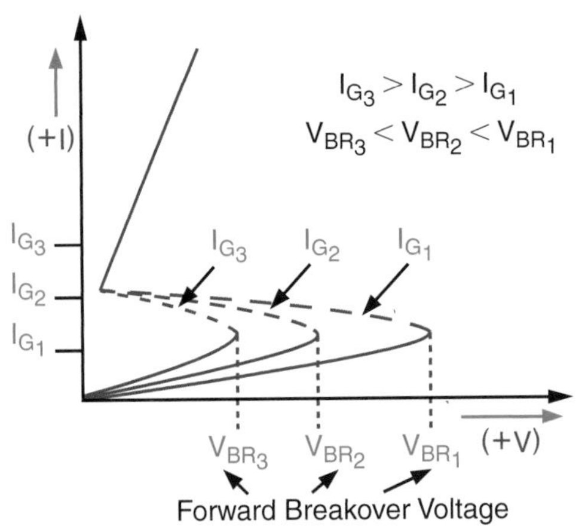

11.6 Commutation

Commutation
The act of turning an "on" thyristor to its "off" state.

The process of turning an SCR off is called **commutation.** Turn-off (commutation) is achieved by reducing the current through the SCR to a value below the holding current level. This is done in practical circuits by one of the following two methods:

1. Opening a series switch that interrupts the current flow in the circuit (Figure 11–9a)
2. Closing a parallel switch that reduces the forward bias to zero (Figure 11–9b)

FIGURE 11–9 Commutation of an SCR: (a) series switch, (b) parallel switch.

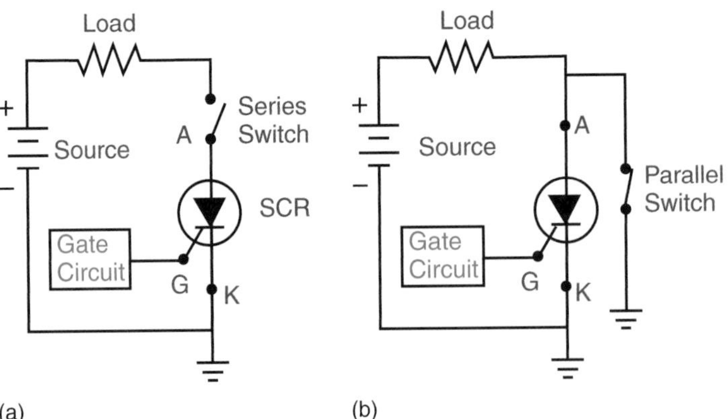

Turning the SCR off is not complete until all carriers in the center junctions of the SCR recombine. Recombination is a process by which free electrons occupy holes (valence shells that have a deficiency of electrons) and eliminate the carrier.

If forward bias is applied to the SCR before recombination is complete, it may turn on. Recombination takes some time.

The **"turn-off" time** is defined as the time that elapses after current flow stops and before forward bias can be applied without turning the device on. A zero bias between the anode and cathode can turn off the SCR. However, the fastest possible turn-off is achieved by applying a reverse bias between the anode and cathode.

Turn-off time
Time that elapses after anode current stops, and before forward bias can be applied without turning the device on.

11.7 SCR Control Circuits

Figure 11–10 shows how an SCR can be used to control AC power applied to a load. Assume that the load is a lamp and analyze how its brightness can be controlled. Because the SCR is a unidirectional device, it will conduct only during half of the AC cycle.

FIGURE 11–10 Half-wave control circuit using an SCR.

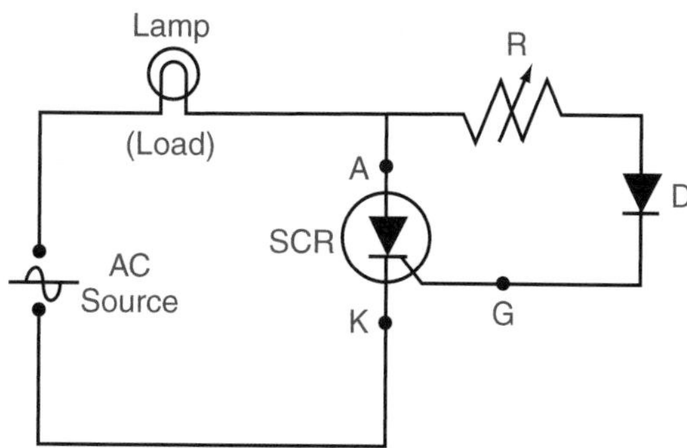

The circuit shown in Figure 11–10 is therefore a half-wave circuit. The value of the variable resistance, *R*, controls the current in the gate circuit and determines at what instant during the conducting half cycle the SCR is turned on. The SCR is reverse biased when the source changes polarity and is turned off. The diode, *D*, keeps the gate current from flowing when the polarity of the source is reversed.

Figure 11–11 demonstrates how controlling the conduction angle can vary the brightness of the lamp. A small conduction angle means that the circuit is on for a small portion of the AC cycle. Conversely, a large conduction angle keeps the circuit on for a large portion of the AC cycle. If the SCR is gated on late during the positive half cycle (small conduction angle), the load power dissipation will be low. On the other hand, if the SCR is triggered earlier during the positive half cycle (larger conduction angle) the load power dissipation increases.

The SCR control circuit is very efficient because most of the power is dissipated in the load. The forward resistance of the SCR is very small, and the power dissipated in the control device (SCR) is only a fraction of the total power.

FIGURE 11–11 Control of power in an AC circuit by varying conduction angles.

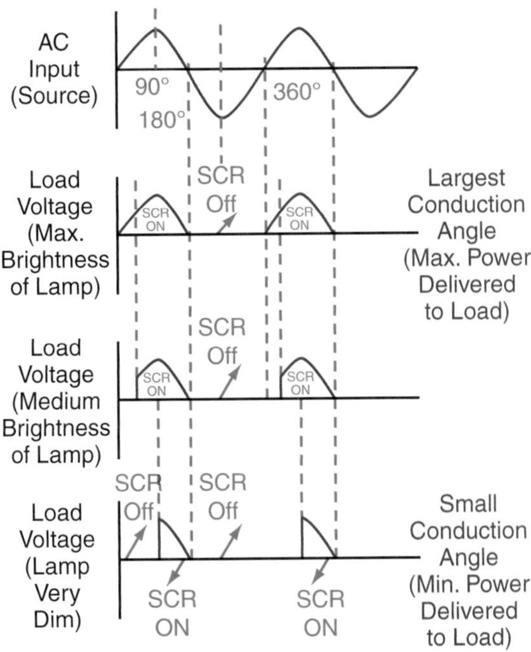

An SCR can be used to obtain full-wave control by combining it with a rectifier circuit. Figure 11–12 illustrates the output of a full-wave rectifier. This output can then be fed to the SCR, as shown in Figure 11–13. When the pulsating DC waveform reduces to zero, there is zero forward bias applied to the SCR, thus dropping the holding current to zero and turning the SCR off.

Figure 11–13 shows a full-wave control circuit using an SCR. A center-tapped full-wave rectifier with diodes D_1 and D_2 is used to obtain full-wave pulsating DC voltage across the SCR. Resistor R sets the gate current, which controls where in the cycle the SCR comes on. Thus, full-wave power control to the load can be achieved by controlling the conduction angle of the SCR. A similar result could be obtained by substituting two SCRs for diodes D_1 and D_2 and triggering them on the alternate half cycles.

FIGURE 11–12 Full-wave pulsating direct current.

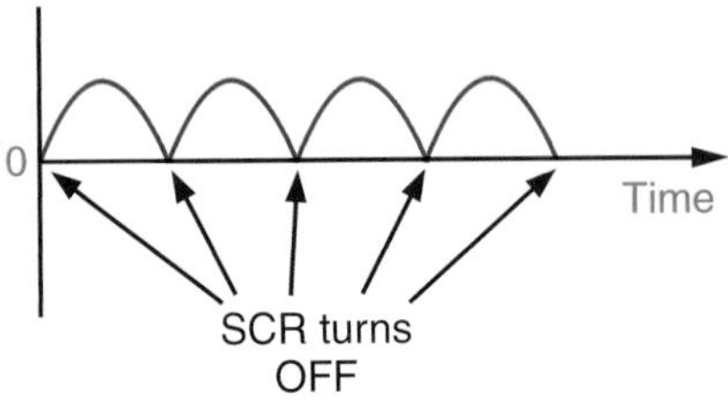

FIGURE 11–13 Full-wave control circuit using an SCR.

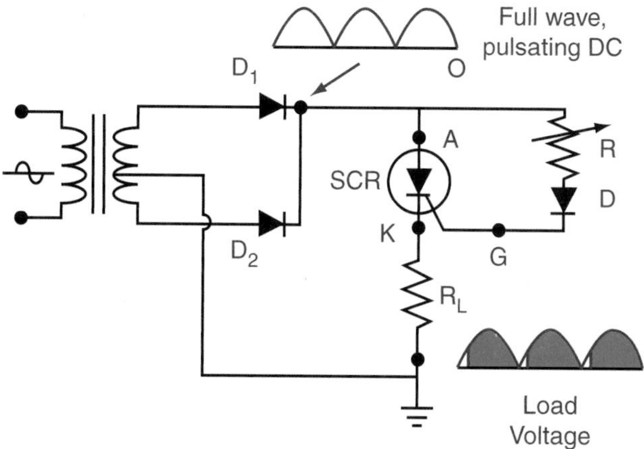

THE TRIAC

11.8 Basic Construction and Operation

The SCR, as discussed in the previous section, is a unidirectional device. A triac (triode AC semiconductor switch) is a four-layered bidirectional device and accomplishes the function of two SCRs. It can be considered two SCRs connected in antiparallel. When one of the SCRs conducts, the other one does not, and vice versa. A triac is therefore a full-wave device and can be viewed as a bidirectional SCR. Both the SCR and triac are thyristors.

Figure 11–14a shows the construction of a triac. The triac connections are called main terminal 1 (MT1), main terminal 2 (MT2), and the gate. The MT2 can be positive or negative with respect to MT1 when triggering occurs. The triggering pulses applied by the gate can be positive or negative with respect to the MT1. This leads to four possible combinations by which the triac can be turned on. They are:

1. MT2 (+), gate (+)
2. MT2 (+), gate (−)
3. MT2 (−), gate (+)
4. MT2 (−), gate (−)

Of the combinations listed previously, 1 and 4 are those generally employed to trigger the triac. Note that when a positive triggering pulse is applied to the gate MT2 is positive. When a negative triggering pulse is applied to the gate, MT2 is negative. By changing the amount of gate current, the instant at which the triac comes on can be controlled.

Figure 11–14b shows the schematic symbol for a triac. Keep in mind that triacs have smaller current and voltage ratings than SCRs. Triacs are suitable for small- and medium-power AC applications, whereas SCRs are capable of handling high-power applications.

FIGURE 11–14 The triac: (a) construction, (b) schematic symbol.

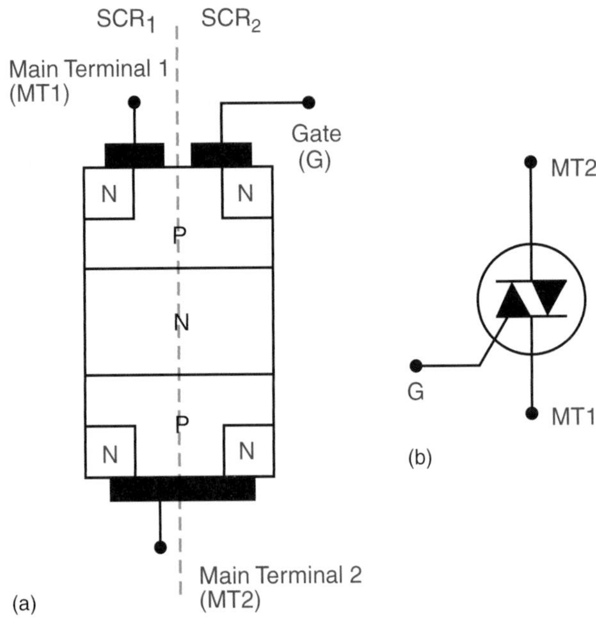

11.9 Triac Control Circuits

Figure 11–15 shows a circuit that uses a triac to control AC power. The gate pulses allow the triac to be triggered during both the positive and negative alteration of the AC source. Just as discussed in SCRs, the timing of these trigger pulses can be used to control the conduction angle and the power delivered to the load.

FIGURE 11–15 Control circuit using a triac.

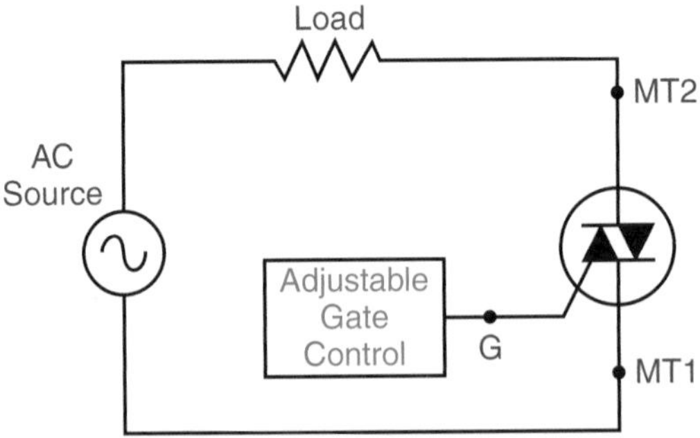

Figure 11–16 illustrates the conduction angle control in a triac control circuit. By comparing Figure 11–16 with Figure 11–11, you can see that the triac is a full-wave device, whereas the SCR is a half-wave device.

FIGURE 11–16 Conduction angle control in a triac control circuit.

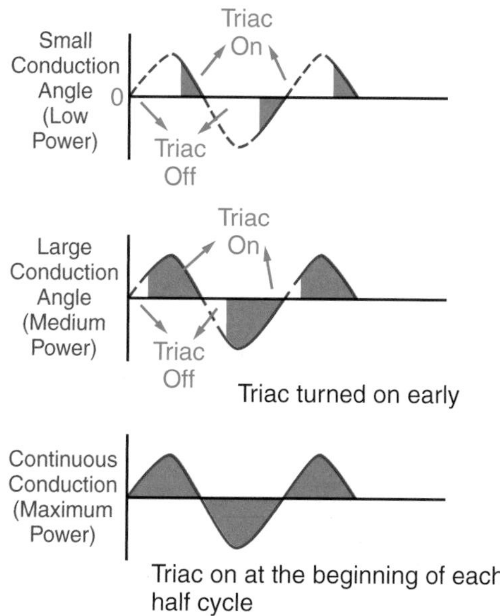

11.10 Static Switching

SCRs and triacs can be used to perform the task of static switching of AC loads. A **static switch** is a switch with no moving parts. Switches that contain moving parts tend to wear out faster and are subjected to contact bounce and arcing. Such problems are eliminated by electronic circuits that use static switches such as triacs and SCRs. The triac functions well as a static switch at lower frequencies (50 Hz to 400 Hz), whereas the SCR is designed to operate up to frequencies of 30 kHz.

Static switch
A switch that uses semiconductor devices such as thyristors to perform the on and off operations.

A three-position static switching circuit is shown in Figure 11–17.

- In position 1, no gate signal is applied and the triac is off.
- In position 2, the presence of the diode causes the gate signal to be applied only during the positive alteration of the AC source. Hence, the triac is on only for half the AC cycle and half power is delivered to the load.
- In position 3, the gate signal is applied to the triac during both alterations of the AC cycle. Hence, the triac conducts during both alterations of the AC cycle and full power is delivered to the load.

Although the three-position switch shown in Figure 11–17 has a mechanical switch, note that it operates in the gate circuit (which carries low currents). As a result, the mechanical switch is not subject to arcing and related contact problems.

FIGURE 11–17 A three-position static switching circuit.

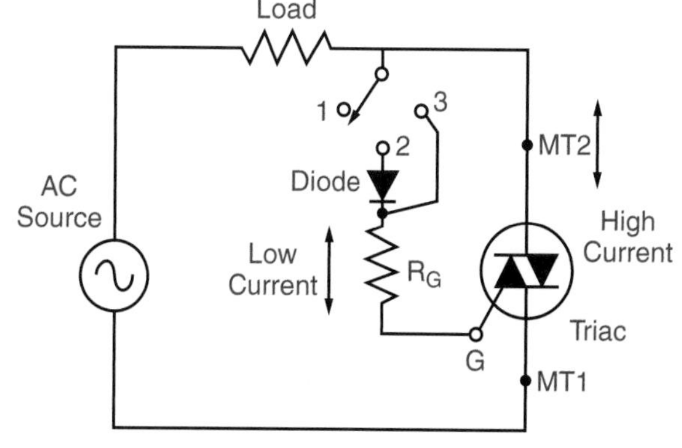

11.11 Commutation

The triac is a bidirectional device, which means that commutation is different from an SCR. When the AC power becomes zero, the triac has to go off. We know that power hits the zero point twice in one AC cycle. If the triac does not turn off at these points, power control cannot be achieved.

When the load is resistive, voltage and current in the circuit are in phase. The zero-power point is reached when the current is zero, the voltage is zero, and the triac is turned off with no problems. However, if the load is reactive (inductive or capacitive) the voltage and current in the circuit are out of phase and turning the triac off may be a problem.

Consider an inductive load such as a motor, where the motor winding offers an inductive reactance. The current in the circuit lags the voltage. When the current becomes zero, the voltage is not zero and is applied across the triac. This voltage may be sufficient to turn the triac on, resulting in false triggering. Thus, commutation is not achieved and power control may be lost.

In addition to inductive loads, transients can affect triacs (and SCRs). The term **transient** is used to refer to a large voltage change occurring for a very short time. The transient voltages are usually associated with noise signals and can be of magnitudes high enough to turn the device on.

Transient
A temporary change in a circuit, usually used when referring to a momentary high-energy pulse.

Thyristors are made of PN junctions, and when not conducting these junctions have a depletion region. The depletion regions are insulators and can act as the dielectric of a capacitor. Hence, we can associate internal capacitances with the nonconducting thyristor. These capacitances respond to sudden changes in voltages caused by transients and draw charging currents. The charging currents may serve as gate currents and turn the device on, thereby resulting in false triggering. Hence, power control will be lost if the thyristor responds to transients.

11.12 Snubber Network

As was discussed in the previous section, inductive loads and transients can trigger the triac on and can cause the power control to be lost. A snubber circuit reduces false triggering in both triac and SCR control circuits. Figure 11–18 shows an RC snubber network connected to a triac control circuit.

FIGURE 11–18 Snubber network.

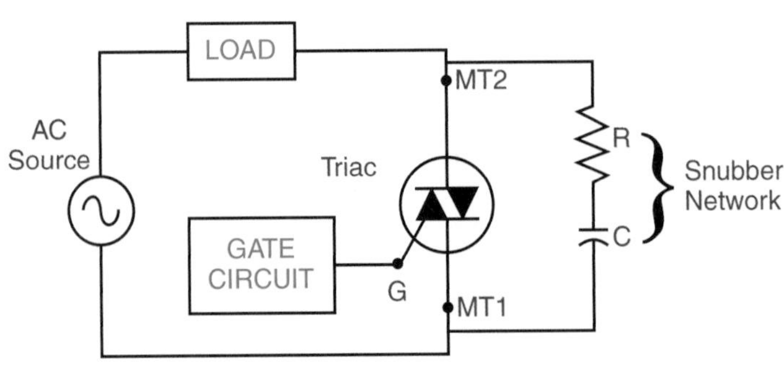

The property of the capacitor to oppose changes in voltage is used in the RC snubber circuit. The RC circuit is connected between MT2 and MT1 of the triac (anode and cathode of the SCR). The snubber network bypasses the charging currents from the nonconducting thyristor. The resistor (R) is included to limit the discharge current of the capacitor (C) when the thyristor turns on.

11.13 Varistors

Most of us have AC outlet extension strips that claim to incorporate transient suppression. One common suppression technique used in these strips utilizes the varistor. The varistor (or MOV, metal-oxide varistor) is a voltage transient suppression device. It is a two-legged, typically coin-size, semiconductor device (Figure 11–19a and Figure 11–19b). Varistor resistance decreases rapidly as terminal voltage rises above its rated RMS voltage. In other words, varistors act like bidirectional high-voltage Zeners.

FIGURE 11–19 (a) Typical varistor compared to a U.S. quarter. (b) Varistor schematic symbols.

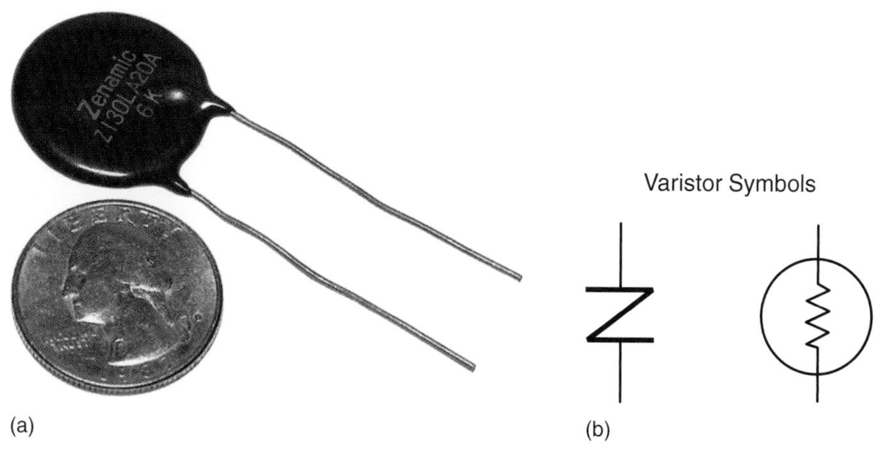

Varistors are intended to prevent voltage spikes from damaging load-switching devices such as SCRs, triacs, relay contacts, and manual switches. Varistors are found connected across power lines, inductive loads, and thyristors. They provide protection in both AC and DC circuit applications, although simple inexpensive high-voltage diodes are good choices for DC circuits.

See Figure 11–20a through Figure 11–20d for typical varistor and snubber wiring. Varistors are often used in conjunction with snubber networks. The major function of the varistor is high-voltage spike suppression. The major function of the snubber is electrical noise suppression.

FieldNote!

In regard to transient suppression, a technician in a maintenance shop was tired from a long day of repairing magnetic contactor relay driver circuit card assemblies (CCAs). He was relieved when his last challenge revealed a shorted SCR. He replaced the SCR with a new component, and before heading home sent the board back to the Field Service department for reinstallation.

Two days later, the same CCA came back. Same problem. Why? Another look at the circuit diagram screamed the answer. There was a varistor across the SCR anode-cathode pins. This time the tech replaced the SCR *and* the varistor. The CCA did not come back from the field again. (See Section 11.19 for the likely varistor failure mode.)

FIGURE 11–20 Noise suppression with snubber across switching device.

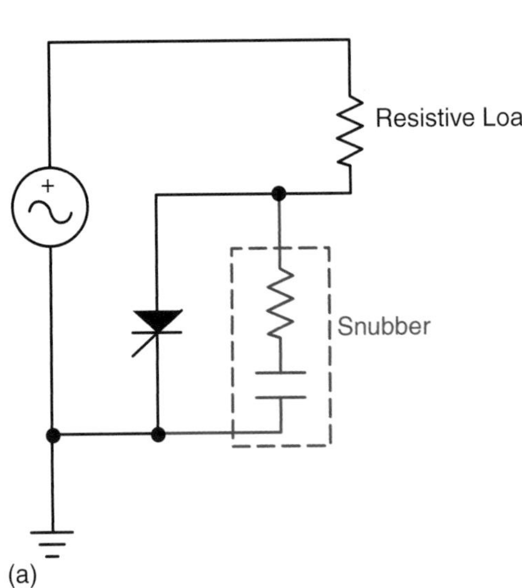

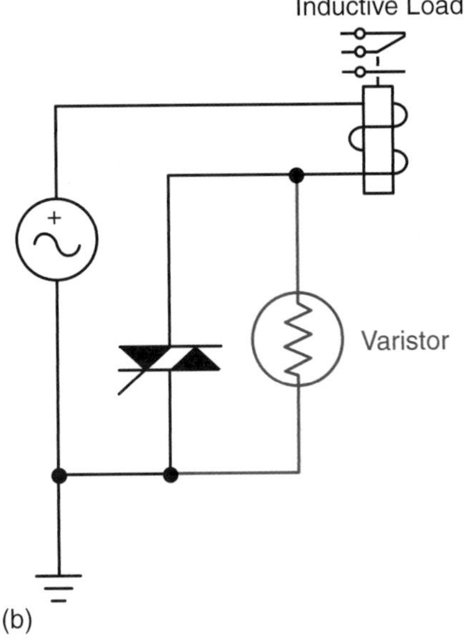

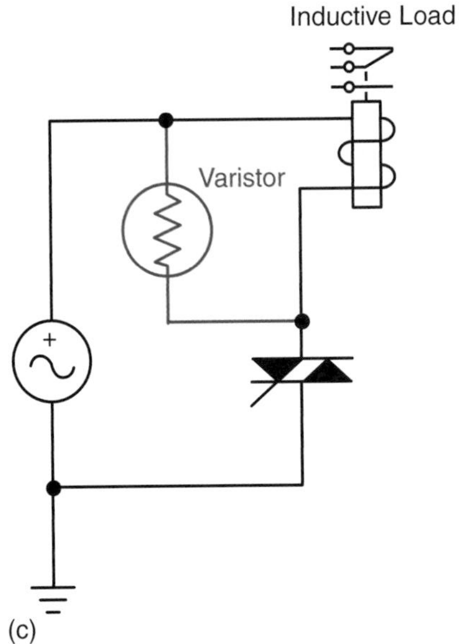

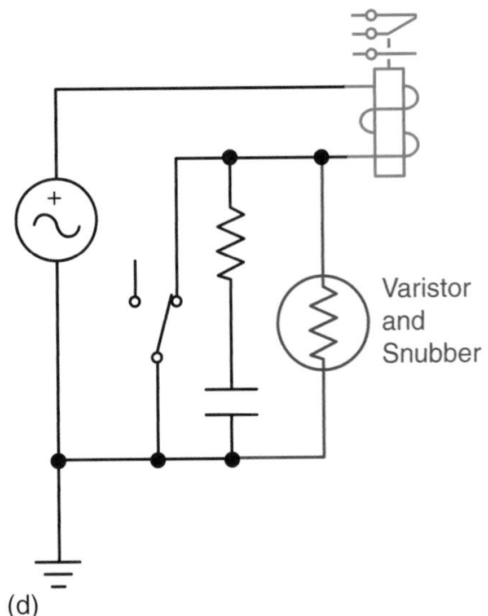

THE DIAC

11.14 Basic Construction and Operation

The diac is a three-layer bidirectional device that has only two external terminals. The construction of the diac is shown in Figure 11–21a, and its schematic symbol is shown in Figure 11–21b. The terminals of the diac are called anode 1 and anode 2 and can be connected interchangeably. The diac's construction is very similar to that of an NPN or PNP transistor. There are, however, crucial differences.

FIGURE 11–21 The diac: (a) construction, (b) schematic symbol.

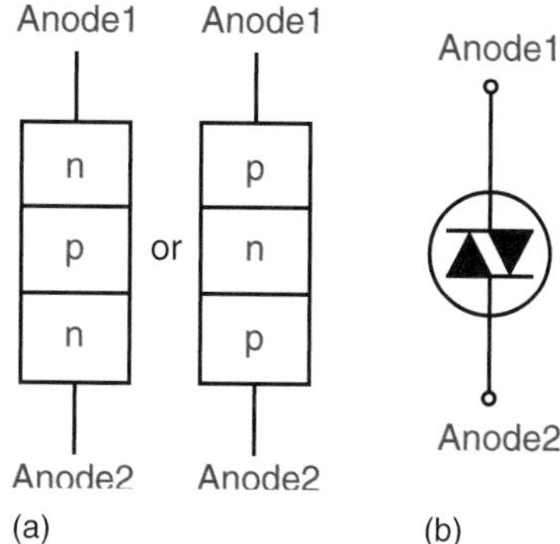

- A diac has no base connection.
- The diac's three regions are identical in size and level of doping. In the BJT, the base is extremely narrow in dimension and very lightly doped.

The operating characteristics of a diac are shown in Figure 11–22. As you can see, the forward operating characteristic is identical to the reverse characteristic. In either direction, the diac remains open (off) until the breakover voltage is reached. At this point, the diac is triggered and conducts (on) in the appropriate direction. The diac continues to conduct until the current drops below the holding current value. The operating principles are the same for an NPN and PNP diac.

FIGURE 11–22 The diac operating characteristic curve.

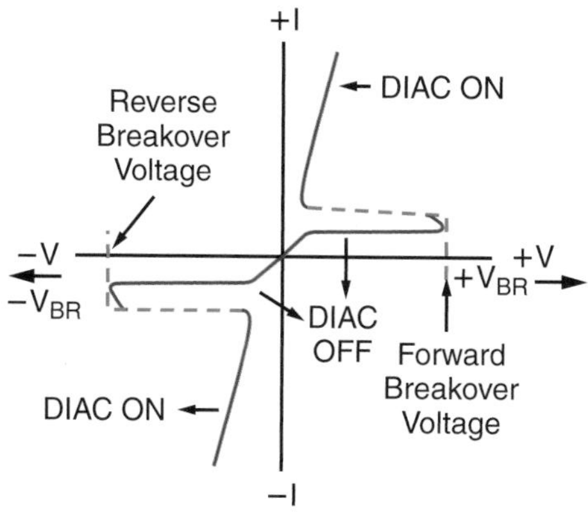

11.15 Diac-Triac Control Circuit

The bidirectional diac is very well suited for triggering the triac. The following sections describe how the two devices work together.

Example

Figure 11–23 shows an example of a diac-triac control circuit. The diac is used in the gate circuit of the triac.

Operation in the Positive Half Cycle

In the positive half cycle (Figure 11–23a), diode D_1 is forward biased and D_2 is reverse biased. A positive voltage appears at point P. When this voltage becomes greater than the breakover value of the diac, the diac conducts. When the voltage is high enough, resistor R passes enough current to trigger the triac. Thus, triggering occurs in the positive alternation of the AC cycle.

Operation in the Negative Half Cycle

During the negative half cycle (Figure 11–23b), diode D_2 is forward biased and D_1 is reverse biased. Point P in the circuit is still positive, but MT2 is negative. This causes the triac to be triggered in the negative alternation of the AC cycle.

The rheostat R is used to change the gate current through the triac, which in turn controls the conduction angle. The load in the circuit is a motor, whose speed of rotation is influenced by the conduction angle. The greater the conduction angle the higher the power delivered to the motor and hence the faster the motor will rotate. Note the presence of RC snubber network R_1C_1 across the diac and R_2C_2 across the triac.

FIGURE 11–23 Diac-triac control circuit.

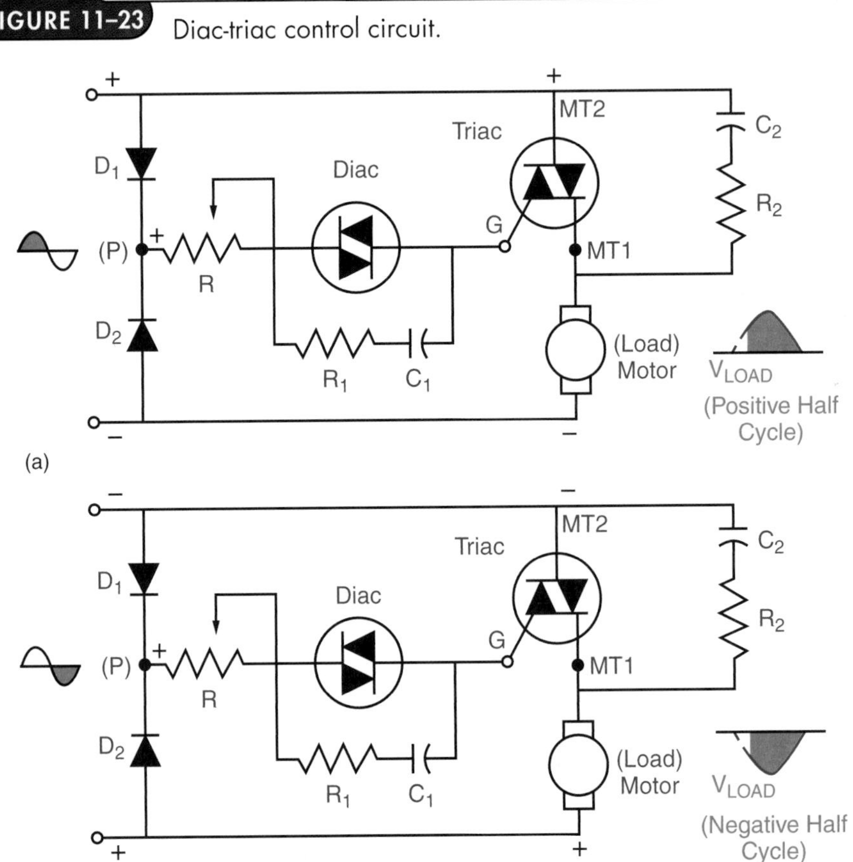

Example

Figure 11–24 shows another example of a diac-triac control circuit. The resistors R_1 and R_2 determine the time it takes for the capacitor C_3 to charge. When C_3 has charged to the breakover point of the diac, the diac conducts, providing a discharge path for the capacitor C_3 through the gate circuit of the triac. The discharge current from C_3 causes the triac to conduct.

The lamp that serves as the load in the circuit can be made to glow with varying brightness by changing the amount of power applied to it. Resistor R_1 determines the timing of the gate current applied to the triac, which in turn controls the conduction angle of the triac.

If the value of R_1 is set to a low value, C_3 charges to the breakover point quickly and the triac comes into conduction earlier in the AC cycle. This will contribute a larger amount of power delivered to the lamp, which will glow with increased brightness. Conversely, decreasing the value of R_1 leads to a chain of events that would reduce the brightness of the lamp.

L_1C_1 is used only for RF filtering. When the LC low-pass filter allows 60 Hz AC line current to pass, it may also allow a wider band of frequencies induced into the wiring from external sources to trigger the triac. The R_1C_2, R_2,C_3 brute force filter provides a narrow frequency trap at the low band end to alleviate this problem and to add triggering stability to the circuit.

R_1, C_2, R_2, and C_3 also form a phase-shift network with the rheostat, forming the primary phase-shift determinant. The resulting phase angle from the RC network delivered to the diac determines the turn-on time of the triac. Recall that phase shift is dependent on the ratio X_C/R. Thus, the smaller R_1 is the greater the phase shift and the less time the triac will conduct.

The L_1C_1 low-pass filter also suppresses radio frequency interference that may be generated in the circuit. When triacs switch from off to on, the sudden increase in current produces harmonic noise. Harmonic energy emitted by the triac circuits can range up to several MHz and may be radiated into the surrounding area by the load wiring. AM radio receivers operating nearby can pick up the signals, and this contributes to noise in the reception.

L_1 and C_1 form a low-pass filter, which (as the name suggests) allows the relatively low AC line frequency signals to pass through to the load. However, signals in the radio frequency range (MHz) will be blocked by the filter and will not reach the load. The presence of the low-pass filter helps eliminate interference to nearby AM radio receivers.

FIGURE 11–24 Another diac-triac control circuit.

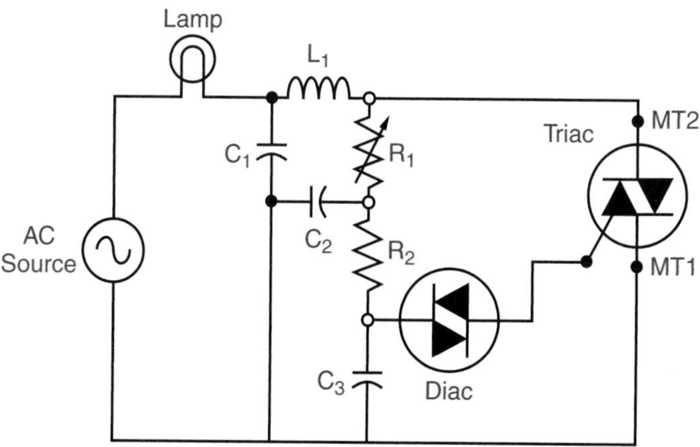

COMPONENT TESTING

11.16 SCR Testing

If an SCR is suspected to be the cause of a malfunction, it should be removed and tested using the ohmmeter test circuit shown in Figure 11–25. The internal battery of the ohmmeter applies different polarities to the terminals of the SCR. The switch SW in the circuit is used to apply or disconnect gate current. The ohmmeter response table shown in Figure 11–25 can be explained as follows:

- Open the SW before connecting the ohmmeter. When switch SW is open, the ohmmeter's battery is not connected to the gate circuit, making the gate current zero and the SCR off. At this setting of SW, the resistance between the anode and cathode should read extremely high, theoretically infinite Ω. This is true no matter what polarity is applied to the cathode.
- When switch SW is closed and the anode is made positive with respect to the cathode, the gate will also be positive through SW. This will turn the SCR on, and 0 Ω (very low resistance) will be measured between the anode and cathode. Open SW at this point. Anode-to-cathode current should continue until the ohmmeter is disconnected. If the anode is made negative with respect to the cathode, the SCR is reverse biased and the ohmmeter should read infinite Ω (or a high resistance) regardless of SW position.

If any one of the ohmmeter readings does not correspond to that listed in the ohmmeter response table of Figure 11–25, the SCR is probably defective and should be replaced.

Note: A DMM may not provide enough holding current to keep the devices turned on once they have been triggered.

FIGURE 11–25 Testing SCRs.

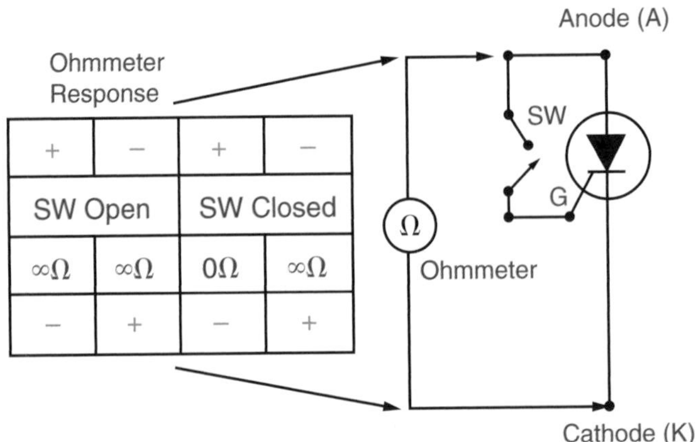

11.17 Triac Testing

If a triac is suspected to be malfunctioning, it should be removed and tested using the ohmmeter test circuit shown in Figure 11–26. As in Figure 11–25, the ohmmeter's internal battery is used to apply different polarities to the terminals. The ohmmeter response table shown in Figure 11–26 can be explained as follows:

- With switch SW open, there is no gate current through the triac and thus the triac is off. Consequently, the resistance between MT2 and MT1 will read very high no matter what polarity is applied between the two.
- When the switch SW is closed, the gate of the triac will receive a trigger current from the battery. Because the triac is bidirectional and operates with either a positive or a negative trigger, the triac should turn on no matter what polarity is applied between MT1 and MT2. Therefore, 0 Ω (low resistance) should be read between MT2 and MT1 either way.

If any one of the ohmmeter readings does not correspond to that listed in the ohmmeter response table of Figure 11–26, the triac is probably defective and should be replaced.

FIGURE 11-26 Testing triacs.

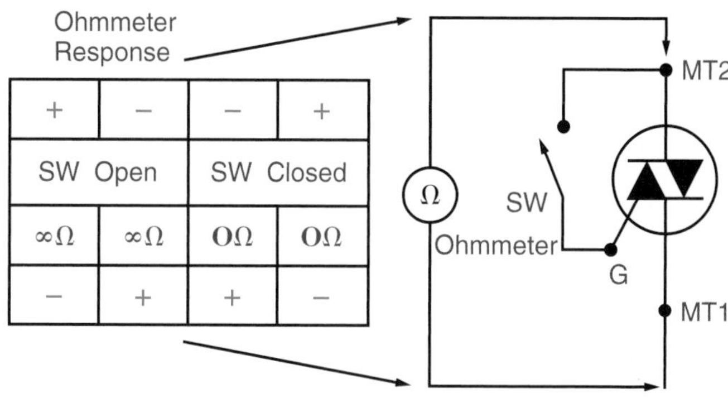

11.18 Diac Testing

If a diac is suspected to be malfunctioning, the diac should be removed and checked with an ohmmeter. The diac is basically two Zener diodes connected back to back in series, and hence the diac should measure ∞ (or very high resistance) between its terminals no matter what polarity is applied between them.

If the ohmmeter reads a low resistance in either of the connections, the diac is probably shorted (see Figure 11–27). Today's DMMs do not present sufficient potential on diode test function, nor resistance test function to cause conventional diacs to reach their reverse breakdown voltage.

FIGURE 11–27 Testing diacs.

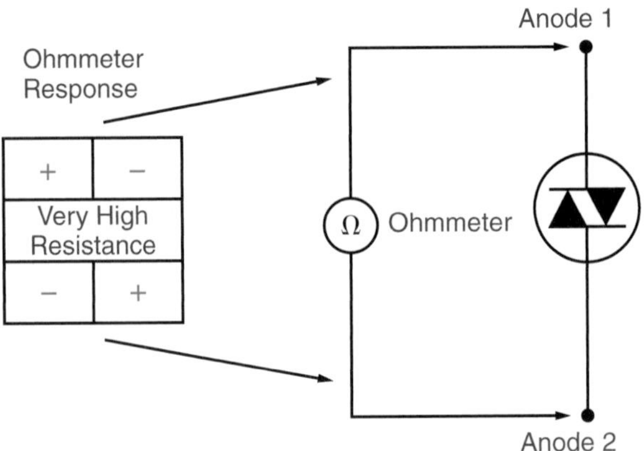

11.19 Varistor Testing

A shorted varistor can easily be identified by observation, or ohmmeter. However, determining an *open* varistor is a different matter. Ordinarily, the varistor may or may not be required to conduct and only exhibits a small capacitive effect. An open varistor *may* not affect routine circuit performance at all. Should a catastrophic voltage spike produce current through the varistor, the varistor may actually exhibit burns or fractures or may even explode.

Sometimes varistors open-circuit after a number of less stressful events. In these cases, the technician may see no evidence that the varistor has opened. Recall from your study of inductors that rapid current changes result in large induced voltages across inductor ends. This would be the case in turn-off of large coil windings. The spike voltage plus the supply voltage would appear across the opening switching device!

The switching device (whether toggle switch, control relay contacts, or solid-state switch) may be stressed to damage or destruction. Excessive contact arcing is a dead giveaway, indicating an open spike suppressor. If the operator has replaced a damaged load, or thyristor switch, he would be wise to replace the varistor as a routine precaution.

PRACTICAL CIRCUITS

11.20 Practical SCR Circuits

Figure 11–28 shows a schematic diagram of a practical elementary crowbar circuit. The purpose of a crowbar circuit is to protect a load circuit from damage in the event the power supply voltage rises above allowable limits. A 15 V power supply is ordinarily allowed to vary by ±1.5 V (10%). Many crowbar circuit configurations can be found. The circuit shown is simple, and is adjustable for a wide range of power supply voltages. Its shortcoming is its slow turn-on. If the 15 V supply exceeds 15 V, the SCR conducts (crowbars), grounding the 15 V power supply output.

The load circuit(s) is thus protected from damage. The fuse blows before damage can be done to the hard-conducting SCR or to the power supply. The 1 kΩ resistor across the fuse allows the 10 kΩ trip-point potentiometer to be adjusted before the fuse is installed. To adjust the potentiometer:

1. Turn off the power supply.
2. Remove the fuse from its socket.
3. Set the potentiometer (rheostat) to 0 Ω.
4. Connect a voltmeter to ±15 V.
5. Turn on the power supply (voltmeter should read slightly less than +15 V).
6. Very slowly rotate the potentiometer wiper until the voltmeter indicates output has dropped to less than +1 V.
7. Move the potentiometer back toward 0 Ω by roughly 20%.
8. Turn off the power supply to reset the SCR.
9. Turn on the power supply, and verify that slightly less than +15 V is present.
10. Turn off the power supply, install the fuse, turn on the power supply, verify +15 V is present, and remove the voltmeter.

FIGURE 11–28 Practical crowbar circuit.

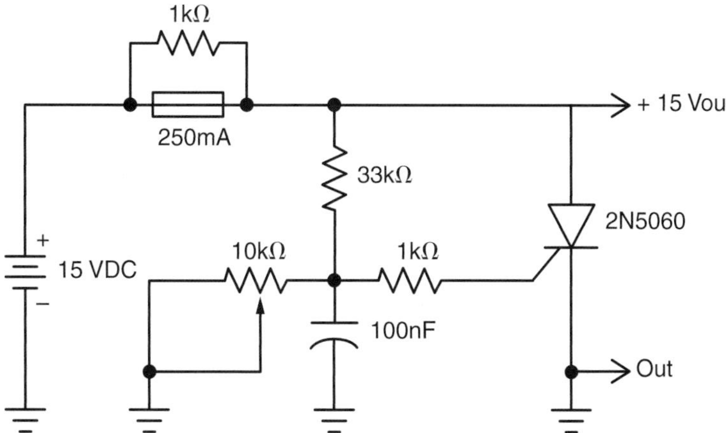

Figure 11–29 shows a schematic diagram of a triple-input SCR control circuit. In this circuit, the load requires a relatively high potential we wish to electrically isolate from the control circuitry. The control circuitry is chosen to be 24 VDC. To turn on the load (a 300 VDC motor in this instance), the relay must be energized. When the motor runs, it represents a danger to the three operators in the vicinity. Therefore, all three operators must simultaneously press momentary switches.

This should guarantee that the workers are at their stations and not in the danger zone when the motor starts. When all three switches are pressed, all three SCRs turn on. The switches may then be deenergized, and the motor will continue to run until the momentary turn-off switch is pressed.

FIGURE 11–29 Practical triple-input SCR control circuit.

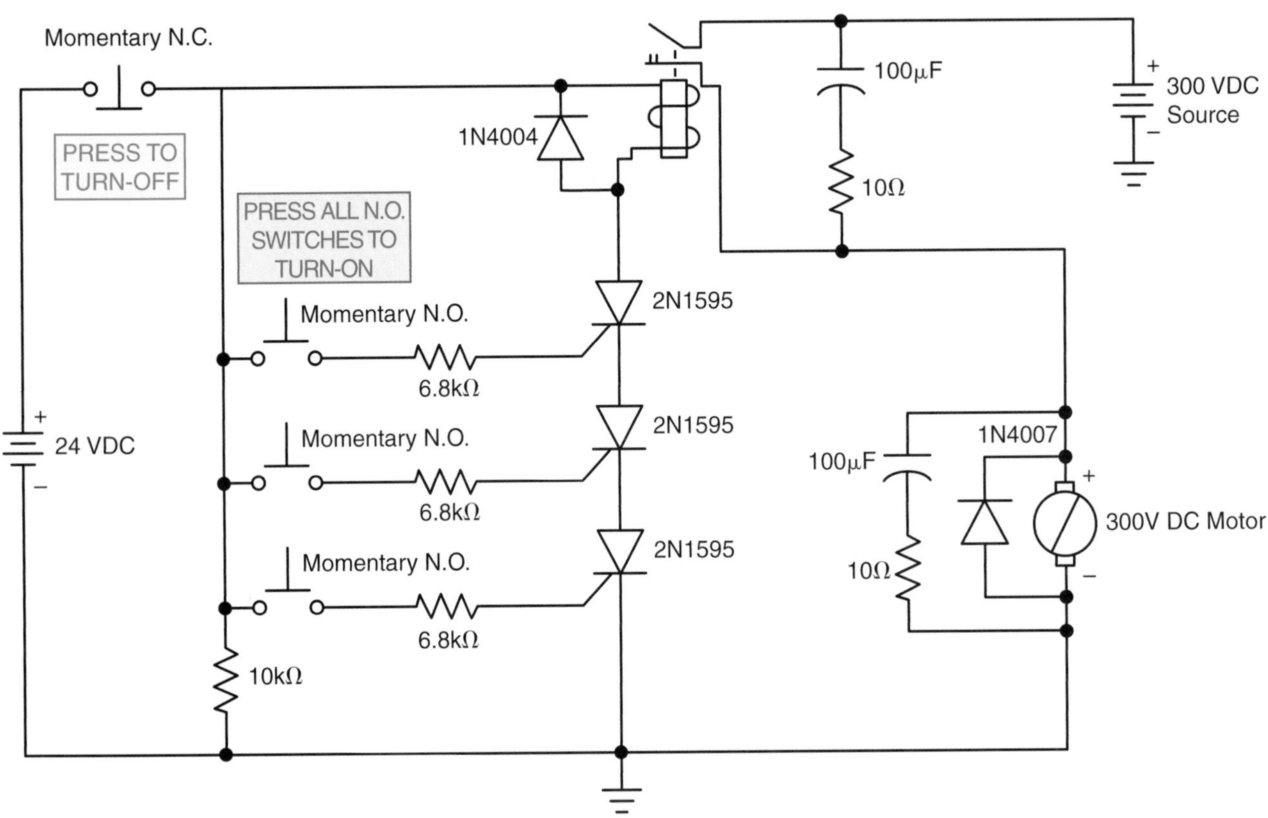

Note the arc suppression snubber across the relay contacts to limit contact arcing and electrical noise. Another snubber is found across the DC motor to absorb CEMF (counter–electromotive force) generated by the motor as the brushes and commutators make/break contact. The diode across the motor also shunts motor CEMF to limit damage to the relay contacts. The diode across the relay coil prevents damage to the SCRs and turn-off switch due to voltage "kickback" when deenergized.

11.21 Practical Triac Circuits

Figure 11–30 shows a schematic diagram of a practical single-phase universal motor controller. Circuits of this type are often found in variable-speed AC hand tools such as drill motors. The universal motor is a series-wound DC motor and will therefore rotate in the same direction regardless of the voltage polarity applied. This means that the motor will run on DC or AC as long as the frequency is not too high.

The triac circuit of Figure 11–30 has a low component count and is very reliable. It has no noise-suppression components and thus noise may be heard from nearby AM radio receivers when the motor turns. Component values will vary depending on the specs of the diac and triac selected.

Figure 11–31 shows a schematic diagram of a circuit commonly used in household lamp dimmers and other low-power single-phase AC devices. Note the similarity to Figure 11–30. The major difference is the noise-suppression capacitors, resistors, and inductor used in the household dimmer.

FIGURE 11–30 Practical universal motor control circuit.

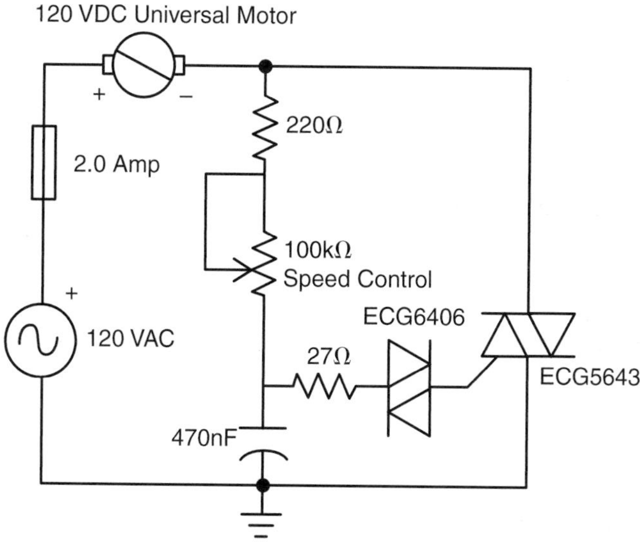

FIGURE 11–31 Practical lamp dimmer circuit.

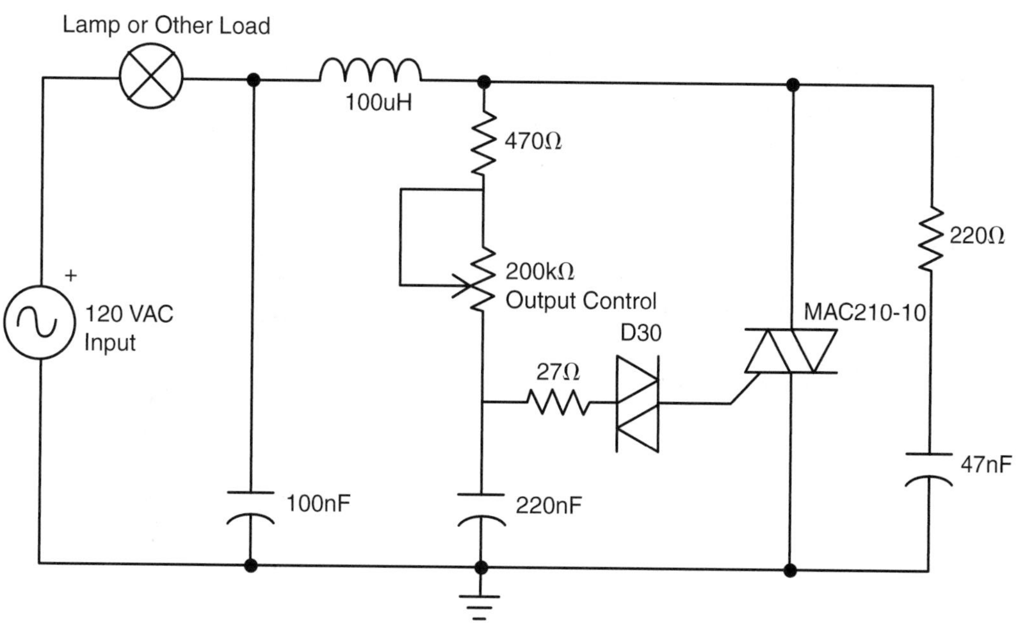

SUMMARY

The discussion in this chapter familiarizes you with electronic control devices (SCR, diac, and triac) and how they can be used in circuits to control the amount of power delivered to the load. This lesson should enable you to identify the schematic symbols of these devices and understand their basic operation.

The lesson also helps you analyze control circuits and appreciate the switching capabilities of thyristors. Finally, the methods of testing an SCR, triac, diac, and varistor (MOV) addressed in the discussion should provide you with techniques to identify malfunctioning components in circuits. Table 11–2 summarizes the semiconductors discussed in this chapter.

TABLE 11–2 Summary of Thyristor Devices

	SCR	Triac	Diac
Symbol	Anode (A) Gate (G) Cathode (K)	Main terminal 2 (MT2) Gate (G) Main terminal 1 (MT1)	Anode 1 (A1) Anode 2 (A2)
Current flow	Unidirectional	Bidirectional	Bidirectional
Common applications	Large current	Modest current Recent improvements allow for increased current capacity.	Triggering device for triacs
Turn on	Forward biased and forward breakover current (which is controlled by the gate current)	Gate pulse in both alterations *Note:* Four combinations of lead termination are possible	Forward and reverse breakover voltage
Turn off (commutation)	(1) Open circuit or (2) Reduce forward bias to 0. *Note:* Recombination has to be complete.	AC power becomes 0 in both positive and negative alteration	Current drops below holding current value
Drawing	K G A SCR	MT2 G MT1 Triac	A2 A1 Diac

REVIEW QUESTIONS

1. What are the advantages of using semiconductor switches for control circuits as opposed to
   a. mechanical switches?
   b. resistor controls?
2. Describe the ohmmeter method for testing SCRs, triacs, and diacs.
3. Examine Figure 11–32. How should resistor R be adjusted to set the load current to zero? To maximum?
4. Which of the circuits discussed in this chapter might be used to make a simple light dimmer?
5. The amount of energy supplied to a load is proportional to the amount of AC waveform passed to it. How does this statement apply to the material covered in this chapter?
6. What type of failure is possible if an "open" varistor is not replaced before powering up a thyristor circuit?

FIGURE 11–32 Full-wave control circuit using an SCR.

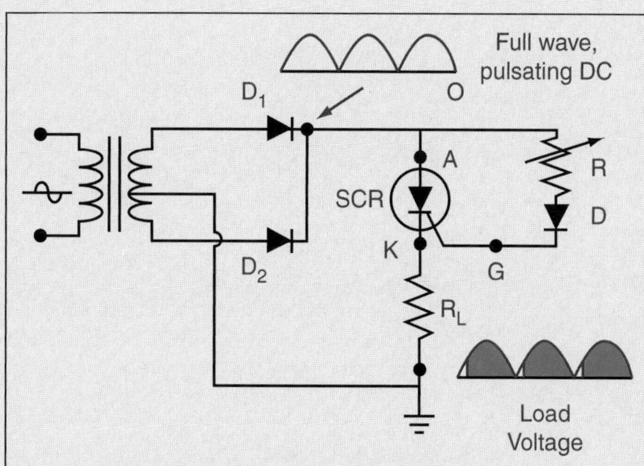

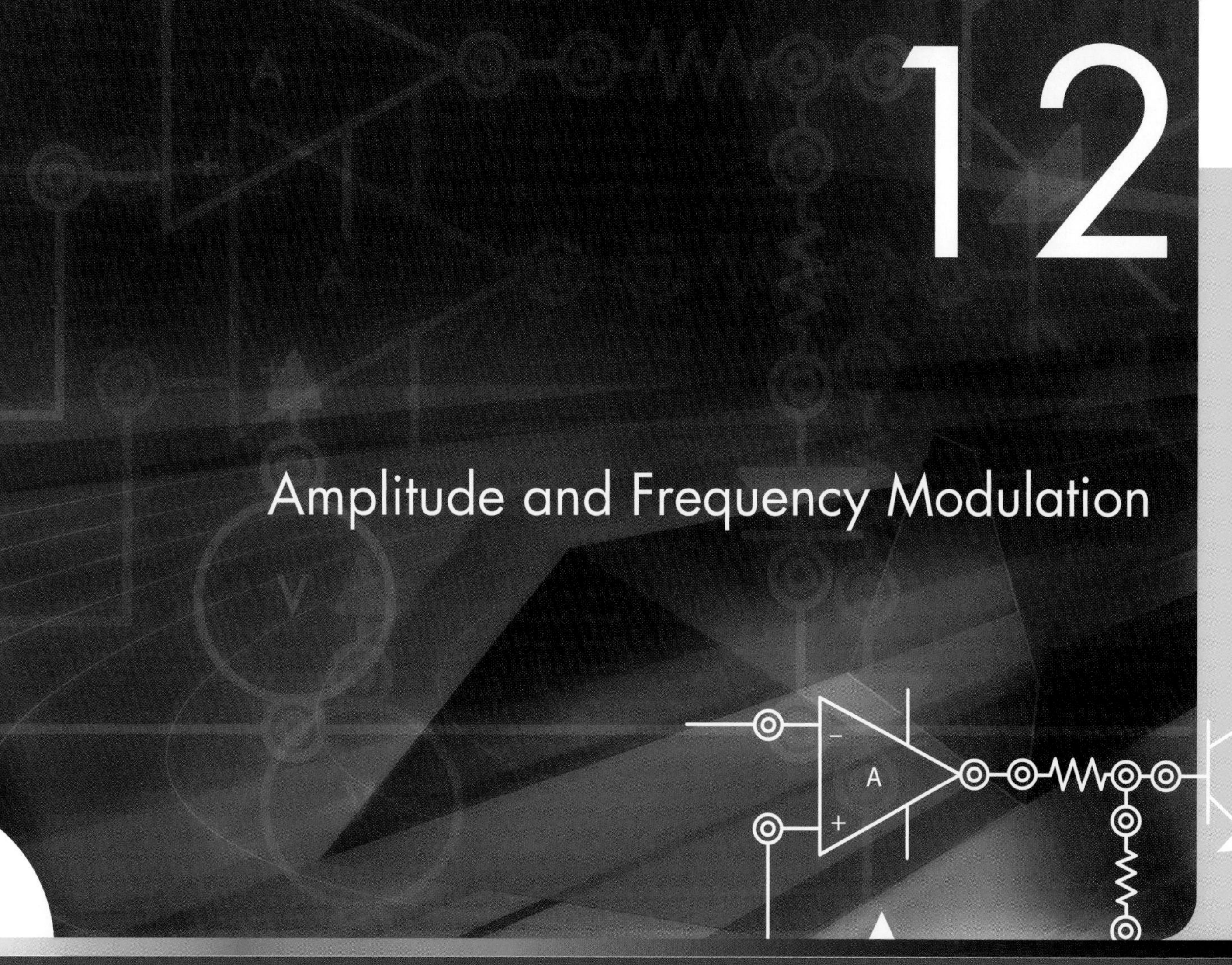

12 Amplitude and Frequency Modulation

OUTLINE

OVERVIEW

The natural desire of people to communicate between two distant points has given birth to modern-day communication systems. The advances in technology today have made it possible to reach practically any point on Earth and distant points in space. This chapter introduces you to the basic concepts of electronic communication systems.

OBJECTIVES

After completing this chapter, the student should be able to:

1. Discuss the purpose of modulation.
2. Identify different types of modulation.
3. Define amplitude modulation and frequency modulation.
4. Explain the operation of modulators and demodulators used in AM and FM.
5. Explain the operation of AM and FM receivers in a block diagram approach.

COMMUNICATION FUNDAMENTALS

12.1 Information Transfer

Communication is the transfer of meaningful information (intelligence) from one point (source) to another (destination). The information to be transmitted can be in the form of audio or video signals, or it could even be digital data. Figure 12–1 illustrates the basic transfer of information from source to destination. Signals can be transmitted from the source to the destination by two means.

1. Wireless communication, where an antenna is used as a transmitting device
2. Line communication, where wave guides, transmission lines, or optical fibers are used as transmitting devices

FIGURE 12–1 Transfer of information.

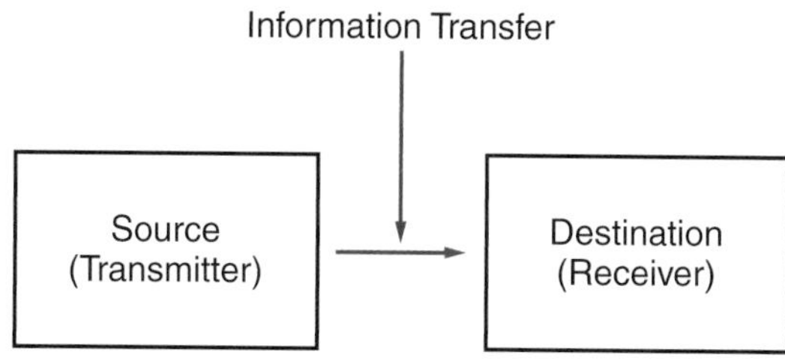

The discussion in this chapter involves wireless communication. An oscillator can be used to produce a high-frequency signal. When the output of this oscillator is fed to a transmitting antenna, it converts the high-frequency alternating current to a radio wave. A radio wave travels at approximately the speed of light, which is 3×10^8 meters/sec.

When this radio wave is incident on another antenna, a high-frequency current is induced in it. This induced current is of a smaller strength but identical in shape to the current in the transmitting antenna. Thus, transfer of electrical energy from one point to the other without the use of wires (wireless communication) is accomplished.

12.2 Modulation and Demodulation

Modulation is defined as the process of impressing information onto a high-frequency signal for the purpose of transmission. The information is in the form of a low-frequency signal and is called the modulating signal, whereas the high-frequency signal is called the **carrier.** The high-frequency signal "carries" the low-frequency signal from the source to the destination. The **modulator,** therefore, has to be present at the source (or the transmitter end) of the communication system.

Demodulation is the process of removing information from the carrier. A **demodulator** recreates the low-frequency modulating signal at the destination (or the receiver end) of the communication system.

Modulation
In electronics and communications, the process of attaching information to a signal or carrier wave.

Carrier wave
The signal used to carry the information after it has been modulated.

Modulator
The electronic circuit that impresses the information signal onto the carrier frequency.

Demodulator
An electronic circuit that recaptures the information signal from the received signal.

You may wonder at this point why the process of modulation or demodulation is required at all. Why not transmit the information directly? The reason is the impracticality of transmitting low-frequency signals. Two important factors contribute to the impracticality.

1. *Antenna length:* For efficient propagation of signals in the air, the length of the antennas is crucial. The antennas length depends on the wavelength of the signal. Wavelength in meters is defined by the equation $\lambda = \dfrac{3 \times 10^8}{f}$.

   High-frequency carriers have small wavelengths, and the antenna required to transmit these signals range over a few feet in length. On the other hand, low-frequency information has a larger wavelength, and the antenna required to transmit these signals may span many miles, rendering transmission highly impractical.
2. *Interference:* If everyone transmitted low frequencies directly, interference between each transmitted signal would make all of the signals ineffective. With modulation, different carrier frequencies can be assigned to transmitters and hence reduce the potential for interference among transmitters.

Table 12–1 shows "bands" of usable frequencies in the electromagnetic spectrum. Atmospheric communications are accomplished in bands ranging from VLF (where earth and water penetration is possible for submarine communication) to EHF, where satellite communication is common.

TABLE 12–1 Radio Frequency Spectrum

Frequency	Designation	Abbreviation
30–300 Hz	Extremely low frequency	ELF
300–3000 Hz	Voice frequency	VF
3–30 kHz	Very low frequency	VLF
30–300 kHz	Low frequency	LF
300 kHz–3 MHz	Medium frequency	MF
3–30 MHz	High frequency	HF
30–300 MHz	Very high frequency	VHF
300 MHz–3 GHz	Ultra high frequency	UHF
3–30 GHz	Super high frequency	SHF
30–300 GHz	Extremely high frequency	EHF

12.3 Basic Communication Systems

There are three basic types of communication systems, based on the characteristics of the carrier that is altered.

1. Systems that use amplitude-modulated (AM) carriers

2. Systems that use angle-modulated modulation, which in turn comprises two variations:
   a. Frequency-modulated (FM) carriers
   b. Phase-modulated (PM) carriers
3. Systems that use digital techniques, normally referred to as pulse modulation

This chapter focuses on amplitude modulation (AM) and frequency modulation (FM) only. Figure 12–2 shows a block diagram of a communication system. The following two sections explain how it works.

FIGURE 12–2 Basic communication system.

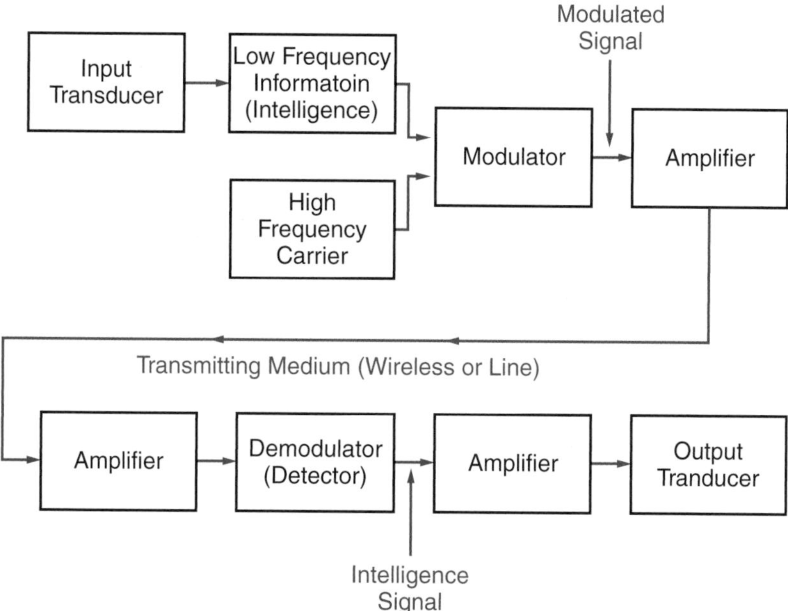

Tansmitting End

An input transducer (such as a microphone or a camera) converts information into an electrical form. This electrical information is the modulating signal. The modulating signal and a high-frequency signal, called the carrier, are injected into the modulator, which produces a modulated signal. The modulated signal is amplified.

Receiving End

The receiver picks up the signal and amplifies it to compensate for the attenuation that occurred during transmission. The output of the amplifier is fed to a demodulator, or **detector,** where the information is extracted from the carrier. The demodulated signal is then fed to another amplifier, which suitably amplifies it so that it can be fed to an output transducer such as a speaker (for audio signals), a monitor (for video signals), or computer (for digital signals). The output transducer converts the information from its electrical form to a physical form such as sound or picture.

Detector
See *Demodulator.*

AMPLITUDE MODULATION

12.4 AM Fundamentals

As the name implies, in an AM system the amplitude of the carrier is altered by the information. In this discussion, we refer to the carrier as radio frequency (RF) signals and the information is audio frequency (AF) signals. Figure 12–3 illustrates the process of amplitude modulation, which produces a resultant amplitude-modulated (AM) wave. It can be seen that the output of the modulator produces an RF signal whose amplitude varies in accordance with the AF signal.

FIGURE 12–3 Process of amplitude modulation.

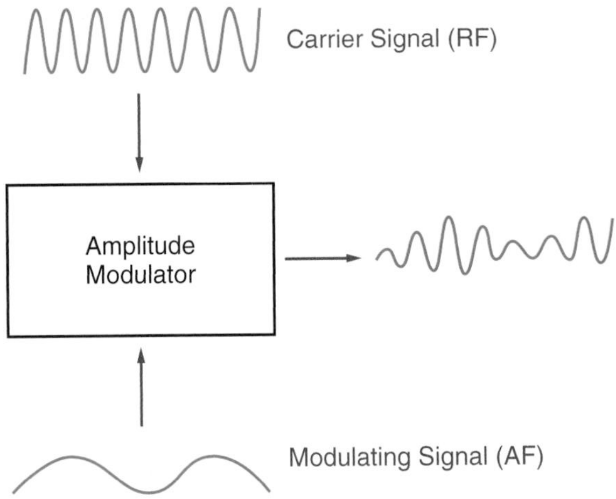

Figure 12–4 shows a 500 watt (input power) high-frequency AM radio transmitter typical of the 1950s. Note the size of the modulator enclosure compared to the RF amplifier sections. Figure 12–5 shows an AM/FM high-frequency radio transceiver of today. Solid-state technology and integrated circuitry, along with improved communication techniques, have collapsed the size and dramatically improved efficiency of radio equipment while reducing cost and improving reliability. An amplitude-modulated signal consists of three frequencies.

1. The carrier frequency that is the original RF oscillator signal
2. The **upper sideband (USB)**, which is carrier frequency plus modulating signal frequency
3. The **lower sideband (LSB)**, which is carrier frequency minus modulating signal frequency

Upper sideband (USB) The carrier frequency plus the modulating frequencies.

Lower sideband (LSB) The carrier frequency minus the modulating frequencies.

The bandwidth of an AM signal is calculated as

Bandwidth (BW) = USB − LSB

or

Bandwidth (BW) = 2 × Modulating signal frequency

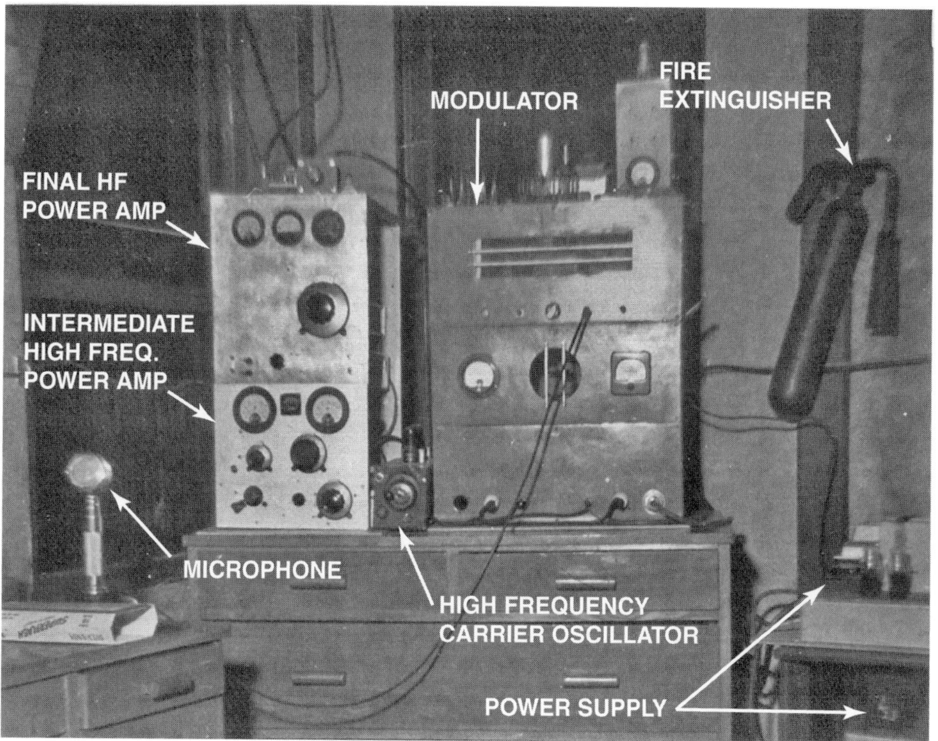

FIGURE 12–4 High-frequency communications transmitter of 1950.

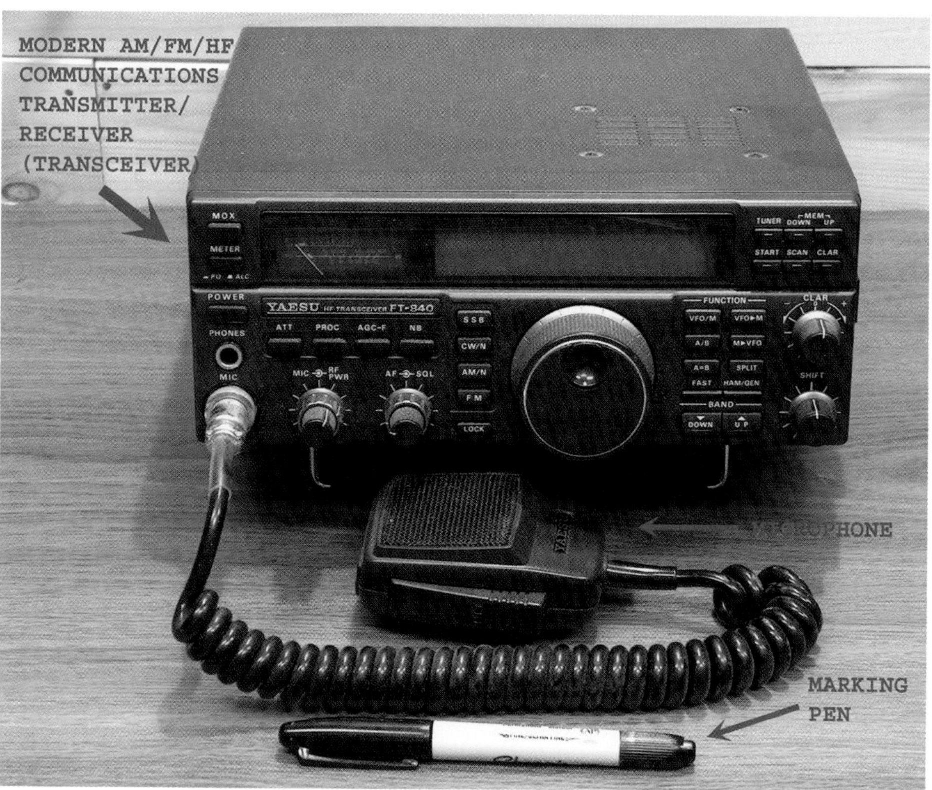

FIGURE 12–5 Modern high-frequency communications transmitter/receiver.

If a 2 kHz audio signal (modulating signal) is used to modulate a 500 kHz RF carrier, the output of the modulator will be made up of three frequencies.

- The carrier frequency is 500 kHz.
- The USB is calculated as (500 kHz + 2 kHz) = 502 kHz.
- The LSB is (500 kHz − 2 kHz) = 498 kHz.

The bandwidth is calculated as
BW = USB − LSB = 502 kHz − 498 kHz = 4 kHz
or
BW = 2 × Modulating frequency = 2 × 2 kHz = 4 KHz.

The AM signal, consisting of its carrier frequency (500 kHz) and its two sidebands (502 kHz and 498 kHz), can be represented with frequency on the X-axis (shown in Figure 12–6). Power in each of the two sidebands cannot exceed 25% of total transmitter output. In wideband applications such as television video, many modulating frequencies are transmitted at the same time. Therefore, useful power (sidebands) may be a very small proportion of the transmitted power.

FIGURE 12–6 AM represented in the frequency domain.

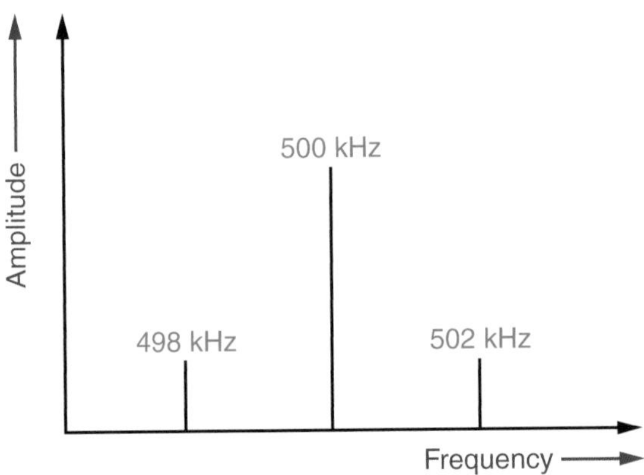

The carrier is largely discarded. An instrument called a spectrum analyzer is used to view an AM signal in the frequency domain. Figure 12–7 shows the AM signal with time on the X-axis. Oscilloscopes are used to view AM signals in the time domain. The observed oscilloscope waveform cannot differentiate sideband signals from carrier signals.

Heterodyne
(1) To mix one signal frequency with another in a nonlinear device to produce their sum and/or difference. (2) The sum or difference signal resulting from the mixing of two or more signals in a nonlinear device.

12.5 A Typical Amplitude Modulator

A typical amplitude modulator is shown in Figure 12–8. The circuit is basically that of a class C amplifier (reverse-base bias). The value of V_{CC} applied to the circuit is varied by the audio signal (information). Class C amplifiers are nonlinear, which means that they will **heterodyne** applied signals. The RF input (carrier) fed to the base of transistor Q_1 via the input circuit formed by resistor R and the capacitor C is one applied signal.

FIGURE 12–7 Modulation and carrier cycles.

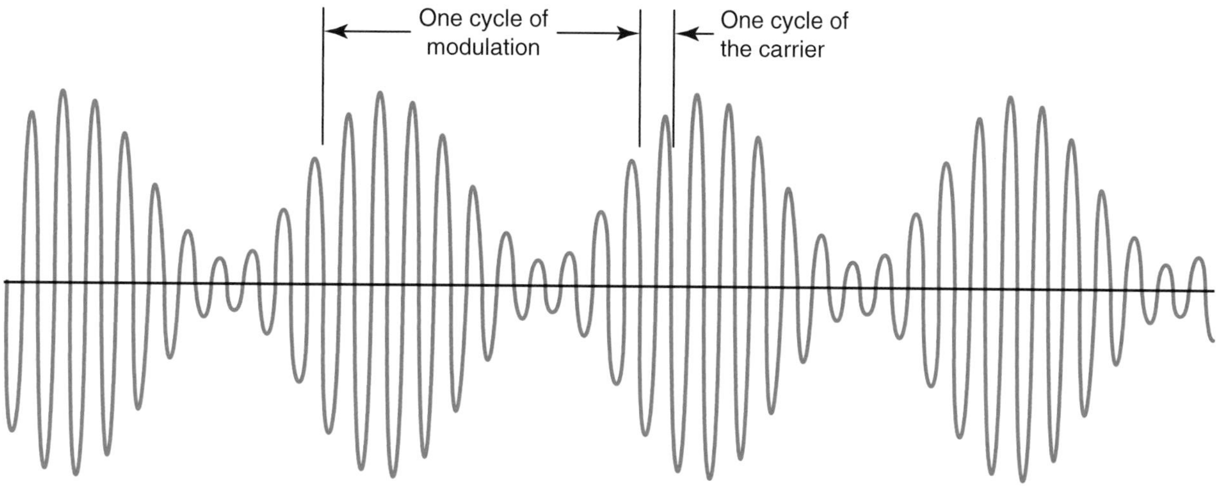

FIGURE 12–8 A typical amplitude modulator.

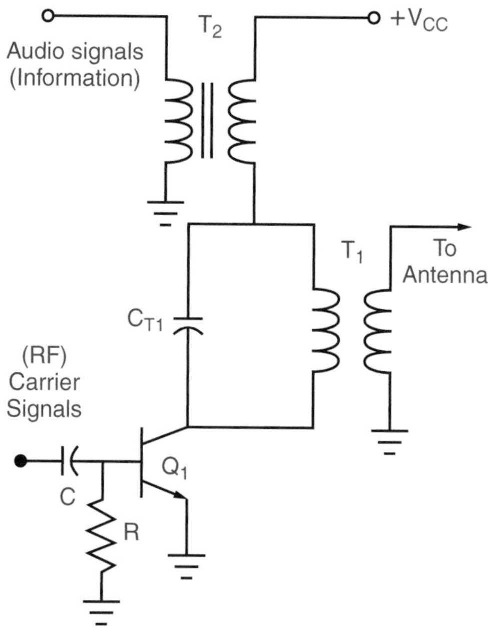

The other applied signal is the audio applied to the collector via T_1's resonant primary winding. Hetrodynes are signals produced in a nonlinear device (including the human ear), which occur at both the sum and the difference of two or more applied signal frequencies. This is why a 500 kHz carrier and a 2 kHz modulating signal produce both 498 and 502 kHz (as illustrated in Figure 12–6).

The primary winding of transformer T_1 and the capacitor C_{T1} form a tank circuit whose resonant frequency is that of the RF input signal. Bandwidth of the LC tuned circuit is sufficient, however, to pass the sideband signals as well as the carrier signal. The transistor functioning in the class C mode is biased beyond the cutoff region. The presence of the RF input causes the transistor to conduct for portions of its positive half cycle, whereas the tank circuit recreates the rest of the RF input. The output signal at the collector of the transistor is, therefore, the RF carrier and sidebands.

As can be seen from Figure 12–8, the audio signal is applied to the primary of transformer T_2. The audio voltage at T_2 secondary either aids or opposes the supply voltage, depending on its polarity. For instance, suppose that the audio signal at the secondary of T_2 is 20 V peak to peak and the supply voltage V_{CC} is 10 V. If the audio input goes to its positive peak (with respect to ground) at 10 V, this voltage opposes the applied V_{CC} and the transistor gets 0 V.

On the other hand, if the audio input goes to its negative peak of 10 V (with respect to ground) this voltage aids the applied V_{CC} and the transistor gets 20 V. Thus, the collector supply for the transistor is not a constant but varies with the audio input. The output signal at the collector of the transistor has amplitude that depends on the supply voltage applied, which in turn depends on the audio signal. Amplitude control is hence achieved.

Because information transmitted using amplitude modulation is contained only in the sidebands, many AM systems filter out the carrier signal before power amplification. This allows almost all transmitted power to be sideband power alone. Little, if any, carrier signal is transmitted.

Because (as in the example of Figure 12–6) both sidebands contain the same information, one of the sidebands can be filtered out along with the carrier. This allows a transmitter to concentrate 100% of its output as useful intelligence, rather than waste a majority transmitting both sidebands as well as the carrier. Television video is an example of single-sideband AM. AM broadcast band radio is an example of double sideband with full carrier.

12.6 AM Detector or Demodulator

An AM detector, or demodulator, recovers information from the modulated signal and is an important part of a receiver. Nonlinear devices such as diodes and transistors (in class C) are at the heart of the detector. A very simple diode detector is shown in Figure 12–9.

In Figure 12–9, the modulated signal is applied to the primary of transformer T. The capacitor C_1 and the primary windings form a resonant circuit that can be tuned to the carrier frequency. At the secondary end of the transformer, the diode D passes only the positive halves of the carrier. The capacitor C_2 acts as a filter, charging and discharging in accordance with the positive peaks of the modulated signal. The output of the detector thus traces the positive peaks, or envelope, of the AM signal, which recreates the information signal.

An AM detector that uses a transistor is shown in Figure 12–10. The base-emitter junction of the transistor acts as a detector diode. The transistor Q and the resistors R_1, R_2, R_C, and R_E constitute a voltage divider biased class B amplifier. The AM input to the circuit is fed through the primary of the transformer T. The capacitor C_1 is tuned to the carrier frequency.

FIGURE 12-9 A simple AM detector using a diode.

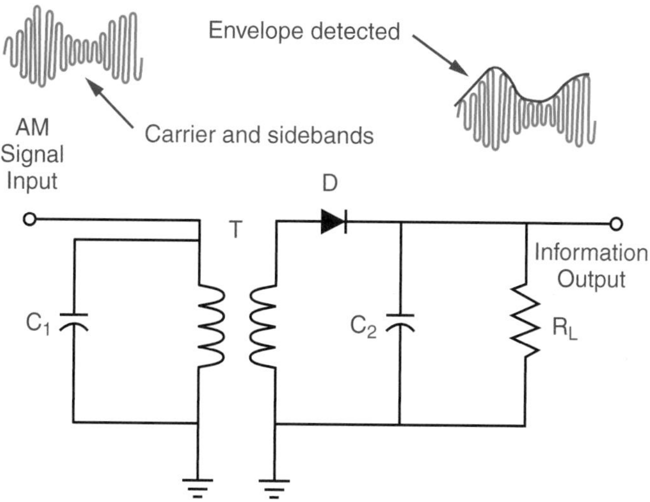

The collector current is the amplified version of the base current, and the value of the base current depends on the modulated signal.

The capacitor C_4 at the collector acts as a low-pass filter and produces a low-frequency output signal, which is the information. The biggest advantage of using a transistor circuit instead of a diode is the fact that it is able to provide a gain to the detected information. It is also a more sensitive detector because the base-emitter junction voltage does not have to be overcome by the signal before conduction can occur. Nonetheless, the transistor detector is seldom used.

FIGURE 12-10 A transistor detector.

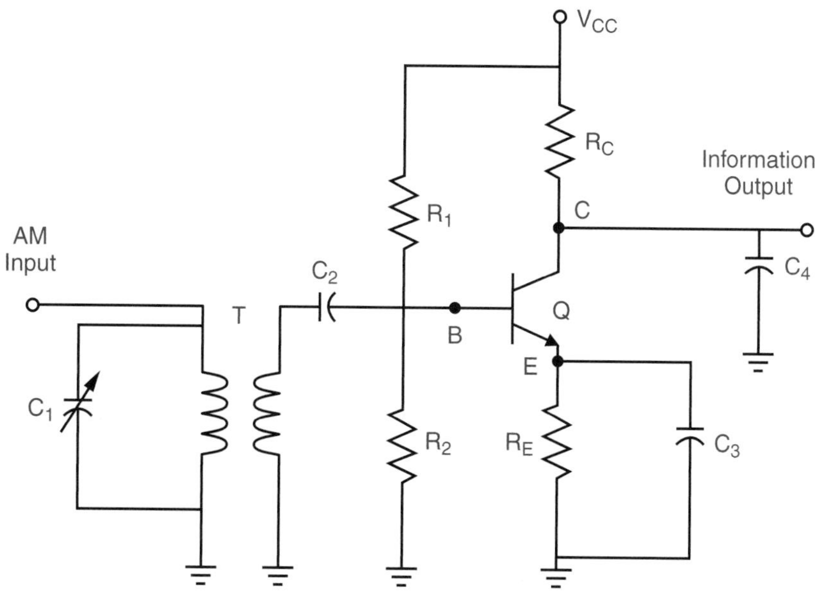

12.7 Bandwidth Considerations

As was discussed previously, an AM signal comprises the carrier and two sidebands, making bandwidth a factor to be considered. For broadcast of AM frequencies without interference, it is important to make sure that the channel bandwidth limits are taken into consideration.

In the standard AM broadcast band, carrier frequencies range from 540 kHz at the lower end to 1,600 kHz at the upper end. Each channel is spaced 10 kHz apart or has a bandwidth of 10 kHz. This means that the modulating frequency has to be limited to a maximum of 5 kHz. When audio signals are considered, 5 kHz is adequate for casual listening of the speaking voice. However, music has frequencies higher than 5 kHz (up to 15 kHz or even higher for better fidelity). For suitable reproduction of high-quality music, a bandwidth of twice the maximum audio frequency (which is 15 kHz $\times$ 2 = 30 kHz) is required. We can therefore conclude that the 10kHz channel spacing is not entirely adequate for good fidelity in an AM broadcast band.

To overcome this, the Federal Communications Commission (FCC) assigns three-channel spacing between carriers in any particular geographic location. This results in a 30 kHz bandwidth and prevents the upper sideband from a lower channel from spilling into the lower sideband of the channel above it.

12.8 Simple AM Receivers

Before we discuss AM receivers, it is important to define and discuss two major characteristics of any receiver.

1. Sensitivity is defined as the minimum amount of input signal the receiver needs to drive an output transducer (such as a speaker) to an acceptable level.
2. Selectivity is defined as the extent to which a receiver is capable of differentiating between one carrier frequency and another. Poor selectivity of a receiver is associated with a large bandwidth, which may result in spilling over of frequencies from other channels, causing interference. On the other hand, a receiver with a small bandwidth may be overly selective, which results in the suppression of desired signals outside the selected bandwidth. An optimal receiver bandwidth based on the bandwidth of the modulating signal is therefore necessary for efficient operation of the receiver.

In its simplest form, an AM receiver consists of an antenna, diode, headphones, and ground. Figure 12–11 illustrates a basic radio receiver. An antenna is needed to capture the radio wave and convert it into an electrical signal. The diode acts as the detector or demodulator. The headphone is a transducer that converts audio signals, in the form of electrical energy, to audible sound.

The earth ground is used to complete the path of current. It needs to be mentioned that in this simple receiver circuit there is no operating power supply. This electronic circuit operates entirely from the induced energy from the antenna. Although the basic receiver shown in Figure 12–11 does work, it is not practical for the following reasons.

- *Poor sensitivity:* The receiver has no component that offers any gain. This contributes to poor sensitivity and failure to pick up weak signals.

Additional electronic stages would need to be added so that the receiver could drive anything larger than a set of headphones.

- *Lack of selectivity:* The receiver has no component with the ability to select among various carrier frequencies. This means that the receiver will "hear" every carrier signal strong enough to drive the headphones. The addition of a tuned LC circuit can resolve this situation.

FIGURE 12–11 A basic radio receiver.

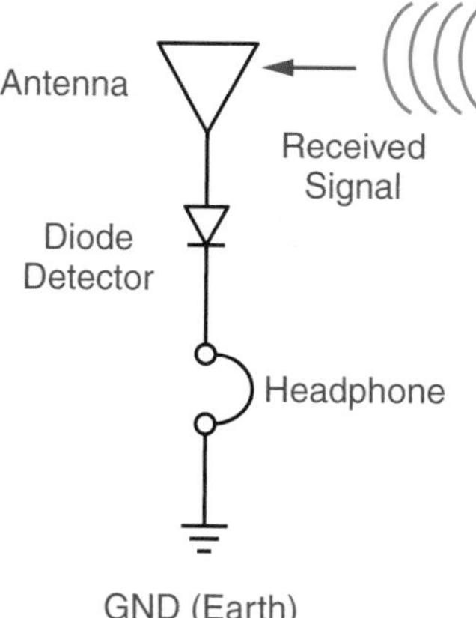

An improved radio receiver is shown in Figure 12–12. Of all frequencies intercepted by the antenna, the desired carrier frequency is chosen using the tuned circuit. The detector (demodulator) recreates the audio signal, which is amplified by the audio amplifier before it is fed to the speaker. It can be seen that in the improved radio receiver sensitivity and selectivity are both improved.

FIGURE 12–12 An improved radio receiver.

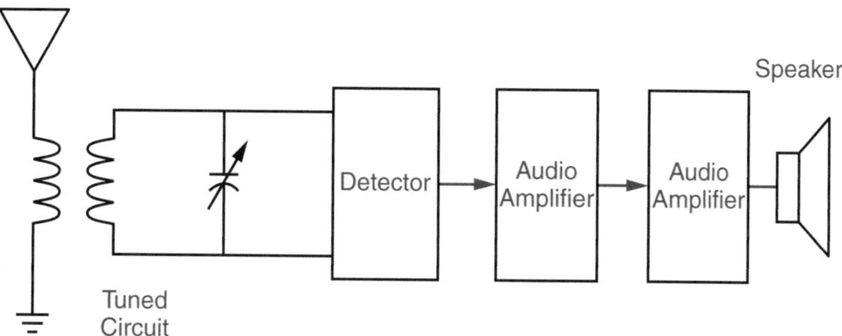

Figure 12–13 shows one of the earliest types of receiver used for commercial radios—the tuned radio frequency (TRF) receiver. Sensitivity and selectivity are greatly improved in a TRF receiver by using more amplifiers and tuned circuits. It employs three tuned circuits (which provide the desired selectivity) and four amplifiers (two radio frequencies and two audio frequencies), which provide the required gain and better selectivity.

FIGURE 12–13 A TRF receiver.

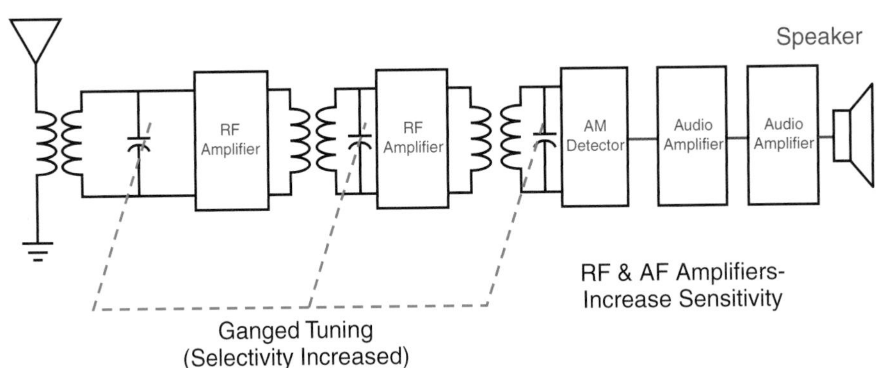

There are, however, a few disadvantages to the TRF receiver. The three tuned circuits are gang tuned, which means that they are tuned in synchrony. The term *tracking* refers to how closely resonant frequencies can be matched at a certain setting of the tuning control. It is very difficult to match (or track) the resonant frequency of each of the tuned circuits. In addition, each tuned circuit can have different bandwidths.

The differences in bandwidth and resonant frequency of the receiver can pose a problem. Moreover, to avoid unwanted oscillation the RF amplifier stages must be electrically and mechanically isolated, and negative feedback must be incorporated. These problems can be largely eliminated by the use of a superheterodyne receiver.

12.9 Superheterodyne Receivers

In the superheterodyne receiver, some of the tuned circuits are confined to a single fixed frequency, eliminating the problem of tracking and changing bandwidths as seen in TRF receivers. This single frequency or fixed frequency is called the intermediate frequency (IF). A superheterodyne receiver converts any received frequency to the IF by the process of mixing or heterodyning. The most commonly used IF for the AM broadcast band is 455 kHz. Different values of IF are assigned to other broadcast bands such as FM and the shortwave bands. Commonly used intermediate frequencies are 455 kHz, 4.5 MHz, 10.7 MHz, and 21 MHz.

Figure 12–14 shows the process of heterodyning. The mixer (heterodyne converter) is fed with two input signals, A and B. The output of the mixer contains the sum of the two inputs ($A + B$), the difference of the two inputs ($A - B$), and the original signals A and B. The difference signal is usually the one used for the IF.

Figure 12–15 shows a block diagram of a superheterodyne receiver. The mixer gets its inputs from an oscillator and the signals coming from the antenna. The oscillator is tuned to a frequency greater than the received frequency by an amount equal to the IF.

FIGURE 12–14 Heterodyne converter, also called a mixer.

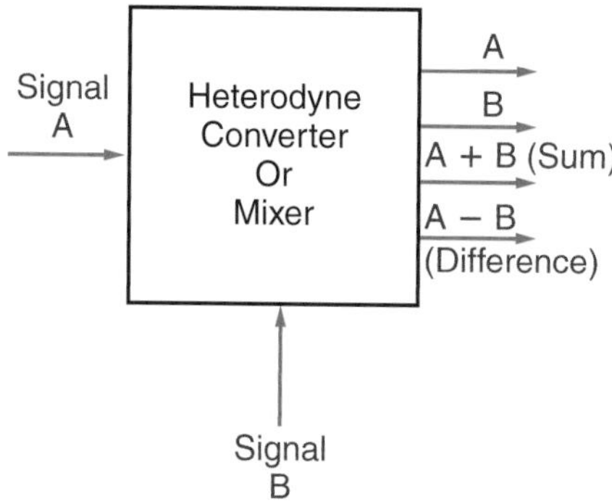

For instance, if a carrier frequency of 1,080 kHz is to be received by an AM broadcast receiver the oscillator should be tuned to a frequency that is equal to

$$\text{Received frequency} + \text{IF} = 1{,}080 \text{ kHz} + 455 \text{ kHz} = 1{,}535 \text{ kHz}$$

The mixer produces the sum and difference frequency at its outputs. The mixer is followed by an IF amplifier stage, which selectively amplifies and passes IF frequencies (those frequencies of about 455 kHz). The output of the mixer produces a sum frequency of 1,080 kHz + 1,535 kHz and a difference frequency of 1,535 kHz − 1,080 kHz = 455 kHz. The sum frequency is rejected by the IF amplifier, whereas the difference frequency (which yields a 455 kHz signal) is passed and amplified by the IF amplifiers. Subsequently, its amplitude variations are detected by the detector.

If while the 1,080 kHz signal is being received a 970 kHz carrier signal is also picked up by the antenna, the mixer will produce a sum frequency of 970 kHz + 1,535 kHz = 2,505 kHz and a difference frequency of 1,535 kHz − 970 kHz = 565 kHz. Both the sum and difference frequencies do not fall in the IF amplifier's pass band, and thus they are both rejected, eliminating interference from other carrier frequencies.

FIGURE 12–15 Block diagram of a superhetrodyne receiver.

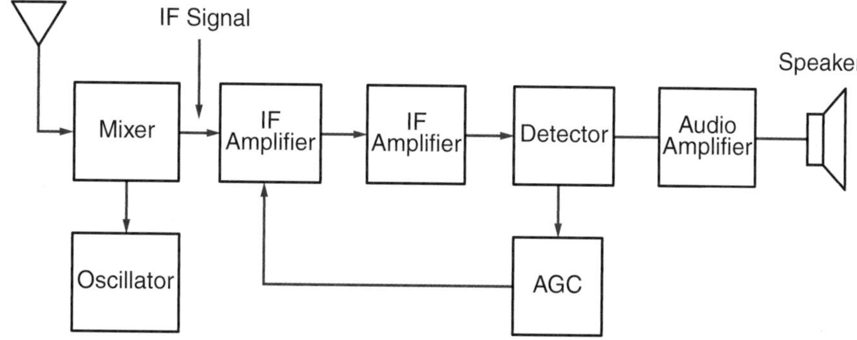

Examine Figure 12–9 again. Note how much it looks like a half-wave rectifier circuit. The filter capacitor C_2 is designed to filter out the RF ripple but not significantly affect the much lower modulation frequency variations. Therefore, a DC voltage (dependent on received carrier level) and modulation AC are present at the information output terminal. The DC component is used to regulate gain of the preceding IF and RF amplifiers to prevent undesirable signal level fluctuations due to atmospheric conditions, distance from the transmitter, and so on.

The DC output of the detector is used for **AGC (automatic gain control)** or **AVC (automatic volume control)**. The purpose of AGC is to maintain a constant signal level at the output of the receiver. If the gain of the IF amplifiers are kept constant when the strength of the signal received by the antenna varies, the output voltage of the receiver will also vary.

Automatic gain control (AGC)
Circuitry incorporated into most AM and FM receivers that automatically limits incoming signal strength to prevent overloading problems such as variable video contrast and sound volume levels.

Automatic volume control (AVC)
See *AGC*. Only used for voice frequency AM radios.

A feedback loop is formed among the detector output, AGC block, and IF amplifier(s). AGC develops a control voltage depending on the strength of the signal at the input of the detector. This control voltage is used to influence the gain of the RF and/or IF amplifier(s). The transistor is employed as the amplifying device in an IF amplifier, and the transistor's gain is varied by the AGC (or AVC) control voltage.

Figure 12–16 shows the AGC characteristics of a transistor. As indicated in the graph, maximum gain is present only at one particular value of collector current (IC). As collector current increases or decreases from this value, gain drops. The AGC control voltage is used to control the bias of the transistor so that the collector current is changed. Thus, if the antenna receives a stronger signal AGC reduces IF/RF amplifier gain, whereas a weaker received signal returns the IF/RF amplifier(s) to the linear portion of their characteristic curves, increasing their gain.

FIGURE 12–16 AGC characteristics of a transistor.

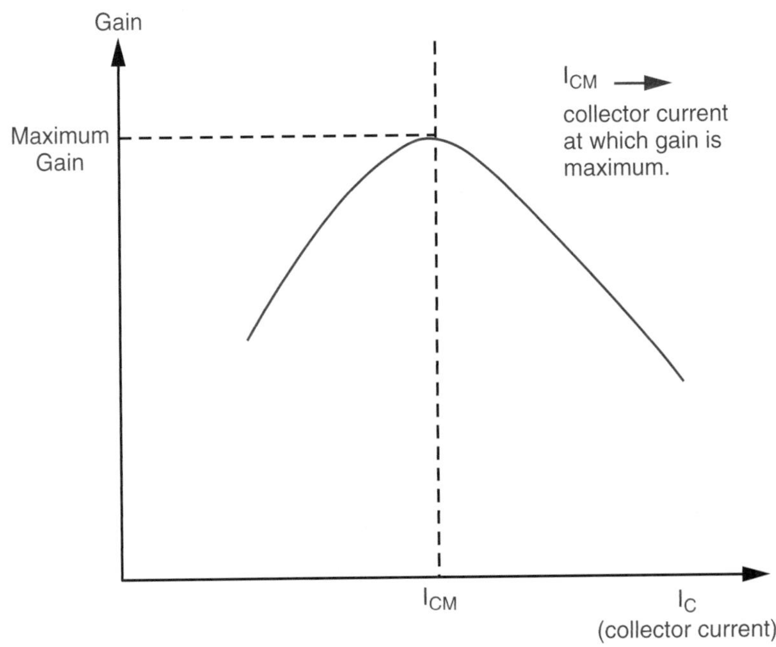

Superheterodyne receivers are vulnerable to image interference. This is because there may be two values of incoming frequencies that can mix with the oscillator to produce IF. For instance, in the case already discussed an oscillator frequency of 1,535 kHz is used to detect a 1,080 kHz signal. The difference between the two frequencies yields 455 kHz. Now let's consider that the antenna picks up a frequency 1,990 kHz. The difference signal available at the output of the mixer will be 1,990 kHz − 1,535 kHz = 455 kHz. This means that the receiver detects a carrier signal of 1,990 kHz, which is considered the image frequency of 1,080 kHz.

The interference produced by the image frequency is called **image interference**. It can be seen that the image frequency is separated from the desired receive frequency of 1,080 kHz by 910 kHz (2 × 455 kHz). Image rejection is improved by having a tuned circuit before the mixer. This tuned circuit can also have an RF amplifier stage to increase the sensitivity of the receiver as well as the selectivity.

Image interference
Interference produced by a signal present on the image frequency.

Another common and highly effective solution to the image rejection problem is dual conversion (Figure 12–17). **Dual conversion** is the use of two intermediate frequencies in a communication receiver. For instance, assume that the first IF is 4,500 kHz and the second IF is 455 kHz. Further assume a desired receive frequency of 1,080 kHz. There is an unwanted signal, as before, at 1,990 kHz. The first local oscillator frequency would be 1,080 kHz + 4,500 kHz = 5580 kHz. The image frequency would be 5,580 kHz + 4,500 kHz = 10,080 kHz. Let's assume an undesired signal *does* also exist at 10,080 kHz. There are now three signals at the 4,500 kHz IF tuned circuits.

Dual conversion
The use of two intermediate frequencies in a communication receiver. These are known as the first IF and second IF.

1. The desired 1,080 kHz signal, now converted to exactly 4,500 kHz (middle of the first IF passband)
2. The undesired 1,990 kHz signal, now converted to 7,570 kHz—way out of the 4,500-kHz first IF passband and thus weak
3. The undesired 10,080 kHz image, which is even further outside the 4,500 kHz first IF passband and thus very weak

Now apply this 4,500 kHz to the second mixer stage, and then to the second IF of 455 kHz, and see what happens. The second local oscillator frequency will be a *fixed* frequency of 4,500 kHz + 455 kHz = 4955 kHz. Let's see what happens to our unwanted images.

- The desired 1,080 kHz signal now becomes 4,955 kHz − 4,500 kHz = 455 kHz square in the middle of the 455 kHz second IF passband.
- The undesired 1,990 kHz signal now becomes 7570 kHz − 4,955 kHz = 2,615 kHz. It is so far out of the 455 kHz passband that it is not detectable.
- The undesired 10,080 kHz signal now becomes 10,080 − 4,955 kHz = 5,125 kHz, which is insanely beyond the 455 kHz second IF passband and is undetectable.

In addition to superb image rejection, dual conversion enhances receiver sensitivity and selectivity over single conversion, direct conversion, and TRF designs.

FIGURE 12–17 Dual conversion.

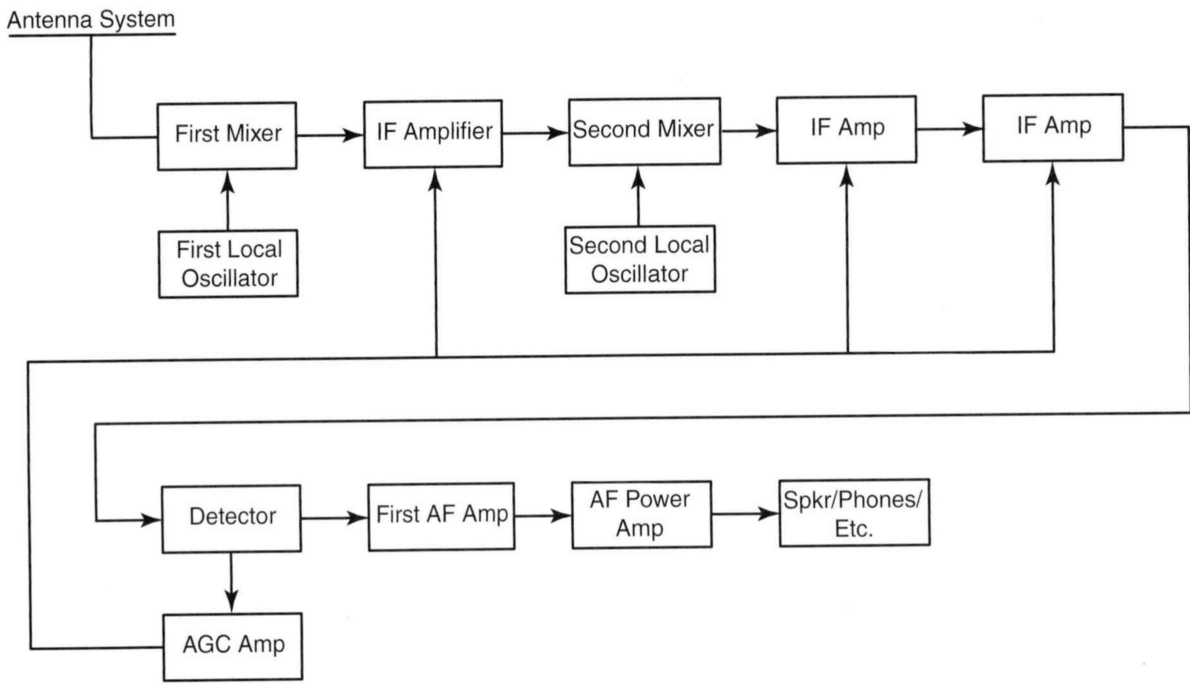

FREQUENCY MODULATION

12.10 FM Fundamentals

Frequency modulation (FM) involves two factors: (1) modulating frequency [for every cycle of modulation the FM carrier moves above, below, and returns to the center (carrier) frequency] and (2) modulating amplitude [the greater the amplitude of the modulating signal the greater the deviation ("swing") of the FM signal above and below center frequency]. Figure 12–18 illustrates a frequency-modulated waveform.

Frequency modulation is preferred over amplitude modulation in commercial broadcasting networks because of its immunity to noise and its wider bandwidth, which allows for better reproduction fidelity. AM is very sensitive to noise. This causes things such as lightning, automobile ignition, or sparking to manifest as amplitude variations on a carrier wave. In AM, amplitude variations carry "loudness" information. Extraneous signal sources algebraically add to the desired signal amplitude and mix in the mixer stages.

These amplitude changes are detected by the receiver as though they were a part of the intended signal, and thus contribute to noise. In the case of FM, the noise signals can be eliminated in the receiver by using limiters that clip off amplitude increases or spikes. This can be done without affecting the detected FM signal because signal amplitude in FM is not a component of modulation frequency or modulation amplitude. In addition, received AM signals at or near the same frequency will interfere with each other, whereas received FM signals at or near the same frequency are far less likely to interfere due to an FM characteristic called "capture effect."

FIGURE 12–18 Frequency-modulated waveform.

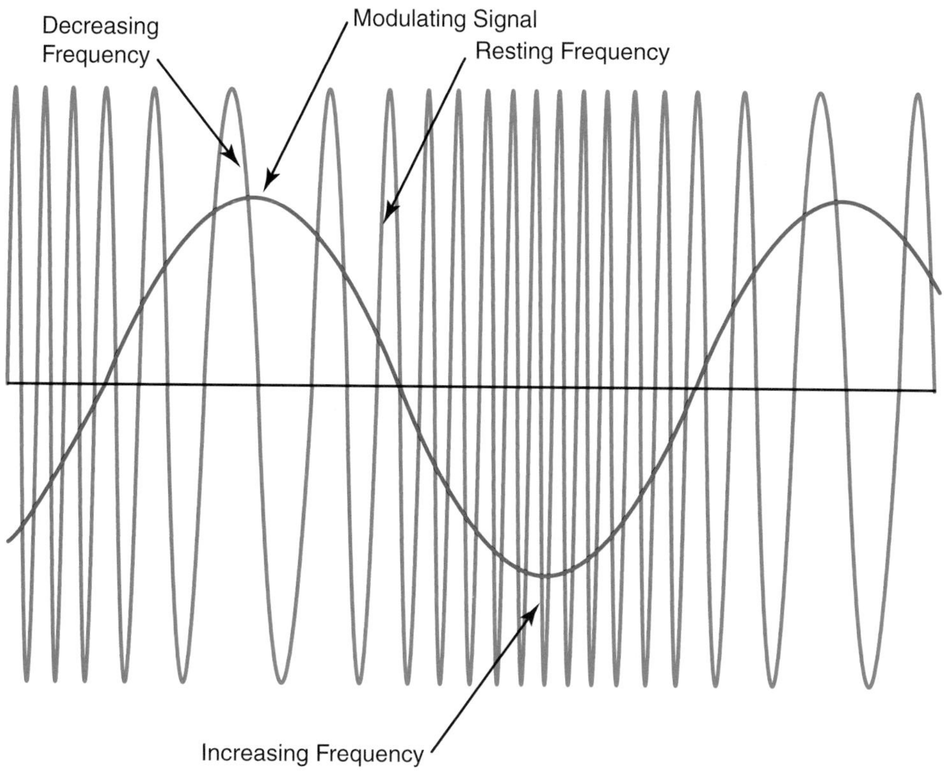

Capture effect causes a receiver to lock onto the strongest signal to the exclusion of weaker ones.

> **Capture effect**
> The tendency of FM receivers to lock onto the strongest of signals occupying the same carrier frequency to the exclusion of weaker signals.

FM produces sidebands, just as AM does. However, FM signals have several sidebands. Consider an FM system with a 200 MHz carrier being modulated by a 10 kHz audio tone. Figure 12–19 shows the sidebands produced as a frequency domain graph.

FIGURE 12–19 FM sidebands.

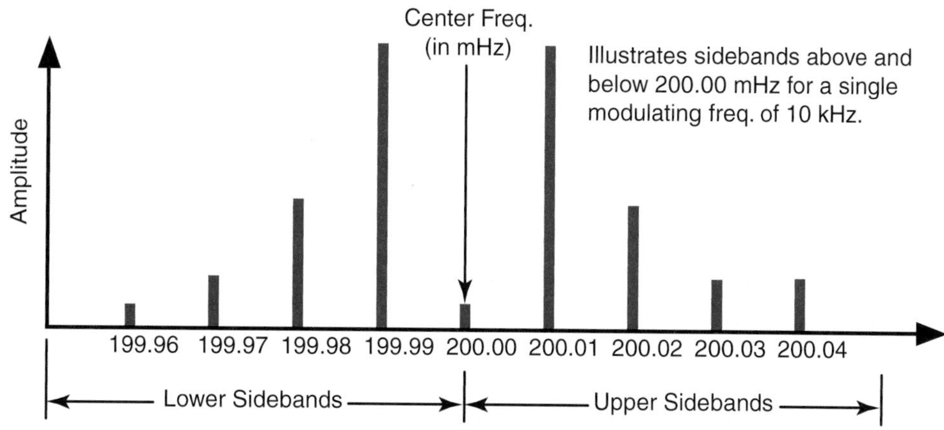

12.11 A Frequency Modulator

In its basic form, an FM transmitter consists of an LC tank circuit used with an oscillator. Figure 12–20 illustrates a basic frequency modulator. The audio input voltage is used to change the instantaneous value of the capacitor. The carrier frequency generated at the output of the oscillator is dependent on the rate at which the capacitance **varies above and below its resting value.** The amplitude of the modulating signal determines the amount of capacitance change, thus dictating **the extent of the carrier frequency deviation above and below the carrier frequency**.

FIGURE 12–20 A basic frequency modulator.

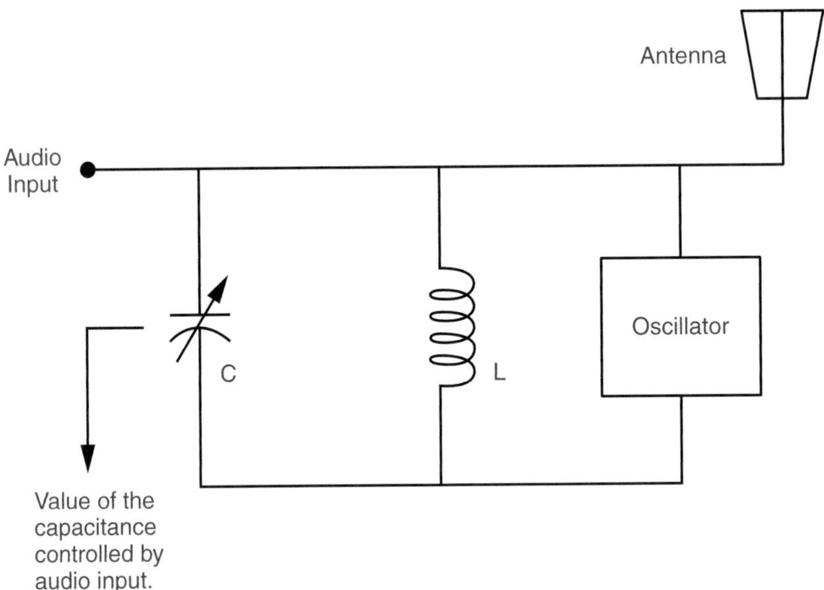

Figure 12–21 shows a detailed frequency modulator. A Colpitts oscillator is used to generate the carrier signal. The values of L and C determine the frequency of oscillation. The diode C_{D1} is a varactor, a voltage controlled variable capacitor. Resistors R_1 and R_2 form a voltage divider network and bias the diode. The value of the diode's capacitance is dependent on the instantaneous audio input. This diode capacitance is in parallel with capacitor C, and hence the oscillator frequency is changed with the net change of capacitance (C_{D1} in parallel with C).

In practice, frequency deviation of the signal from the modulator is very small. To be of practical use, deviation must be increased to standards set by the FCC. Increasing deviation is usually accomplished by multiplying the FM signal in RF amplifiers known as multipliers, usually doublers or triplers. Thus, a modulator with a 5 MHz carrier and ±100 Hz deviation can be multiplied to a 200 MHz FM carrier with a deviation of ±4 kHz.

Recall that class C is the only nonlinear operation mode. Therefore, class C is needed for multiplication to occur in amplifier stages between the modulator and power amplifier. RF power amplification may use any class of operation, although for efficiency class C is usually chosen.

FIGURE 12–21 A complete frequency modulator.

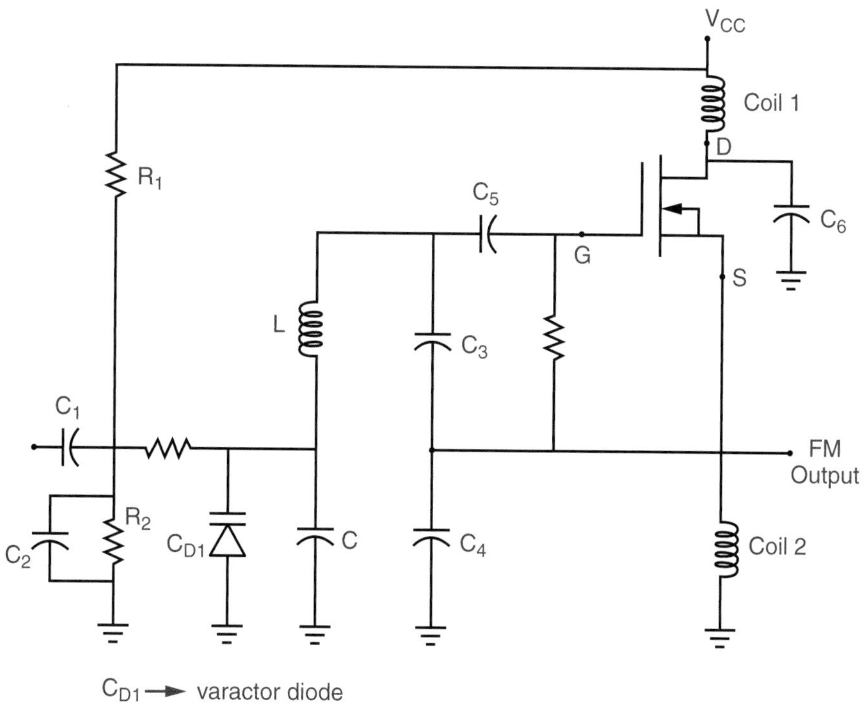

12.12 FM Detector or Discriminator

FM is detected by using a circuit called a discriminator. A discriminator reproduces the information (modulating signal) at the output when fed with an FM signal. A discriminator is shown in Figure 12–22. Tank circuits L_1C_1 and L_2C_2 are present on the secondary side of the transformer. The frequency response curve for the two tank circuits is shown in Figure 12–23. As indicated in the figure, the resonant frequencies for the two tank circuits are above and below the carrier frequency f_O.

When the nonmodulated carrier, which is at the resting frequency, is fed to the primary side of the transformer the tank circuits have identical voltages because they are operating at the carrier frequency. The identical voltages drive the diodes D_1 and D_2 to conduct in equal amounts.

Currents through the resistors R_1 and R_2 are identical in magnitude but opposite in direction because of the direction of the diodes. This causes the drops across the resistors R_1 and R_2 to be equal in magnitude but opposite in sense. Hence, the output voltage measured across R_1 and R_2 amounts to zero. The discriminator thus produces no output voltage for the resting frequency or the nonmodulated carrier frequency.

When the frequency of the carrier increases due to modulation, the tank circuit L_2C_2 has a higher voltage than L_1C_1. This causes diode D_2 to conduct more than diode D_1, and hence the voltage across resistor R_2 is more than resistor R_1. The result is a positive output voltage. As frequencies increase toward the resonant frequency of L_2C_2, the output voltage continues to be more positive.

FIGURE 12–22 A discriminator circuit.

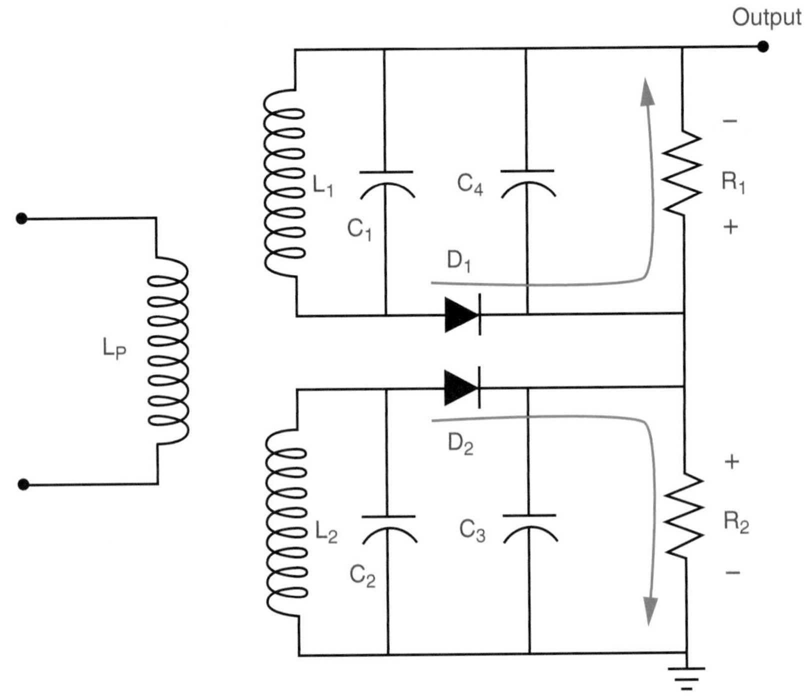

FIGURE 12–23 Frequency response curve for tank circuits.

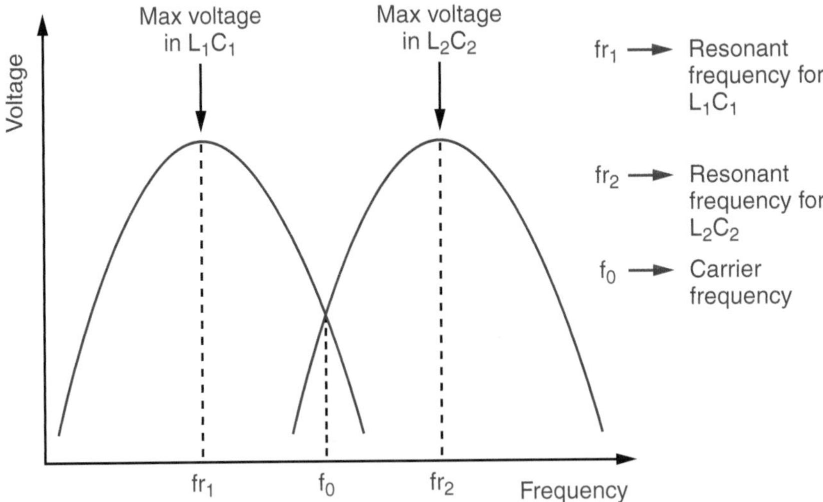

On the other hand, when the frequency of the carrier decreases due to modulation the tank circuit L_1C_1 has a higher voltage than L_2C_2. This causes the diode D_1 to conduct more than D_2, and hence the voltage across resistor R_1 is more than resistor R_2. The result is a negative output voltage.

As frequencies decrease toward the resonant frequency of L_1C_1, the output voltage continues to be more negative. This is summarized as follows:

- The output of the discriminator is zero when the carrier frequency is at rest.
- The output voltage of the discriminator increases in the positive direction as the frequencies increase.
- The output voltage of the discriminator increases in the negative direction as the frequencies decrease.

The output of the discriminator is thus a function of both the rate at which the signal deviates above and below center frequency and the range of that deviation.

12.13 FM Receiver

The block diagram of a typical FM superheterodyne receiver is shown in Figure 12–24. Mixing or heterodyning is achieved just as described in the case of an AM receiver. The intermediate frequency is 10.7 MHz for the FM broadcast band. The output of the IF amplifier is fed to a limiter. The limiter is provided to eliminate noise signals that occur as spikes or amplitude variations. Limiting clips the output, which means that the noise components are removed. In doing so, the FM signal does not lose any information components in the carrier because information is coded in the form of frequency variations. Figure 12–25 illustrates the operation of a limiter.

FIGURE 12–24 Block diagram of an FM superheterodyne receiver.

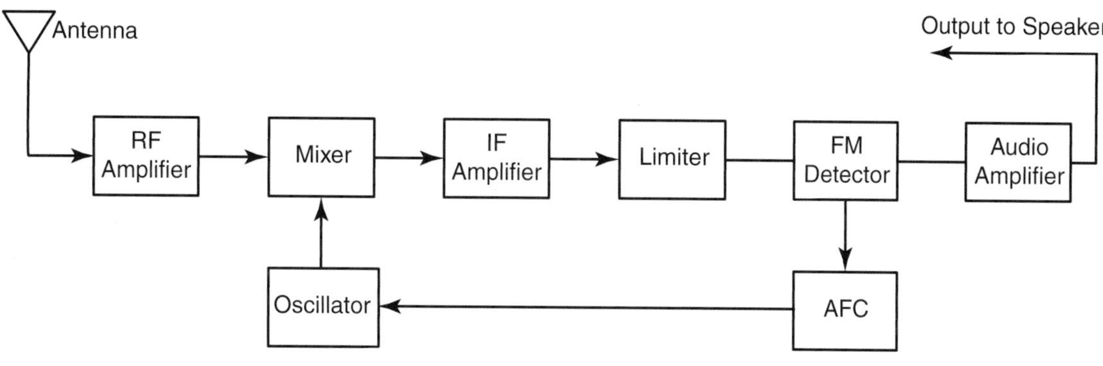

FIGURE 12–25 Operation of a limiter.

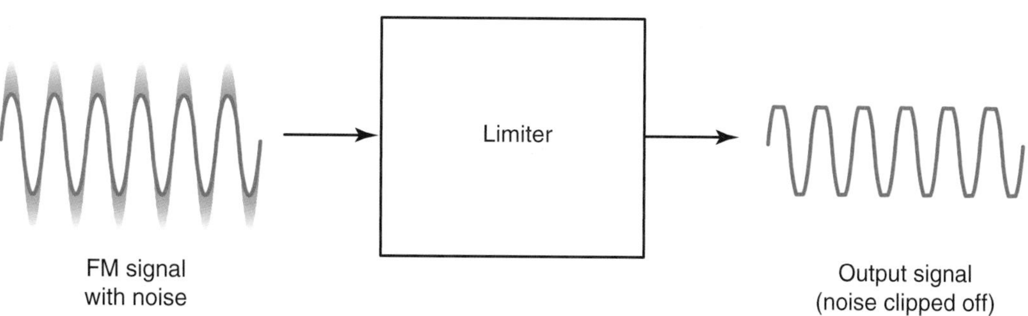

Note that the process of limiting or clipping to eliminate noise could not be used in AM. In AM, the information is coded in the form of amplitude variations. The noise signals also appear as amplitude variations, and if they are clipped relevant information may be lost.

The output of the limiter is fed to the detector or discriminator. The detector feeds its signals to an audio amplifier and a block named AFC, which stands for **automatic frequency control.** AFC is used to keep the local oscillator frequency from drifting. This is what happens. If the incoming carrier frequency is properly tuned, average detector output is 0 V. However, if the receiver's local oscillator drifts off frequency, average detector DC voltage will tend to be either above or below zero volts. This voltage drift is used as a correction voltage to shift the local oscillator frequency back where it belongs. This technique is called automatic frequency control.

Automatic frequency control (AFC)
Circuitry incorporated into most frequency modulation receivers to keep the receiver local oscillator from drifting off frequency.

SUMMARY

Information of various types often needs to be transmitted from one location to another. For example:

- Voice and music from a studio to your home via a radio
- Voice, music, and/or video from a remote location to your television receiver
- Digital data from one computer to another
- Process information from an industrial process to a controller
- Control signals from a controller to a process

In each of these cases, the information can be converted to an electric signal that varies in amplitude according to the information. For example, digital data can be encoded as an "on" or "off" signal. This is a form of amplitude modulation (AM). In other cases, the information may be of a continuous or analog nature, such as audio or video. In yet other situations, the information is converted from analog to digital.

Whatever the format of the information signals, they are usually fairly low frequency, on the order of a few kilohertz up to a maximum of 100 kHz or so. For a variety of reasons, these low frequencies do not lend themselves to long-distance transmission means such as radio waves or light waves through fiber-optic cable.

To remedy this situation, the information is often impressed on a higher-frequency carrier signal. This action is called modulation. This combination of the two signals is readily transmitted for long distances, depending on the type of transmission path (e.g., fiber-optic cable, radio, coaxial cable, twisted pair, and so on). This situation is the basis for most radio and television transmissions. In this type of transmission, a transmitter impresses information onto the carrier wave and the result is then applied to an antenna and broadcast into the atmosphere. At the other end, a receiver demodulates the received signal and converts the information back into the original format.

Modern radio systems use either AM or frequency modulation (FM). In AM, the information is used to vary the amplitude of the carrier signal. In FM, the information is used to vary the frequency of the signal. Other offshoots of these two fundamental types have been developed. However, they are all based in one way or another on the two mainstays.

REVIEW QUESTIONS

1. The choice of carrier frequency makes a large difference in how far a transmitted signal will carry. FM stations in the United States use frequencies in the vicinity of 100 MHz, which is line-of-site communication. Why? What problems could occur if the signals traveled farther than line of site?
2. Why is a higher-frequency carrier signal required for radio transmission? Discuss the pros and cons of such a system.
3. Using Figure 12–26, walk through a block-by-block explanation of a basic communication system.

FIGURE 12–26 Basic communication system.

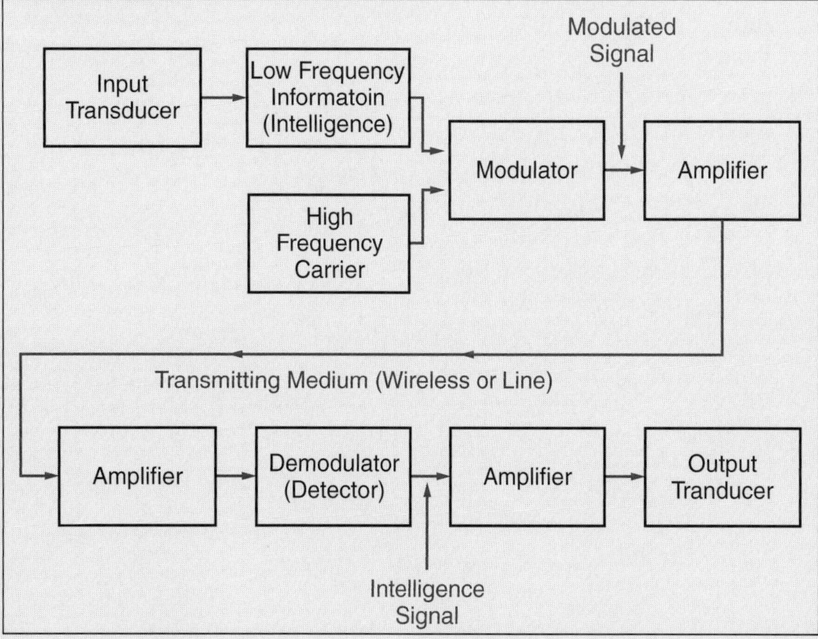

4. Bandwidth is the difference between the lowest frequency and the highest frequency in a transmitted signal. For example, the bandwidth of an AM signal with a carrier frequency of 1,600 kHz and an applied audio signal of 4 kHz is 8 kHz (1,604 kHz − 1,596 kHz). Consider what would happen to a signal if after modulation the carrier signal and the lower portion of the sideband were removed. What would be the bandwidth of this single-sideband signal?
5. Assume that the circuit of Figure 12–27 is being used to generate an AM signal for a radio broadcast. What will happen if the audio input is adjusted so that the output is clipped? (This is called overdriving.) What will happen if the audio input is decreased to a very small value?

FIGURE 12–27 A typical amplitude modulator.

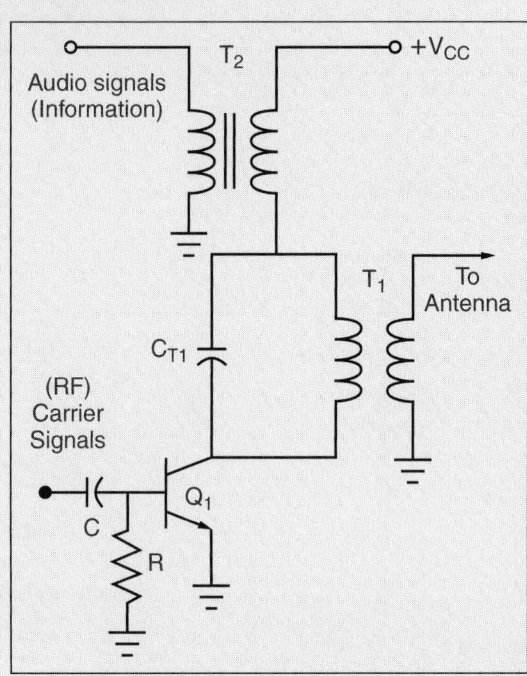

6. Figure 12–28 shows an improved version of the basic receiver shown in Figure 12–11. Note the additional components. Discuss how the additional components might help overcome the two problems of Figure 12–11 that were discussed in the chapter.
7. How might the receiver shown in Figure 12–15 be improved for additional selectivity and/or sensitivity?
8. Most of the modern music radio stations use FM. Why?
9. A certain receiver is being used to tune a 600-kHz radio station. The intermediate frequency for the receiver is 455 kHz. Another radio station is very close and broadcasts on a frequency of 1,510 kHz. Is there any possible problem? Why or why not?
10. In the circuit of Figure 12–26, the diode rectifies the input so that only half the waveform is passed through to the capacitor and the diode. The capacitor bypasses the high RF and allows the audio voltage to develop a signal across the headphones. How does the circuit of Figure 12–11 work?

FIGURE 12–28 Improved basic receiver (crystal set).

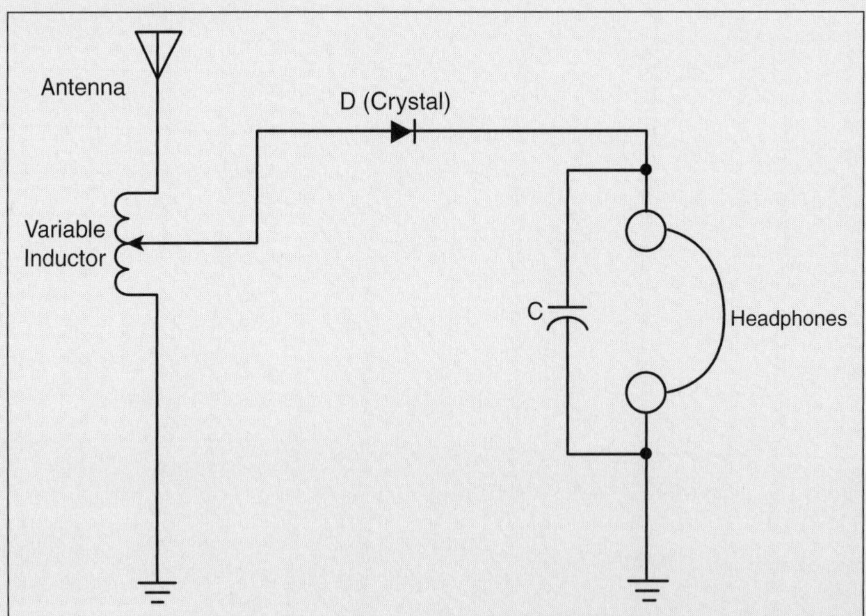

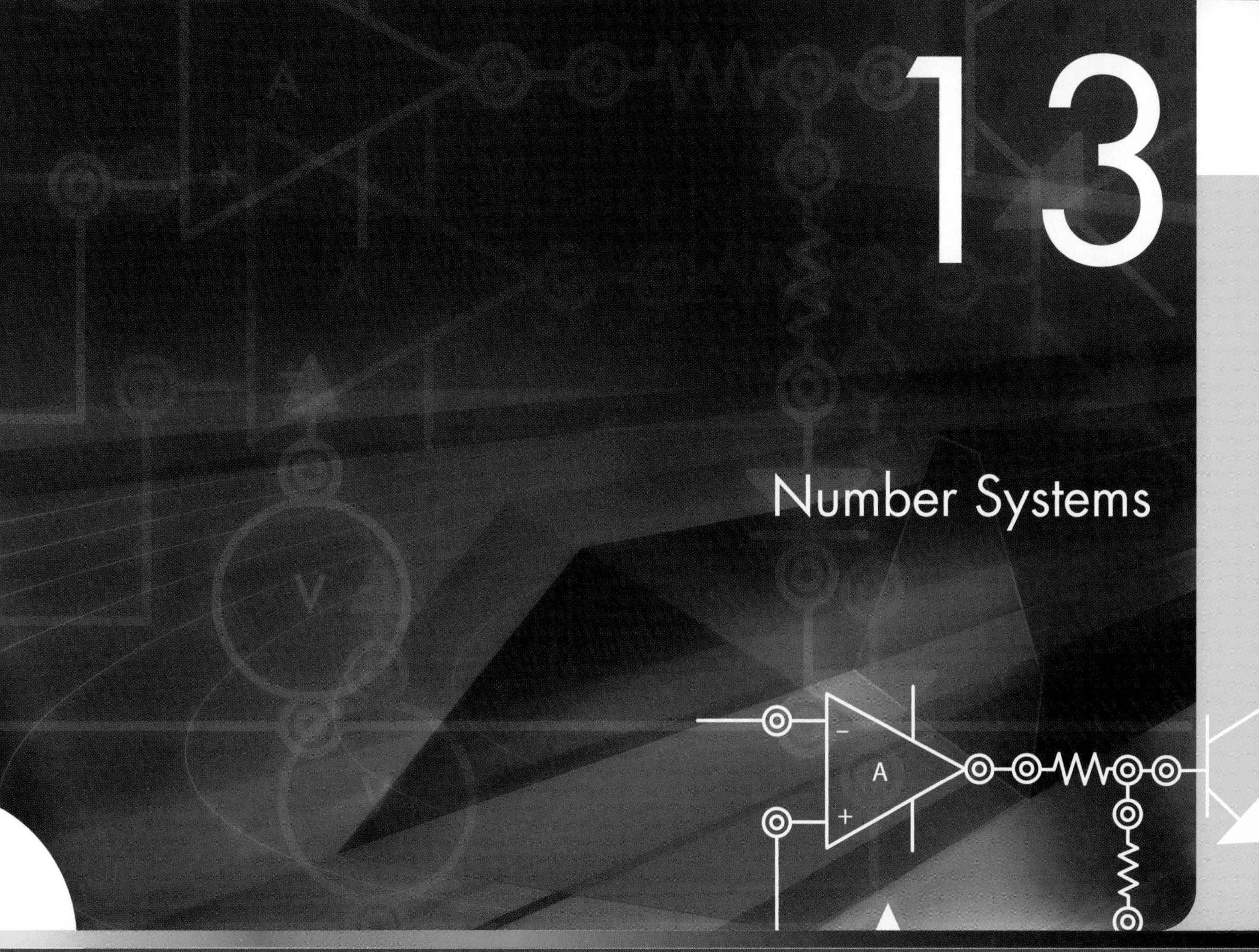

13

Number Systems

OUTLINE

OVERVIEW

Since before recorded history, humans have needed to count things. Primitive civilizations used what modern mathematicians call one-to-one correspondence. Pebbles were collected to keep count of the number of goats in the herd. A bag of these pebbles represented the entire herd. Other methods of keeping count included making marks and tying knots in leather thongs. In time, humans began using words to represent these numbers.

With the advent of writing and record keeping came the need for symbols to represent certain numbers of items. Figure 13–1 shows the hieroglyphics used for numbers by the Egyptians about 3000 B.C. Although crude compared to our modern notation, the Egyptian method offered some conveniences. For example, the symbols did not have to be listed in any particular order. The bottom two lines of hieroglyphics both represent the same number: 221,222.

Modern number systems are codes that differ from each other in the number of digits and symbols that might be used. For each discrete value there is an assigned symbol. Once the code is memorized, common arithmetic functions such as adding, subtracting, multiplying, and dividing can be performed.

The number system with which you are most familiar—because you learned it from kindergarten on, is based on the number 10, and is called the decimal system. The decimal system was developed primarily by Arab cultures with contributions from others. Computers use a number system based on the number 2, primarily because an electric circuit is often thought of as being either off or on. Other number systems have been developed over the years to assist us in performing calculations.

In this chapter, you will learn about the four most commonly used numbering systems: the binary (base 2), the octal (base 8), the decimal (base 10), and the hexadecimal (base 16). Knowledge of these systems will be extremely helpful to you in your career, especially if your work takes you into computers and modern digital systems.

OBJECTIVES

After completing this chapter, the student should be able to:

1. Explain the number systems with bases of 2, 8, 10, and 16.
2. Convert numbers among the bases 2, 8, 10, and 16.
3. Use binary-coded decimals.

FIGURE 13–1 Egyptian hieroglyphics used for counting.

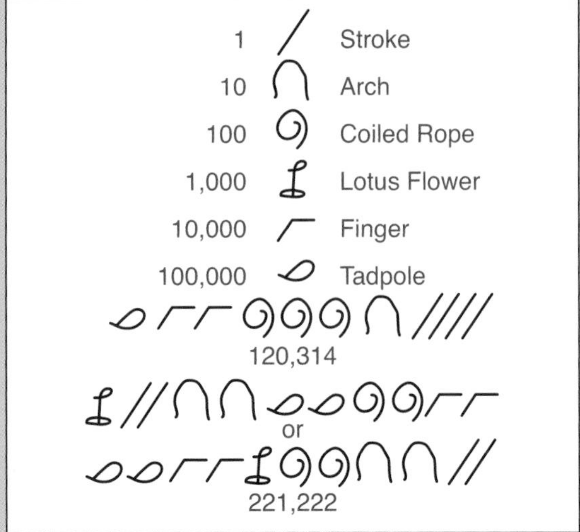

DECIMAL AND BINARY NUMBER SYSTEMS

13.1 The Decimal System

The number of digits or symbols used in a number system is called its **base** or **radix.** The decimal system we use every day has 10 digits. It is the most familiar to most of us. Table 13–1 compares the number of strokes to the symbols used in the decimal system.

For example, the number 9,652 is actually shorthand for a value equal to a quantity of 1s, 10s, 100s, and 1,000s. This number can be written according to its positional weight.

$$(9 \times 10^3) + (6 \times 10^2) + (5 \times 10^1) + (2 \times 10^0) = 9{,}000 + 600 + 50 + 2 = 9{,}652$$

This is because the 9 is in the third position, the 6 is in the second position, the 5 is in the first position, and the 2 is in the zero position. Even though we do not think about it, what we do with each of the shorthand numbers is multiply each number by 10 raised to the number's position weight and then add all numbers.

The previous example considers only whole numbers, numbers to the left of the decimal point. It is often necessary to work with fractional numbers. Decimal fractions are numbers whose positions have weights that are negative powers of 10. The numbers to the right of the decimal point start in the position called the −1 position. Thus:

$$10^{-1} = \frac{1}{10^1} = 0.1$$

$$10^{-2} = \frac{1}{10^2} = 0.01$$

$$10^{-3} = \frac{1}{10^3} = 0.001$$

$$10^{-4} = \frac{1}{10^4} = 0.0001$$

$$10^{-5} = \frac{1}{10^5} = 0.00001$$

$$10^{-6} = \frac{1}{10^6} = 0.000001$$

Note that in each case the numeral 1 falls in the same position as the value of its power. For example, for 10^{-6} the number is the decimal followed by five zeroes and then the number. A decimal point for base 10 separates the integer and the factional parts of the number. The whole part of the number is left of the decimal, and the fractional part is right of the decimal. To illustrate this, consider the number 328.75.

$$(3 \times 10^2) + (2 \times 10^1) + (8 \times 10^0) + (7 \times 10^{-1}) + (5 \times 10^{-2}) =$$

$$300 + 20 + 8 + 0.7 + 0.05 = 328.75$$

In this example, the numeral to the far left (3) carries the greatest weight and is designated the **MSD (most significant digit).** The numeral to the far right (5) is the **LSD (least significant digit)** because it has the lowest weight in determining the overall value of the number.

Base
The number of digits or symbols used in a number system.

Radix
The base of a system of numbers, such as 2 in the binary system and 10 in the decimal system. Also called the base.

TABLE 13–1

Comparison of Strokes to Symbols in the Decimal Number System

Strokes	Symbols
None	0
/	1
//	2
///	3
////	4
/////	5
//////	6
///////	7
////////	8
/////////	9

Most significant digit (MSD)
The digit in any number that carries the most weight (i.e., represents the biggest part). For example, in the number 4,256.79 the numeral 4 is the MSD.

Least significant digit (LSD)
The digit in any number that carries the least weight (i.e., represents the smallest part). For example, in the number 4,256.79 the numeral 9 is the LSD.

The convention is to write only the digits and deduce the corresponding powers of 10 from their position. A decimal number with a decimal point is represented by a string of coefficients:

$$\ldots A_5\, A_4\, A_3\, A_2\, A_1\, A_0\, A_{-1}\, A_{-2}\, A_{-3} \ldots$$

Each A_n is one of the 10 digits used in the decimal system, with the subscript n providing the position of the digit. Therefore, the coefficient A is multiplied by 10^n for each position, and each total is added to obtain the number they represent.

13.2 The Binary System

The decimal number system is a wonderful system that has served us well in our everyday lives. Digital and electrical systems, however, are natural binary number systems where only two states exist. Either the system is on or it is off.

Digital systems need to accept a variety of number systems, depending on the type of measurement or input instruments. These numbers are processed within the digital system, and the proper number system is selected for the output device. The input number system, the digital number system, and the output number system may all be the same, or they may all be different. Figure 13–2 illustrates this concept.

FIGURE 13–2 Digital system.

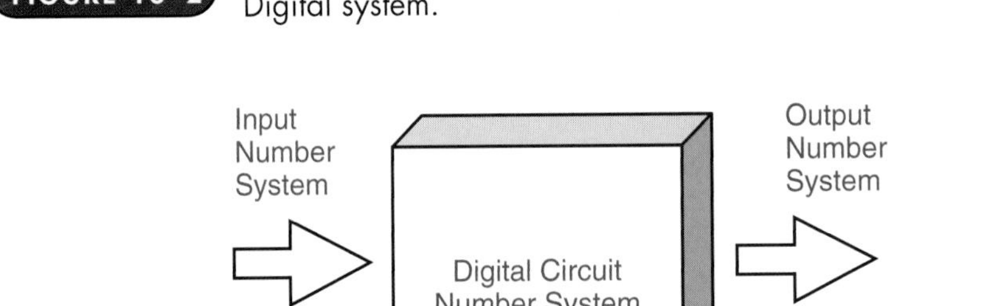

The binary number system is the simplest of those that use positional notation. The two digits used in the binary system are 0 and 1. Table 13–2 indicates the number of strokes associated with the symbols for the binary digits.

TABLE 13–2 Comparison of Strokes to Symbols in the Binary Number System

Strokes	Symbols
None	0
/	1

The binary number system has a radix of 2 (N_2) and therefore uses only the digits 0 and 1 to represent any numerical value. The number is expressed with a string of 1s and 0s and a possible binary point.

The value of every digit in a binary number is determined by its position in the binary number string.

Table 13–3 shows three ways to count to nine by comparing the number of strokes to the binary and decimal systems of numbers.

TABLE 13–3 Comparison of Strokes and Digits Between the Binary and Decimal Number Systems

Strokes	Binary Number	Decimal Number
None	0	0
/	1	1
//	10	2
///	11	3
////	100	4
/////	101	5
/////	110	6
///////	111	7
////////	1000	8
/////////	1001	9

To evaluate the total value of the number, the specific bits and the weights of their positions must be considered. As in the decimal system, the first digit to the left of the binary point has a power of 0, with sequential powers to the left of 1, 2, 3, 4, and so on. The difference exists in the base number. The base is now 2 instead of 10 used in the decimal system. Remember that any number raised to the zero power is equal to 1. A condensed listing of the powers of 2 is given in Table 13–4, with equivalent numbers in the decimal system.

TABLE 13–4 Decimal Values of the Powers of 2

$2^0 = 1_{10}$	$2^6 = 64_{10}$
$2^1 = 2_{10}$	$2^7 = 128_{10}$
$2^2 = 4_{10}$	$2^8 = 256_{10}$
$2^3 = 8_{10}$	$2^9 = 512_{10}$
$2^4 = 16_{10}$	$2^{10} = 1{,}024_{10}$
$2^5 = 32_{10}$	$2^{11} = 2{,}048_{10}$

Thus, the binary number 101010 can be evaluated using the same approach we used earlier for decimal numbers.

$$101010_2 = 1 \times 2^5 + 0 \times 2^4 + 1 \times 2^3 + 0 \times 2^2 + 1 \times 2^1 + 0 \times 2^0 =$$

$$32 + 0 + 8 + 0 + 2 + 0 = 42_{10}$$

In the binary number system, the radix point is called the binary point, as opposed to the decimal point in the decimal system. Fractional binary numbers are expressed as negative powers of 2, just as fractional decimal numbers are expressed as negative powers of 10. Following is a partial list of the values of the binary number to the right of the binary point, with their decimal equivalents.

$$2^{-1} = \frac{1}{2^1} = 0.5_{10}$$

$$2^{-2} = \frac{1}{2^2} = 0.25_{10}$$

$$2^{-3} = \frac{1}{2^3} = 0.125_{10}$$

$$2^{-4} = \frac{1}{2^4} = 0.0625_{10}$$

$$2^{-5} = \frac{1}{2^5} = 0.03125_{10}$$

$$2^{-6} = \frac{1}{2^6} = 0.015625_{10}$$

$$2^{-7} = \frac{1}{2^7} = 0.0078125_{10}$$

$$2^{-8} = \frac{1}{2^8} = 0.00390625_{10}$$

The decimal equivalent of the binary number 0.1011 is calculated as

$$(1 \times 2^{-1}) + (0 \times 2^{-2}) + (1 \times 2^{-3}) + (1 \times 2^{-4}) =$$

$$0.5_{10} + 0 + 0.125_{10} + 0.0625_{10} = 0.6875_{10}$$

There is another way to evaluate this fraction. If the binary point is ignored, the number 1011_2 is equal to

$$11_{10}\{1011_2 = (1 \times 2^3)_{10} + (0 \times 2^2)_{10} + (1 \times 2^1)_{10} + (1 \times 2^0)_{10} =$$

$$8_{10} + 0_{10} + 2_{10} + 1_{10} = 11_{10}\}$$

Four bits are possible, providing a maximum count (0 through 15_{10}) of $2^4 = 16$ distinct values.

$$\frac{11_{10}}{16_{10}} = 0.6875_{10}$$

13.3 Decimal-to-Binary Conversion

Even though most scientific calculators will easily convert from one number system to another, it is helpful to understand the methods used. For example, one method of finding the binary value of a decimal number is to determine a set of binary weight values whose sum is equal to the decimal number. Consequently, the number 14 can be expressed as the sum of binary weights.

$$14 = 8 + 4 + 2$$

By placing a 1 under the appropriate weight, the binary equivalent number is calculated, as shown in Table 13–5. Using this procedure, 14_{10} is found to equal to 1110_2. A more systematic method for converting from a decimal number to a binary number is to use the repeated division-by-2 process. This method is illustrated in Figure 13–3.

TABLE 13–5 Converting (Encoding) Binary Number 1110_2 from Decimal Number 14_{10}

Power position	2^3	2^2	2^1	2^0
Decimal (14)	8+	4+	2+	0
Power of 2	$(1 \times 2^3)_{10}$	$(1 \times 2^2)_{10}$	$(1 \times 2^1)_{10}$	$(0 \times 2^0)_{10}$
Binary	1	1	1	0

FIGURE 13–3 Repeated division-by-2 method to convert from decimal to binary number.

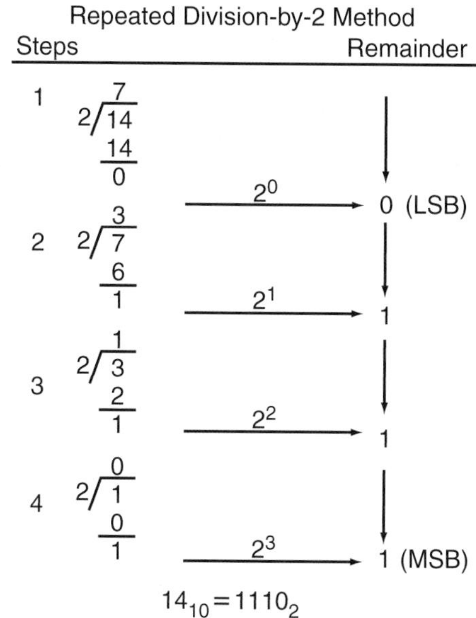

Refer to Figure 13–3. To convert the decimal number 14 to binary, the 14 is divided by 2. The remainder is carried over as the least significant binary digit. In this position (2^0) it represents 0. The quotient, 7, from the first division by 2 is then divided by 2 in step 2. Seven divided by 2 is 3, with a remainder of 1. The second binary digit in the 2^1 position is therefore 1. In step 3, the quotient 3 is divided by 2, yielding a quotient of 1 and a remainder of 1 in the 2^2 position.

In step 4, the quotient 1 from step 3 is divided by 2, yielding a quotient of 0 and a remainder of 1 in the 2^3 position. The results of this process show that 14_{10} is equal to 1110_2.

Decimal fractions can be converted to binary fractions using the sum-of-weights method. For example:

$$0.875_{10} = 0.5 + 0.25 + 0.125 = 2^{-1} + 2^{-2} + 2^{-3} = 0.111_2$$

Decimal fractions can also be converted to binary fractions using a more systematic process called the repeated multiplication-by-2 method. Figure 13–4 shows this process.

FIGURE 13–4 Repeated multiplication-by-2 method to convert from decimal to binary number.

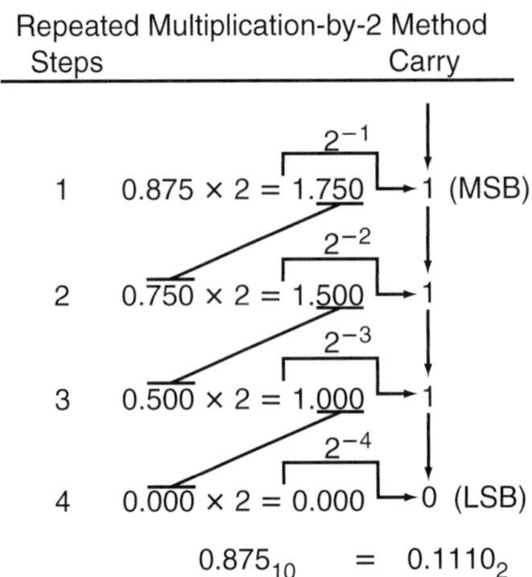

In step 1, 0.875 is multiplied by 2. Because the result is equal to or greater than 1, the 1 carries as the most significant fractional digit (2^{-1} position of the binary fraction). The remaining fraction, after the 1 is removed, is then multiplied by 2 in step 2. Once again, the result is equal to or greater than 1 and thus the 1 carries into the 2^{-2} position of the fraction. The remaining fraction, 0.500, is again multiplied by 2 in step 3 to yield 1.000. The 1 carries to the third position of the fraction, 2^{-3}. The result of the exercise is that $0.875_{10} = 0.111_2$.

13.4 Binary Number Sizes

Bit
Single binary digit that can represent 0 or 1.

Nibble
Four-bit (XXXX) binary value.

Byte
An 8-bit (XXXXXXXX) binary number (word). Note that this is also the amount of storage space required to store an 8-bit number.

A **bit** is a single binary digit. A bit can represent two discrete values (0 or 1). Binary bits can be conveniently grouped into sets of four. A 4-bit binary value is called a **nibble.** A nibble can represent 16 discrete values (0 through 15). An 8-bit binary number (two nibbles) is called a **byte.** A byte can represent 256 discrete values (0 through 255).

Another term we use in digital electronics is the *word.* The length of a binary word is usually some multiple of bytes. Thus, a word may be 1 byte (8 bits), 2 bytes (16 bits), 4 bytes (32 bits), or even more. Although unusual, a word may be as short as 2 bits. Digital equipment normally uses a fixed word size. The size of the word determines the magnitude and resolution of numbers processed with the system. The number of bits in a word determines the number of discrete states that can exist and the maximum decimal number value that can be calculated.

In the binary system, the number of states can be calculated with the following formula:

$$N = 2^n$$

where, N is equal to the total number of states and n is equal to the number of bits in the word. For example, a 3-bit word will have 8 states and a 4-bit word will have 16 states.

$$N = 2^3 = 8 \text{ and } N = 2^4 = 16$$

The maximum value of a number that can be represented is 1 less than the number of bits in a word. The largest numbers that can be represented by 3 and 4 bits are:

$$N = 2^3 - 1 = 7\ (111_2 = 7_{10})$$

$$N = 2^4 - 1 = 15\ (1111_2 = 15_{10})$$

Note that 2^4 still represents a total of 16 numbers, including $0 \rightarrow 0$, 1, 2, 3, 4, 5, 6, 7, 8, 9, 10, 11, 12, 13, 14, and 15. The number of bits in the binary system needed to represent a given number (N) can be determined using the following equation:

$$B = 3.32 \times \log_{10}(N + 1)$$

The common logarithm can be obtained from a set of tables or from almost any scientific calculator. For example, to calculate the maximum number of bits when the largest number that needs to be processed is 700 you would use the following:

$$N = 3.32 \times \log_{10}(700 + 1)$$

$$N = 3.32 \times \log_{10} 701$$

$$N = 3.32 \times 2.8457 = 9.448$$

Because fractional bits cannot be implemented, the number of bits required will be the next highest number. Thus, to represent the number 700 you need 10 bits. This can be confirmed with the previously given equation for determining the largest number that can be calculated with a fixed number of bits.

$$N = 2^{10} - 1 = 1{,}023$$

If 9 bits had been used, the maximum number would had been

$$N = 2^9 - 1 = 511$$

13.5 Binary-Coded Decimal

You have seen that numbers can be represented by binary digits. Letters and other symbols can also be represented by 1s and 0s. Combinations of binary digits that correspond to numbers, symbols, or letters are called digital codes. In many applications, special codes are used to indicate error detection and corrections.

Consider the arrangement shown in Figure 13–5. When switch A is up and all others down, the number represented is 4. In sequence, switch B alone will input a 3, switch C alone a 2, and switch D alone a 1. The numerals 0, 1, 2, 3, and 4 can be sent to the computer using these switches. Zero occurs when all switches are in their down positions.

FIGURE 13–5 Numerical values switches (4, 3, 2, 1).

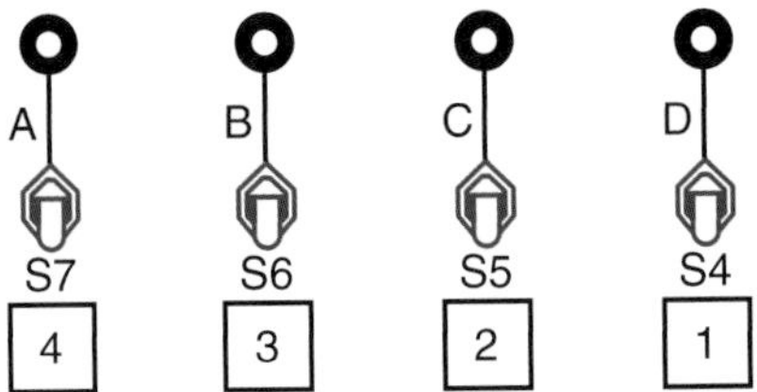

With a single switch, the highest number is 4. Suppose the number 5 is required. This could be obtained by putting switch A (4) and switch D (1) in their up positions. It is also possible to obtain a 5 by putting switches B (3) and C (2) up. This could cause some confusion in a computer system where more than one combination can produce a common result.

For the number 3, switches C (2) and D (1) in their up positions could be used. The number 3 can therefore be eliminated from the code. This means that switch A has a value of 4 and switches C and D have values of 2 and 1, respectively. Unfortunately, this combination can count only from 0 to 7. To be compatible with the decimal system, it is necessary that the count go to 9 to include all integers. To fix this problem, the value of switch A is transferred to the vacant switch B. In place of the 4 value for switch A, the new value of 8 is assigned. The new coding for the switches is shown in Figure 13–6.

FIGURE 13–6 Binary-coded decimal switches (8, 4, 2, 1).

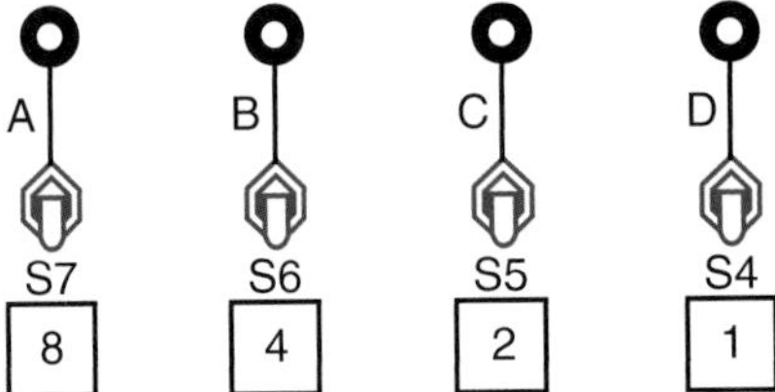

Binary-coded decimal (BCD) means that the binary code represents all numbers used in the decimal code (0–9). It is designated the 8421 code, where the position of the digit is weighted in terms of these numbers. The 8421 code is the predominant BCD code.

With 4 bits, 16 numbers (2^4) can be represented. In the 8421 code, only 10 of these numbers are used, and 1010_2, 1011_2, 1100_2, 1101_2, 1110_2, and 1111_2 are invalid. To express any number in BCD, simply replace each decimal digit with the appropriate 4-bit code.

Figure 13–7 shows a schematic diagram of a decimal-to-BCD converter you can build. The light-emitting diodes (LEDs) can be any size or color, the 300 Ω resistors can be 0.25 watt or larger, and the battery may be a 9 V transistor radio battery. The switches may be pushbutton or toggle switches. The signal diodes may be any generic rectifier diode, such as 1N4001. Trace the schematic diagram to see how the diode matrix works.

FIGURE 13–7 Practical decimal to BCD converter circuit.

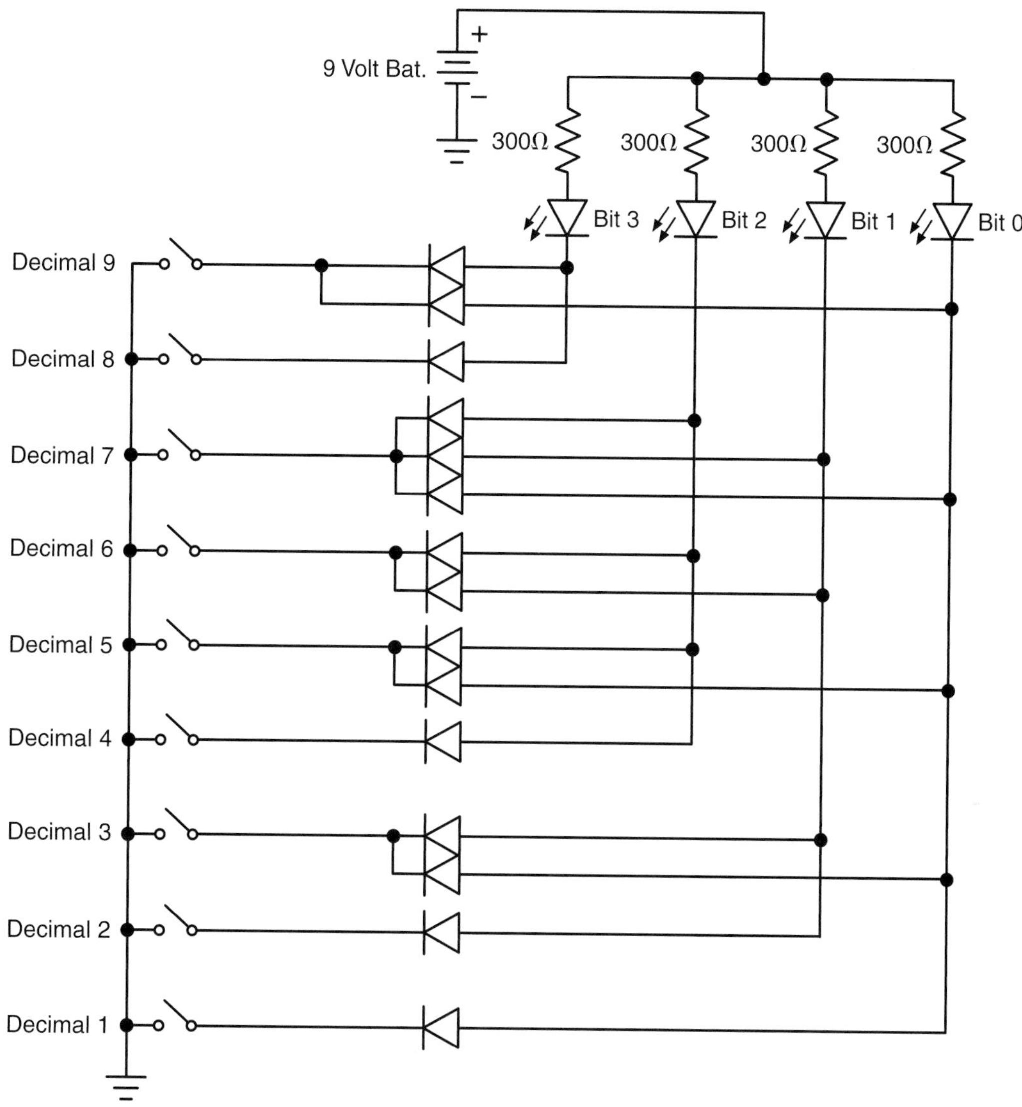

THE OCTAL NUMBER SYSTEM

Octal numbers are often used with microprocessors and have a radix of 8. The primary application of octal numbers is representation of binary numbers. The digits 0, 1, 2, 3, 4, 5, 6, and 7 are used in the octal number system. These digits have the same numerical values as the decimal system.

As with the decimal and binary number systems, each digit in the octal number system carries a positional weight. The weight of each position is determined by some power of 8. For example, the octal number 234.018 can be written

$$234.01_8 = (2 \times 8^2) + (3 \times 8^1) + (4 \times 8^0) + (0 \times 8^{-1}) + (1 \times 8^{-2}) =$$

$$128 + 24 + 4 + 0 + 0.015625 = 156.015625_{10}$$

The decimal value of the octal number is determined by multiplying each digit by its positional weight and adding the results. The octal (radix) point separates the integer from the factional part of the number. The decimal values of the positional integers from the octal point left are:

$$8^0 = 1_{10}$$

$$8^1 = 8_{10}$$

$$8^2 = 64_{10}$$

$$8^3 = 512_{10}$$

$$8^4 = 4{,}096_{10}$$

$$8^5 = 32{,}768_{10}$$

$$8^6 = 262{,}144_{10}$$

$$8^7 = 2{,}097{,}152_{10}$$

Fractional values in the decimal system from the octal point right are:

$$8^{-1} = \frac{1}{8^1} = 0.125_{10}$$

$$8^{-2} = \frac{1}{8^2} = 0.015625_{10}$$

$$8^{-3} = \frac{1}{8^3} = 0.001953125_{10}$$

$$8^{-4} = \frac{1}{8^4} = 0.000244140625_{10}$$

Each octal number can be represented by a 3-bit binary code. The binary, octal, and decimal equivalent values are outlined in Table 13–6.

TABLE 13–6 Decimal, Octal, and Binary Equivalents

Decimal	Octal	Binary
0	0	000
1	1	001
2	2	010
3	3	011
4	4	100
5	5	101
6	6	110
7	7	111

To convert an octal number to a binary number, simply replace each octal digit with the equivalent 3-bit binary code. Figure 13–8 shows examples of this approach. Figure 13–9 shows examples of converting from a binary number to an octal number. Begin at the binary point and break the binary number into groups of 3 bits, and then convert each group to the equivalent octal digit. With binary whole numbers, the binary point is understood to be to the right of the least significant bit (LSB or sometimes LSBit).

FIGURE 13–8 Octal-to-binary conversion.

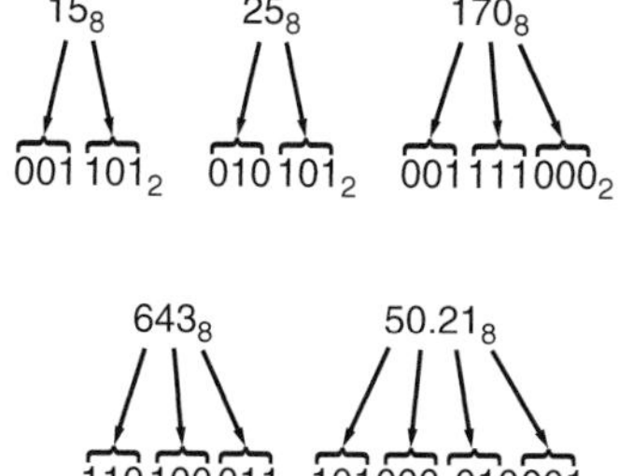

FIGURE 13–9 Binary-to-octal conversion.

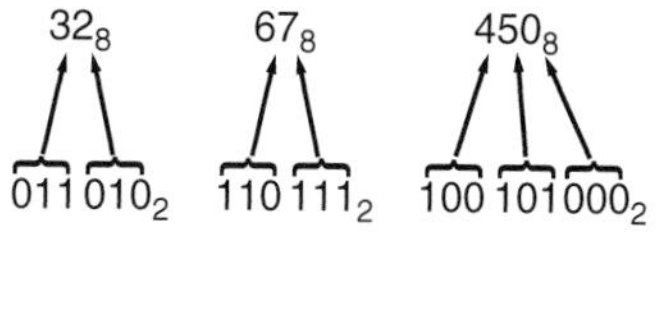

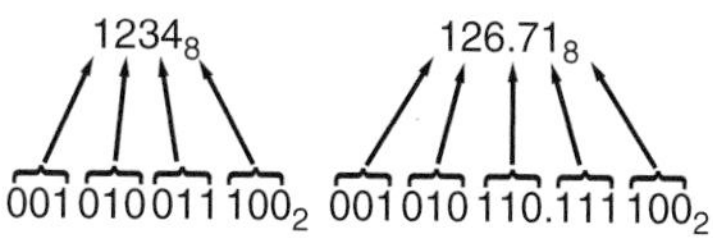

Zeros can be added with the most significant bit (MSB) and LSB where needed to complete the group of three digits. These zeros are sometimes not in the original numbers. They do not affect the values of the number when added. The repeated division-by-8 system can be used to convert a decimal number to an octal number. This is illustrated in Figure 13–10, with the conversion of decimal number 500_{10} to its octal equivalent 764_8.

FIGURE 13–10 Decimal-to-octal conversion.

Repeated Division-by-8 Method

Steps			Remainder
1	$500 \div 8 = 62$; $500 - 48 \cdot 10$… 48, 20, 16, 4	8^0	4 (LSD)
2	$62 \div 8 = 7$; 56, 6	8^1	6
3	$7 \div 8 = 0$; 0, 7	8^2	7 (MSD)

$500_{10} = 764_8$

In Figure 13–10, the number 500_{10} is divided by 8. The remainder, 4, is the LSD in the octal number system. The quotient of the first division, 62, is then divided by 8, with a remainder of 6, which is put in the 8^1 position of the octal number. The quotient of the second division, 7, is then divided by 8, with a remainder of 7, the MSD in the octal system for the decimal number 500. The 7 is in the 8^2 position of the octal number.

THE HEXADECIMAL (16) NUMBER SYSTEM

As microprocessors have become faster, the use of the octal number system has decreased and is being replaced by the hexadecimal number system. This allows for the processing of more bits at the same time. Because the hexadecimal system has a base of 16, it is composed of 16 digits and characters. Most digital systems process binary data in groups that are multiples of 4 bits. This makes the hexadecimal system convenient because each four-bit number represents one hexadecimal number.

The hexadecimal number system, also referred to as the alphanumeric hexadecimal number system, uses decimals 0 through 9 and alphabet letters A through F. The letters are used because it is necessary to represent 16_{10} different values with a single digit for each value. The letters A through F represent 10_{10} through 15_{10}, respectively. Table 13–7 outlines the corresponding values of the decimal, hexadecimal, and binary number systems.

TABLE 13–7 Decimal, Hexadecimal, and Binary Equivalents

Decimal	Hexadecimal	Binary
0	0	0000
1	1	0001
2	2	0010
3	3	0011
4	4	0100
5	5	0101
6	6	0110
7	7	0111
8	8	1000
9	9	1001
10	A	1010
11	B	1011
12	C	1100
13	D	1101
14	E	1110
15	F	1111

It might seem awkward using letters for numbers, but after you become familiar with the notation it becomes much easier. As with the previous number systems, each digit position of the hexadecimal number carries a positional weight.

Positional values left of the hexadecimal point are:

$16^0 = 1_{10}$

$16^1 = 16_{10}$

$16^2 = 256_{10}$

$16^3 = 4{,}096_{10}$

$16^4 = 65{,}536_{10}$

$16^5 = 1{,}048{,}576_{10}$

$16^6 = 16{,}777{,}216_{10}$

Positional values to the right of the hexadecimal point are:

$$16^{-1} = \frac{1}{16} = 0.0625_{10}$$

$$16^{-} = \frac{1}{16^2} = 0.00390625_{10}$$

$$16^{-3} = \frac{1}{16^3} = 0.000244140625_{10}$$

$$16^{-4} = \frac{1}{16^4} = 0.0000152587809625_{10}$$

A decimal number can be converted to a hexadecimal number by the repeated division-by-16 process. This is shown in Figure 13–11. To convert 700_{10} to hexadecimal, you first divide by 16. The remainder of 12 from the first division is the LSD, represented by C.

FIGURE 13–11 Decimal-to-hexadecimal conversion.

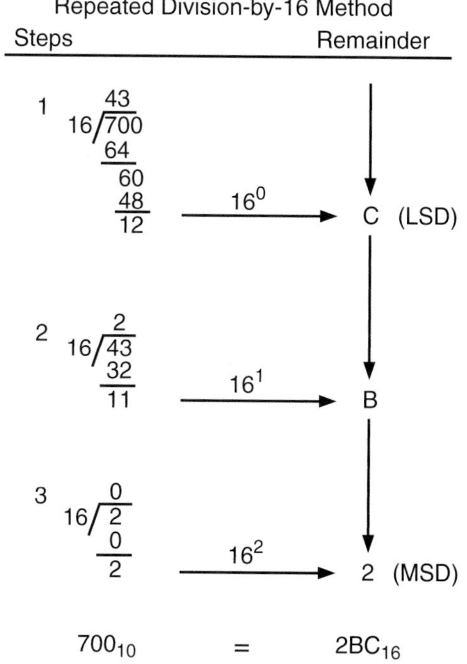

The quotient of the first division, 43, is now divided by 16, with a remainder of 11. B represents 11 in the hexadecimal number system and is placed in the 16^1 position. The quotient 2 from the second division is now divided by 16, with a remainder of 2. The numeral 2 is placed in the most significant digit position, giving $2BC_{16}$ for the decimal number 700_{10}.

Binary numbers can easily be converted to hexadecimal values. The binary number is converted to groups of four, starting with the binary point. Zeros can be added to the most significant group so that four digits are available without changing the value of the binary number. Each of these groups is then replaced with the corresponding hexadecimal value (see Figure 13–12).

Hexadecimal numbers can be changed to binary numbers in a similar manner. Each of the hexadecimal numbers is changed to an equal-value four-digit binary number, starting at the hexadecimal point. Four examples are shown in Figure 13–13. Counting with the hexadecimal system is similar to the decimal system.

0 1 2 3 4 5 6 7 8 9 A B C D E F 10 11 12 13 14 15 16 17 18 19

1A 1B 1C 1D 1E 1F 20 21 22 23 24 25 26 27 28 29 2A 2B 2C 2D

2E 2F 30 31 32 33 . . .

FIGURE 13–12 Binary-to-hexadecimal conversion.

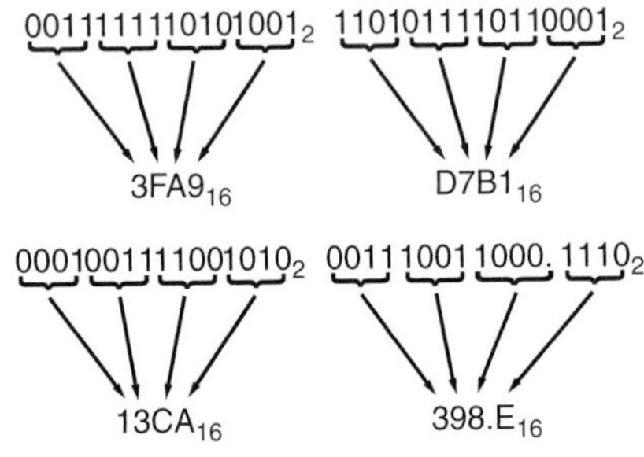

FIGURE 13–13 Hexadecimal-to-binary conversion.

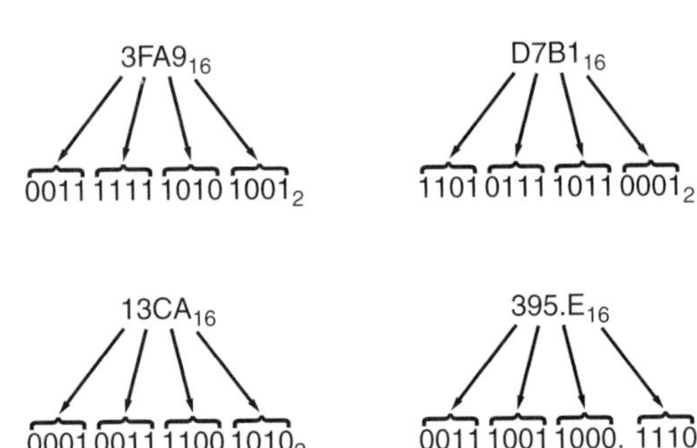

Two hexadecimal digits allow a count to FF_{16}, which is 255_{10}. A count beyond this decimal number requires additional hexadecimal digits. For example, 100_{16} equals 256_{10}, 101_{16} equals 257_{10}, and so on. The maximum three-digit hexadecimal number is FFF_{16}, or $4{,}095_{10}$. The maximum four-digit hexadecimal number, $FFFF_{16}$, is equal to $65{,}535_{10}$.

Occasionally, you may need to convert from hexadecimal to decimal. One way of making this conversion is to convert to binary and from binary to decimal. Conversion can be made directly to a decimal number by multiplying the number in each position by the power of 16 for that position and then adding the numbers. An example of the process follows for the number $C53_{16}$.

$$C53_{16} = (12 \times 16^2) + (5 \times 16^1) + (3 \times 16^0) = 3{,}072 + 80 + 3 = 3{,}155_{10}$$

Example

A digital-to-analog converter (DAC) integrated circuit is configured to convert an 8-bit input to a DC output voltage ranging from 0 V to 10 V. Figure 13–14 shows a simplified DAC sample circuit. When switches close, the related bit becomes logic level 1.

a. Find the output voltage step size.
b. Find the minimum DC output voltage.
c. Find the maximum DC output voltage.
d. Find the DC output voltage if the digital input is $F2_{16}$.
e. Find the binary input if the DC output is 4.15 V.

Solution:

a. Given that this circuit uses 8 bits, the number of steps will be:

$$N = 2^8 - 1 = 255_{10}$$

Then, 10 V divided into 255 steps results in:

$$\text{Step size} = \frac{10\text{ V}}{255\text{ Steps}} = 0.0392\text{ V/Step}$$

b. Minimum output voltage will appear at the output when no input bits are selected:

$$\text{Output} = \text{Step size} \times \text{Steps} \Rightarrow 0.0392\text{ V} \times 0\text{ Steps} = 0\text{ V}$$

c. Maximum output voltage will appear at the output when all input bits are selected (close all switches):

$$\text{Output} = \text{Step size} \times \text{Steps} \Rightarrow 0.0392\text{ V} \times 255_{10}\text{ Steps} = 10.0\text{ V}$$

d. First convert $F2_{16}$ steps to decimal steps:

$$F2_{16} = (15_{10} \times 16_{10}{}^{1}) + (2_{10} \times 16_{10}{}^{0}) = 240_{10} + 2_{10} = 242_{10}$$

Then find the output, as in step c:

$$\text{Output} = \text{Step size} \times \text{Steps} \Rightarrow 0.0392\text{ V} \times 242_{10}\text{ Steps} = 9.49\text{ V}$$

e. From Table 13–7, the hex value $F2_{16}$ can easily be expanded:

		F				2			HEX NUMBER
1	1	1	1	/	0	0	1	0	EQUIV. BINARY
SW7	SW6	SW5	SW4	/	SW3	SW2	SW1	SW0	SWITCH NAME
on	on	on	on		off	off	on	off	SWITCH POS.

Examine Figure 13–15. This figure is a practical application of counter, comparator, and inverter integrated circuits. Digital pulses are applied to the circuit at the count input. The pulses are added in counter U2, and can be read at U2's Q_D, Q_C, Q_B, and Q_A as a binary number, with Q_D the most significant bit and Q_A the least significant bit.

FIGURE 13–14 Simplified DAC sample circuit.

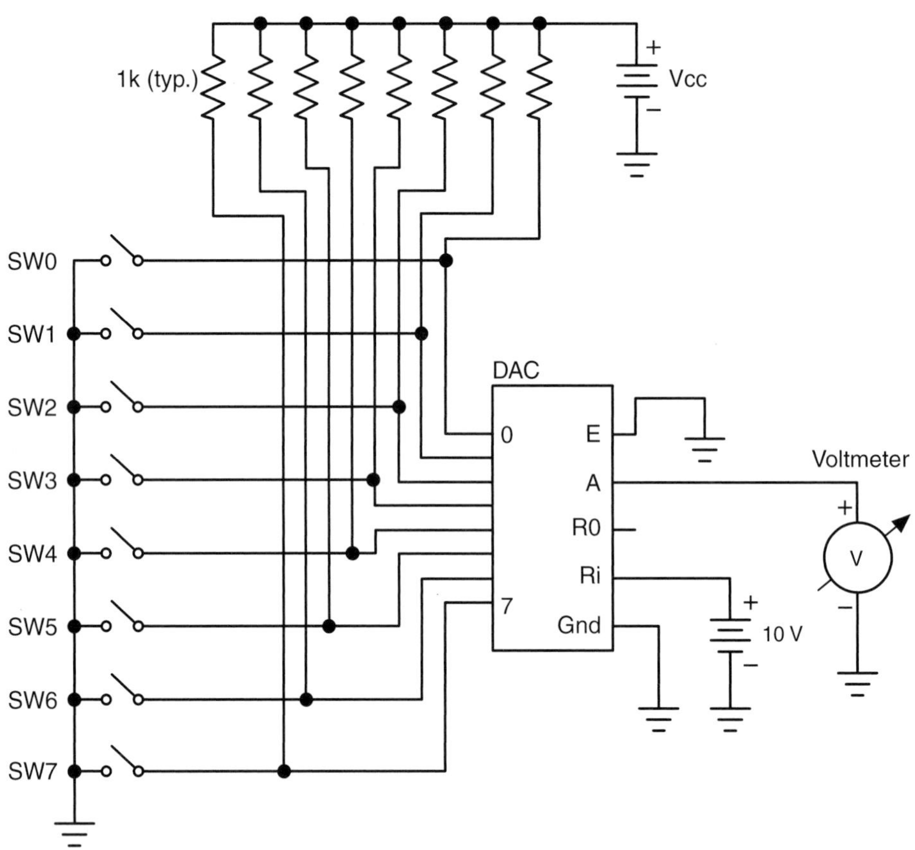

Once the accumulated value at U2 reaches 1111_2 (F_{16}), the next count received will cause the following:

- Output Q_D through Q_A will all roll over to 0000_2, and continue counting up.
- CO (carry output) will change states, causing U1 to increment from 0000_2 to 0001_2.

Output from U1 will increment only once for every 16_{10} pulses received at the count input. U1 and U2 then combine to form an 8-bit counter circuit. When pressed, the CLEAR THE COUNT N.C. switch will reset the counters U1 and U2 to zero.

U3 is an integrated circuit called a comparator. The digital signal applied at U3 A_3, A_2, A_1, and A_0 is compared to the digital signal applied at B_3, B_2, B_1, and B_0, respectively. If the A nibble is less than the B nibble, the inverter IC (U4) will illuminate the upper red LED. If the A nibble is equal to the B nibble, the inverter U5 will illuminate the green LED. If the A nibble is greater than the B nibble, U6 will illuminate the lower red LED. In practice, the LEDs could be replaced by audible signals, or provide closed-loop feedback to control the device(s) generating the pulses.

Example

In Figure 13–15, note that switch 6 at U3, B1 is closed. Switches 4, 5, and 7 are open. What range of count inputs will illuminate the green LED? Express the answer in binary, hex, and decimal.

Solution:

The comparator (U3) responds to U1 output, and U1 is the most significant nibble. Switches at B3 through B0 are configured: B3 = Logic Low, B2 = Logic high, B1 = Logic low, and B0 = Logic low, or 0100_2.

U1 NIBBLE	U2 NIBBLE
0 1 0 0	X X X X

Therefore, the green LED will illuminate for any count input ranging from $0100\ 0000_2$ to $0100\ 1111_2$—or expressed in hexadecimal from 40_{16} to $4F_{16}$. Conversion from hex to decimal is achieved as follows.

$$N_{10(\text{minimum})} = (4 \times 16^1) + (0 \times 16^0) = 64_{10}$$

$$N_{10(\text{maximum})} = (4 \times 16^1) + (15 \times 16^0) = 79_{10}$$

FIGURE 13–15 Practical event counter comparator circuit.

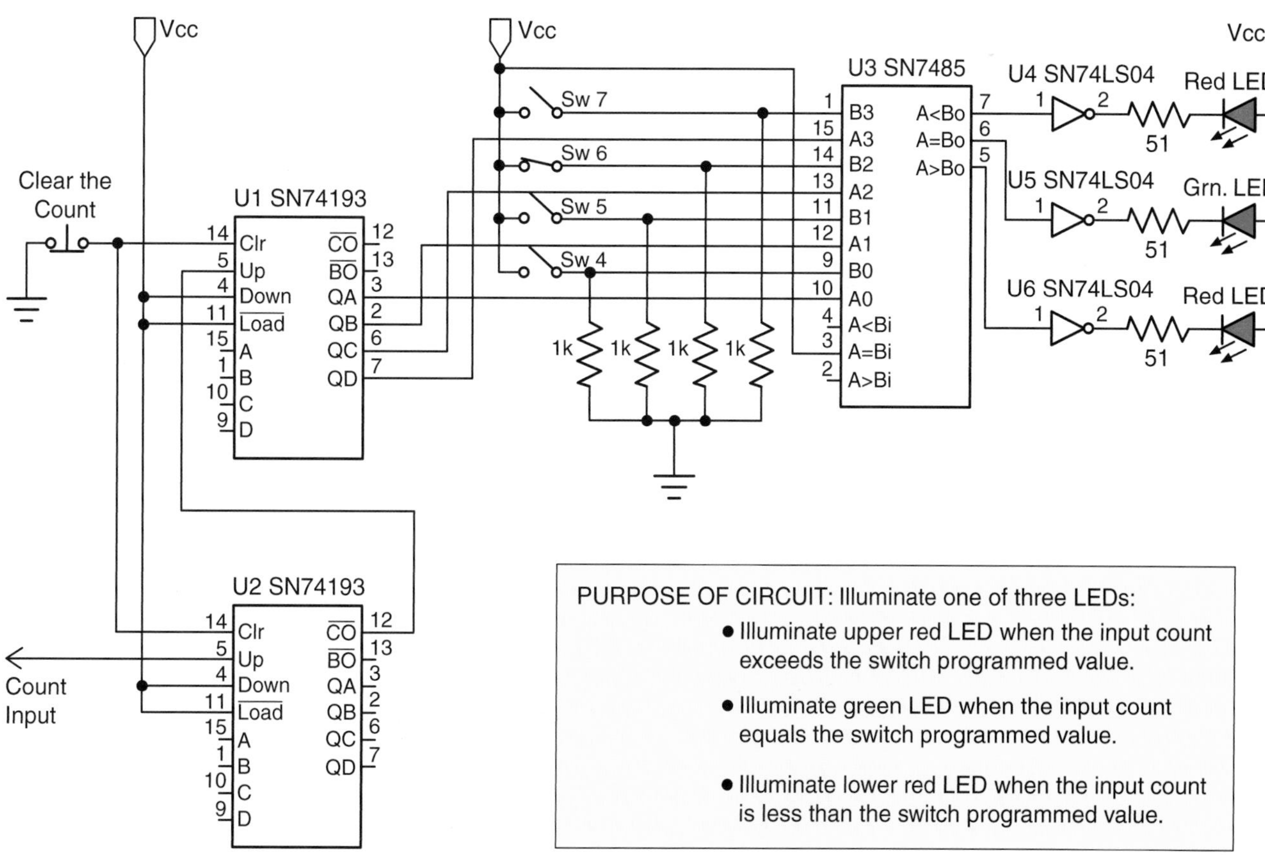

SUMMARY

The 10-digit decimal numbering system we use every day has limitations when dealing with modern digital and electrical systems. Basically, only two states exist with the digital numbering system: on and off. The requirement for the use of a variety of numbering methods, depending on the type of measurement or input needed, has led to the development of additional numbering systems for the digital world.

The simplest form we have studied is the binary numbering system, containing only two digits. Because of the large number of digits needed by the binary system, the octal and hexadecimal systems are often used instead. These allow larger numbers using smaller numbers of digits.

REVIEW QUESTIONS

1. Convert the number $353{,}652_{10}$ to
   - a. Binary
   - b. Octal
   - c. Hexadecimal
   - d. Egyptian
2. Consider the concept of the LSD and MSD. How do these affect computations such as multiplication, division, subtraction, and addition?
3. What is the octal point? The hexadecimal point? The binary point? How do they relate to the decimal point?
4. When would you use the repeated division-by-2 method? When would you use the repeated multiplication-by-2 method?
5. The most often-used binary code for computer applications is called the American Standard Code for Information Interchange, usually abbreviated as ASCII. The most fundamental form for ASCII uses a 7-bit binary word. So-called extended ASCII uses an 8-bit binary word.
   - a. How many different characters or words can ASCII represent?
   - b. How many different characters or words can extended ASCII represent?
6. Use the following table for the indicated switch positions of Figure 13–16.
   - a. What binary and decimal numbers are represented?
   - b. Which, if any, of the positions are "not allowed"?

Check your answers by converting from decimal to binary and back again.

7. In some electrical system protection equipment, settings are applied by the use of a 32-bit binary word. To shorten the amount of storage space, the binary word is usually abbreviated using an 8-bit hexadecimal word. What binary number is represented by $FC34A7A8_{16}$?
8. In each of the various number systems, it takes 2 bits to represent the radix. For example, in the binary system it is the number $2 = 10_2$, in the octal system the number $8 = 10_8$, and so on. Explain why you always need 2 bits to express the radix of a given number system.
9. How many characters can ASCII (see question 5) represent? Express your answer in hexadecimal.
10. Convert the number $8{,}461.203_{10}$ to octal.

FIGURE 13–16 Binary-coded decimal switches (8, 4, 2, 1).

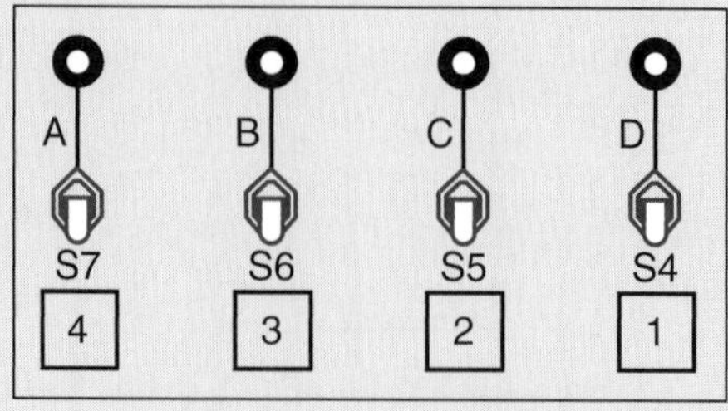

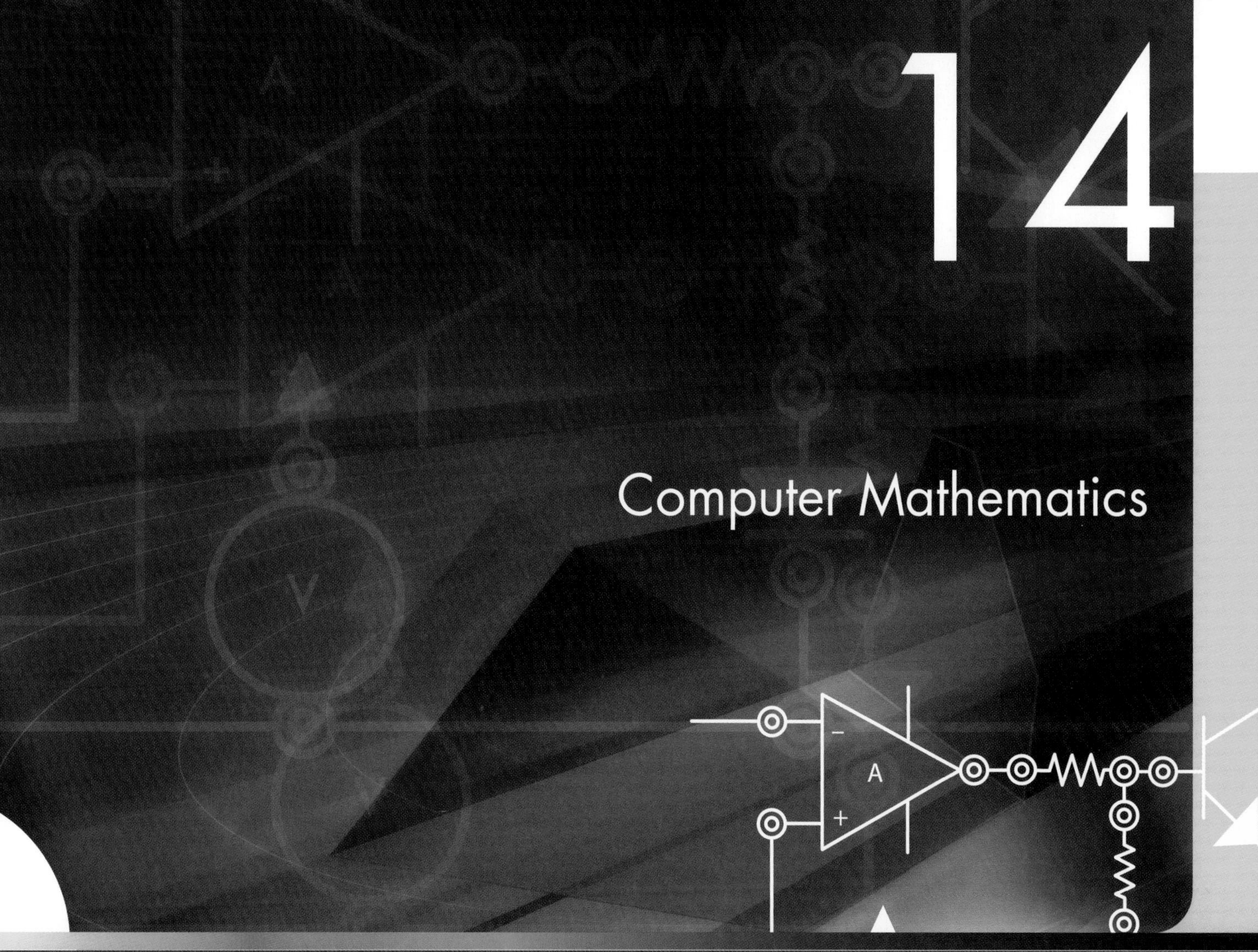

14

Computer Mathematics

OUTLINE

OVERVIEW

Virtually all operations within a number system can be simplified to addition and subtraction. Multiplication, division, exponentiation, and all other such mathematical operations are simply extensions of addition and subtraction. Throughout your education, you have been taught to perform all of these operations in the decimal system.

However, binary arithmetic is fundamental to all digital systems. To understand digital systems, you must learn the basics of binary arithmetic. In this chapter, you will learn how to add, subtract, multiply, and divide using binary numbers. Decimal mathematics is reviewed as a guide to understanding the decimal system better.

OBJECTIVES

After completing this chapter, the student should be able to:

1. Add binary numbers
2. Subtract binary numbers
3. Multiply binary numbers
4. Divide binary numbers

ADDITION

14.1 Decimal Addition

Addition is a manipulation of numbers that represent the combining of common physical quantities. For example, in the decimal system 3 + 4 = 7 symbolizes combining /// strokes with //// strokes to get /////// strokes. Binary addition is much like decimal addition. Using two decimal numbers, if $54{,}678_{10}$ is added to $69{,}142_{10}$ a sum of $123{,}820_{10}$ is obtained. Table 14–1 indicates how this decimal number is calculated.

TABLE 14–1 Decimal Addition

Column number	6	5	4	3	2	1
Column power	10^5	10^4	10^3	10^2	10^1	10^0
Carry	1	1	0	1	1	—
Augend	—	5	4	6	7	8
Addend	—	6	9	1	4	2
Sum	1	2	3	8	2	0

Example

Find the sum of the **addends** 54,678 and 69,142 using Table 14–1.

Solution:

First column (the one to the right, the 1s column or the 100 column)

Adding the rightmost column (10^0) gives (8 + 2 = 10). This is expressed in the sum as the digit 0 under 8 and 2, with a carry of 1 above the 7 of the **augend.**

Second column

The carry is then added in the second column, 1 + 7 + 4 equals 12. This is expressed as a 2 in the sum in the 10^1 column, with a carry of 1 to be added in the third column.

Third column

Adding the third column gives 1 + 6 + 1 = 8. The 8 becomes part of the sum in the 10^2 column. There is no carry in this case, and thus a 0 is placed above the 4 in the 10^3 column.

Fourth column

Adding the fourth column gives 0 + 4 + 9 = 13. The 3 becomes part of the sum and the 1 is the carry to the 10^4 column above the 5.

Fifth column

Adding the fifth column gives 1 + 5 + 6 = 12. The 2 becomes part of the sum in the 10^4 column and the carry of 1 is added in the 10^5 column.

Addend
Any of a set of numbers to be added.

Augend
The first in a series of addends.

14.2 Binary Addition

The same basic operations are performed with binary numbers, using the following four rules.

$0 + 0 = 0$

$0 + 1 = 1$

$1 + 0 = 1$

$1 + 1 = 10$ (0 with a carry of 1)

Note that the first three rules result in a single bit, 0 or 1. In the fourth case, the addition of two 1s results in a binary 2 (10_2).

Example

Use Table 14–2 to add 011_2 and 001_2.

Solution:

First column

In the 2^0 column, adding $1 + 1$ results in a 0 in the sum of that column (with a carry of 1).

Second column

Adding the 2^1 column gives $1 + 1 + 0 = 10$. The 0 remains in the 2^1 column and the 1 is carried to the 2^2 column.

Third column

Adding the 2^2 column gives $1 + 0 + 0 = 1$. The sum therefore is 100^2, which is equal to 4_{10}. This checks because $011_2 = 3_{10}$ and $001_2 = 1_{10}$.

When there is a carry digit, there is a situation in which 3 bits are being added. The rules for handling each of the possible sums are:

Carry + Augend + Addend = Sum
$1 + 0 + 0 = 01$ (1 with a carry of 0)
$1 + 1 + 0 = 10$ (0 with a carry of 1)
$1 + 0 + 1 = 10$ (0 with a carry of 1)
$1 + 1 + 1 = 11$ (1 with a carry of 1)

TABLE 14–2 Binary Addition

Column number	3	2	1
Column power	2^2	2^1	2^0
Carry	1	1	—
Augend	0	1	1
Addend	0	0	1
Sum	1	0	0

[a] All digits are base 2.

Example

To illustrate these possibilities, add 1011^2 to 1001^2 using Table 14–3. In this binary addition example, 11_{10} is added to 9_{10} to obtain a sum of 20_{10}. If four 1s are to be added, cumulatively add two numbers each time:

$1 + 1 = 10$
$10 + 1 = 11$
$11 + 1 = 100$

TABLE 14–3 Another Example of Binary Addition

Column number	5	4	3	2	1
Column power	2^4	2^3	2^2	2^1	2^0
Carry	1	0	1	1	–
Augend	–	1	0	1	1
Addend	–	1	0	0	1
Sum	1	0	1	0	0

Example

Add 111111_2 to 11001100_2 using Table 14–4.
Evaluate this addition in terms of base 10 numbers.

$$\begin{array}{r} 11001100_2 = 204_{10} \\ +00111111_2 = 63_{10} \\ \hline 100001011_2 = 267_{10} \end{array}$$

TABLE 14–4 A Third Example of Binary Addition

Column number	9	8	7	6	5	4	3	2	1
Column power	2^8	2^7	2^6	2^5	2^4	2^3	2^2	2^1	2^0
Carry	1	1	1	1	1	1	0	0	–
Augend	–	1	1	0	0	1	1	0	0
Addend	–	0	0	1	1	1	1	1	1
Sum	1	0	0	0	0	1	0	1	1

14.3 Computer Addition

There are several methods by which computers add numbers. One thing common to all computers is that numbers are added in pairs only. If there is a sum of three numbers, a pair will be added to obtain the first sum and the remaining addend will then be added for the final sum.

Methods for addition are classified as either parallel or serial. The difference is the method in which the numbers are transmitted to and from the computer or processed within the computer.

Parallel Representation

In the parallel method, a number's code is transmitted on a set of leads, one for each digit in the number code. The number 1001_2 would require four leads. The first and fourth lead would be high, and the second and third lead would be low to transmit the number 9_{10} ($1001_2 = 9_{10}$). The parallel methods of addition will add two numbers by considering all positions in the numbers at once and producing a set of signals on a third set of leads that corresponds to the sum. Figure 14–1 shows parallel processing of the binary word 10110010. Note that there is one lead for each digit.

FIGURE 14–1 Parallel processing of the binary word 10110010.

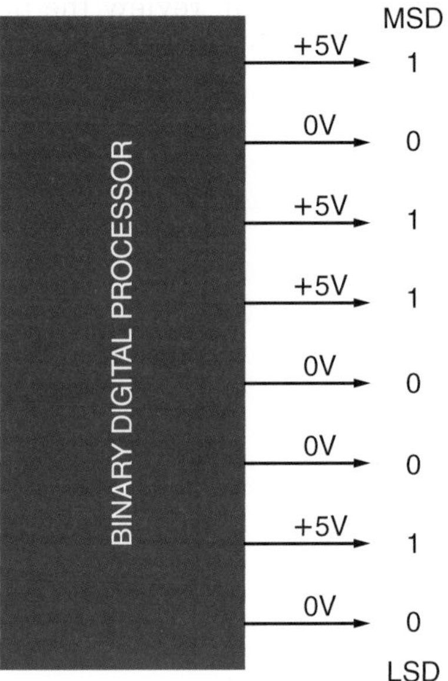

Serial Representation

With the serial method, numbers are transmitted to the machine in sequence, one digit at a time, beginning with the most significant digit (MSD). As the sequence of signals is fed into the processing unit, the digits are combined progressively and the sum is given as a sequential output. Figure 14–2 shows the serial transmission of the binary word 10110010. Note that only a single transmission line is necessary because the bits are separated by time.

The parallel method is faster but requires more apparatus. The serial method requires less apparatus but generally requires more time for processing. Many factors go into the selection of one system over the other. New innovations in protocol, such as the universal serial bus (USB) and IEEE 1394 protocol, have made serial transfer extremely fast.

FIGURE 14–2 Serial processing of the binary word 10110010.

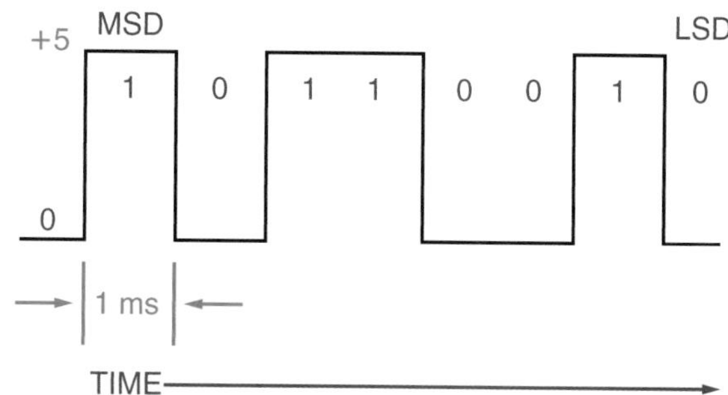

SUBTRACTION

14.4 Decimal Subtraction

Binary subtraction is performed in the same manner as decimal subtraction. Before attempting binary subtraction, review the decimal method of subtracting 2768_{10} from 7605_{10}. See Table 14–5.

TABLE 14–5 Decimal Subtraction

Column power	10^3	10^2	10^1	10^0
Minuend after borrow	6	15	9	15
Minuend	7	6	0	5
Subtrahend	2	7	6	8
Difference	4	8	3	7

Because the digit 8 in the **subtrahend** is greater than **minuend** 5, 1 is borrowed from the next higher-order digit in the minuend. If the digit is 0 (as in the example), 1 is borrowed from the next higher order that contains a number other than 0. That number is reduced (6 to 5 in the example), and the digit skipped in the minuend becomes a 9. This is the equivalent of subtracting 1 from 20, with 19 being the difference. In the decimal system, the digit borrowed has the value of 10, and thus the 5 in the 10^0 place is now 15, as shown in the top row of numbers ($15 - 8 = 7$).

Continuing: $9 - 6 = 3$ for the 10^1 position. Because 5 is less than 7 in the 10^2 position, 1 is borrowed to make the number 15 and the 7 in the 10^3 position is reduced to 6. $15 - 7 = 8$ and $6 - 2 = 4$ completes the problem.

Subtrahend
A quantity or number to be subtracted from another. In the equation $50 - 16 = 34$, the subtrahend is 16.

Minuend
The quantity from which another quantity, the subtrahend, is to be subtracted. In the equation $50 - 16 = 34$, the minuend is 50.

14.5 Binary Subtraction

In the decimal system, the borrowed number has the value of the radix (or 10). In the binary number system, the borrowed number will also have the value of the base (in this case, 2). When one binary number is subtracted from another, the same method described for the decimal system is used.

The basic rules for binary subtraction are:

$0 - 0 = 0$

$1 - 1 = 0$

$1 - 0 = 1$

$10 - 1 = 1$ (0 − 1, with a borrow of 1)

$10 - 1 = 1$ can be better understood when using strokes

$$// - / = /$$

To subtract larger binary numbers, subtract column by column from the 2^0 position to 2^n, where n is the power of the MSD. In Table 14–6, 101_2 is subtracted from 111_2.

TABLE 14–6 Binary Subtraction

Column power	2^2	2^1	2^0
Minuend after borrow	—	—	—
Minuend	1	1	1
Subtrahend	1	0	1
Difference	0	1	0

In this case, it was not necessary to borrow a digit. The minuend is 7_{10}, and the subtrahend is 5_{10}. The difference is 2_{10}. This verifies the value of the difference obtained using the binary system. Table 14–7 outlines a second example.

TABLE 14–7 A Second Example of Binary Subtraction

Column power	2^3	2^2	2^1	2^0
Minuend after borrow	—	0	10	—
Minuend	1	1	0	1
Subtrahend	1	0	1	0
Difference	0	0	1	1

In this example, it was necessary to borrow a digit in the 2^1 position. In this case, $10 - 1 = 1$. To confirm the results, the decimal solution to the problem is $13 - 10 = 3$. Like decimal numbers, binary numbers can also be negative. Subtracting 111_2 from 100_2 will result in a negative value, as indicated in Table 14–8.

TABLE 14–8 A Negative Binary Number

Column power	2^2	2^1	2^0
Minuend after borrow	—	—	—
Minuend	1	0	0
Subtrahend	1	1	1
Difference	0	1	1

Note that as with decimal numbers we subtract the smaller number from the larger and then prefix the sign of the larger of the two numbers. In this case, it is the sign of the subtrahend. The decimal values verify the difference obtained in the binary subtraction: 4 − 7 = −3. Two more examples of binary subtraction follow, with verification using decimal numbers.

Minuend after borrow:		0	1	1	1	10	1	10
Minuend:	1	1	0	0	0	1	0	0
Subtrahend:	−0	0	1	0	0	1	0	1
Difference:	1	0	0	1	1	1	1	1

The decimal solution for the preceding problem: 196 − 37 = 159

Minuend after borrow:	0		10	10				
Minuend:	1	1	1	0	1	1	1	0
Subtrahend:	−1	0	1	1	1	0	1	0
Difference:	0	0	1	1	0	1	0	0

The decimal solution for the preceding problem: 238 − 186 = 52

Note that when a borrow is required 1 is obtained from the next higher-order bit that contains a 1. That bit becomes zero, as illustrated in the 2^6 position of both problems. All bits skipped that had 0 become 1, as best illustrated in the first problem. For example, if 1 is subtracted from 1000_2 the result is 0111_2.

MULTIPLICATION

14.6 Decimal Multiplication

Multiplication is nothing more than a short method of addition. For example, the decimal multiplication problem 3 × 2 = 6 is the same as 3 + 3 = 6. 3 × 3 = 9 is the same as 3 + 3 + 3 = 9. The addition process is fine for small numbers, and most of us probably tried to remain with addition when we were first introduced to the multiplication process. However, we soon found that this was futile when numbers such as 324 × 467 were introduced.

Multiplicand:	324
Multiplier:	× 467
First partial product:	2268
Second partial product:	1944
Third partial product:	1296
Carry:	02110
Final product:	151,308

Multiplicand
The number to be multiplied by another. In 8 × 32, the multiplicand is 8.

Multiplier
The number by which another number is multiplied. In 8 × 32, the multiplier is 32.

To add 324 to itself 467 times would be quite a task. Using the short form of multiplication, the multiplicand is multiplied by the multiplier one digit at a time to obtain a partial product for each. The partial products are then added. Note that the carries for each row of addition have been included in the example.

14.7 Binary Multiplication

Binary multiplication is much easier than decimal multiplication. In binary multiplication, we have to deal with only two digits (0 and 1) and there are only four rules.

$0 \times 0 = 0$

$0 \times 1 = 0$

$1 \times 0 = 0$

$1 \times 1 = 1$

Numbers with several bits are processed just like decimal numbers. Partial products are obtained and are added. Examine the following examples. The results using binary multiplication methods are verified by substituting decimal numbers for the binary numbers.

Multiplicand:	11
Multiplier:	1
Final product:	11

The decimal equivalent of the binary multiplication: $3 \times 1 = 3$

Multiplicand:	11
Multiplier:	11
First partial product:	11
Second partial product:	11
Carry:	10
Final product:	1001

The decimal equivalent of the binary multiplication: $3 \times 3 = 9$

Multiplicand:	111
Multiplier:	101
First partial product:	111
Second partial product:	000
Third partial product:	111
Carry:	11100
Final product:	100011

The decimal equivalent for the preceding problem: $7 \times 5 = 35$

Multiplicand:	1111
Multiplier:	1010
First partial product:	0000
Second partial product:	1111
Third partial product:	0000
Fourth partial product:	1111
Carry:	1111
Final product:	10010110

The decimal equivalent for the preceding problem: $15 \times 10 = 150$

Keep in mind that just as in decimal multiplication you must keep track of any zeros by setting a zero product under the 0 bit in the multiplier. This is very important when the zero occupies the least significant digit (LSD).

DIVISION

14.8 Decimal Division

Division is the reverse of multiplication. It is the process of determining how many times one number can be subtracted from another. In the decimal number system, most students are probably familiar with short and long division. In short division, only a single digit is the **divisor**, whereas in long division two or more digits are in the divisor. See the examples that follow.

Divisor
The quantity by which another quantity, the dividend, is to be divided. In 45 ÷ 3 = 15, 3 is the divisor.

Example

Short division

```
   202
 8)1616
```

Where 8 is called the divisor, 1616 is called the **dividend** and 202 is called the **quotient.**

Dividend
A quantity to be divided. In 45 ÷ 3 = 15, 45 is the dividend.

Quotient
The number obtained by dividing one quantity by another. In 45 ÷ 3 = 15, 15 is the quotient.

Example

Long division

```
     457
 19)8683
   -76
    108
   -95
    133
   -133
    000
```

Where 19 is the divisor, 8683 is the dividend, 457 is the quotient, and 000 is the remainder.

In division, the most significant digit in the dividend is examined to determine if the divisor is smaller or greater in value. In our short division example using radix 10, 8 is greater than 1 and thus the quotient is 0 for this position. In these cases where the divisor is a single digit, the two most significant digits are selected. Eight is divided into 16 to give a quotient of 2 in the 10^2 position. The third MSD is 1 and is smaller than the divisor of 8. The quotient in the 10^1 position is 0, with the 1 carrying for the next round (8 into 16 is 2).

The same process is used for the long division problem. The divisor 19 is larger than the MSD of 8, giving a quotient of 0 in the 10^3 position. The 8 carries, and thus in the second round 19 is divided into 86 (with a quotient of 4 in the 10^2 position). Four is multiplied by 19 to give a product of 76, which is subtracted from 86. The remainder is 10. The 8 in the 10^1 position is brought down to provide a remainder of 108. One hundred and eight divided by 19 equals 5, with a remainder of 13. The 3 in the 10^0 position is brought down to provide a remainder of 133. One hundred and thirty-three divided by 19 equals 7, with no remainder.

14.9 Binary Division

Binary division is performed much like decimal division. Binary division, however, is a simpler process because only two numbers are used rather than 10 (as in the decimal number system). See the examples of binary division that follow.

Example

Long division

```
     110
  10)1100
    -10
      10    Remainder
     -10
      00    Remainder

       111
  101)100011
     -101
       111    Remainder
      -101
        101   Remainder
       -101
        000   Remainder
```

When using long division, the dividend is examined (beginning with the MSD) to determine the number of bits required to exceed the value of the divisor. In the first example, this occurs above the 2^2 position of the dividend. A 1 is placed in this position and multiplied times the divisor. The value of the divisor is subtracted from the dividend. This results in a remainder of 1.

Bring down the 0 in the 2^1 position to the remainder. Place a 1 above the 2^1 position of the dividend and multiply times the divisor. Subtract the product from the remainder. This leaves zero remainder. Put a 0 above the 2^0 position. The results of the first binary division can be confirmed using the decimal number system: $1100_2 = 12_{10}$ and $10_2 = 2_{10}$.

$$\frac{1100_2}{10_2} = \frac{12_{10}}{2_{10}} = 110_2$$

You may have noticed that division of a binary number by 2 can be accomplished simply by moving the binary point one place to the left. Division by 4 only requires movement of the binary point two places left, and so on. For multiplication, move the binary point to the right.

The same approach is taken in the second example. The dividend is examined (beginning with the MSD) to determine how many places are necessary to just exceed or equal the value of the divisor. This is true at the 2^2 position of the dividend. A 1 is placed above this position and multiplied times the divisor. Subtract 101 from 1000 in the dividend. This results in a remainder of 11, which is less than 101 of the divisor.

Like decimal mathematics, this remainder must be less than the divisor or an error has been committed. Bring down 1 from the dividend to make the remainder greater than the divisor. Place 1 in the quotient above the 2^1 position of the dividend and multiply times the divisor. Subtract 101 from 111, leaving a remainder of 10. Bring down the 1 from the dividend and again divide by the divisor. Place a 1 in the quotient above the 2^0 position and multiply times the divisor. Subtract the product from the remainder. The result is zero. Confirming with decimal arithmetic yields:

$$\frac{100011_2}{101_2} = \frac{35_{10}}{5_{10}} = 7_{10} = 111_2$$

14.10 Hexadecimal Operations

Manual addition, subtraction, multiplication, and division in base 16 is a bit more difficult due to the large span of values (0 through 15). However, because hexadecimal numbers can be very easily converted to and from binary the following example will illustrate an easy process for hex math manipulations when operators are only a couple of bytes wide.

Example

Add two hexadecimal numbers.

- **Augend** $\mathbf{B6_{16}}$ Converted to equivalent binary = $1011\ \ 0110_2$
- **Addend** $\mathbf{1B_{16}}$ Converted to equivalent binary = $0001\ \ 1011_2$

$$\begin{array}{r} 1011\ \ 0110 \\ +0001\ \ 1011 \\ \hline 1101\ \ 0001_2 \end{array}$$

Converted to equivalent hexadecimal **Sum =** $\mathbf{D1_{16}}$

The same rules apply in math operations in hexadecimal as are employed in the decimal system. Table 14–9 outlines a simple hexadecimal subtraction of 55_{16} from BA_{16}.

TABLE 14–9 First Example of Hexadecimal Subtraction

Column power	16^3	16^2	16^1	16^0
Minuend after borrow	—	—	—	—
Minuend	—	—	B	A
Subtrahend	—	—	5	5
Difference	—	—	6	5

Table 14–10 illustrates a simple hexadecimal subtraction of $C7_{16}$ from $E2_{16}$.

TABLE 14–10 Second Example of Hexadecimal Subtraction

Column power	16^3	16^2	16^1	16^0
Minuend after borrow	—	—	D	12
Minuend	—	—	E	2
Subtrahend	—	—	C	7
Difference	—	—	1	8

Table 14–11 illustrates a simple hexadecimal subtraction of CF_{16} from $9A_{16}$. When the minuend is smaller than the subtrahend, subtract the smaller number from the larger number and change the sign.

TABLE 14–11 Third Example of Hexadecimal Subtraction with a Negative Result

Column power	16^3	16^2	16^1	16^0
Minuend after borrow	—	—	—	—
Minuend	—	—	9	A
Subtrahend	—	—	C	F
Difference	—	—	3	5

Table 14–12 illustrates hexadecimal multiplication of $3F_{16}$ times $A2_{16}$.

TABLE 14–12 Example of Hexadecimal Multiplication

Column Power	16^3	16^2	16^1	16^0
Multiplicand	—	—	3	F
Multiplier	—	—	A	2
First partial product	—	—	1	E
Second partial product	—	—	6	—
Third partial product	—	9	6	—
Fourth partial product	1	E	—	—
Carry	1	—	—	—
Product	2	7	D	E

The following illustrates hexadecimal division of CC_{16} by C_{16}:

$$\begin{array}{r} 11 \\ C\overline{)CC} \\ \underline{-C} \\ 0C \\ \underline{-C} \\ 0 \end{array}$$

SUMMARY

Binary arithmetic is much like decimal arithmetic. The basic mathematic operations of adding, subtracting, multiplying, and dividing can be performed using binary numbers. These processes are easier than the decimal process because only two numbers, 0 and 1, are used. Results of the binary processes can be confirmed using the decimal number system. Material covered in this chapter really does have daily applications for many electrical/electronics technicians, as illustrated by the following example:

Example

Temp Corporation has a controlled-temperature chamber in use daily. The operator is to insert several product samples into the chamber, and following a written test procedure set the oven temperature to 50_{10}°C above its current temperature. The keypad has 20 buttons labeled 0 through F, ENTER, CLEAR, HEAT ON, and HEAT OFF. The digital display indicates 21_{16}°C.

What is the desired hexadecimal temperature increase, and what hexadecimal value should be punched into the keypad?

1. Convert the desired 50_{10}-degree increase to hexadecimal (hex). Refer to Figure 12–10 for the manual procedure, or use your pocket calculator. Temperature increase desired: $50_{10}°C = 32_{16}°C$
2. Add the current chamber temperature and the desired increase: $21_{16}°C + 32_{16}°C = 53_{16}°C$
3. Punch in hex 53, ENTER, and HEAT ON.

PRACTICE PROBLEMS

1. Addition
   - a. 101110 + 11011 = ?
   - b. 1111 + 1011 = ?
   - c. 001110001 + 100011 = ?
   - d. 100001 + 101101 = ?
2. Subtraction
   - a. 100100 − 10110 = ?
   - b. 101101 − 0101 = ?
   - c. 101 − 011 = ?
   - d. 111111 − 100001 = ?
3. Multiplication
   - a. 1001 × 11 = ?
   - b. 0110 × 100 = ?
   - c. 1100 × 101 = ?
   - d. 11111 × 101110 = ?
4. Division
   - a. 1001 / 11 = ?
   - b. 110011011 / 111010 = ?
   - c. 1110011 / 111 = ?
   - d. 11011011 / 1011 = ?

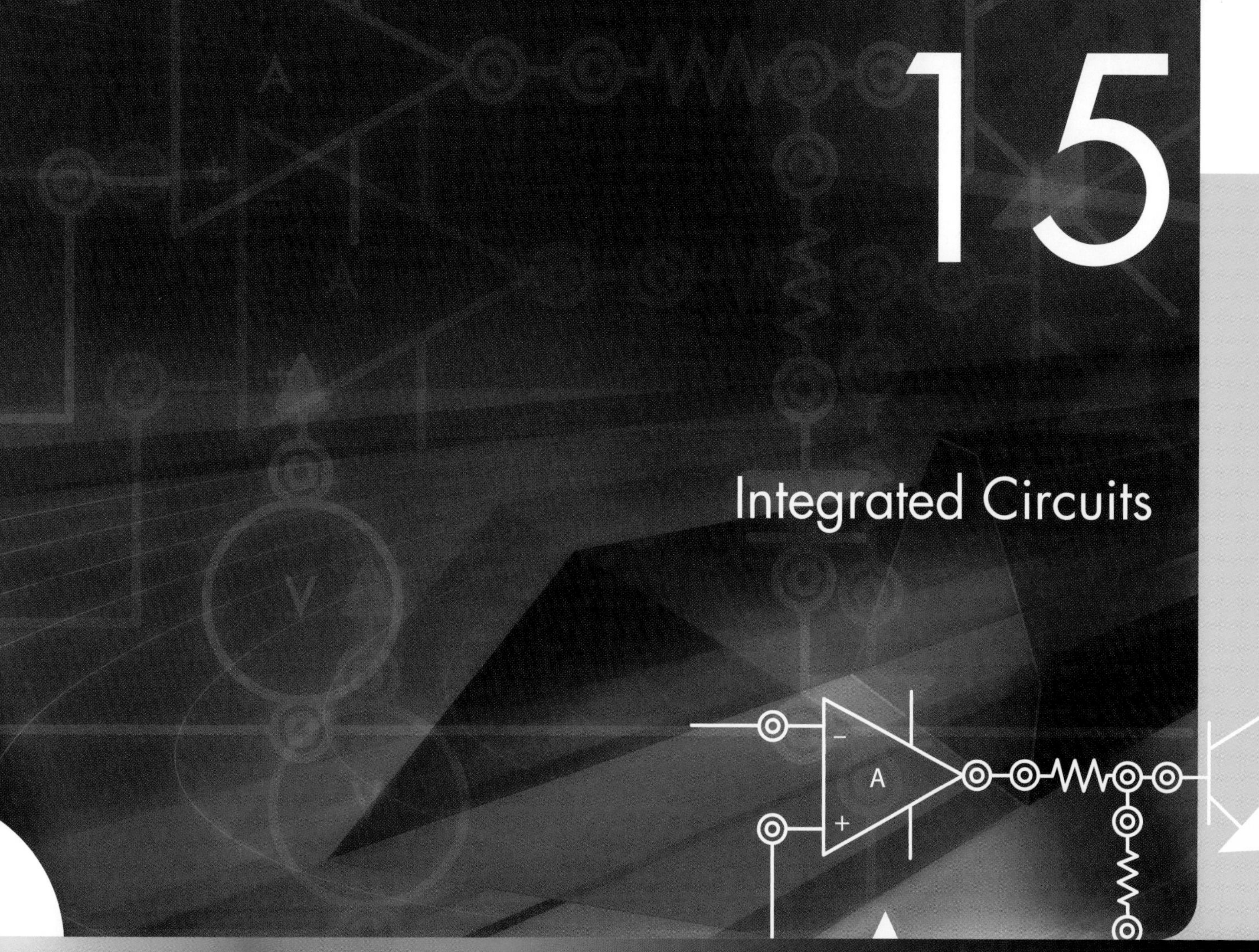

15 Integrated Circuits

OUTLINE

OVERVIEW

From the earliest days of electrical technology until the late 1950s, electronic circuits consisted of individual components installed on a chassis or framework and wired together. Then, in 1958 a Texas Instruments engineer named Jack Kilby invented a device that has become known as the integrated circuit (IC). Kilby's invention uses one piece of semiconductor material with many devices etched into it.

Resistors, capacitors, and various types of semiconductor devices are most commonly found in an IC. In the past 15 years, the IC has reached a level of sophistication where literally millions of components can be etched onto one chip that is smaller than a postage stamp.

Since ICs were introduced in 1958, electronics has advanced at a quick pace. The use of ICs makes possible applications that would otherwise be too large or too costly or require too much cooling. This chapter introduces you to ICs and some of their more common implementations.

OBJECTIVES

After completing this chapter, the student should be able to:

1. List and define the various types of IC timers.
2. Identify the internal parts of the IC timers.
3. Describe the operations for circuits covered.
4. Describe digital signal processing.
5. Explain functions of phase-locked loop and their building blocks.
6. Perform the calculations introduced in this lesson.

INTRODUCTION TO INTEGRATED CIRCUITS

15.1 Discrete Circuits Versus Integrated Circuits

The differences between discrete circuits and ICs can be summarized as follows:

- Discrete circuits use individual resistors, diodes, transistors, capacitors, and other devices to complete the circuit function. These individual parts are usually mounted on some sort of a circuit board and then interconnected using wires or solder traces.
- To accomplish the same function, discrete circuit assemblies are usually more expensive than an equivalent IC.
- Although IC assemblies do not eliminate the need for circuit boards, assembly, soldering, and testing, they do allow the same circuit to be produced for a lower cost—and a more complex circuit to be produced for the same cost.
- With ICs, the number of discrete parts can be reduced.
- ICs are smaller and use less power, in addition to costing less to manufacture.
- Because the electronics involved in IC assemblies often require fewer alignment steps, ICs can be set up and calibrated for a lower cost.
- Because ICs allow for fewer discrete parts in a piece of equipment, IC systems tend to be more reliable.

TechTip!

In regard to pin identification, determination of the correct IC pins to probe is often confusing. Three generations ago there were no transistors, no ICs, and no circuit boards. Radio repairmen worked with vacuum tubes (tubes, or valves) and point-to-point wiring. Tube sockets were accessible only from the underside of the radio chassis.

Clock hands advance "clockwise." Therefore, it was natural that pins be numbered clockwise from a mark (key). IC pin numbering retains the same standard numbering technique (Figure 15–1b). IC pins are counted clockwise from a mark. ICs, however, can often be probed from either top or bottom of the circuit card. Thus, we say if **clockwise from the bottom then counterclockwise from the top.** Be sure to start counting from the mark (key)!

15.2 Schematics

The internal features for ICs are seldom shown in schematics. The technician does not usually need to know the circuit details inside the IC. It is more important to know what the IC is supposed to do and how it works as a part of the overall circuit. Figure 15–1 shows two typical ways (a and b) of illustrating an IC. The schematic and a few voltage specifications are all a technician usually needs to verify proper operation of the IC.

FIGURE 15–1 Normal way of showing an IC on a schematic.

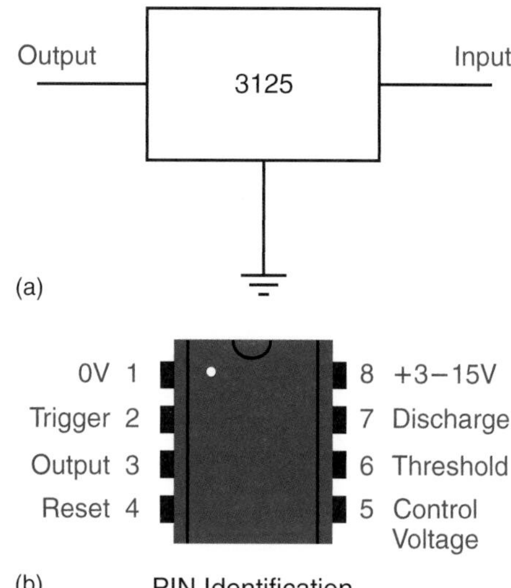

15.3 Fabrication

Photolithography
Creation of ICs using a photographic system.

Isolation diffusion
Part of the IC fabrication process that creates insulating barriers.

The manufacture of ICs starts in a radio frequency furnace. The P-type silicon wafers are processed using **photolithography.** This process creates the desired P-type and N-type zones in the substrate (wafer). Later in the process, the numerous circuit functions in the IC are electrically insulated from each other by **isolation diffusion** (forming an area designed to block current flow).

THE 555 TIMER

15.4 Introduction

Duty cycle
Usually, duty cycle is that percentage of the input cycle period when a load is driven by its source.

The NE555 IC timer is very popular among circuit designers because of its low cost, versatility, and stable time delays. It allows timing intervals from microseconds to hours. Depending on the preferred output waveform, the oscillator mode requires three or more external components. Frequencies from less than 1 Hz to 500 kHz with **duty cycles** from 1% to 99% can be achieved.

15.5 Internal Circuitry and Operation

Flip-flop
See bistable multivibrator.

Bistable
Able to operate steadily in either one of two states. Will not leave one state until triggered to do so.

Bistable multivibrator
A multivibrator in which either of the two active devices may remain conducting (with the other nonconducting) until the application of an external pulse. Also known as Eccles-Jordan circuit, Eccles-Jordan multivibrator, flip-flop circuit, or trigger circuit.

The major components of the 555 timer are shown in Figure 15–2. It houses two voltage comparators, a **flip-flop** (also called a **bistable multivibrator**), a resistor divider network, a discharge transistor, and an output amplifier with up to 200 mA current capability. The three divider resistors are of equal value. This network sets the trigger comparator at one-third of V_{CC} and the threshold comparator trip point at two-thirds of V_{CC}. V_{CC} may range from 4.5 to 16 V.

In Figure 15–2, assume that $V_{CC} = 12$ V. Given this, the trigger point will be 4 V ($\frac{1}{3} \times 12$ V) and the threshold point will be 8 V ($\frac{2}{3} \times 12$). When pin **2** falls below 4 V, the trigger comparator output changes states and sets the flip-flop to the high state, and output pin **3** goes high.

If pin **2** returns to a value higher than 4 V, the output stays high because the flip-flop knows that it was set. On the other hand, if pin **6** climbs above 8 V the threshold comparator changes states and resets the flip-flop to its low state. This causes the output (pin **3**) to go low and the discharge transistor to be turned on.

Note that the output of the 555 timer is digital; that is, either high or low. When it is low, it is near ground potential, and when it is high it is close to V_{CC}. In Figure 15–2, pin **6** is usually connected to a capacitor that is part of an external RC timing network. If the capacitor voltage exceeds $\frac{2}{3}$ V_{CC}, the threshold comparator will reset the flip-flop to the low state. This turns on the discharge transistor, which can be used to discharge the external capacitor in preparation for the next timing cycle.

The reset, pin **4**, allows for direct access to the flip-flop. This pin overrides the various timer functions and pins. The reset can be used to stop a timing cycle. The reset pin is a digital input, and when it is taken low it resets the flip-flop, turns on the discharge transistor, and drives output pin **3** low. The reset function is not usually needed, and thus pin **4** is normally tied to V_{CC}.

FIGURE 15–2 Diagram of NE555 IC timer.

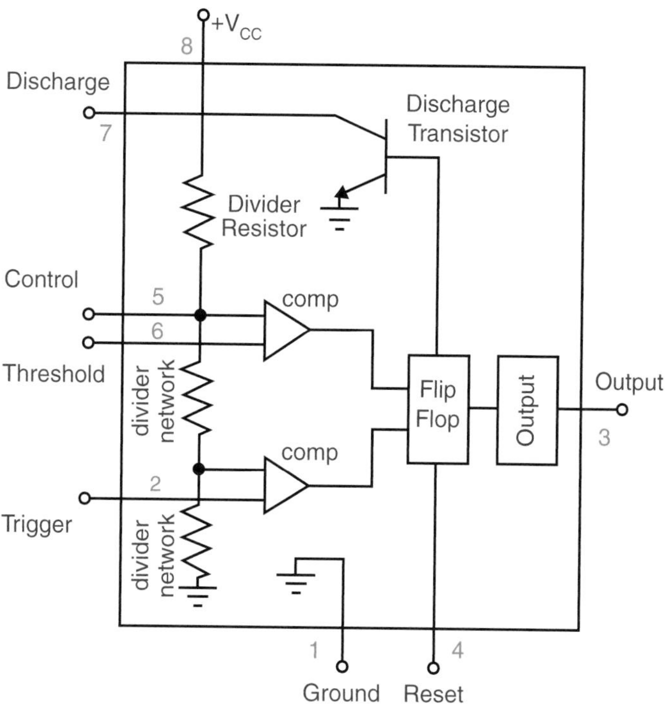

15.6 Monostable Mode

The IC timer is connected for the **monostable** (one-shot) mode in Figure 15–3. This mode produces an RC-controlled output pulse that goes high when the device is triggered. The timer is considered negative-edge triggered, which means that the timing cycle begins at t_1 when the trigger input falls below $\frac{1}{3}\ V_{CC}$.

Monostable
Having only one stable state.

Once the trigger input is greater than $\frac{1}{3}\ V_{CC}$, the time-out period begins, which means that the trigger pulse cannot be wider than the output pulse. When the trigger pulse is wider, the trigger input will need to be AC coupled. Using Figure 15–4, you can see that the 0.1 μF coupling capacitor and the 5 kΩ resistor differentiate the input trigger pulse. The effective width of the trigger pulse is decreased by this pulse differentiation (AC coupling).

The width of the output pulse is RC controlled in the monostable (one-shot) circuit, and the timing capacitor begins charging through the timing resistor when the timer is triggered. Recall that as the capacitor voltage reaches $\frac{2}{3}\ V_{CC}$ the flip-flop is reset. This turns on the discharge transistor, and the capacitor is emptied in preparation for the next cycle. This results in the output pulse width equaling 1.1 time constants. The formula for finding the output pulse is $t_{ON} = 1.1\ RC$.

TechTip!

Pin **5** of the 555 timer (the control voltage input) is seldom utilized. However, in some situations brief voltage transients can cause inadvertent triggering of the 555. Even nearby lightning can induce sufficient voltage into the timer to cause false triggering. This problem is generally overcome by addition of a noise filter capacitor between pin **5** and ground. This precaution does not otherwise affect timer operation. Figure 15–3 includes a noise filter capacitor, as explained.

FIGURE 15–3 Monostable mode.

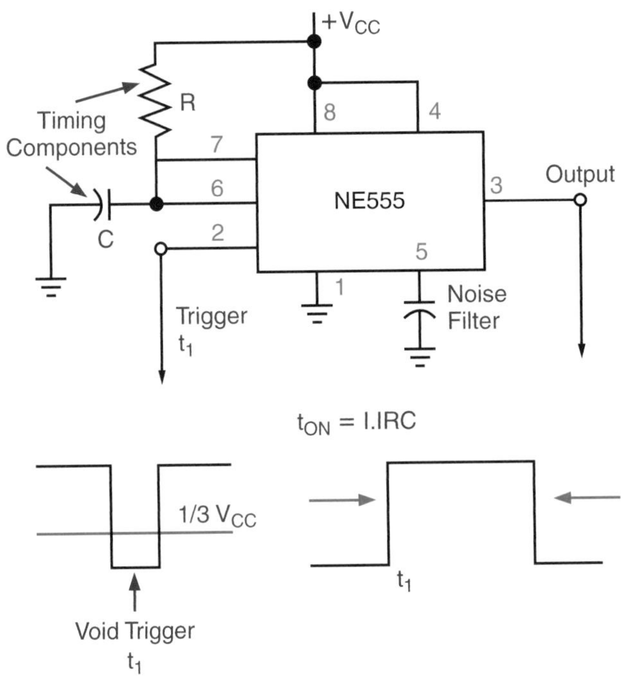

FIGURE 15–4 AC-coupled trigger pulse.

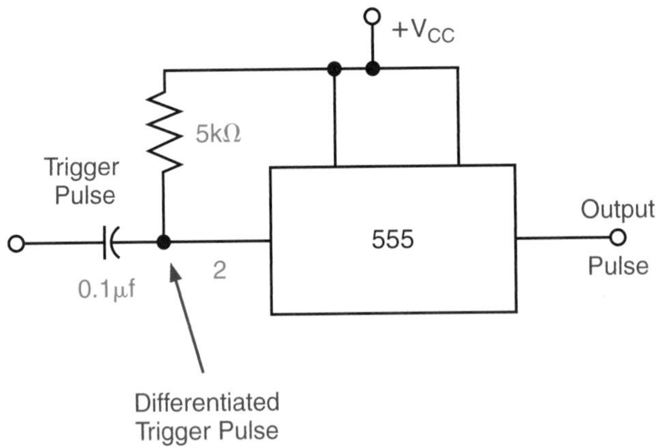

Example

Find the output width for Figure 15–4. $R = 5$ kΩ and $C = 0.1$ μF. The pulse width is equal to 1.1 time constants.

$$t_{ON} = 1.1 \times R \times C = 1.1 \times 5\ \text{k}\Omega \times 0.1\ \mu\text{F} = 0.55\ \text{m sec}$$

Figure 15–5 shows a typical application for a 555 connected as a monostable (one-shot) timer. This circuit may be used in applications such as an automatic shutoff timer for a photographic exposure lamp. When the momentary N.O. switch is pressed and released, the 555 timer output at pin 3 goes high. The 555 output drives an inexpensive solid-state relay block (a triac switch may be built from discrete components, or a suitable relay may be used), which turns on the exposure lamp.

FIGURE 15–5 Practical exposure lamp timer circuit.

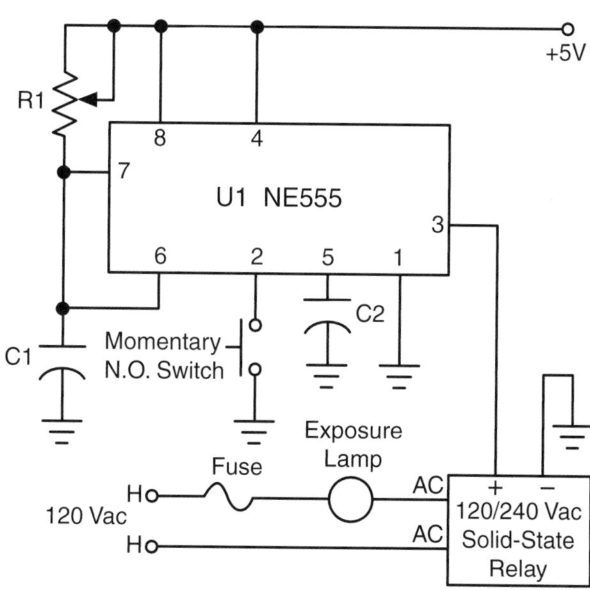

When the 555 one-shot output drops low (times out), the exposure lamp turns off. The potentiometer at 555 pin **7** may be calibrated, and it may be labeled in minutes and seconds so that the operator may select the lamp-on time, press the switch, and walk away. Long time constants often require electrolytic capacitors. Accuracy of electrolytic capacitor values is notoriously poor. Therefore, calibration is required. In the circuit of Figure 15–5, time constant values must be high because exposure times may need to be quite high. If we select a maximum time of 15 minutes and a capacitance of 1,000 μF, the potentiometer value would be calculated as follows.

$$t = 1.1 \times RC \Rightarrow R = \frac{t}{1.1 \times C}$$

$$R = \frac{15 \text{ minutes}}{1.1 \times 1{,}000\ \mu\text{F}} = \frac{15 \text{ min} \times 60 \text{ sec/min}}{1.1 \times 1{,}000\ \mu\text{F}} = 818\ \text{k}\Omega$$

The potentiometer, R_1, may be rounded up to 1 megohm. C_2, used for stability, may be any convenient value from 0.001 μF to 0.1 μF.

15.7 Astable Mode

Astable
A circuit that alternates automatically and continuously between two unstable states at a frequency dependent on circuit constants; for example, a blocking oscillator.

An example of the timer configured in the **astable** (free-running) mode is shown in Figure 15–6, with the trigger (pin **2**) being tied to the threshold (pin **6**). As the circuit is turned on and the timing capacitor is discharged, the timer begins charging through the series combination (R_1 and R_2).

When the capacitor voltage attains $\frac{2}{3}$ V_{CC}, the output drops to low and the discharge transistor turns on. At this point, the capacitor discharges through R_2. The output changes to high and the discharge transistor turns off as the capacitor reaches $\frac{1}{3}$ V_{CC}. Now the capacitor starts charging through R_1 and R_2. This cycle will continuously repeat as the capacitor charges and discharges.

FIGURE 15–6 Free-running or astable mode.

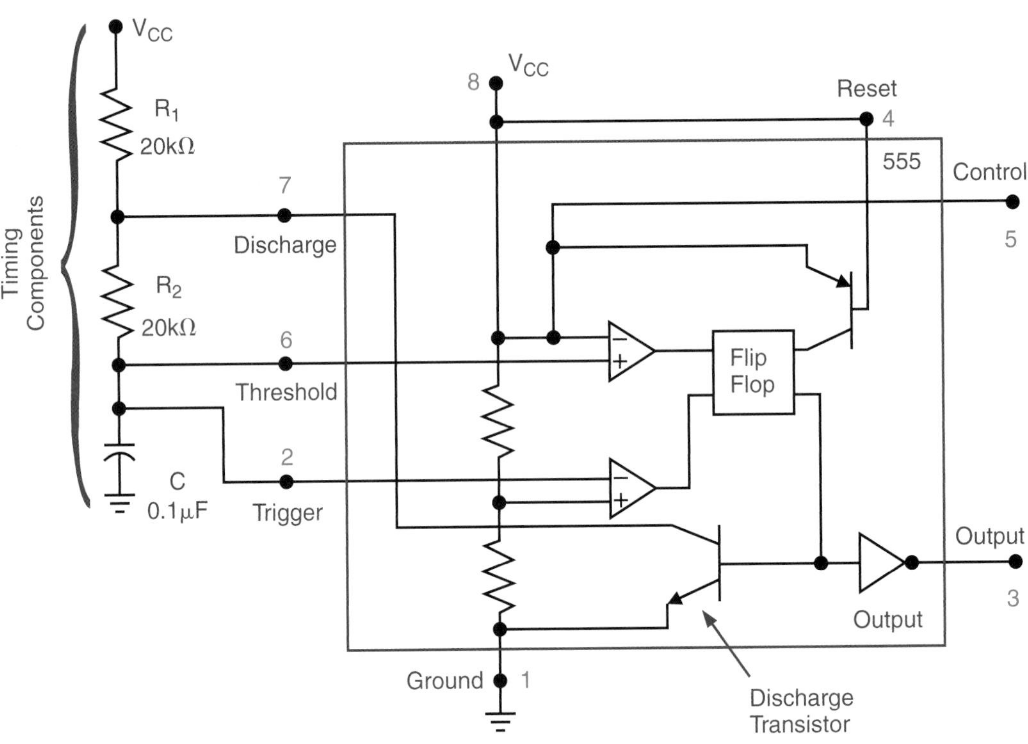

The following formula calculates the time the output is held high in an astable circuit through two resistors:

$$t_{high} = 0.69(R_1 + R_2)C$$

Example

Say that both timing resistors in Figure 15–6 are 20 kΩ and the timing capacitor is 0.1 μF. Find the time the output will remain high.

$$t_{high} = 0.69(R_1 + R_2)C = 0.69(20\text{ k}\Omega + 20\text{ k}\Omega)0.1\ \mu\text{F} = 2.76\text{ m sec}$$

Because the discharge path is passing through only one resistor (R_2), the time the output is held low is found by using a similar formula:

$$t_{low} = 0.69(R_2)C$$

Example

Using the same data from the previous example, with the 20 kΩ resistor and the 0.1 μF timing capacitor find the time the output will remain low.

$$t_{low} = 0.69(R_2)C = 0.69 \times 20\ \text{k}\Omega \times 0.1\ \mu\text{F} = 1.38\ \text{m sec}$$

Because the resistors are equal, the time held low is half the time held high. Because the times held high and low are different, the output waveform is nonsymmetrical. When you add t_{high} and t_{low}, the result is the total period (which is needed to calculate the output frequency). The output frequency is equal to the reciprocal of the total period, which can be found by using the following formula:

$$f_O = \frac{1.45}{(R_1 + 2R_2)C}$$

Example

Again using the data from the previous example, where the resistors are 20 kΩ and the timing capacitor is 0.1 μF, find the output frequency.

$$f_O = \frac{1.45}{(R_1 + 2R_2)C} = \frac{1.45}{(20\ \text{k}\Omega + 40\ \text{k}\Omega)0.1\ \mu\text{F}} = 241.67\ \text{Hz}$$

The duty cycle is the portion (percentage) of time the output is high. This is found by taking the time the output is high and dividing it by the total period of the waveform. The formula is:

$$D = \frac{R_1 + R_2}{R_1 + 2R_2} \times 100$$

Example

Using the information from Figure 15–6 and two 20 kΩ timing resistors, calculate the duty cycle of this rectangular waveform.

$$D = \frac{20\ \text{k}\Omega + 20\ \text{k}\Omega}{20\ \text{k}\Omega + 2(20\ \text{k}\Omega)} \times 100 = 66.7\%$$

Study Figure 15–6. Because the timing capacitor charges through the two resistors and discharges only through R_1, this circuit can produce only a rectangular wave. A square wave, which is a rectangular wave with a 50% duty cycle, cannot be produced with this circuit.

As R_1 gets smaller relative to R_2, the duty cycle will get closer to 50% but will not reach 50%. Making R_1 equal to 0 Ω in the formula will result in a 50% duty cycle. However, this would damage the IC because there would be no current limiting for the internal discharge transistor.

An alternative circuit that allows duty cycles of 50% or less is shown in Figure 15–7. A diode, which bypasses R_2 in the charging circuit, has been added in parallel with R_2. The timing capacitor will still discharge through R_2 but only charges through R_1. The previous formulas are modified slightly for the altered circuit.

$$t_{high} = 0.69(R_1)C$$

$$t_{low} = 0.69(R_2)C$$

$$f_O = \frac{1.45}{(R_1 + R_2)C}$$

$$D\% = \frac{R_1}{R_1 + R_2} \times 100$$

FIGURE 15–7 Bypass diode around R_2.

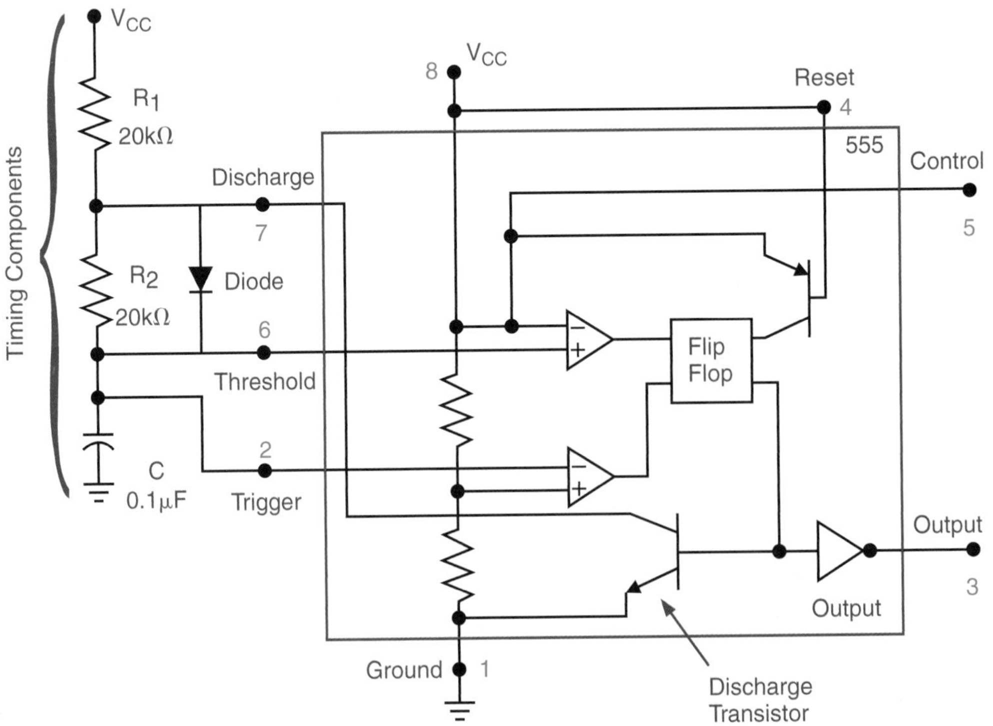

It is now possible to find resistor values that will produce square waves.

Example

Find the resistor values that will produce a 2 kHz square wave given a timing capacitor of 0.01 μF.
Solution:
Begin with the formula for output frequency.

$$f_O = \frac{1.45}{(R_1 + R_2)C}$$

Manipulate the formula to find $R_1 + R_2$, the total resistor value.

$$R_1 + R_2 = \frac{1.45}{f_O \times C} = \frac{1.45}{2\text{ kHz} \times 0.01\ \mu\text{F}} = 72.5\text{ k}\Omega$$

Because a square wave has a 50% duty cycle, each resistor must be the same. Therefore, each resistor must be half of 72.5 kΩ.

$$R_1 = R_2 = \frac{72.5\text{ k}\Omega}{2} = 36.25\text{ k}\Omega$$

Figure 15–8 illustrates a common application of the 555 timer connected as an astable oscillator. In this circuit, the 555 output from pin **3** drives an infrared light-emitting diode (IR-LED). An IR-LED, phototransistor or photodiode located across an opening is directed toward the IR-LED. If an object blocks the infrared path, the IR-LED detector circuit will trigger and halt a motor, snap a photo, sound an alarm, increment a counter, and so on.

FIGURE 15–8 Practical IR-LED transmitter circuit.

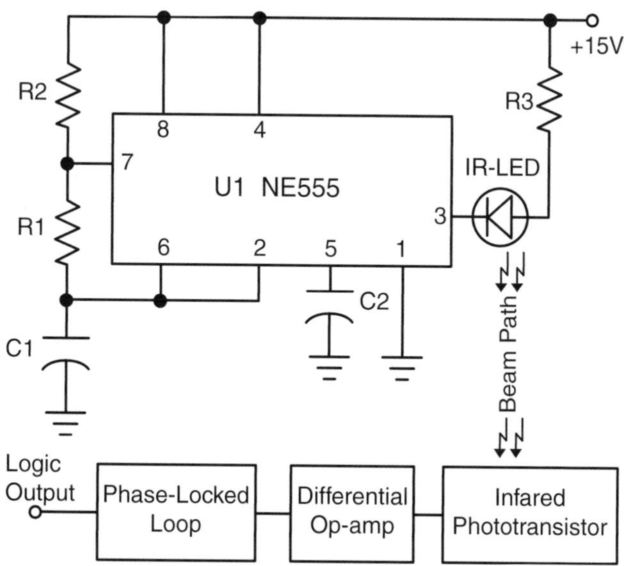

The IR-LED is pulsed on and off by the astable 555 so that other sources of IR (the sun, for example) will not cause accidental triggering of the detector circuit. An additional reason for pulsing, especially at a low duty cycle, is that the power rating of the IR-LED (an average measurement) may be extended, expanding range and improving reliability of the system. Solving for the component values is accomplished as follows:

Select the pulse frequency: 1 kHz

Find the period: $t = \frac{1}{f} \Rightarrow t = 1{,}000$ microseconds

Select the "output low" time: $t_{(low)} = 100$ microseconds

Select the "output high" time: $t_{(high)} = 900$ microseconds

Select C_1: 0.10 µF

Find R_1: $t_{(low)} = 0.69(R_1)C_1 \Rightarrow R_1 = \frac{100\ \mu S}{0.69 \times C_1} = 1449\ \Omega$

Find R_2: $t_{(high)} = 0.69(R_1 + R_2)C_1 \Rightarrow R_1 + R_2 = \frac{t_{(high)}}{0.69 \times C_1} \Rightarrow$

$$R_2 = \frac{t_{(high)}}{0.69 \times C_1} - R_1 \Rightarrow R_2 = \frac{t_{(high)}}{0.69 \times C_1} - R_1 \Rightarrow$$

$$R_2 = \frac{900\ \mu S}{0.69 \times 0.1\ \mu F} - 1449\ \Omega = 13{,}043\ \Omega - 1449\ \Omega = 11{,}594\ \Omega$$

Select IR-LED current per specifications: 50 mA at 1.2 V.

Find R_3: $V_{R3} = V_{CC} - V_{IR\text{-}LED} \Rightarrow 15\ V - 1.2\ V = 13.8\ V$

$$R_3 = \frac{V_{R3}}{I_{R3}} \Rightarrow R_3 = \frac{13.8\ V}{50\ mA} = 276\ \Omega$$

Round-off R_3 to 270 Ω

$P_{R3} = I^2 \times R_3 \Rightarrow P_{R3} = 50^2\ mA \times 270\ \Omega = 0.69\ W$

Round-off P_{R3} to 1 W

Note that the IR-LED in Figure 15–8 is driven only when 555 pin **3** is low. The duty cycle is, then:

$$\text{Duty cycle\%} = \frac{t_{(low)}}{t_{(low)} + t_{(high)}} \times 100 \Rightarrow \frac{100\ \mu S}{1000\ \mu S} \times 100 = 10\%$$

If the frequency is critical, a potentiometer could be substituted for R_2. When the IR light beam crosses to the receiver, it is detected by the phototransistor, amplified by the op amp, and decoded to produce a digital low at the output of the phase-locked loop (PLL). The PLL, shown as part of the detector block, is discussed later in this chapter.

15.8 Time-Delay Mode

The versatility of 555 applications is shown in Figure 15–9. This configuration creates a stable timing output with the addition of a transistor and two diodes to the RC timing network. The frequency can be varied over a wide range while maintaining a constant 50% duty cycle.

FIGURE 15–9 Time-delay mode.

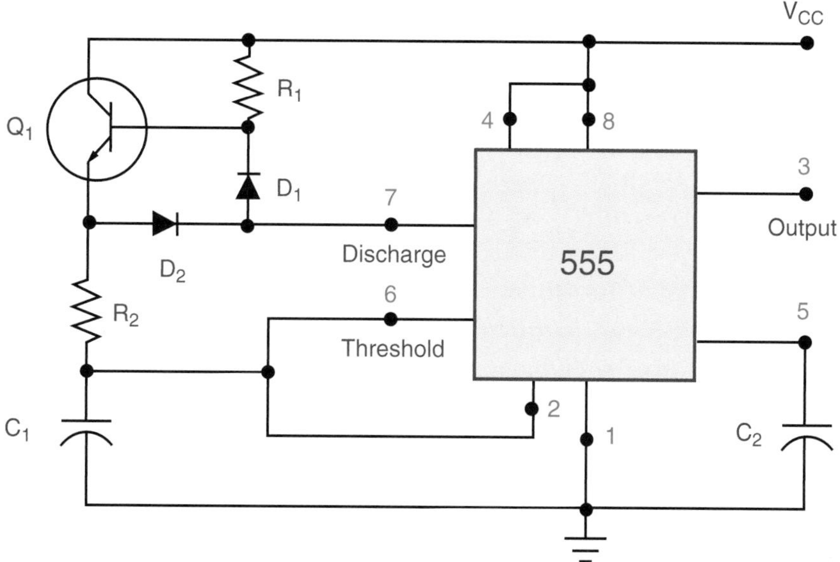

When the output is high, the transistor is biased into saturation by R_1 so that the charging current passes through the transistor and R_2 to C_1. When the output goes low, the 555 internal discharge transistor (pin **7**) cuts off the transistor Q_1 and discharges the capacitor through R_2 and D_2. The high and low periods are equal. The values of capacitor (C_1) and resistor (R_1) are dependent on the type or length of timing desired.

Example

If $R = 260\ \text{k}\Omega$ and $C = 25\ \mu\text{F}$, the time delay is calculated using the following formula:

$$t_{delay} = 1.1 \times R \times C = 1.1 \times 260\ \text{k}\Omega \times 25\ \mu\text{F} = 7.15\ \text{sec}$$

In this mode, if the trigger signal goes high before the IC times out the output will not go low. Security alarms take advantage of this feature by allowing time to clear an area before the alarm is armed.

OTHER INTEGRATED CIRCUITS

In earlier chapters, you were introduced to some of the most widely used ICs, such as differential amplifiers and operational amplifiers. ICs can also replace transistor stages such as an intermediate frequency (IF) amplifier. The use of an IC (with multiple transistors) will normally provide improved performance, such as more gain, better selectively, greater noise rejection, and so on.

Linear ICs that provide more than one function are usually called subsystem ICs. An example of a subsystem IC is a television sound system containing a limiter, an FM detector, an IF amplifier, a regulated power supply, and an electronic volume control. This type of IC reduces the number of parts needed in system applications. With fewer parts, cost, circuit board real estate, power dissipation, and maintenance are reduced, whereas reliability and operating speed are improved.

DIGITAL SIGNAL PROCESSING

One application for digital signal processing (DSP) involves changing AM analog signals (such as music) into a stream of numbers. The circuit then performs various arithmetic operations on those numbers and changes the resulting numbers back into an improved or enhanced signal. With DSP, many different things can be achieved. Some examples are:

- Eliminate unwanted signal characteristics or components (such as echoes, noise, or interference)
- Compress or expand a signal
- Enhance a signal to provide special effects (such as surround sound)
- Balance the different frequency components
- Demodulate an encoded signal

Analog circuits can also accomplish these effects. However, the DSP is preferred because the technology is small and inexpensive compared to analog equipment. Other benefits of the DSP include the following:

- Changes can be made through software adjustments rather than component changes. This reduces the cost and time spent making modifications.
- Digital designs are more reliable, with maintenance and calibration being easier to perform and less costly.
- They are not sensitive to environmental changes (temperature and humidity) and are not affected by aging, as are analog circuits.

This DSP system includes the following stages (Figure 15–10):

1. Amplifier
2. Low-pass input filter, also call an antialiasing filter
3. Sample-and-hold circuit
4. Analog-to-digital converter
5. Memory
6. Microprocessor
7. Digital-to-analog converter
8. Low-pass output filter

FIGURE 15–10 Diagram of a digital signal processing system.

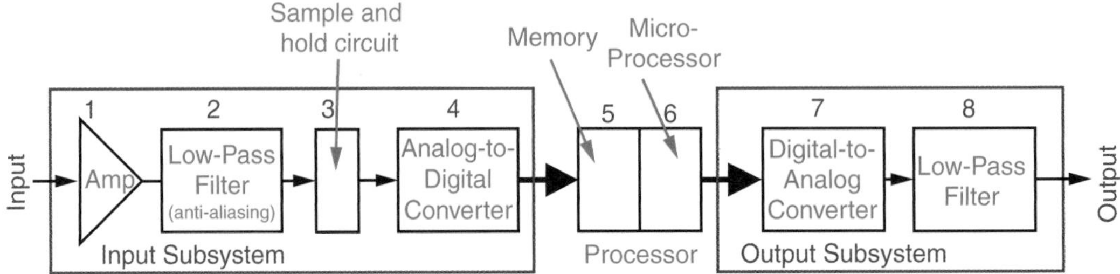

Today, DSP has many applications in addition to audio signal processing. For example, a DSP system may be configured to convert one video format to another. Simple digital instructions may program the generation of multiple signal waveforms simultaneously, or make voltage and frequency measurements at multiple test points. DSP ICs can be configured by software to perform a seemingly endless variety of functions. The technician is generally not trained nor equipped to troubleshoot a DSP system.

In the rare event of DSP failure, replacement is required, and an identically programmed DSP system must be made available. These components, with all of their advantages, are usually very expensive. Thus, the technician needs to be confident that his analysis of a defective DSP system is correct before proceeding with replacement.

ANALOG-TO-DIGITAL CONVERSION

Analog-to-digital (A/D) and digital-to-analog (D/A) conversion is required in many IC applications. The digital signal processor mentioned earlier has both in its circuitry. Analog signals are continuous; that is, their voltage changes smoothly over time. On the other hand, a discrete signal can change in voltage value at any point in time.

To make digital processing possible, the continuous signal must be changed into a digital signal. Changing the continuous signal into a discrete signal uses a process called sampling. Basically, "snapshots" (samples) of the signal are taken at different points in time. The higher the sampling rate the better the discrete signal will represent the continuous signal. Conversely, the discrete signal will be worse as the sampling rate becomes lower. This will result in a poor quality of sound or picture.

Before the signal is received for the sample-and-hold process, it first passes through an antialiasing filter, which helps remove interference and high-frequency noise. After sampling, the signal is sent to the A/D converter. An A/D converter converts each signal sample into a string of bits (*bit* combines the words *binary* and *digit*).

Bits are binary values. The value 0 is represented by a voltage. The value 1 is usually represented by a more-positive voltage than the value of 0. The digital logic IC family dictates the voltage limits for 1s and 0s. It is common today to find a variety of logic levels on a single circuit card assembly. The old standard **TTL** family of ICs define 0s as −0.5 V through +0.8 V and 1s as +2.4 V through +5.5 V. This is generally called "5 volt logic."

Rather than discussing voltage levels, we equate the term *1* with the term *HI* and the term *0* with the term *LO*.

Bit
A binary 1 or 0.

TTL
A standard abbreviation for the solid-state transistor family of logic devices known as Transistor–Transistor Logic.

Example

How can the decimal number 253 be defined in TTL voltage terms?
From the number systems chapter, $253_{10} = FD_{16} = 1111\ \ 1101_2$. Because 1s are about 5 V in TTL and 0s are about 0 V in TTL, 1111 1101 can be stated as 5 V, 5 V, 5 V, 5 V, 5 V, 5 V, 0 V, 5 V. How can 253 be defined as HIs and LOs?

$$253_{10} = 1111\ \ 1101_2 = \text{HI, HI, HI, HI, HI, HI, LO, HI}$$

Byte
An 8 bit number.

The smallest increment in a binary number is a single bit. The next larger grouping of values used in the binary system is a grouping of four bits, called a nibble. The combination of two nibbles is called a **byte.** Bytes can be grouped into words. Words are defined by the system in which they are used.

A word can be 1 byte or more. It is common to find words of 2 bytes (16 bits), 4 bytes (32 bits), and even more. It is also possible to encounter words with an odd number of bits, such as 10. A small A/D (or ADC), D/A (or DAC), or memory may use 10 or 12 bits to a word.

Some examples of bytes are:

- $0000\ 0000_2$ (this byte represents the decimal 0)
- $0000\ 0100_2$ (this byte represents the decimal 4)
- $1000\ 0111_2$ (this byte represents the decimal 135)
- $1111\ 1111_2$ (this byte represents the decimal 255)

After the A/D has converted the signals to bits, bytes, and words, they are sent to the memory and then (when called upon) to the microprocessor, which performs the calculations specified by the software program. From there, it is sent to the DAC converter and then on to a low-pass filter. A good example of this application is the telephone answering machine most of us have at home.

PHASE-LOCKED LOOP

The PLL, discrete circuitry known for many years, began to find massive application in the early 1980s with the popularity of single-chip ICs such as the Signetics NE565, 567, and others.

15.9 Introduction

A PLL is an assemblage of four basic electronics blocks consisting of a phase detector, filter, DC amplifier, and voltage-controlled oscillator (VCO) (Figure 15–11).

FIGURE 15–11 Block diagram of NE665 and 567 PLLs.

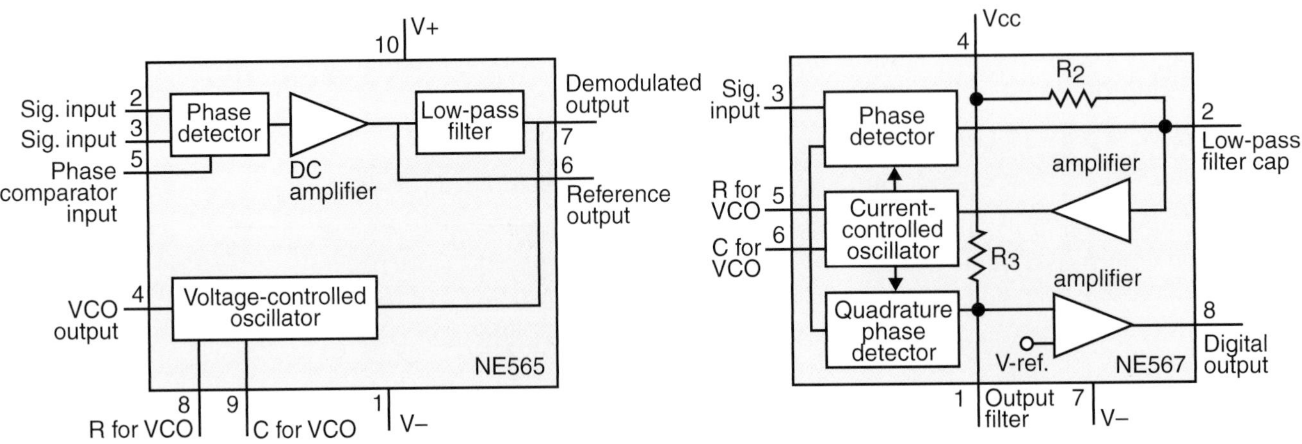

Dramatic reductions in circuit board real estate, component count, and cost are benefits of the single-IC PLL.

PLLs can perform a wide variety of functions, including frequency synthesis, FM (phase) detection, frequency multiplication and division, tone decoding, and signal regeneration. The manner in which a minimum of external components are applied in conjunction with the PLL determines the function and operating characteristics of the PLL circuits.

15.10 FM Detection

Study Figure 15–12, which shows a PLL FM detector. Pin **2** is the capacitively coupled single-ended input to the PLL detector. Pin **3**, the differential input pin, is grounded through a matching 560 Ω resistor. Audio output is taken from pins **6** and **7**. Pins **10** and **1** are the differential power points. The free-running VCO frequency is determined by the capacitor at pin **9**, and resistance at pin **8**, with the following formula.

Select $R = 5.5\ \text{k}\Omega$ and $IF = 455\ \text{kHz}$
Find C

$$f_O = \frac{1.2}{4RC} \Rightarrow C = \frac{1.2}{4 \times 5.5\ \text{k}\Omega \times 455\ \text{kHz}} = \frac{1.2}{1 \times 10^{10}} = 120\ \text{pF}$$

FIGURE 15–12 Practical PLL FM detector circuit.

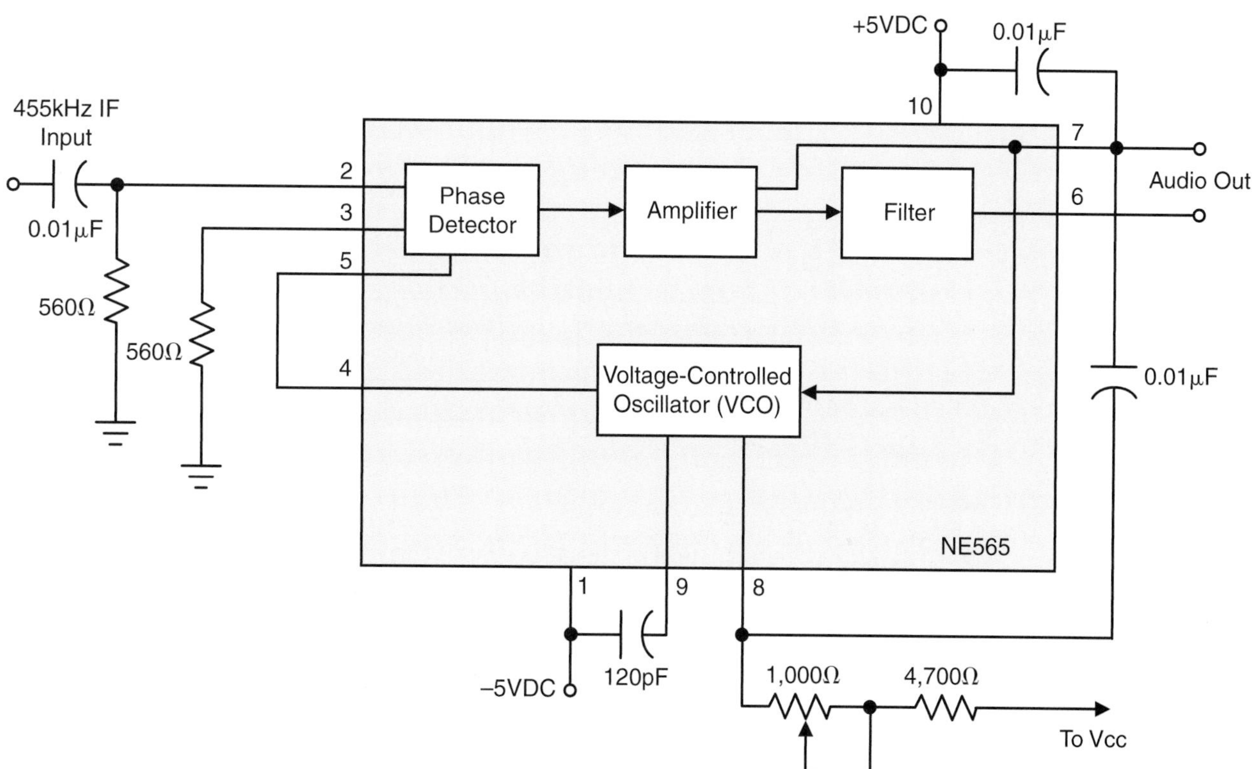

The series-connected 1 kΩ variable resistor shown in Figure 15–12 allows first-time fine-tuning of the VCO to the 455 kHz center frequency. This is necessary because the components are not sufficiently accurate to ensure 455 kHz resonance without trimming.

15.11 Frequency Synthesis

The block diagram shown in Figure 15–13 is a PLL application of tremendous importance. It shows a PLL and auxilliary devices that form a frequency synthesizer. Synthesizers are highly accurate signal frequency sources employing as few as one crystal oscillator, but capable of producing multiple stable output frequencies. In Figure 15–13, a crystal oscillator of 2.000 MHz drives a frequency divider. The frequency divider produces

$$\frac{2.000\text{ MHz}}{20{,}000} = 100\text{ Hz}$$

FIGURE 15–13 PLL frequency synthesizer block diagram.

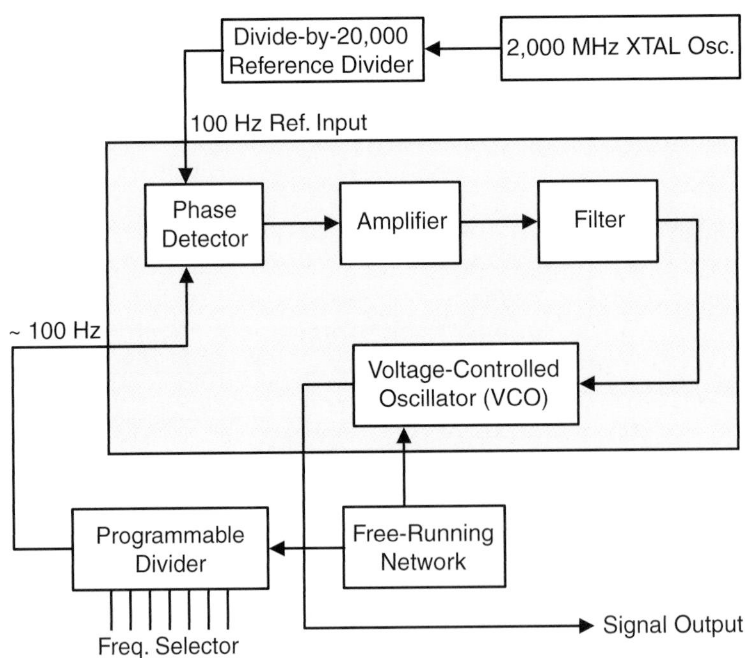

This constant 100 Hz signal is one of the two signals introduced into the PLL phase detector. The VCO output is routed to two points: the synthesizer output and a programmable divider circuit. The programmable divider can be externally controlled to divide by a variety of combinations. Assume the divider in Figure 15–13 can furnish divide-bys of 9,950 through 20,750 in incremental steps of 100.

Further assume that the programmable divider is currently set to divide by 12,250. Programmable divider output is delivered to the remaining phase detector input. The free-running network is simply the R&C components that set the VCO free-running frequency to approximately the middle of the range of desired synthesizer output frequencies. If the VCO is free running at approximately 1,535 kHz, the programmable divider output will be:

$$\text{Programmable divider output} = \frac{f_{VCO}}{\text{Programmable divide-by}}$$

$$\frac{1{,}535 \text{ kHz}}{12{,}250} = 125.306 \text{ Hz}$$

Because the two-phase detector inputs are not equal (100 versus 125.306 Hz), the difference will be amplified, filtered, and delivered to the VCO as a DC voltage to shift the VCO output frequency down until the programmable divider output is equal to 100 Hz. This occurs very quickly. From that point on the synthesizer output will equal:

$$f_{VCO} = \text{Programmable divider output} \times \text{Programmable divide-by} \Rightarrow$$

$$100 \text{ Hz} \times 12{,}250 = 1{,}225 \text{ kHz}$$

Using the same calculation for the proposed programmable divide-by limits, the band of signal frequencies produced will be:

$$100 \text{ Hz} \times 9{,}950 = 995 \text{ kHz}$$

through

$$100 \text{ Hz} \times 20{,}750 = 2{,}075 \text{ kHz}$$

If the IF of a superhetrodyne radio receiver is 455 kHz and the local oscillator operates between 995 (above the carrier) and 2,075 kHz (above the carrier), the received frequency band would be 995 kHz − 455 kHz = 540 kHz through 2,075 kHz − 455 kHz = 1,620 kHz. In other words, the hypothetical PLL synthesizer of Figure 15–13 would be appropriate for a digitally tuned AM broadcast band receiver. The same configuration is suitable for use as the master oscillator of a radio transmitter.

15.12 Tone Decoding

The 567 tone decoder of Figure 15–14 finds application where recognition of one or more tones is desired. Pin **8** of the 567 drives from high to ground when PLL VCO lock occurs with an incoming signal. Typical applications of the 567 tone decoder PLL include touch-tone decoding, wireless intercoms, and ultrasonic control.

Input to the PLL of Figure 15–14 can easily be connected to the audio circuit of a broadcast band receiver. It is preferable that input signal level be greater than 200 mV. Only when the Emergency Broadcast System (EBS) activates the dual-tone alert signal (853 and 960 Hz) will the circuit of Figure 15–14 produce a LOGIC HIGH output from pin **3** of the 4001 NOR logic gate.

Component values for the EBS alert tone decoder circuit are calculated as follows:
Assign 853 Hz to PLL U1

$$f_{O1} \simeq \frac{1.1}{R_1C_1} \Rightarrow R_1 \simeq \frac{1.1}{f_{O1} \times C_1}$$

Given: $f_{O1} = 853$ Hz

Assume $C_1 = 0.22$ μF

$$R_1 \simeq \frac{1.1}{853 \text{ Hz} \times 0.22 \text{ μF}} \simeq 5{,}861 \text{ kΩ}$$

From the BANDWIDTH CHARACTERISTIC CURVE:

$$\text{For 5\% Bandwidth: } f_O \times C_2(\text{μF}) = 10{,}000 \Rightarrow C_2(\text{μF}) = \frac{10{,}000}{853 \text{ Hz}} = 11.7 \text{ μF}$$

$$C_3 = 2 \times C_2(\text{μF}) = 2 \times 11.7 \text{ μF} = 23.4 \text{ μF}$$

R_1 will have to be trimmed, using a potentiometer and frequency counter on pin **6** to center the VCO on 853 Hz. Round off C_2 to 12 μF. Round off C_3 to 22 μF
Assign 960 Hz to PLL U2

$$f_{O2} \simeq \frac{1.1}{R_2C_4} \Rightarrow R_2 \simeq \frac{1.1}{f_{O2} \times C_4}$$

Given: $f_{O2} = 960$ Hz

Assume $C_4 = 0.22$ μF

$$R_2 \simeq \frac{1.1}{960 \text{ Hz} \times 0.22 \text{ μF}} \simeq 5.208 \text{ kΩ}$$

From the BANDWIDTH CHARACTERISTIC CURVE:

$$\text{For 5\% Bandwidth: } f_O \times C_5(\text{μF}) = 10{,}000 \Rightarrow C_5(\text{μF}) = \frac{10{,}000}{960 \text{ Hz}} = 10.4 \text{ μF}$$

$$C_6 = 2 \times C_5(\text{μF}) = 2 \times 10.4 \text{ μF} = 20.8 \text{ μF}$$

R_1 will have to be trimmed, using a potentiometer and frequency counter on pin **6** to center the VCO on 960 Hz.
Round off C_2 to 10 μF.
Round off C_3 to 22 μF

The input coupling capacitor is common to both PLLs. We use the rule of thumb for coupling and bypass capacitance that the capacitive reactance should be no greater than 10% of the load impedance. The specification sheet on the 567 states that the input resistance is 20 kΩ. Therefore:

$$X_{C(Input\ Coupling\ Cap.)} = \frac{1}{10} \times R_{input} = \frac{1}{10} \times 20 \text{ kΩ} = 2 \text{ kΩ}$$

$$C_{(Input\ Coupling\ Cap.)} = \frac{1}{2 \times \pi \times f \times X_C} = \frac{1}{2 \times \pi \times 853 \text{ Hz} \times 2 \text{ kΩ}}$$

$$C_{(Input\ Coupling\ Cap.)} = 0.093 \text{ μF}$$

Round off the input coupling capacitor to 0.1 μF

FIGURE 15–14 Practical PLL EBS alert tone decoder circuit.

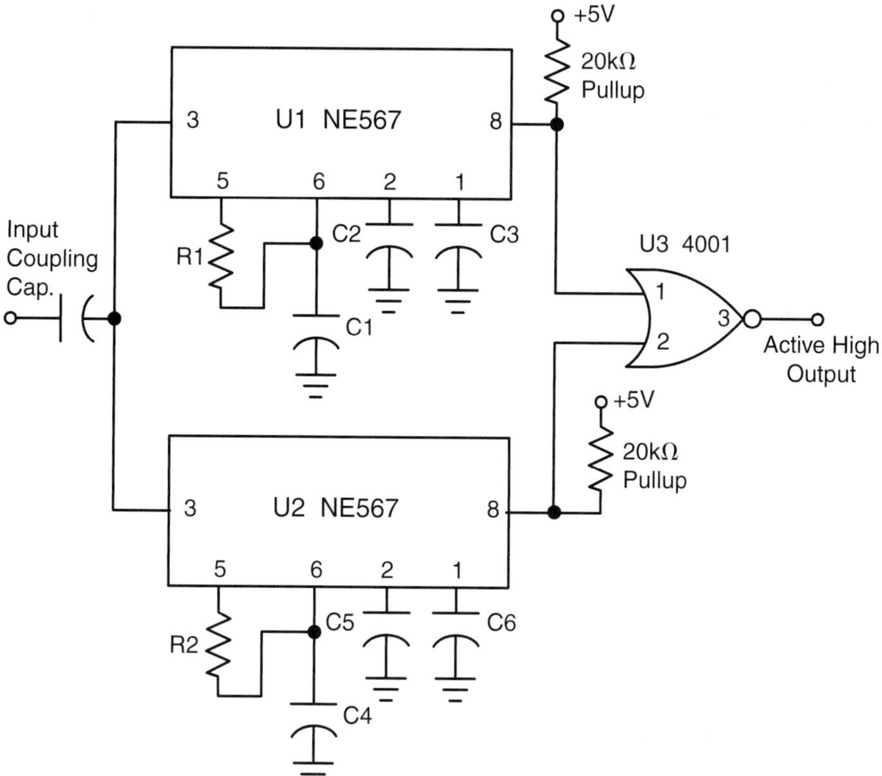

SUMMARY

The invention of the IC changed the world of electronics forever. Circuits and systems are routinely developed using ICs that would have required an impossible number of discrete components. ICs you have already learned about include differential amplifiers, operational amplifiers, and Darlington pairs.

The major components of the NE555 IC timer are voltage comparators, a bistable flip-flop, a resistor divider network, a discharge transistor, and an output amplifier. The network trigger comparator is set at $\frac{1}{3}$ V_{CC}, and the threshold comparator trip point is set at $\frac{2}{3}$ V_{CC}. The V_{CC} ranges from 4.5 to 16 V. The output at pin **3** is digital.

In this chapter, you learned the various components and modifications of three different modes of the 555 timer. The three modes are the monostable mode, the astable mode, and the time-delay mode. Digital signal processing was introduced, as well as some benefits of this system compared to analog equipment. During the A/D conversion phase of the illustrated DSP, the analog signal is converted into a string of bits (termed *words*). Words are usually a multiple of bytes. A byte is 8 bits. A bit is a binary 1 (HI) or 0 (LO).

REVIEW QUESTIONS

1. What are the advantages and disadvantages of ICs compared to discrete component circuits?
2. Describe, in simple terms, how an IC is fabricated.
3. Use the following terms in sentences:
   a. Bit
   b. Word
   c. Nibble
   d. Byte
4. Which of the three operating modes should be used (NE555) for the following?
   a. Steady square-wave output
   b. A switch that can trigger a given action after a fixed time delay
   c. A circuit that can be used to determine how long a particular operation requires
5. What are the advantages of digital signal processing compared to analog processing?
6. Using Figure 15–15, describe the basic purpose and operation of each of the blocks in the digital signal processor.
7. In a group discussion, identify and discuss at least five other electronic applications that could be enhanced by the use of ICs.

FIGURE 15–15 Diagram of a digital signal processing system.

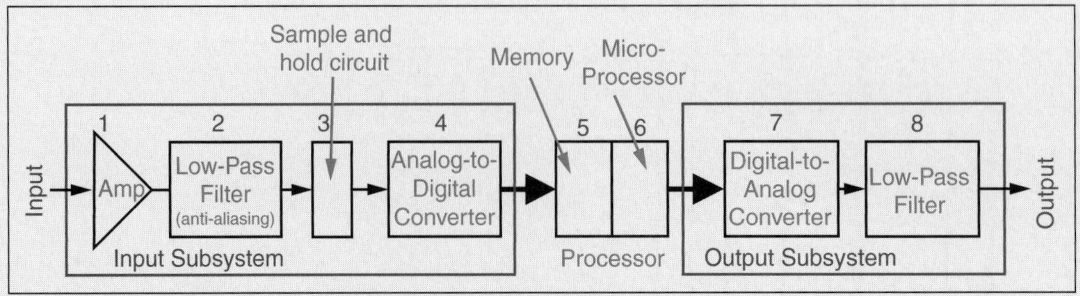

PRACTICE PROBLEMS

The following questions refer to the NE555 timing circuit.

1. In the monostable mode, what is the time duration for $R = 10$ kΩ and 0.5 μF?
2. In the monostable mode, if the pulse width is 2 m sec and the capacitor is 0.1 μF, what is the value of R?
3. In the astable mode, you have the following data: $t_{high} = 4$ m sec and $t_{low} = 1.5$ m sec. What is the duty cycle as a percentage?
4. Using the solution from problem 3, if $R_1 = 6$ kΩ find the value of R_2.
5. Using the values from problems 3 and 4 and a value of $C = 0.2$ μF, what is f_O?
6. Find the values for two resistors that will produce a 2 kHz square wave with a 70% duty cycle. The timing capacitor is 0.01 μF.
7. You wish to use an NE555 to build an 8-second time-delay circuit. You have a 20 μF capacitor. What size resistor do you need?
8. You are analyzing an NE555 circuit (Figure 15–16) in an astable mode. The resistors are $R_1 = 10$ kΩ, $R_2 = 30$ kΩ, and $C = 2$ μF. Answer the following:
   a. D = ??
   b. t_{high} = ??
   c. t_{low} = ??
   d. f_O = ??

FIGURE 15–16 Free-running or astable mode.

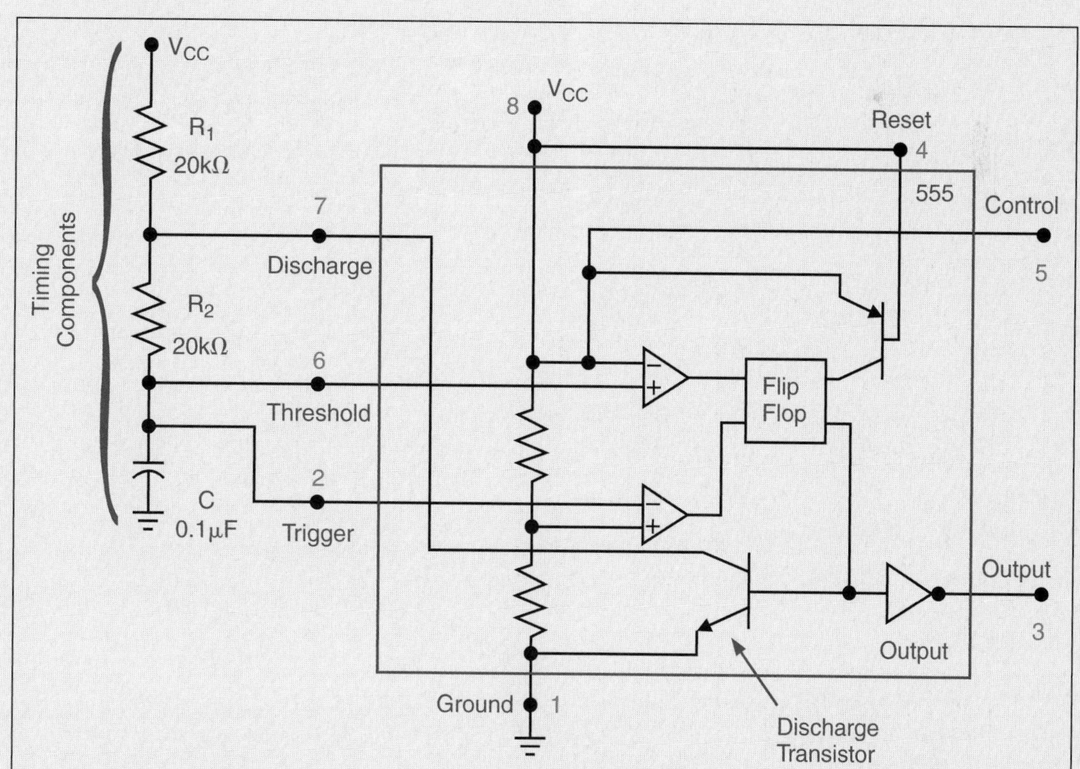

9. Answer problem 8 using Figure 15–17.

a. D = ??

b. t_{high} = ??

c. t_{low} = ??

d. f_O = ??

FIGURE 15–17 Bypass diode around R_2.

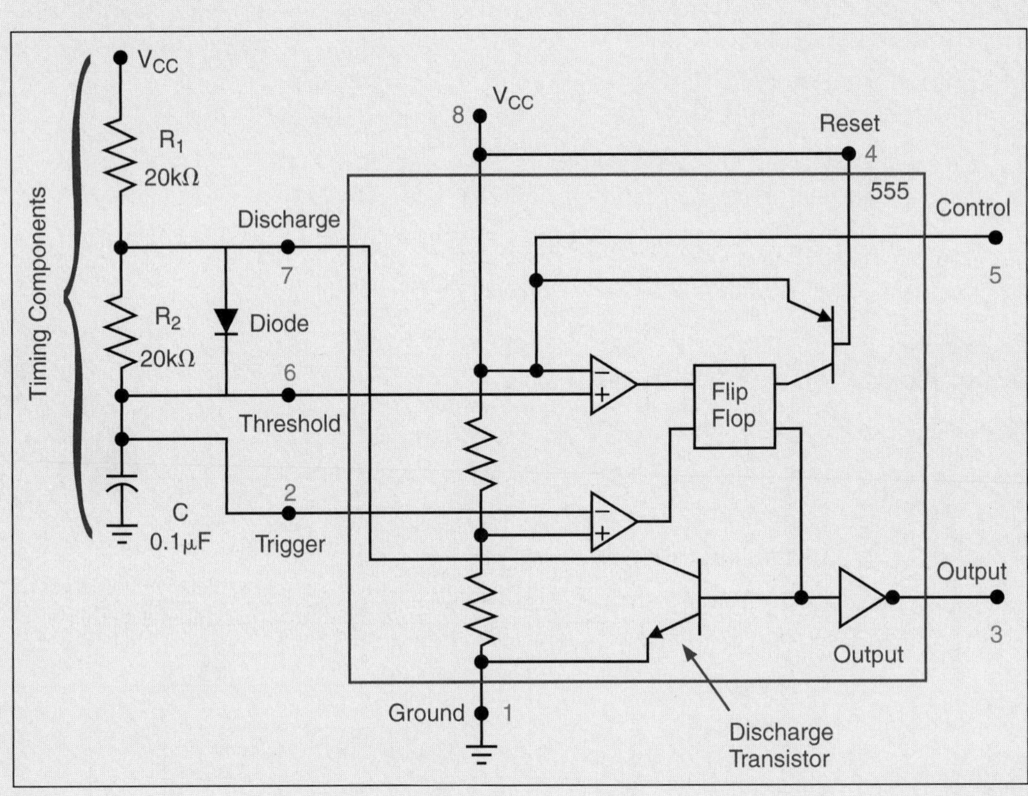

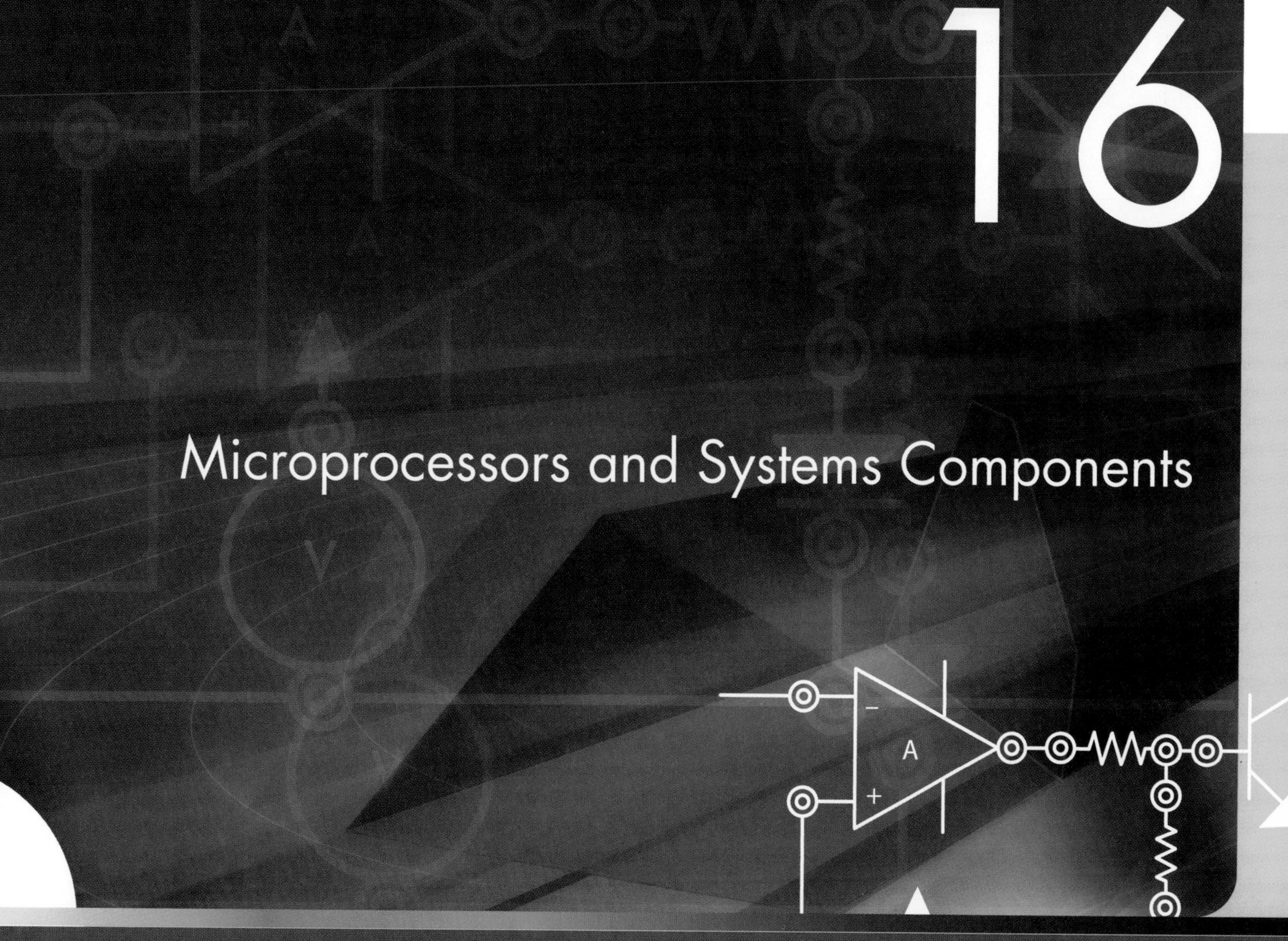

16

Microprocessors and Systems Components

OUTLINE

OVERVIEW

Early computers used vacuum tubes, were extremely large and expensive, and had limited capabilities. Today's **microcomputers** are smaller, less expensive, and much faster than older **mainframes** or **minicomputers.** These improvements have been brought about primarily by the advent of the microprocessor.

Probably the most significant use of integrated circuits has been in the development of the microprocessor, an integrated circuit that has the central processing unit (CPU) and possibly other peripheral components all on one chip. Since the initial circuits were introduced in the early 1970s, microprocessors have continued to grow in power and capability, even though the basic electrical structure and operation of the CPU has not changed significantly. At the same time, they have been decreasing in size and increasing in density.

Microprocessors composed of millions of transistors and other circuit components are now commonplace. Truly, the small laptop and handheld computers we use today have become many times more powerful than the room-sized mainframe computers of the 1950s and 1960s.

As you work through the material in this chapter, keep in mind that computer technology is the most dynamic technology in history. It is changing and advancing so fast that almost any publication about the state of the art is out of date before it hits print. This chapter presents many of the fundamentals upon which the computer technology of today is based.

OBJECTIVES

After completing this chapter, the student should be able to:

1. List the major components of the digital computer and CPU.
2. Explain how the major components of the CPU work together to enable a program to function.
3. Explain how multiplexers function.
4. Describe the operating principles of various types of computer memory.
5. Explain the difference between volatile and nonvolatile memory.

Microcomputer
A microprocessor combined with input/output interface devices, some type of external memory, and the other elements required to form a working computer system. It is smaller, lower in cost, and much faster than a minicomputer. Also known as a micro.

Mainframe
A large computer. Older models are made of discrete components and often require cabinetry that fills several rooms.

Minicomputer
A relatively small general-purpose digital computer intermediate in size between a microcomputer and a mainframe.

DIGITAL COMPUTERS

16.1 History and Modern Architecture

In use roughly two thousand years ago, the abacus (a mechanical mathematical calculator using beads) has remained a staple of the marketplace in the Far and Middle East into the second half of the twentieth century and first part of the twenty-first century. Through intervening centuries, advances in mathematics paved the way for the mechanical calculators of the last two hundred years. Electricity led to switches, motors, and relays, enhancing mechanical calculator speed, power, and ease of operation.

The vacuum tube amplifier/switch and World War II converged to produce the first electronic **computers**, the British Colossus and American ENIAC. The transistor, born at Bell Labs in 1947 was followed by the integrated circuit in 1960. Huge improvements in speed, low power consumption, and economy of space opened the floodgates to computer advances that will continue into the future. At the heart of all modern computers is the **microprocessor.**

The arithmetic/logic unit (ALU) and the control unit (CU), the two major elements of the digital computer, are usually combined to form what is known as the **central processing unit (CPU)**. The memory, which stores operating instructions and results, and the input/output (I/O) buffers are the other two major components of the computer.

Although the memory and I/O buffers are sometimes contained on chips separate from the CPU, usually all functions of the modern microprocessor, including a clock oscillator (the timing standard of the microprocessor), are included on one chip. See Figure 16–1. Vast amounts of additional memory are usually found external to the microprocessor, along with the relatively small storage space available "onboard" the microprocessor.

As its name suggests, the (CU) controls the flow of information through the other components, whereas the ALU manipulates the data based on the instructions of the program. The I/O buffers may be internal or external to the microprocessor. The buffers are controlled by microprocessor control lines that time transmission or reception of data into and out of the microprocessor, something like a series of coordinated traffic lights.

Examples of data leaving the microprocessor are address lines, data output, and read/write lines. Examples of data received from the outside world by the microprocessor are data input, interrupts, and resets. The data bus (which carries and transfers data), address bus (which controls memory), and control bus (which carries instruction and housekeeping information) make up the three types of data that move digital signals about a microprocessor-based computer.

Computer
A machine into which an operator may apply numerical data and instructions, and from which an operator may retrieve logical numeric results.

Microprocessor
An integrated circuit that today normally contains a CPU, processor clock, some memory, buffered address, data input/output lines, and control signals.

Central processing unit (CPU)
The part of a computer containing the circuits required to interpret and execute instructions.

FIGURE 16–1 Major elements of a digital digital computer.

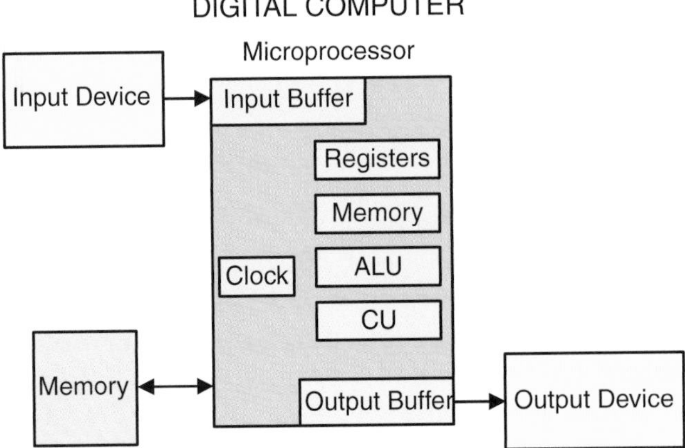

16.2 The Central Processing Unit

Although all CPUs contain arithmetic, logic, and control functions, they are not all the same. A microprocessor is a CPU along with peripheral logic functions contained on a single chip. A modern CPU, or microprocessor, IC may look like the example shown in Figure 16–2. More complicated computer systems may involve more than one CPU, and multiple circuit boards.

FIGURE 16–2 Motorola MC68040 Microprocessor IC.

All CPUs must contain the following fundamental elements:

- The logic, control, and arithmetic functions capable of performing the program instructions
- A program counter, which points to the location in memory where the next instruction code is to be found
- At least one register (accumulator), often a one-word memory location, where data taken from and returned to memory will be stored
- A data counter, which points to the next location in memory where data can be stored and retrieved
- An instruction register, which determines the operation to be executed as defined by the program

In general, the CPU carries out the manipulations and the bits of information are fetched from and returned to memory. The instructions, which determine how the bit patterns are manipulated, are also in memory. These manipulations are performed in the section of the CPU called the ALU, which performs the following functions:

- Boolean operations
- Binary addition
- Complementation
- Shift of data to the right or left

Whereas Boolean operations allow for logical decisions, addition, complementation, and data shifting provide for any of the mathematical operations (addition, subtraction, multiplication, and division). Other functions are often included for speed, but the four functions listed allow for almost any type of bit manipulation.

Most simple inexpensive microprocessors handle 8-bit (1 byte) to 16-bit (2 bytes) word lengths. Larger computers can handle up to 128-bit word lengths or even more. All parts of the CPU must handle the same word lengths. The CU determines the order in which the ALU operates. The instruction register, which is controlled by the program, sends the instructions to the CU. The CU decodes the instructions (bit patterns) from the instruction register. With these instructions, the CU sends the appropriate sequence of signals to control the ALU logic and flow of data through the ALU.

The data bits are transferred between the ALU and the registers by way of an internal data bus. Some computers have extra registers to allow for the holding of intermediate results and for the taking in of additional information. The purpose of the additional registers is quicker operational speed and simpler programming. Table 16–1 shows five status flags produced by the CPU to indicate the status or result of ALU operations. These status flags are contained in a register called the condition code register.

TABLE 16–1 CPU Status Flags

Flag	Purpose
Carry	The carry flag is used when the entire word cannot be handled in a single operation. An example is a microprocessor that handles only 8-bit (1-byte) words but needs to add two 16-bit (2-byte) words. Adding the two lower-order bytes together and the two higher-order bytes together can accomplish this. If the two lower-order bytes produce a carry when added, the carry bit is set. If a carry is produced, one will be added to the least significant bit of the higher-order byte addition.
Interrupt	The interrupt flag helps get data into and out of the computer while ensuring that the correct sequence of events takes place.
Zero	The zero flag is set if a result is a zero.
Sign	The sign flag indicates whether a number is positive or negative. The sign flag is 0 for positive and 1 for a negative value.
Overflow	The overflow flag is used with the sign flag when multiple-byte words are processed.

16.3 Arithmetic/Logic Unit

Figure 16–3 shows a schematic diagram including the SN74181 ALU IC. Although the 74181 has been a popular ALU for many years, it is an effective tool to illustrate the purpose of the ALU. The ALU allows two 4-bit numbers to be operated upon. A0 through A3 and B0 through B3 are the inputs for the 4-bit digital words, and S0 through S3 are the control functions. Outputs are channeled through F0 through F3. As is the general rule, A0, B0, and F0 are the least significant bits (LSBs), whereas A3, B3, and F3 are the most significant bits (MSBs).

FIGURE 16–3 The SN74181 IC.

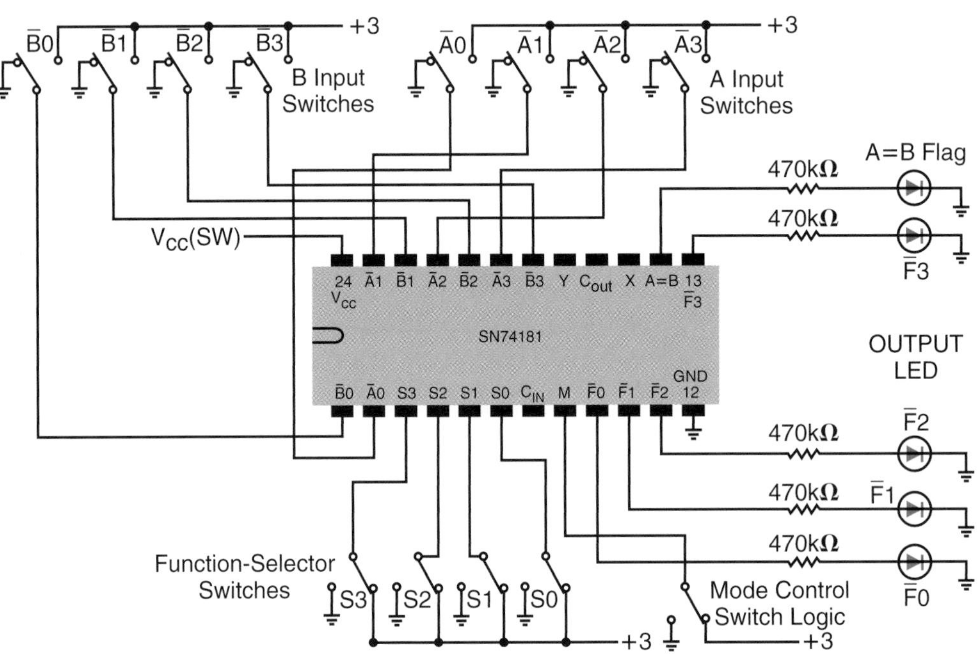

The ALU can be set to perform either arithmetic or logic operations by the setting of control line M. When control line M is low, arithmetic operations can be performed. When control line M is high, logic operations can be performed. In this ALU, a flag is used when A = B. Although the 74181 is only a 4-bit device, it can be chained with others to handle 8-bit-wide or 16-bit-wide data words.

Table 16–2 outlines the arithmetic functions available for A and B with high levels active. By changing the state of the selector switches (S0 through S3), you can choose the desired arithmetic functions to be performed. The desired logic outputs can also be obtained by having the mode selector switch in the logic position. The logic functions table is not shown here.

Note that if S3 through S0 are set at 0001 and A is set at 0010 and B is set at 1001, the output at F3-F0 will be 1011. Although this ALU handles only 4-bit words, it is similar to ALUs found in more complex computers.

TABLE 16–2 Table of Arithmetic Operations

Function Select				Output Function	
S_3	S_2	S_1	S_0	Low Levels Active	High Levels Active
L	L	L	L	F = A MINUS 1	F = A
L	L	L	H	F = AB MINUS 1	F = A + B
L	L	H	L	F = AB MINUS 1	F = A + B
L	L	H	H	F = MINUS 1 (2's complement)	F = MINUS 1 (2's complement)
L	H	L	L	F = A PLUS (A + $\overline{B}$)	F = A PLUS AB
L	H	L	H	F = AB PLUS (A + $\overline{B}$)	F = (A + B) PLUS A$\overline{B}$
L	H	H	L	F = A MINUS B MINUS 1	F = A MINUS B MINUS 1
L	H	H	H	F = A + B	F = AB MINUS 1
H	L	L	L	F = A PLUS (A + B)	F = A PLUS AB
H	L	L	H	F = A PLUS B	F = A PLUS B
H	L	H	L	F = AB PLUS (A + B)	F = (A + B) PLUS AB
H.	L	H	H	F = A + B	F = AB MINUS t
H	H	L	L	F = A PLUS At	F = A PLUS At
H	H	L	H	F = AB PLUS A	F = (A + B) PLUS A
H	H	H	L	F = AB PLUS A	F = (A + B) PLUS A
H	H	H	H	F = A	F = A MINUS 1

16.4 Multiplexers

Although a common bus is used to transfer data between registers and the ALU, a **multiplexer** (as illustrated in Figure 16–4) can be used for interconnections as well. The data bits emanating from the outputs of keyboards, adders, counters, and registers must be connected to other counters, registers, adders, and output devices at the correct time. The multiplexer routes these pulse trains along a shared path. The CU directs the multiplexer selector switch.

Multiplexer
An electrical or electronic device that allows the transmission of two or more signals along a smaller number of pathways.

FIGURE 16–4 Multiplexer.

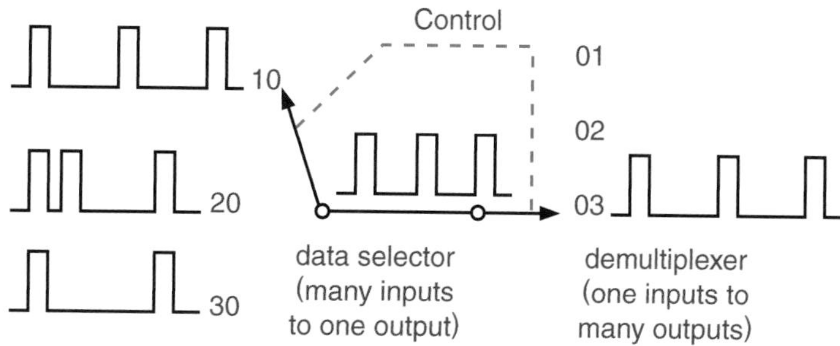

Demultiplexing is the inverse operation. The demultiplexer receives inputs from a single data line. It might first carry a train of pulses from the source to destination 3, the following pulse train to destination 1, a third train to destination 2, and so on. Figure 16–4 illustrates not only multiplexer function but demultiplexer function.

A simple four-line to one-line multiplexer comprising one NOR gate, four INVERTERS, and four AND gates is shown in Figure 16–5. Any of the four data input lines can be routed to the single output, depending on the logic states of the two data-select inputs (as shown in Table 16–3).

FIGURE 16–5 Four-input multiplexer.

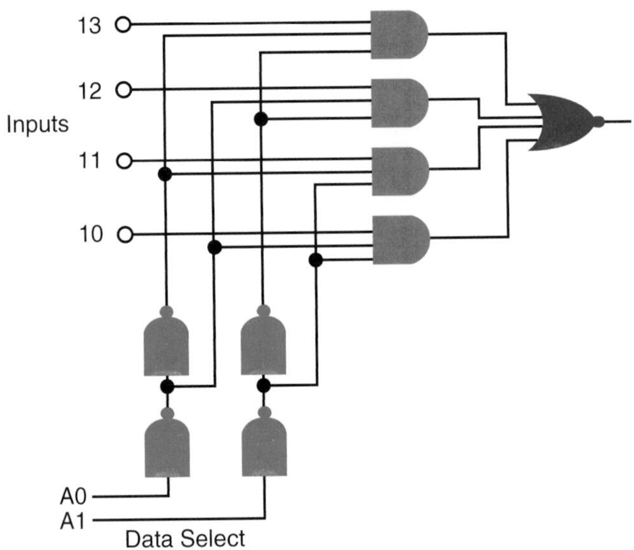

TABLE 16–3 I/O Table for Figure 16–5

Data Inputs		
A0	A1	Input Appearing at Output
1	0	11
1	1	13
0	0	10
0	1	12

This multiplexer is also known as a 4:1 multiplexer, where any one of four lines can be selected. If the number of data-select inputs were increased to three, it would become an 8:1 multiplexer, where eight different lines can be multiplexed onto a single line. This could be expanded to accommodate any number of lines to be multiplexed to a single output line.

The reverse could be done with a demultiplexer. This would allow a single pulse train to be distributed to as many outputs as permitted by the data selects. Under the direction of the CU and using the common bus structure, multiplexing provides that the correct interconnections between all CPU parts occur at the right times.

MEMORY TYPES

16.5 Introduction

Electronic memory is complex, normally being packed into specialized integrated circuits for the retention of thousands, millions, and even billions of data bytes. Every data byte in memory is assigned a unique address. Although exotic memory storage formats exist, memory is generally assigned as follows.

If a memory block has only one address line (call it A_0), it will have two storage locations: A0 HI and A0 LO. From your study of binary arithmetic, you can conclude that an eight-line memory address bus (A_7 through A_0) would allow 2^8 (256_{10}) memory locations (0 through 255 decimal). A 16 bit wide memory (A_{15} through A_0) can contain 65,536 bytes. Memory can also be accessed serially, using multiplexing. Each memory address may point to data. A memory address may contain only a single bit (1 or 0). On the other hand, a memory address may contain a nibble or a byte. Refer to memory IC specifications.

The storage location of electronic memory can be reached by using the address bus. Aside from the address lines and data lines, two or more control lines are required. Memory also requires at least two additional lines: a read/write control line and a chip select/enable control line (Figure 16–6). It is important to understand that the content of memory remain unchanged when read. The memory will change only when different or additional information is written into it.

FIGURE 16–6 Bus types.

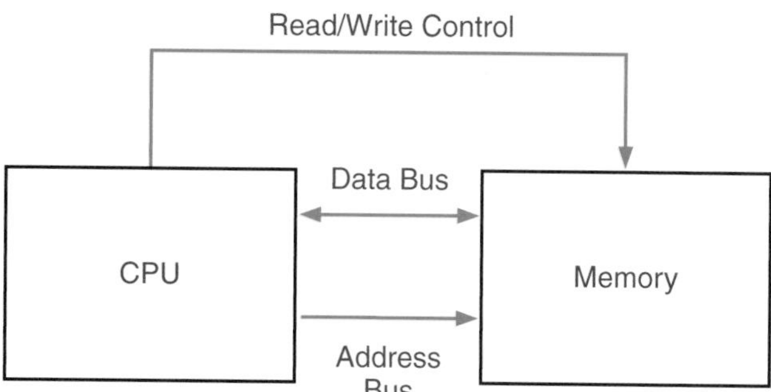

Memory can be either volatile or nonvolatile. Volatile memory is that which cannot retain data once power ("refresh") circuits have been discontinued. Nonvolatile memory is that which does not lose stored data when powered down. This is not to say that nonvolatile memory cannot be purposely changed. Some nonvolatile memory types can be factory or "field" programmed, but the equipment operator does not normally have that capability. Whether volatile or nonvolatile memory is utilized depends on its intended use.

It is possible to control many different memory locations because the data bus and address bus are separate. For example, assume that a certain microprocessor CPU handles data word lengths of 1 byte (8 bits) and has addressing capabilities of 2 bytes (16 bits). This means that there are 2^{16}, or 65,536, addressable locations. Each location is capable of holding one 8-bit word. This 65,536-word memory is commonly termed 64-K memory. Standard memory size is a rounded-off figure. It is not unusual today to find a USB (universal serial bus) memory stick with a capacity of 1 GB (gigabyte; 1,000 million bytes).

The terms *read* and *write* can be easily understood. Both terms apply to the direction data moves into and out of the microprocessor. Read and write are always viewed **from the perspective of the CPU,** or more generally from the microprocessor. When the microprocessor needs to take in data from the outside world, it reads data. When the microprocessor is ready to send data to the outside world, it writes data.

The microprocessor uses a control line known as the read/write (R/W) line. Usually, the microprocessor drives the R/W line HI for reading and LO for writing. A memory IC must then have a corresponding R/W input pin. When the microprocessor places a HI on the R/W line, the memory will make its data available to be read into the microprocessor.

16.6 Internal Memory: Core

Larger, older computers used strictly core memory. Core memory is extremely reliable, read/writable, and random accessible. Core memory still finds application today, although due to space, cost, and speed requirements it is unlikely to be found in smaller business and personal computers. Tiny rings of magnetic oxide are mixed with ceramic material, used to strengthen the product, to form the core memory.

Each ring contains 1 bit of data. The rings will store a 0 when magnetized in one direction and a 1 when magnetized in another direction. The cores are built in rectangular patterns (arrays). The length of a core is the same as the word length for the computer. It will be as wide as possible. This length and width produce what is called a plane (see Figure 16–7), and these planes are stacked to increase the memory size.

FIGURE 16–7 Core planes.

As seen in Figure 16–8, wires go through each core element. These wires allow for the reading, writing, and addressing functions. Depending on the direction of the current in the X and Y lines, the core can be read or written into. These lines also allow for addressing. The sense and inhibit lines ensure the core has an output and control. Although core memory is nonvolatile, its data must be rewritten as it is read out. Because this restoring of data is done internally (inside the core), it appears that the data has been retained. Some advantages of using core memory are:

- Useful in larger computers because they can hold huge amounts of data
- Extremely reliable
- Nonvolatile

However, disadvantages of using core memory are:

- Expensive
- Complex circuit boards
- Operating speed is limited

FIGURE 16–8 Magnetic core memory.

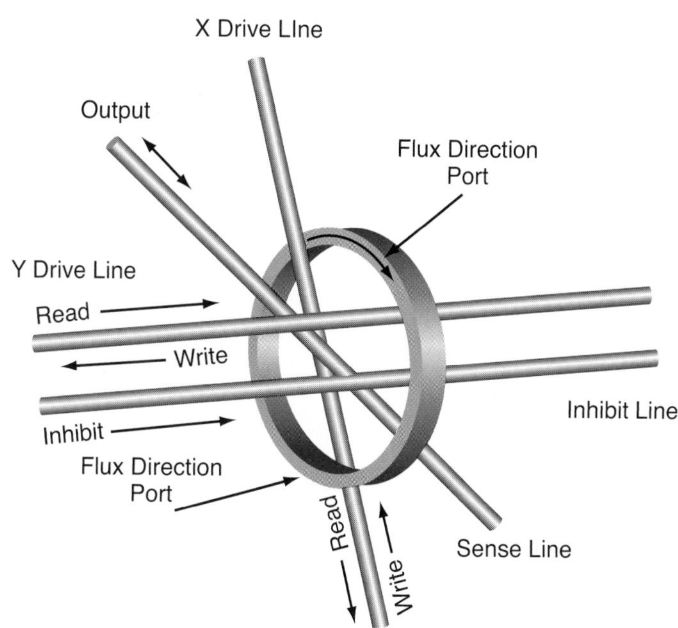

16.7 Internal Memory: Static and Dynamic Random-Access Memory

Random-access memory (RAM) chips are made of capacitors and transistors and work by storing electronic charges. The capacitor stores the charge, and the transistor can turn the charge on or off. In RAM chips, the state of the charges (i.e., bit values) can be changed, whereas ROM (read-only memory) data bits are either permanently on or off.

DRAM (dynamic RAM) is the standard main memory and can be thought of as a rectangular array of cells. Each transistor holds a single bit of data. The capacitor holds the charge temporarily, and thus must be refreshed. This means that the value is periodically read and rewritten. The refresh speed is expressed in nanoseconds (nS). Refresh cycles necessarily slow down data access.

Unlike DRAM, SRAM (static RAM) can store data without the automatic refresh process. Static comes from the fact that refresh is not needed. Data is overwritten when a write command is performed. The data remains valid until the power source is removed. A battery is often used to ensure that data is not lost when normal operating power is removed. It is not unusual for some SRAM to be supplied with backup power from a large capacitor rather than from a battery.

SRAM has two transistors per bit. SRAM has the advantage of being much faster, although more expensive, than DRAM. An example of a 4,096 × 1 bit SRAM is shown in Figure 16–9. As seen in the diagram, memory is a 64 × 64 array, which gives us 4,096 (64 × 64) bits. The two sets of buses address the array. A0 through A5 decode the row, and A6 through A11 decode the column.

FIGURE 16–9 Block diagram of a 4,096 bit SRAM.

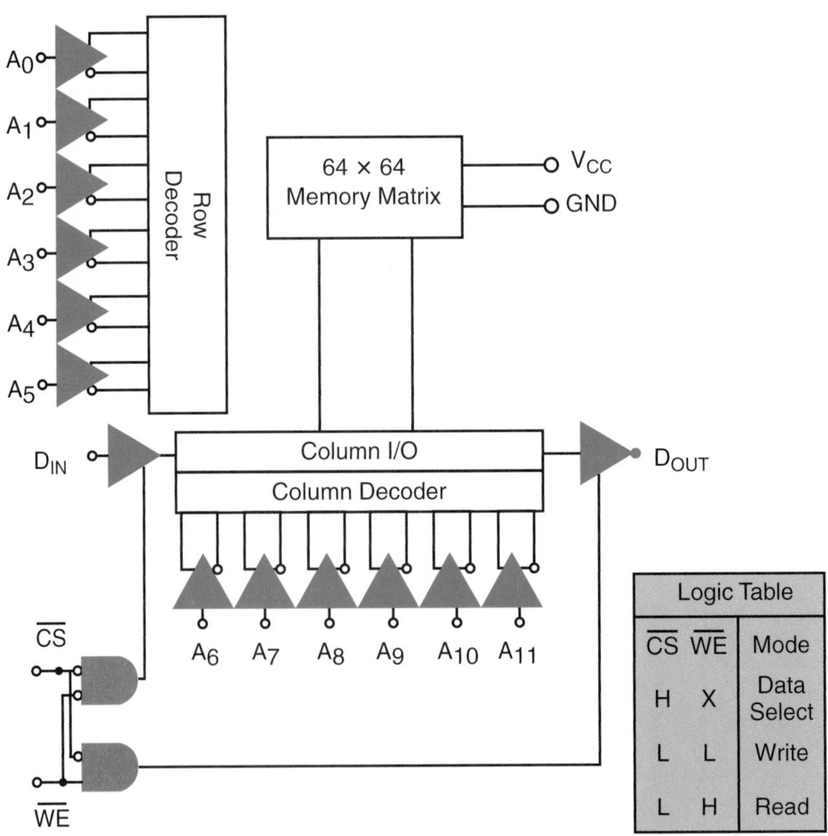

To access the element in 2,1, lines A1 and A6 would be high and the others low. To access the element in 16, 32, lines A4 and A11 would be high. The column I/O forms the output, which appears at D_{out}

The CS (chip select) remains high until data is needed from the memory. When the CS is low, data can be read or written into, depending on the state of WE (write enable). Data can be written into a location, selected by A0 through A11, when WE is low. When WE is high, stored data can be read and becomes available at D_{out}

16.8 Internal Memory: Read-Only Memory

ROM (read-only memory) is a misleading expression. The ROM is also random accessible memory. Data can be retrieved from any location simply by applying the desired memory address and reading the output. ROM is used to store data for a variety of purposes. For example, it may be a computer boot-up program that will never need to be rewritten. It may be a long series of numbers representing voltages that end up being music from a speaker. Because ROM does not change (at least not at the operator's discretion), backup power is not needed.

ROMs consist of a memory matrix, as shown in Figure 16–10, with each cell containing 1 bit of permanent data. Although the types of memory cells vary, assume here that the cells are made up of diodes and switches. When the switch is open, the electric signal cannot get through (low state). If the switch is closed and the circuit is complete, the electric signal can get through (high state).

FIGURE 16–10 Structure of read-only memory.

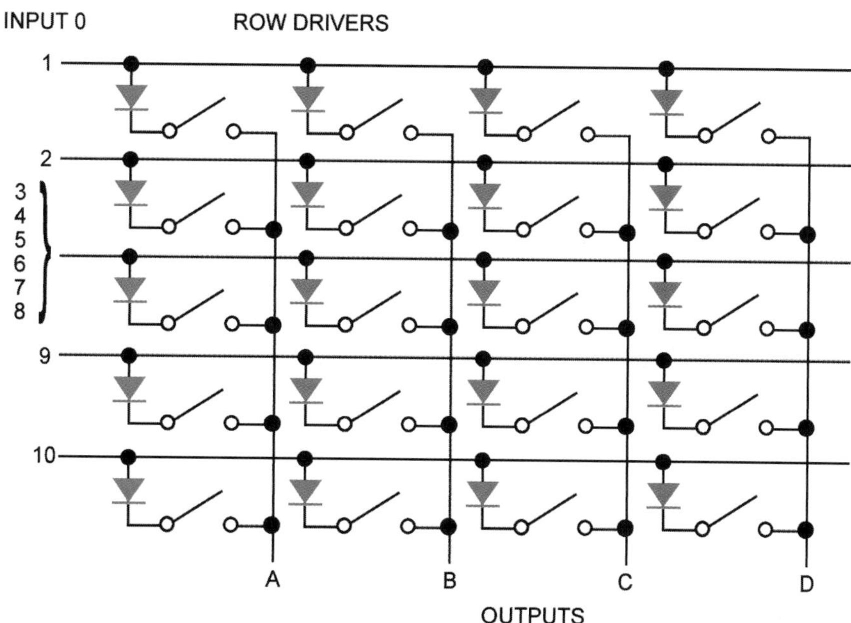

The switch is a link placed there during the fabrication process. Therefore, the high or low (0 or 1) state of each cell is permanent because it is set according to the needs of the program. To read the content of any of the rows in Figure 16–10, you must address that row. The row is addressed by applying a high state to that row. Only the content of the row addressed is seen at the output lines. The content of the other rows will not affect the output.

TechTip!

Occasionally the technician must replace a socketted IC, especially a microprocessor, memory, **microcontroller,** and so on having many pins. Successful insertion of the ICs depends on correct pin alignment. Be sure the IC is positioned correctly. Study Figure 16–2. Pin 1 should be matched with socket position 1. Do not try to install the IC until you are sure of the correct orientation. Can you spot at least three bent pins? Be sure all pins are straight.

Bent pins usually prevent proper operation of the circuit, cause additional damage to the circuit, and often lead to broken pins. Bent microprocessor or memory pins may be very difficult or impossible to spot without removing them from their sockets. Not all microprocessors are socketted. Some have in excess of 500 pins, and are soldered onto the circuit card assembly from beneath. These are referred to as ball grid array ICs.

If switches A and D in row 9 are high, the word 1001 will appear at the ABCD outputs lines.

In this arrangement, 14 connections were needed to access the 40 individual memory elements (10 words of 4 bits each). If the same arrangement were used for a typical ROM, this would lead to an impractical number of connections and wires. A typical ROM has 512 × 8 elements, which would require 4,096 (512 × 8) connections.

Therefore, an arrangement using decoders is used to access each element in an array. Figure 16–11 shows a 4 bit BCD-to-decimal decoder connected to the input lines of the previously shown 10 × 4 memory. A 2-bit decoder is used to select any of the four outputs. As is shown, this decoder is connected to the output with AND circuits.

FIGURE 16–11 Decoders allow for fewer connections.

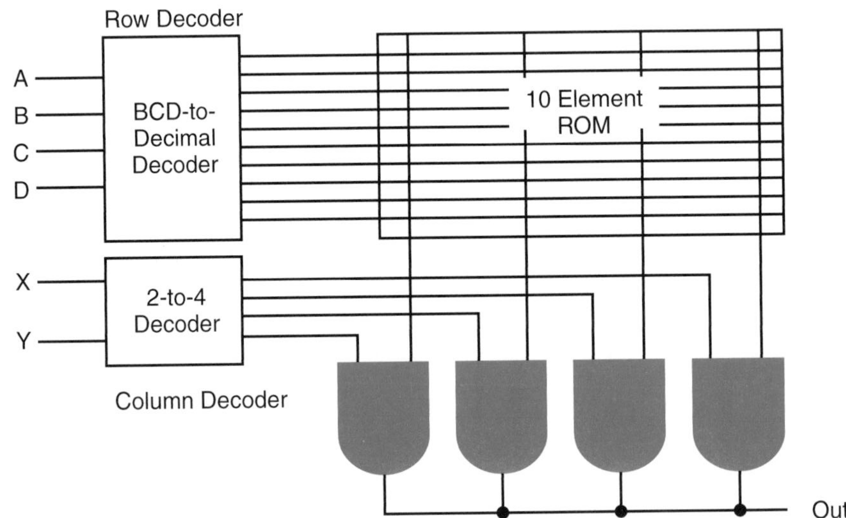

Applying the 4-digit and 2-digit codes to each decoder will result in a 1 at the chosen row and column. The stored pattern, whether it is a 0 or 1, will be sent to each output line when a 1 is applied to the selected row. Then a 1 is applied to the selected column AND gate, which energizes only that AND gate. If the matrix output is 0, the output is 0. A 1 will produce an output of 1.

With this new arrangement, the 10 × 4 memory needs only 6 connections instead of the original 14. Using this organization, it becomes practical to use ROMs with as many memory cells as required. This same structure is used with RAM.

Microcontroller
A small electronic device, usually containing a microprocessor, which based on a fixed set of input data and instructions produces a limited number of predetermined outputs for the accomplishment of specific tasks.

16.9 Internal Memory: Programmable Read-Only Memory

Programmable read-only memories (PROMs) have the same properties as ROMs, with the exception that they can be written on once. Once they have been written on, or programmed, they are just like ROMs. The benefit of PROMs is that they can be economical in small quantities.

The memory cells of the PROM are manufactured in the closed state (1) by using a switch contact consisting of a thin alloy layer. This layer is something like a fuse. The programmer applies a current through any memory cell and blows the fuse element so that the cell is now in the open state (0). The EPROMs (erasable PROMs) do not use permanently open or closed switches. The memory can be changed electrically by exposing the chip to ultraviolet light.

16.10 External Memory: Mass Storage

Magnetic Storage

External memory is used in addition to internal memory when data is not needed for the moment but must be kept for an extended period of time. Data and programs that can be accessed quickly can be stored in external memories. Most computer systems use an external mass storage system. Digital tape is one form of external memory. Because of the slow access to the information on the digital tape, it is used mainly for long-term storage that will be eventually read into internal memory.

The most popular external storage is the magnetic disk system, which allows for almost immediate access to data stored on it. Data can be stored onto the disks almost as quickly. They allow for large amounts of data to be stored externally. Modern magnetic disk systems can be large storage or small. Disks with capabilities up to a terabyte are becoming available. Sizes in the hundreds of gigabytes are common.

The disk or (in the case of disk systems) disks are coated with a magnetic material. The read/write heads of the computer can read stored data or they can write data onto the disk. They can range in size from the common floppy disk to the very large hard disks used with mainframe computers.

Optical Storage

Modern storage media such as **CD** and **DVD** are becoming more and more common. In fact, CDs and DVDs are quickly replacing the 3.5-inch magnetic diskette external magnetic storage in microcomputer systems. These two storage devices use a plastic disk. The information is stored digitally on the disk by etching the information into the material using a laser. A large variety of these two devices is available, including both permanent and rewritable types.

CD
Compact disk. An optical media that stores information using laser-imprinted plastic media. CDs can store up to almost a gigabyte of data.

DVD
Digital video disk or digital versatile disk. A modern optical storage media that can store multiple gigabytes of data on a disk.

PROCESSORS IN INDUSTRY

Computers are found almost everywhere. They are found on desktops at the office, on home kitchen tabletops, at construction sites, in delivery trucks, in spacecraft, and on today's battlefields. Computers are not the only devices that use microprocessors, however. The factory floor, in addition to computers, is heavily populated by a wide variety of other microprocessor-based devices.

Highly reliable and economical programmable logic controllers (PLCs) are manufactured by many companies around the world. They are programmed using a variety of software languages to accomplish specific tasks, and may respond to hundreds of input variables to control manufacturing processes and other applications. Many PLCs provide easy-to-understand graphic displays. They enable operators to quickly respond to real-time situations, analyze problems, and make component troubleshooting and replacement far easier than ever before. Digital communications can tie in multiple PLC systems and allow system supervision from halfway around the Earth when required.

SUMMARY

The major elements of the microprocessor are the control unit (CU), arithmetic/logic unit (ALU), input/output buffers, and memory. The CU and the ALU are usually housed together to form what is known as the CPU. The CU directs how the information flows through the other components. The data is manipulated in the ALU, and the results and other data are stored in memory. All of this information is carried through the bus lines, either the address bus or the data bus.

All CPUs must contain five basic elements. They must be able to perform the instructions of the program using the logic, control, and arithmetic functions; have a program counter; have at least one register; have a data counter; and have an instruction register. The ALU must include four basic functions, but it can include more. It must be able to perform Boolean operations, binary addition, complementation, and shift of data. The CU controls the sequence in which the ALU performs operations.

Multiplexers are electronic devices that alternate a larger number of signals onto a smaller number of signal paths within a system. Memory can be categorized as either volatile (permanent) or nonvolatile (temporary). Some of the memories discussed in this chapter are:

- Static and dynamic random access memory (SRAM and DRAM)
- Read-only memory (ROM)
- External storage (magnetic tape or disks and optical disks)

Programmable logic controllers (PLCs) are microprocessor-based devices commonly used to control and monitor industrial processes.

REVIEW QUESTIONS

1. Name the elements found in a microprocessor.
2. What is the purpose of:
   a. The CPU?
   b. The arithmetic/logic unit?
   c. The control unit?
3. What are the four basic functions carried out in the CPU?
4. How does a multiplexer work? (Use a four-input multiplexer for your explanation.)
5. Explain the use of five flags in a CPU.
6. What type of memory is a:
   a. Computer floppy disk?
   b. A computer hard drive (fixed disk)?
   c. An SRAM chip?
   d. A CD?
7. What is the purpose of an EPROM?
8. What is the purpose of a PROM?
9. When your personal computer first boots up, what type of memory establishes its initial start-up parameters?
10. What is the purpose of I/O buffers in a CPU?

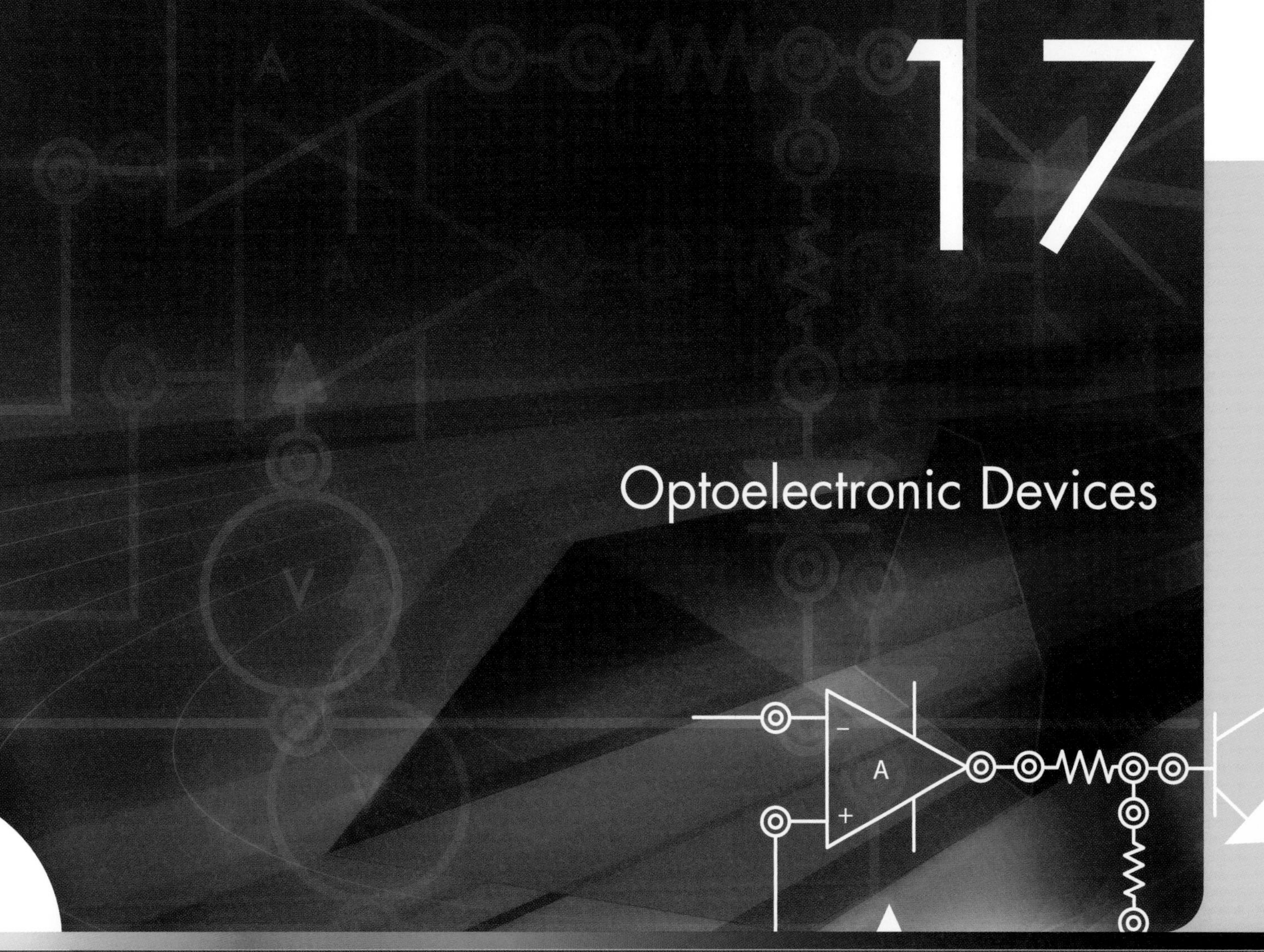

17

Optoelectronic Devices

OUTLINE

OVERVIEW

In this chapter, we will be looking at a class of devices whose operation combines optics and electronics. These devices are, in general, classified under optoelectronic devices. Optoelectronic devices can be divided into two categories.

1. Light-generating devices: Those that create and/or modify light
2. Light-detecting devices: Those whose characteristics are changed or controlled by light

In today's world, solid-state optoelectronic devices have widespread applications in the areas of fiber-optic communication, optical sensors, alarm systems, and so on. However, before we look at the devices themselves it is important to first examine the basic principles of light.

OBJECTIVES

After completing this chapter, the student should be able to:

1. Identify the schematic symbols of optoelectronic devices.
2. Define the operation of optoelectronic devices.
3. Discuss applications of light-emitting and light-detecting devices.
4. Describe steps to test LEDs.

THE OPTICAL SPECTRUM

Light is a form of electromagnetic radiation. If measured in frequency **(hertz)**, the optical light spectrum extends from 30 THz to 3 PHz. As illustrated in Figure 17–1, the **optical spectrum** encompasses three bands—only one of which is visible to the naked eye.

Hertz
The SI unit of frequency. It replaced cycles per second.

Peta (P)
The unit prefix indicating 10^{15}. For example: 30 PHz = 30×10^{15} Hz.

Tera (T)
The unit prefix indicating 10^{12}. For example: 30 THz = 30×10^{12} Hz.

Optical spectrum
The range of electromagnetic frequencies that fall into the region we call light.

FIGURE 17–1 The optical spectrum.

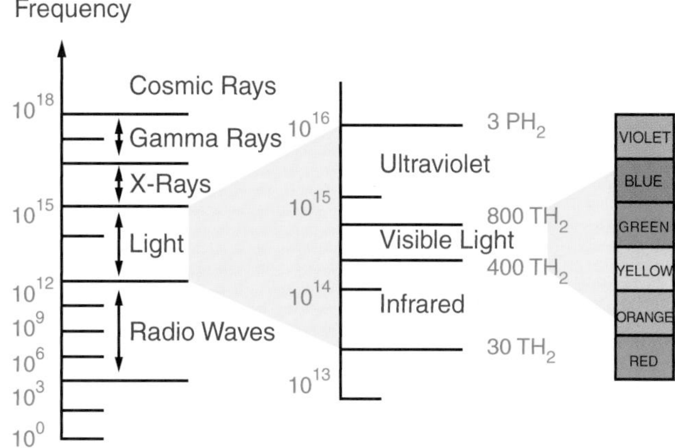

The frequencies range from infrared (not visible) to visible light to ultraviolet (not visible). (1) The infrared band (30 THz to 400 THz) comprises frequencies lower than the human eye can detect. (2) The visible light band (400 THz to 800 THz) consists of frequencies the human eye can detect. This band can be further subdivided into the colors of the rainbow: red, orange, yellow, green, blue, and violet. (3) The ultraviolet band (800 THz to 3 PHz) is made up of frequencies above those the human eye can detect.

Consider the highlighted electromagnetic cycle shown in Figure 17–2. This figure shows that it is possible to express frequency as wavelength, and vice versa. Wavelength is the length of one cycle as it travels through space. It is represented by the Greek letter lambda (λ) and is related to frequency by the following formula:

$$\lambda = \frac{c}{f}$$

Here:
λ = wavelength in meters (m)
c = the speed of light in meters/second ($c = 3 \times 10^8$ m/sec)
f = frequency in Hz

Definitions of a few more quantities associated with light will better explain optoelectronic devices.

Luminous flux (light flux)
The rate at which visible light energy is produced or received at a source. It is measured in lumen (lm).

Luminous intensity (light intensity)
The luminous flux per unit area, measured as lumens/meter^2 (lm/m^2).

FIGURE 17–2 One sinusoidal cycle of an electromagnetic wave.

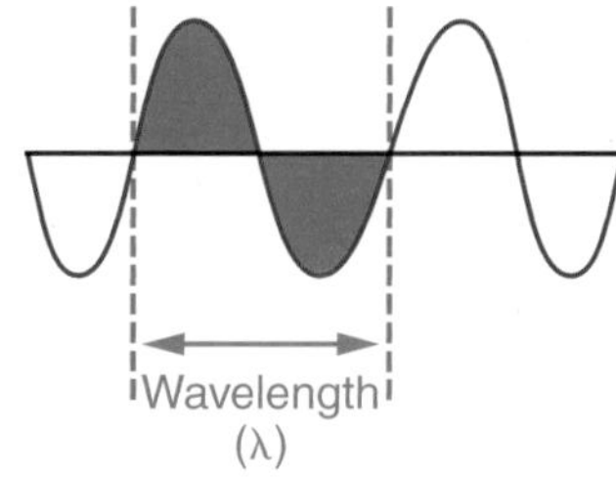

LIGHT-EMITTING DIODE

17.1 Principles of Light Generation

A light-emitting diode (LED), as the name suggests, is a semiconductor device that generates light. Light generation begins at the atomic level. The classic model of the atom (Figure 17–3) is made up of protons, neutrons, and electrons. Protons and neutrons cluster in the center to form the nucleus of the atom. The electrons orbit the nucleus in somewhat the same way that planets orbit the sun.

FIGURE 17–3 The Bohr model of the atom.

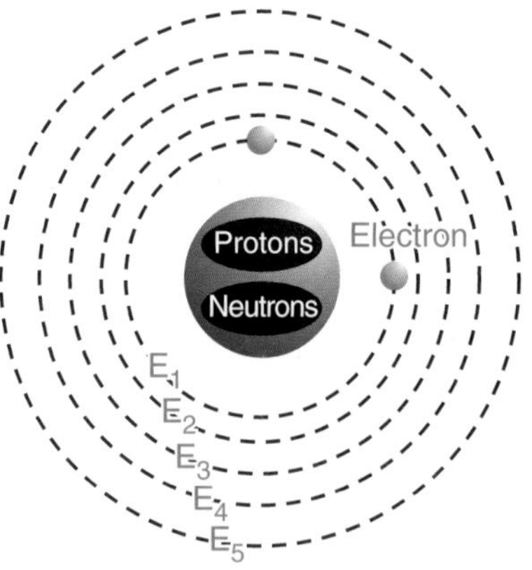

Figure 17–3 is the Bohr atomic model. This diagram shows the orbital paths of electrons. Each orbital is associated with a specific energy level. The orbital an electron occupies depends on the amount of energy the electron possesses. Electrons with lower energy occupy the lower orbits, whereas electrons with higher energy levels occupy the higher orbits. Thus, electrons in the orbit E_1 are at a lower energy level than electrons in the orbit E_2.

The natural state in which an electron exists is said to be its ground state. Energy may be imparted to an atom by heat, electrical force, or collision with other particles.

Upon absorption of energy, an electron can jump to a higher orbit (as shown in Figure 17–4a). In this state, the atom is said to be excited. This is an unnatural state, and the electron will remain in this higher orbit for only a very short time before reverting back to its natural ground state.

Because the electron now possesses too much energy to exist at the ground state, it must shed the excess energy. The electron sheds the excess energy in the form of electromagnetic radiation. In some materials, this energy is released at the frequencies of visible light. The process of energy release is shown in Figure 17–4b.

FIGURE 17–4 Electron energy absorption and emission of light energy: (a) absorption, (b) emission.

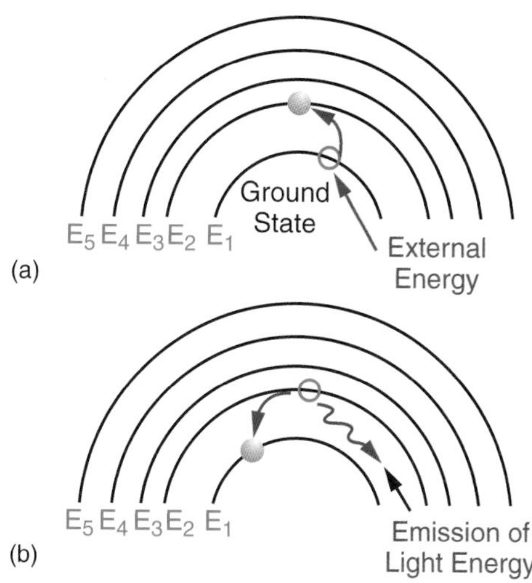

17.2 Basic Operation

An LED is a PN junction diode of special construction. Like all semiconductor diodes, its behavior is influenced by biasing conditions. Figure 17–5 shows the basic operation of an LED. The PN junction is forward biased when the cathode (N-type region) is negative with respect to the anode. Under forward bias, the electrons from the N-type region and the holes in the P-type material move toward the PN junction.

At the junction, the holes and electrons combine. Electrons are at a higher energy level than the holes. The electrons literally fall into the holes as they combine, and in doing so release the excess amount of energy in the form of light. There is a current through the LED in the forward direction, just as in the case of an ordinary diode.

When the PN junction is reverse biased, the LED does not conduct and there is no release of light. Hence, the LED emits light when forward biased and does not emit light when reverse biased. The schematic symbol of the LED is shown in Figure 17–6. The LED operates at forward voltages typically ranging from 1.2 V to 4.3 V. The reverse breakdown voltage is in the range of −3 V to −10 V.

FIGURE 17–5 Basic operation of a light-emitting diode (LED).

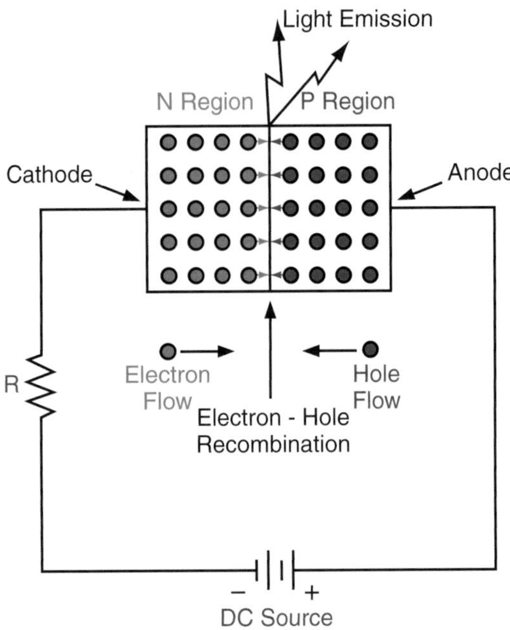

FIGURE 17–6 Schematic symbol of an LED.

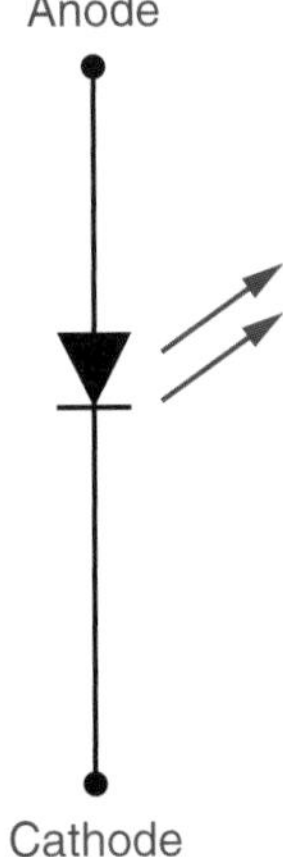

17.3 LED Construction

If you are wondering why the LED emits light and an ordinary diode does not, it is because ordinary diodes are made of silicon, which is opaque or impenetrable as far as light is concerned. Energy in ordinary diodes is therefore released in the form of heat.

LEDs are made of semiconductor materials that are semitransparent to light. Energy released is in the form of light, which can escape to the surroundings. LEDs are normally made of gallium-arsenide phosphide (GaAsP), which emits visible red light.

Gallium phosphide (GaP) produces visible green light. LEDs that produce yellow and blue light are also available.

Figure 17–7 shows an LED chip and a typical LED package. The LED chip is attached to the cathode and anode leads through thin wires. The plastic case serves as a lens that conducts light away from the LED and also acts as a magnifier. The entire case may be dyed or tinted with a color that enhances the on/off contrast of the LED. LED leads are identified in one of the following three ways (see Figure 17–8):

1. The leads may have different lengths, and the shorter of the two leads is the cathode (Figure 17–8a).
2. The cathode lead is flattened (Figure 17–8b).
3. One side of the case may be flattened. The lead closest to the flattened side is the cathode (Figure 17–8c).

FIGURE 17–7 The LED: (a) the chip, (b) typical package.

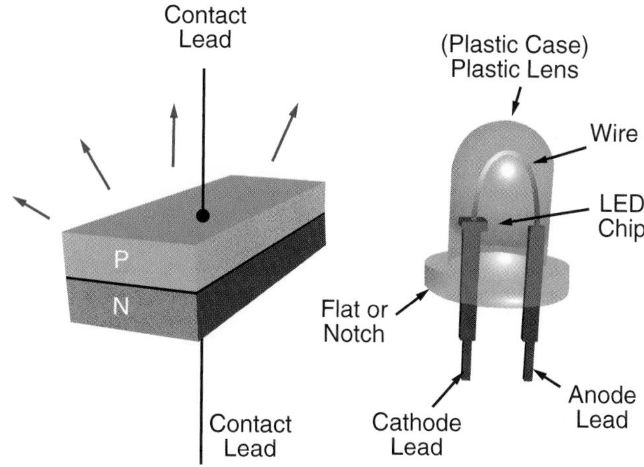

FIGURE 17–8 Identification of LED leads.

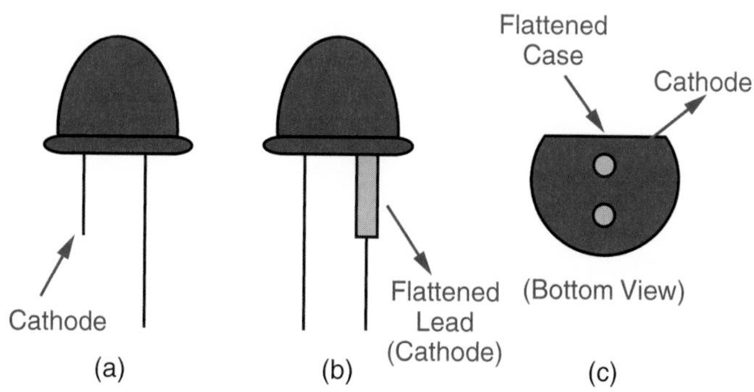

17.4 Applications

LEDs have replaced small incandescent light bulbs as display devices. The long life expectancy, ruggedness, and much lower power consumption have contributed to the choice of LEDs over incandescent light bulbs. Various types of liquid crystal displays have also replaced LEDs in some applications because they use much less energy than LEDs.

LED Indicator Circuits

An individual LED is commonly used as a power level indicator in a circuit. Figure 17–9 shows how an LED can be used as a power level indicator in a computer, a CD player, a stereo amplifier, or other such device. The LED is connected in parallel to the system's internal power supply, along with a series resistor R_S. When the power switch is on, the diode is forward biased and emits a light.

FIGURE 17–9 LED power indicator.

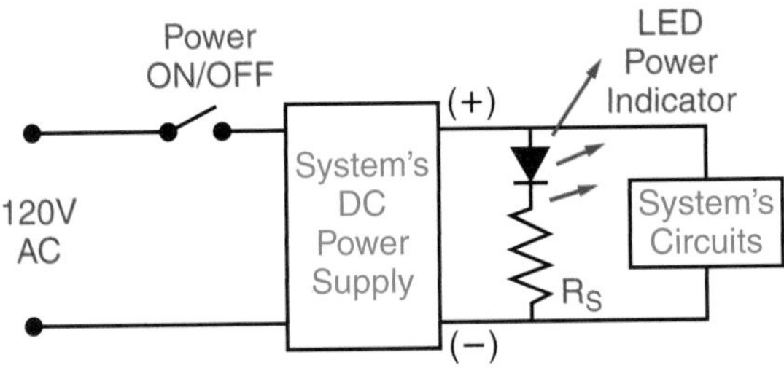

Series resistor R_S limits the current to the LED and is often referred to as a "current limiter." Most LEDs have a forward-biased voltage drop between 1.2 V and 2.5 V, although high intensity LEDs can have higher forward–biased voltage drops. Note that this is much higher than the normal silicon diode (0.7 V) or germanium diode (0.3 V) introduced in Chapter 1. The resistance value and wattage rating of R_S can be calculated with Ohm's law. R_S resistance value is found by subtracting the LED's voltage drop from the source and then dividing by the LED's designed current. To find the wattage rating for R_S, simply apply the power formula $P = I^2R$.

Example

What is the resistance and wattage rating of a limiting resistor in a 12 VDC circuit for an LED that requires a 2 V forward bias with a 20 mA current rating?

$$R_S = \frac{V_S - V_D}{I_D} = \frac{12\text{ V} - 2\text{ V}}{20\text{ mA}} = 500\ \Omega$$

and

$$P = I^2R = (20\text{ mA})^2 \times 500\ \Omega = 200\text{ mW}$$

Therefore, a standard-size 0.25 watt resistor is required. (Note: 0.25 watt = 250 mW.)

Multicolor LEDs

Multicolor LEDs are available that will:

- Emit one color light when the supply voltage is one polarity
- Emit a second color when the polarity is reversed
- Emit a third color when the bias polarity is rapidly switched

Figure 17–10 shows the schematic symbol for multicolor LEDs. Multicolor LEDs are usually two LEDs connected in antiparallel. That is, the anode of each diode is connected to the cathode of the other. Each LED can emit light only when forward biased. Thus, when voltage of either polarity is applied one LED is forward biased and emits its native color.

The two LEDs most commonly used are red and green in color. The green LED is normally used to indicate whether something is functioning properly, and the red LED is used to indicate that there is a problem. If the multicolor LED is rapidly switched between the two polarities, the red/green LED appears to produce a third color (yellow).

FIGURE 17–10 Multicolor LEDs.

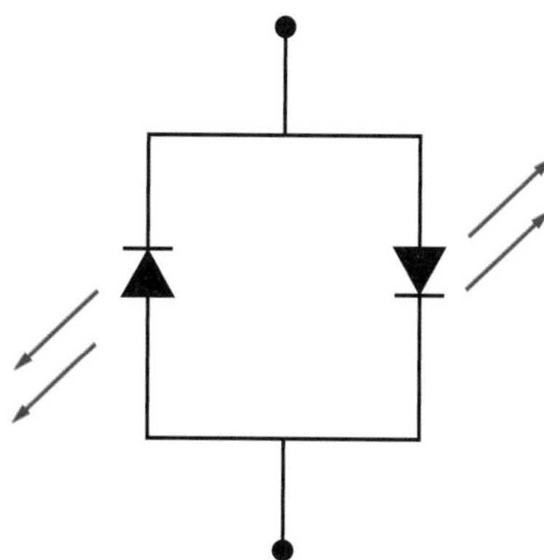

Infrared-Emitting Diode

As discussed earlier in this chapter, the infrared band (30 THz to 400 THz) falls below the frequencies the human eye can detect. Diodes made of gallium arsenide release energy by way of heat and infrared light. Such a diode is called an infrared-emitting diode (IRED). IREDs are used for home electronics such as remote controls, fiber-optic communications, discriminating organic solvents in the field of medicine, and other such applications.

Multisegment LED Displays

LEDs are very widely used in multisegment displays. Figure 17–11 shows the most commonly used multisegment display. Its seven segments are labeled *a*, *b*, *c*, *d*, *e*, *f*, and *g*. The LED labeled *dp* is used to display the decimal point. By lighting a combination of different LEDs, any number from 0 to 9 can be displayed.

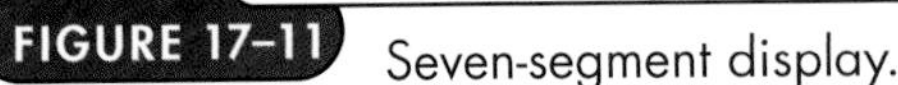

FIGURE 17–11 Seven-segment display.

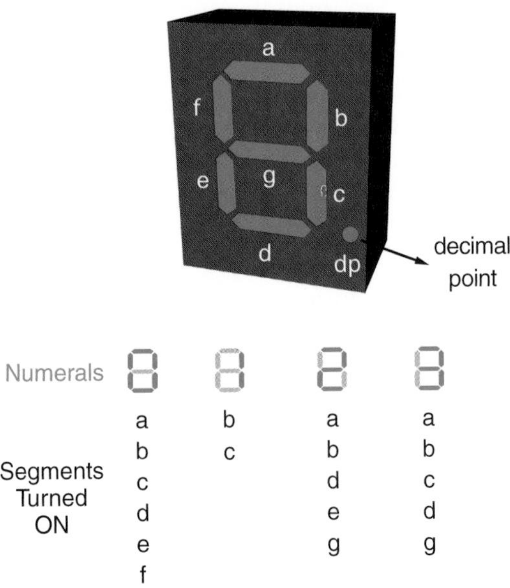

The seven-segment display cannot be used very efficiently to display all alphabets, and thus other multisegment displays have been developed to satisfy these requirements. The 16-segment display is illustrated in Figure 17–12, and the 5 × 7 dot matrix display is shown in Figure 17–13.

FIGURE 17–12 Sixteen-segment display.

FIGURE 17–13 5 x 7 dot matrix display.

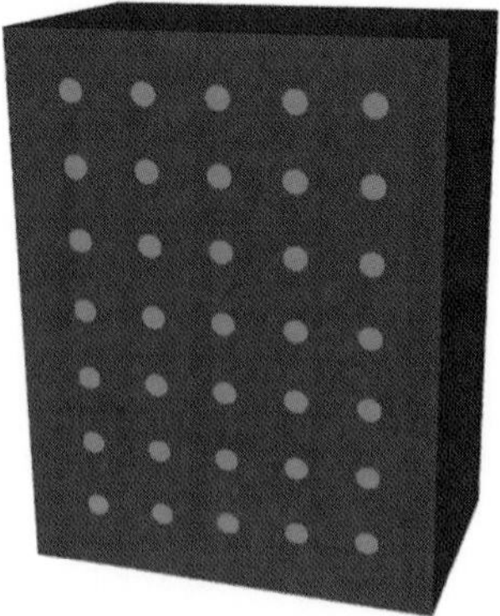

17.5 LED Testing

LEDs are usually damaged by excessive current flowing through them. Such damaged LEDs can be identified by a discoloration on the casing of the LED, due to the burned junction. The LED can be tested with a DMM (digital multimeter). In the diode check position, the DMM supplies about 2.5 V at very low current.

If at this setting the cathode of the LED is connected to the negative lead of the DMM and the anode of the LED is connected to the positive of the DMM, the supplied 2.5 V should be sufficient to forward bias the LED and cause it to glow dimly, and the forward breakover voltage should be indicated on the meter display. Reversing the meter leads should, just as a conventional diode, read open (∞).

LASER DIODE

17.6 Principles of Laser Generation

LASER is an acronym of light amplification by stimulated emission of radiation. Most light sources are multichromatic. That is, they comprise many different colors. Each color corresponds to a different frequency of light. In an incoherent light source, most of these waves are out of phase with one another. Figure 17–14a shows waves of various lengths. Note how these waves (A, B, and C) add to and strengthen each other at some points, while opposing and weakening each other at other points. Most ordinary light sources diffuse, or spread out in various directions, and thus the light beam cannot travel very far.

Laser light sources, on the other hand, are monochromatic (have only a single color associated with a wavelength). All waves are in phase (are coherent). Laser light sources emit light in only one direction. To summarize, laser light requires emitted light to be monochromatic, coherent, and unidirectional. Figure 17–14b illustrates waveforms A, B, and C to be of one frequency (wavelength) and in phase. Note that waveform D of Figure 17–14b agrees with A, B, and C and should be interpreted as their amplified sum. This type of emission of energy by the electron is termed **stimulated emission.** When a large number of electrons are subjected to stimulated emission, the amplification of light (as demonstrated in coherent light) takes place.

Stimulated emission The emission of radiation from a system as an excited electron energy level transitions to a lower energy level.

FIGURE 17–14 Light sources: (a) incoherent, (b) coherent.

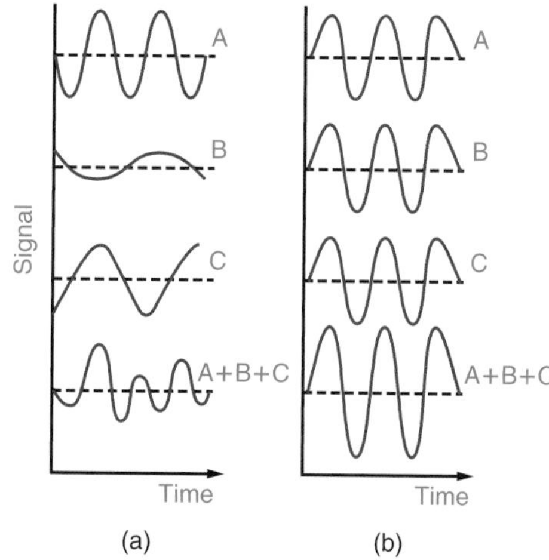

FIGURE 17–15

A typical injection laser diode (ILD) package.

17.7 Basic Operation and Construction

Laser diodes are light-emitting devices, capable of emitting laser beams. One of the commonly used laser diodes is the injection laser diode (ILD). A typical ILD package is shown in Figure 17–15. The laser diode works exactly like an ordinary diode. Figure 17–16 shows the structure of an injection laser. The P-type and N-type materials are made of AlGaAs (aluminium gallium arsenide).

FIGURE 17–16 Structure of the injection laser diode (ILD).

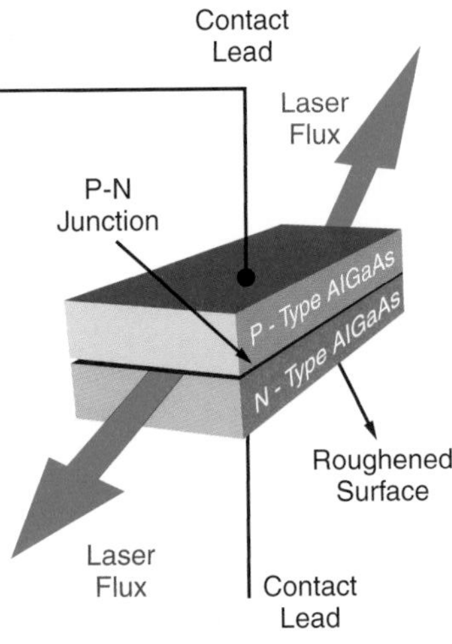

When forward biased, light is emitted at the PN junction. However, unlike the LED the emitted light is coherent and monochromatic. In the LED, the light emitted is scattered in all directions. In the laser diode, however, the light emitted can escape only from the end faces because the edges are roughened. In most laser diodes, one of the end faces is coated with a reflective material so that the radiation of the laser is emitted only in one direction.

The laser diode emits light when forward biased and can withstand only relatively small reverse-biased voltages. A high reverse voltage can damage or destroy the laser diode. The schematic symbol for the laser diode is shown in Figure 17–17. Note that the light emission arrows are zigzagged rather than straight, as in the symbol for an ordinary LED.

FIGURE 17–17

Schematic symbol for a laser diode.

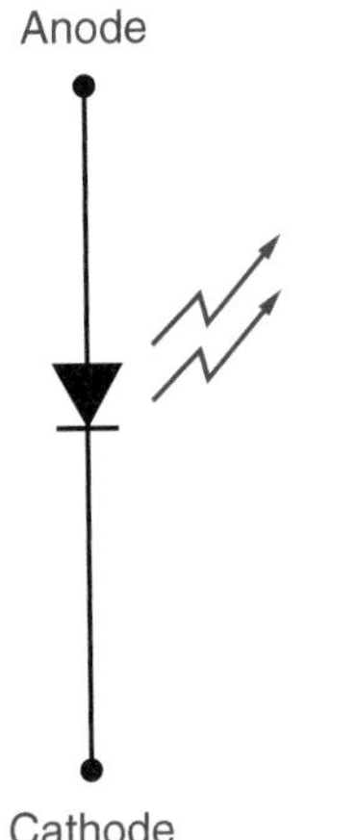

17.8 Applications of the Laser Diode

Laser diodes are used primarily in fiber-optic communications. It is the only device capable of producing optical energy high enough in concentration to pass through lengthy fiber-optic cables. However, it also has some major disadvantages compared to LEDs.

- They cost 10 times more than LEDs.
- Their life expectancy is 10 times shorter than that of LEDs.
- They require elaborate power supplies and consume much more power than LEDs.

Laser diodes do not compete with LEDs on a cost or reliability basis. Therefore, LEDs are used as much as possible in fiber-optic systems, and laser diodes are used only when absolutely necessary.

Some applications of LASER diodes are:

- Fixed product scanners such as found in most grocery stores in the United States.
- Handheld barcode scanners such as used in inventory control in most large warehouses.
- Handheld laser pointers used in meeting room presentations.
- Laser projection devices being adopted by surgeons, mechanics, aviators, and technicians. Laser scanners project low-power images directly onto the user's retina as an overlay in conjunction with the real-world image.
- Land survey range-finding accuracy has been revolutionized by the use of laser diodes. Laser transmission from the total station (reflected from a distant point) is detected, time-delay measured, and converted to an equivalent distance. Figure 17–18 shows a Leica model TC407 Total Station that incorporates laser range finding.
- Document printers and scanners often incorporate lasers.
- Lasers are used in construction to paint line and angle projections, saving time and nearly eliminating the potential for manual measurement errors.

FIGURE 17–18 Leica model TC407 Total Station. Copyright: Leica Geosystems AG.

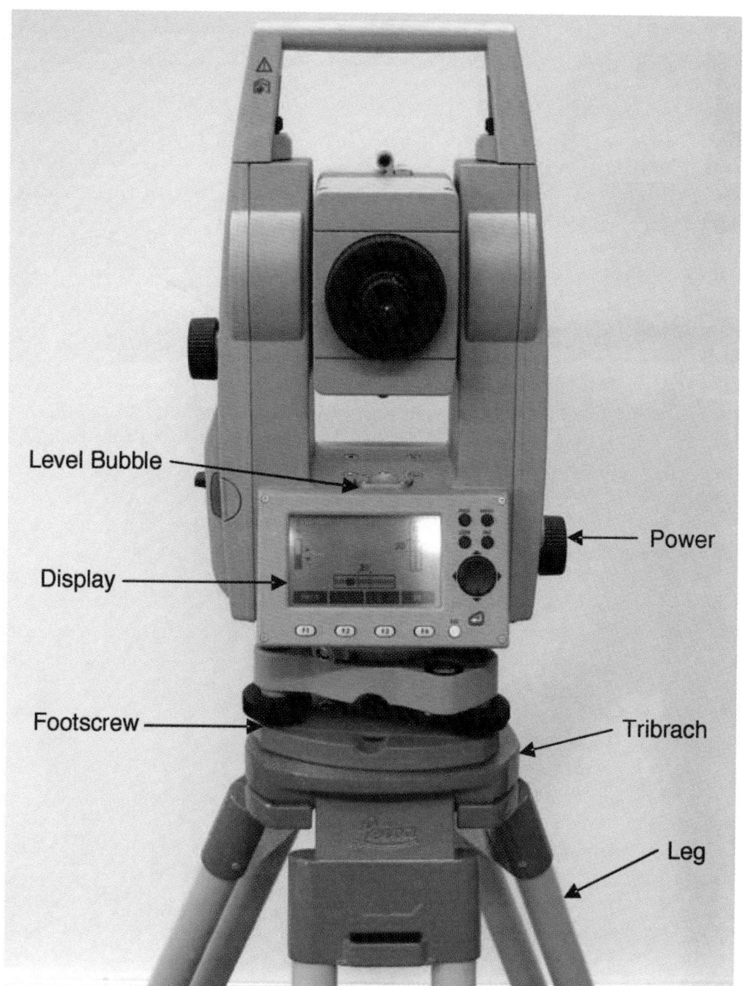

LASERS are categorized into five class types:

- *Class I*: Laser output between 0.221 and 0.39 milliwatts. Class I lasers are eye safe.
- *Class II*: Laser output between 0.39 and 1.0 milliwatts. Class II lasers are eye safe with *caution*.
- *Class IIIa*: Class IIIa applies to visible spectrum laser output between 1.0 and 5.0 milliwatts. These lasers are considered eye safe with *danger*.
- *Class IIIb*: Class IIIb applies to visible spectrum laser output between 5.0 and 500 milliwatts, as well as to invisible spectrum laser output of less than 500 milliwatts. These lasers produce sufficient output to cause damage to the retina, and must be labeled DANGER.
- *Class IV*: Class IV applies to visible and invisible lasers of 500 mW and higher. These devices must be labeled DANGER.

Laser pointers fall within Class II or Class IIIa. Note the Class IIIa Danger label on the laser pointer body shown in Figure 17–19.

FIGURE 17–19 Class IIIa laser pointer.

PHOTODIODE

17.9 Basic Operation and Construction

A photodiode is a light-receiving device that contains a semiconductor PN junction. Figure 17–20 shows a typical photodiode package. A glass window or convex lens allows light to enter the case and strike the photodiode mounted within the glass case.

FIGURE 17–20 Typical package of a photodiode.

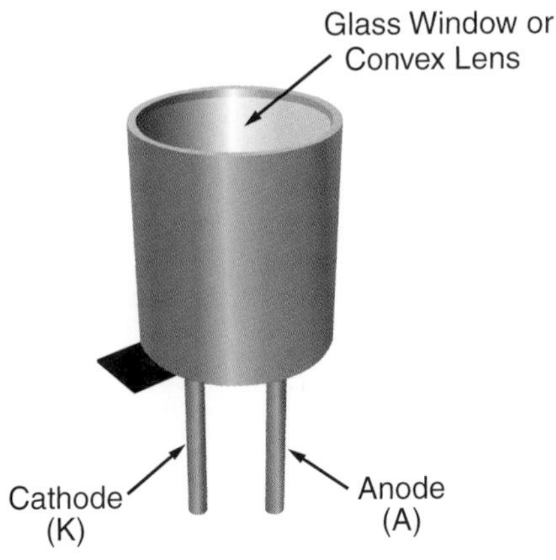

Figure 17–21 shows the construction of a photodiode. It is constructed basically with a P-type region that is diffused into the N-type region. The metal base makes connection between the cathode terminal and the N-type region, and the metal ring makes contact between the anode and the P-type region. Light enters the photodiode through a hole in the metal ring, which accommodates the glass or convex lens. Figure 17–22 shows the most commonly used schematic symbol of a photodiode. Note that the two arrows point toward the photodiode, indicating that it responds to light. Photodiodes can operate in two modes.

1. Photovoltaic mode
2. Photoconductive mode

FIGURE 17–21 Construction of the PN photodiode.

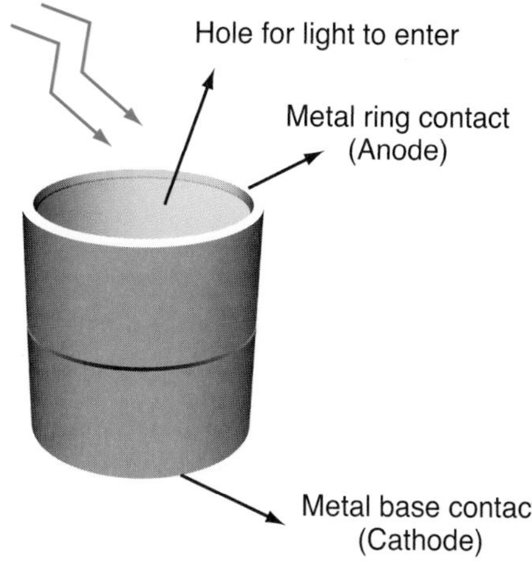

FIGURE 17–22 Schematic symbol of a photodiode.

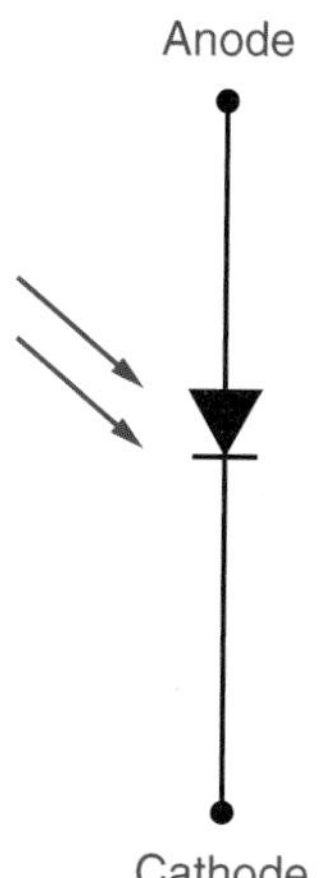

Photovoltaic Mode

When operating in the photovoltaic mode, the photodiode generates a voltage in response to light. The incidence of light on the photodiode creates electron-hole pairs. The electrons generated in the depletion region are attracted to the positively charged ions in the N-type material, and the holes are attracted to the negatively charged ions in the P-type material. This creates a separation of charges, and a small voltage drop of about 0.45 V is developed across the diode.

Figure 17–23 shows the photovoltaic mode of operation. In this mode of operation, the photodiode acts as a solar cell. If a load resistor is connected across the voltage source, a small current will flow from the cathode to the anode.

FIGURE 17–23 Photovoltaic mode of operation.

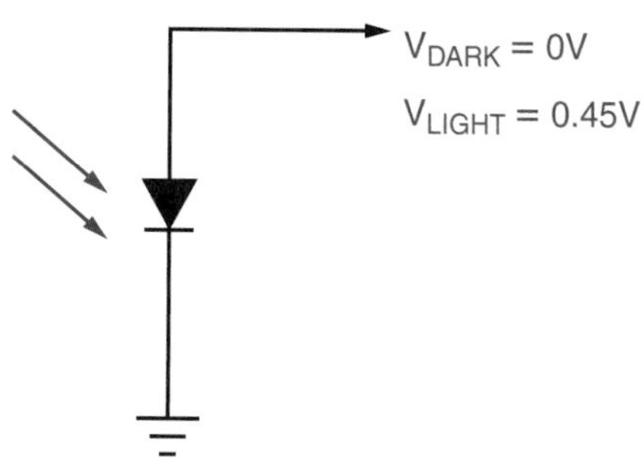

Photoconductive Mode

In the photoconductive mode, the conductance of the diode changes when light is applied. In this mode, the photodiode is reverse biased. Figure 17–24 shows a reverse-biased photodiode in the photoconductive mode. The depletion region of the reverse-biased photodiode is very wide and the resistance of the diode is high, and hence there will be only a small reverse current through it. This reverse current that flows through the diode when there is no light being applied is called dark current (I_D).

FIGURE 17–24 Photoconductive mode of operation.

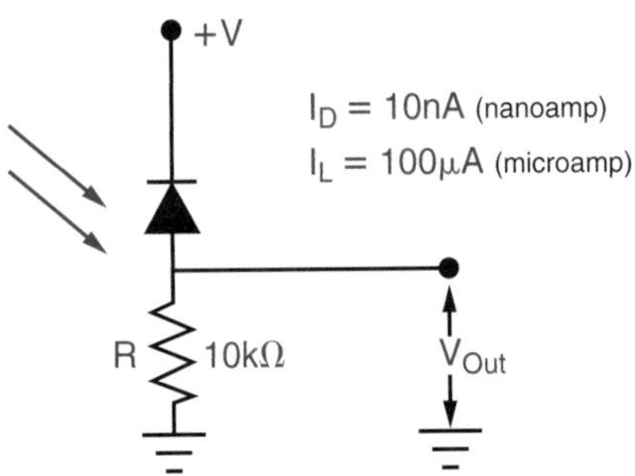

When light is applied, electron-hole pairs are generated. The electrons are attracted to the positive bias voltage, and the holes are attracted to the negative bias voltage. This movement of electrons and holes causes a considerable reverse current to flow through the photodiode. The resistance of the photodiode is very low when light is applied. If the intensity of light is increased, the resistance decreases and therefore the reverse current increases. The current that passes through the photodiode when light is being applied is called the light current (I_L).

The conductivity of the photodiode is low when there is no light applied, and the conductivity increases as the intensity of light increases. Consequently, the magnitude of the dark current is very much smaller than that of the light current. Consider the example shown in Figure 17–24. Assume that the dark current (I_D) flowing through the diode is 10 nA and the light current (I_L) is 100 mA. The output voltage (V_{OUT}) is dependent on the amount of current flowing in the circuit. With no light present,

$$V_{OUT} = I_D \times R = 10 \text{ nA} \times 10 \text{ k}\Omega = 100 \ \mu\text{V}$$

With light present,

$$V_{OUT} = I_L \times R = 100 \ \mu\text{A} \times 10 \text{ k}\Omega = 1 \text{ V}$$

17.10 Applications

Solar Cells

A solar cell is a photodiode operated in its photovoltaic mode. In its most common application, it is used to convert light energy to electrical energy and thereby serves as a DC voltage source.

In solar power applications, solar cells are connected in series/parallel arrays. Parallel connections increase the total current supplied by the array, whereas series connections increase the voltage.

For example, in Figure 17–25 if each solar cell is capable of delivering about 0.8 A of current at 0.5 V of voltage the entire 5 × 4 array would be capable of supplying a total of 4 A of current at a voltage of 2.0 V. The current generated in the solar panels can also be used to charge batteries during the day, so that the stored power can be consumed when it is dark. Solar cells thus can be used to store energy during the day to light streetlamps at night.

FIGURE 17–25 Solar panel.

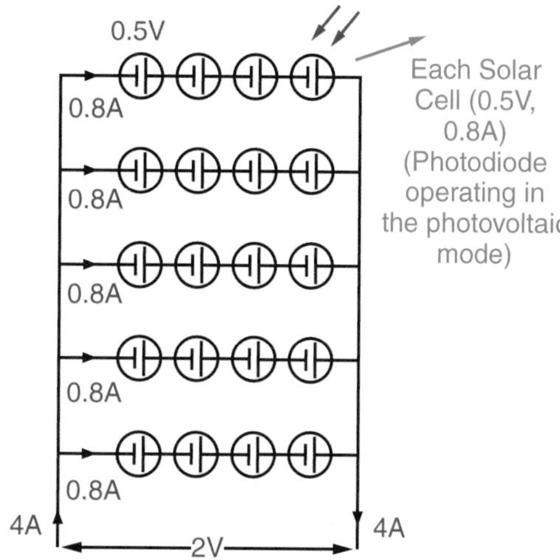

Optocouplers

Optocouplers are devices that use light to optically couple a signal between two electrically isolated points. Optocouplers are also called optoisolators. These devices include a light-emitting device and a light-sensing device, both available in one package. Figure 17–26 shows an optocoupler that uses an LED and a photodiode. The two devices are placed together in an opaque case. The photodiode operates in the reverse-biased photoconductive mode.

FIGURE 17–26 An optocoupler used to control load current.

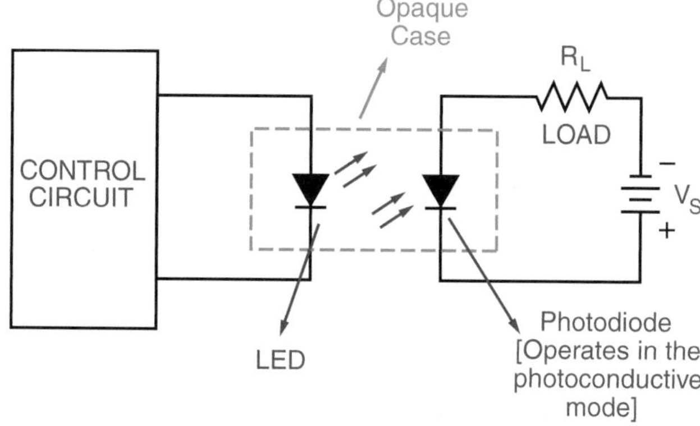

A control circuit forward biases the LED. When the LED lights up, the light strikes the photodiode, increasing its conductivity. This results in current flowing through the load resistor R_L. When the LED is turned off, the conductivity of the photodiode is low and there is no current through the load. The load current is controlled by the control circuit even though there is no direct electrical connection. Such a system is particularly useful for isolating control circuit noise from a sensitive electronic load circuit.

PHOTORESISTOR

17.11 Basic Operation and Construction

FIGURE 17–27

Schematic symbol of a photoresistor.

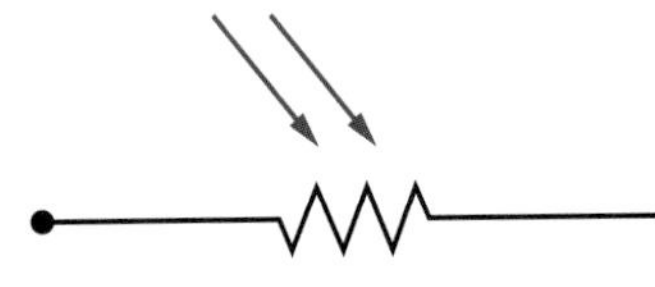

A photoresistor is a light-detecting device and is also called a photoconductive cell or light-dependent resistor. It is a passive device composed of a semiconductor material that changes resistance when its surfaced is exposed to light. The schematic symbol for a photoresistor is shown in Figure 17–27.

In conventional semiconductor devices, electron-hole pairs are created by heat energy. In photoresistors, the semiconductor material is light sensitive and free electrons are created by light energy. With the creation of free electrons, the resistance of the material drops. The greater the intensity of light the greater will be the number of free electrons and the smaller will be its resistance.

Photoresistors are the simplest and the least expensive in the class of optoelectronic devices. A typical construction is shown in Figure 17–28. The light-sensitive semiconductor material is arranged in a zigzag strip whose ends are attached to external terminals. The semiconductor material is either cadmium sulfide (CdS) or cadmium selenide (CdSe). A glass or transparent cover is attached for light to pass through to the device. The resistance of a photoresistor is largest when it is not conducting and is called the dark resistance.

FIGURE 17–28 Structure of a photoconductive cell.

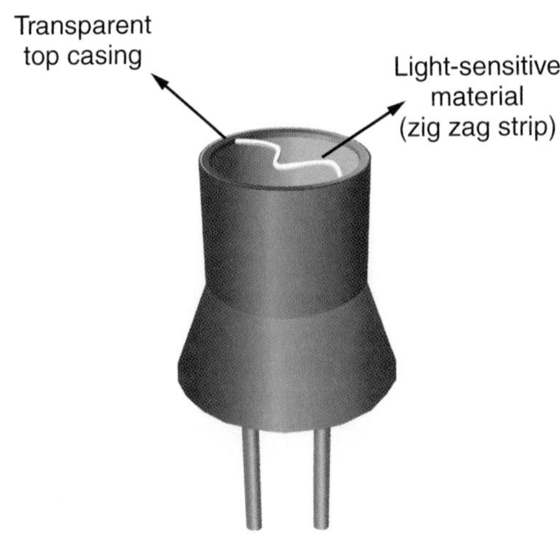

17.12 Applications

Photoresistors are used in lighting controls, automatic door openers, alarm-activating circuits, and other such applications. An example of a photoresistor streetlight control circuit is shown in Figure 17–29. The opening and closing of the relay contact of Figure 17–29 is controlled by the photoresistor. A minimum amount of energizing current flowing through the relay coil is required to activate the switching.

FIGURE 17–29 Photoresistor used to control a streetlight.

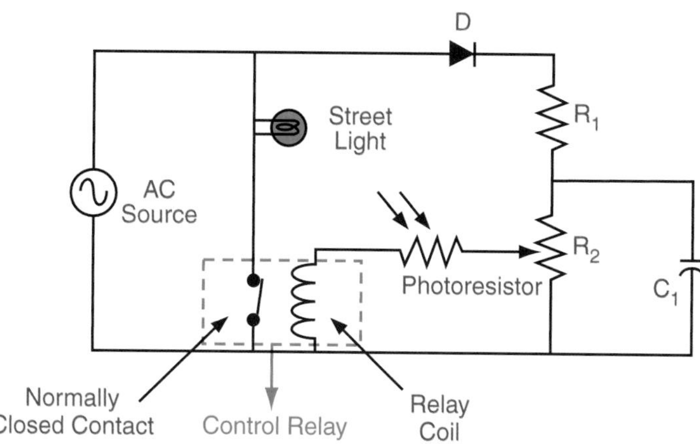

The relay contact is normally closed. At night, when there is no light incident on the photoresistor, its resistance is large, making the current through the photoresistor and the coil very small. The relay is therefore not activated, and the contact remains in its normally closed position. This causes the streetlight to be connected to the AC source.

However, during daylight (when there is sufficient light striking the photoresistor) its resistance drops, increasing the current through the coil. This current is sufficient to energize the relay coil, and therefore the relay contact changes its position from the normally closed position to the open position. This results in the streetlight being cut off from the source. Note that the diode D, resistors R_1 and R_2, and capacitor C_1 form a half-wave rectifier circuit with filter that supplies a DC voltage to the photoresistor and relay circuit.

OTHER OPTOELECTRONIC DEVICES

17.13 Phototransistor

A phototransistor is a light-detecting device that is also called a photo sensor. It is a transistor whose base current is supplied by the carriers generated due to the incident or striking light. The collector current of the transistor is thus controlled by the intensity of light. The schematic symbol of a phototransistor is shown in Figure 17–30.

FIGURE 17–30 Schematic symbol of a phototransistor.

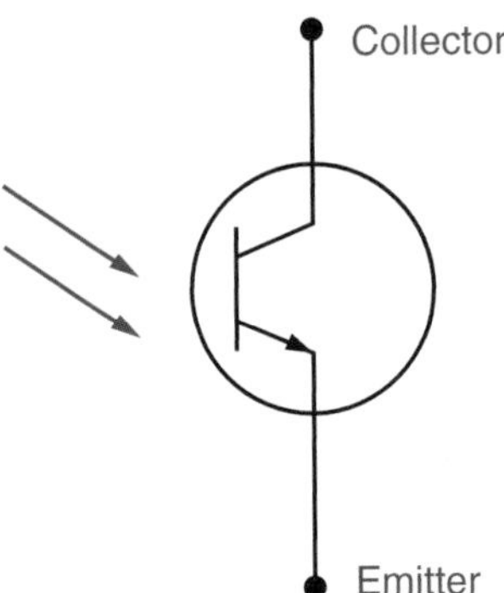

Although the phototransistor is a two-terminal device (like the photodiode), it produces a higher output current than the photodiode. The photodiode, however, has a faster response time. Thus, the phototransistor is preferred in high-current applications and the photodiode is used in high-speed operations.

17.14 Photodarlington

A photodarlington is a phototransistor packaged with another transistor connected in the Darlington configuration. The schematic symbol of the photodarlington is shown in Figure 17–31. Because of its large current gain, the photodarlington produces greater output current than either the photodiode or phototransistor. However, the response of the photodarlington is slower than both the photodiode and the phototransistor.

FIGURE 17–31 Schematic symbol of a photodarlington.

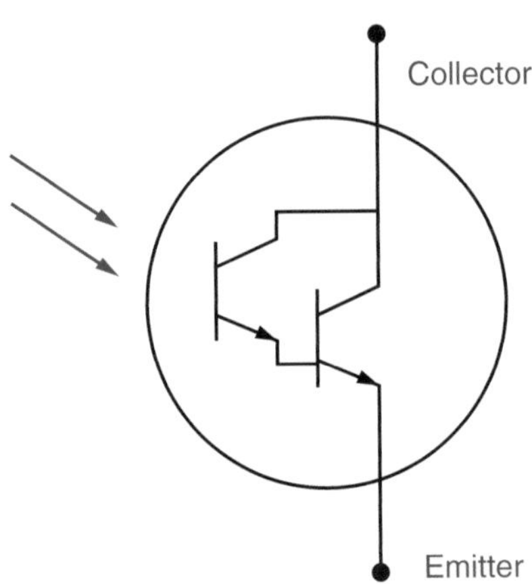

17.15 The LASCR

LASCR is an acronym of light-activated silicon-controlled rectifier. The incoming light strikes the photosensitive surface of the device, which serves as the gate signal that triggers the LASCR. The LASCR can be used in an optically coupled phase control circuit. Figure 17–32 shows the schematic symbol of the LASCR.

FIGURE 17–32 Schematic symbol of the LASCR.

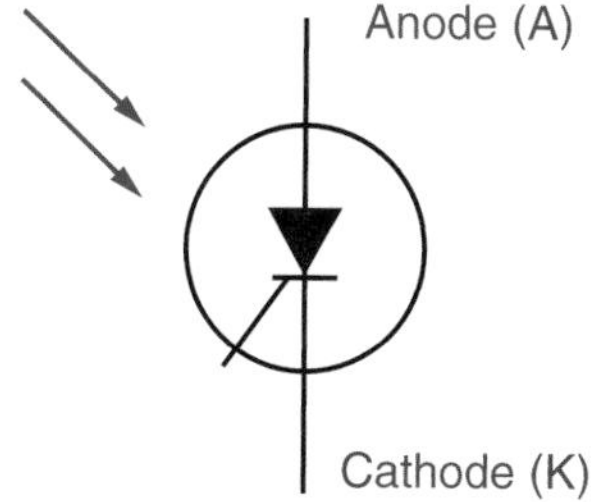

SUMMARY

This chapter introduces you to some of the most commonly used optoelectronic devices. The lesson should enable you to identify optoelectronic devices and describe their basic operations. The discussion in this chapter also acquaints you with applications of light-emitting and light-detecting devices. Some of the key topics in this chapter include:

- Light-generation devices: Those that create and/or modify light
- Light-detecting devices: Those whose characteristics are changed or controlled by light
- Light is a form of electromagnetic radiation or electromagnetic waves
- Wavelength is defined by the formula $\lambda = \frac{c}{f}$
- Light-emitting diode basic operation, construction, and identification
- Multicolor LED operation based on the polarity of the biasing
- Seven-segment, sixteen-segment, and 5×7 dot matrix displays
- LASER (light amplification by stimulated emission of radiation) operation and construction.
- Incoherent and coherent light sources
- Photodiode in both photovoltaic and photoconductive modes
- Voltage and current summation of solar cell arrays
- Input isolation of a control circuit by means of optocoupler devices
- Photoresistors and how their resistance varies with the intensity of light
- Phototransistor and photodarlington transistors versus photodiodes
- LASCR (light-activated silicon-controlled rectifier)

REVIEW QUESTIONS

1. What is the optical light spectrum?
   a. Frequency and wavelength
   b. Visible versus invisible
   c. Infrared and ultraviolet
2. The infrared band is given as 30 to 400 THz. What is this in wavelength? Answer the same question for visible light and ultraviolet light.
3. Define, in your own words, the differences between coherent and incoherent light.
4. Discuss the operation of the LED.
   a. How does it emit light?
   b. When does it emit light?
5. A certain piece of equipment shows a green light when it is working properly and a red light when it is not working properly. The light appears to come from the same device. How might this work?
6. A particular fiber-optic application will use a very long length of fiber-optic cable. Would this application use an LED or a laser diode? Why?
7. Discuss the two modes of operation of a photodiode.
8. A certain control circuit is being used to control the operation of a high-power motor starter. The motor starter generates enormous amounts of noise that might interfere with the control system. Discuss how the circuit might be designed to isolate the noise from the controls.

9. Redraw the circuit of Figure 17–33 so that it works with a normally open contact instead of normally closed.

10. You wish to operate a door opener with a light source. Draw a block diagram of how this might be done.

FIGURE 17–33 Photoresistor used to control a streetlight.

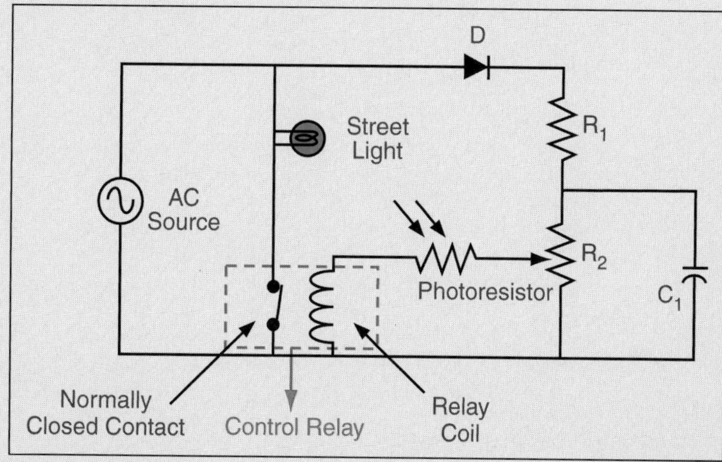

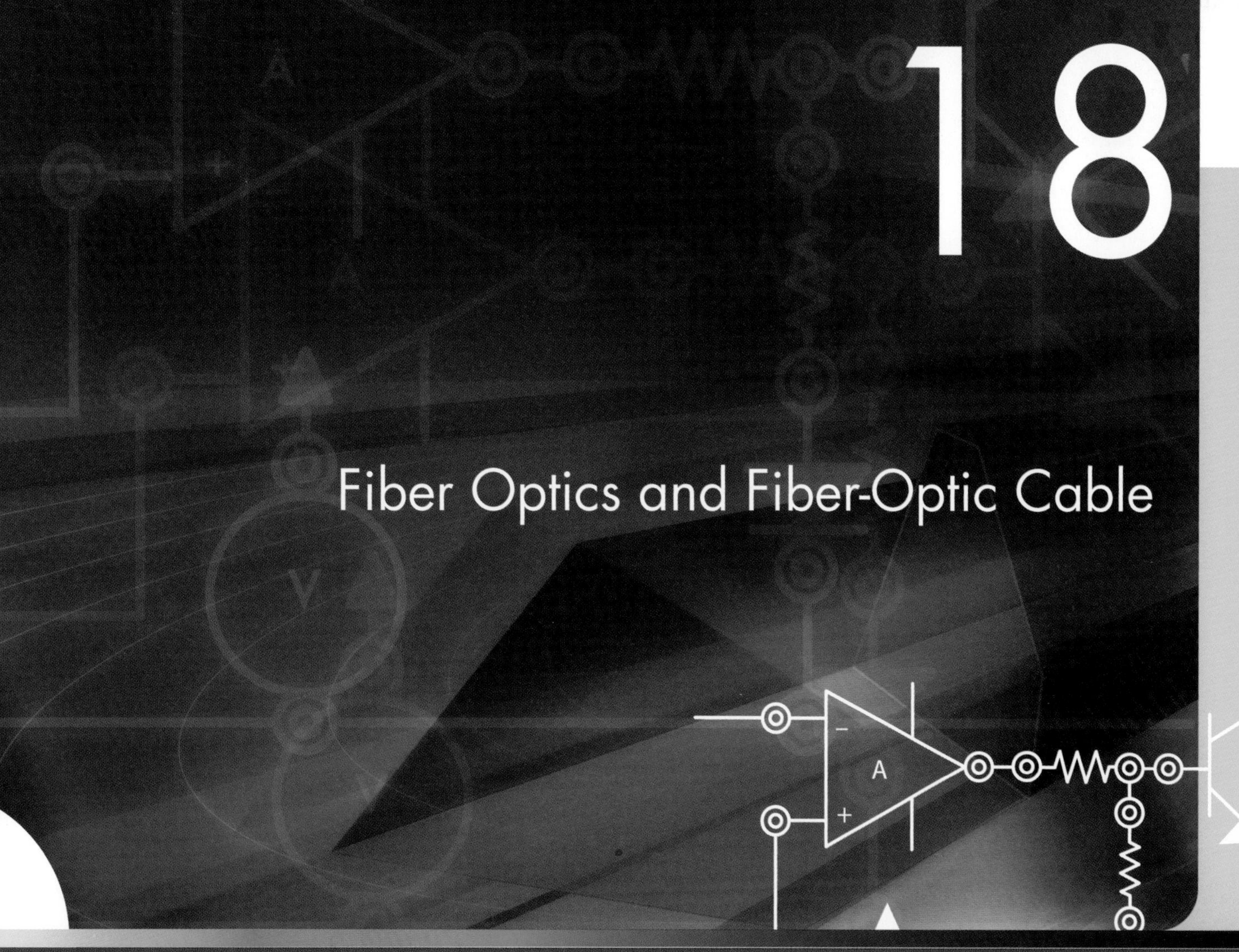

18

Fiber Optics and Fiber-Optic Cable

OUTLINE

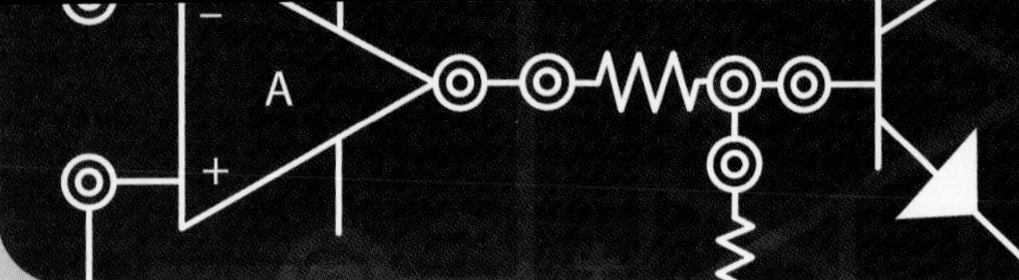

OVERVIEW

Light travels in a straight line when not bothered by an outside influence such as water, oil, or reflective surfaces. Microscopes, telescopes, and cameras are examples of devices that use the "straight line" properties of light. Other applications, such as periscopes, require that light be bent around corners or refracted at an angle.

William Wheeler patented the idea of "piping" light in 1880. His idea was to use a bright arc light and pipes with highly reflective inside surfaces. Figure 18–1 shows how he "piped" light to different rooms in a house using a single light source. Ideas and uses for fiber optics developed from these first inventions. However, until recently the use of light to carry information or data was limited to flashers and other such direct visual devices.

Then, in 1977 GTE and AT&T started using fiber cables to carry telephone signals on a specially modulated light beam. Today, fiber optics are used in virtually all areas of health, business, industry, communications, and government. Uses vary from internal patient examinations (**endoscopy**), machine process control, and monitoring, to satellite communications.

A fiber-optic cable is made of spun glass or a special transparent plastic. It does not conduct electricity, but rather it conducts "electrically encoded" light signals, such as those generated by a laser diode. Fiber-optic signals are virtually immune to electrical interference and noise caused by high-power applications or highly magnetic fields. Because they do not use an electrical signal, the shock hazard that might be present with copper wire is virtually eliminated.

OBJECTIVES

After completing this chapter, the student should be able to:

1. Describe the components of a fiber-optic system.
2. Explain the mechanism used by fiber optics to transmit electrical signals.
3. Identify types of fiber-optic cables and connectors.
4. Explain the uses and applications of fiber optics in an electrical system.

Endoscopy
A minimally invasive diagnostic medical procedure used to evaluate the interior surfaces of an organ by inserting a small scope in the body, usually through a natural body opening.

FIGURE 18–1 Patented light-piping system for lighting a house.

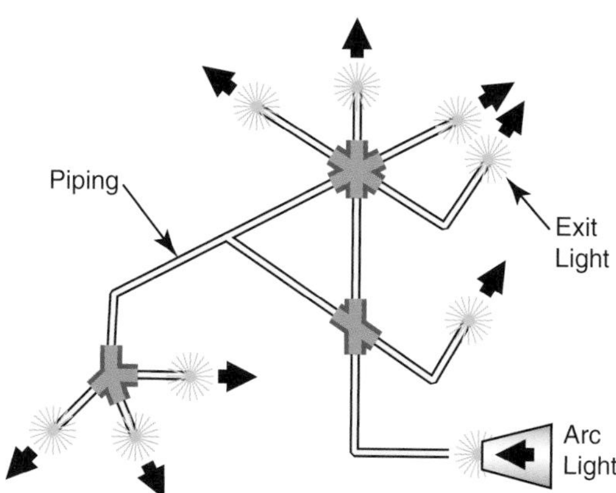

FIBER-OPTIC FUNDAMENTALS

18.1 Principles of Operation and Construction

Reflection and Refraction

In free space, light waves travel in a straight line at approximately 300,000,000 m/s (or about 186,000 mi/s). When light traveling in one medium (e.g., air) strikes a different medium (e.g., water or glass), some of it is rejected by the new medium and **reflects** back. The remainder of the light passes into the new medium. However, because light travels at a different speed in the new medium it is refracted (bent).

Reflect
To throw or bend back (light, for example) from a surface.

Consider Figure 18–2. An incident light beam is directed at the surface of a swimming pool. When the light beam strikes the water, some of it is reflected back, as shown by the reflected light beam. Note that the angle of incidence (45°) is equal to the angle of **reflection.**

Reflection
The act of reflecting or the state of being reflected.

FIGURE 18–2 Properties of reflected and under refracted light.

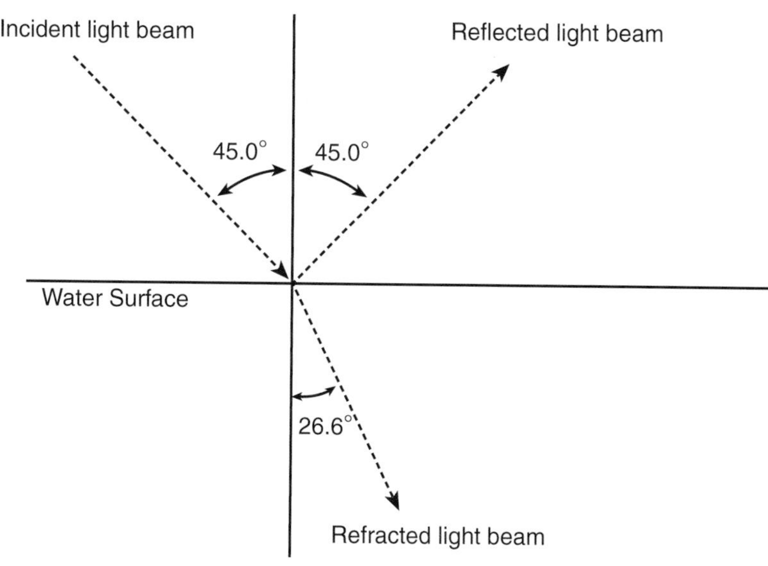

Refraction
The turning or bending of any wave, such as a light or sound wave, when it passes from one medium into another of different density.

Some of the light passes into the water. However, because the light travels more slowly in water than in air it is bent. The angle of **refraction** (26.6°) is different than the angle of incidence. The actual angle of refraction depends on the materials, the angle of incidence, and the refractive index of the material. [The refractive index (n) is given by the formula $n = \frac{c_v}{c_m}$, where c_v is the speed of light in a vacuum and c_m is the speed of light in the material it is entering.] For an example of refraction, think about looking at a fish from the docks of a clear lake. The actual location of the fish is deceiving to your eyes because light is refracting in the water.

Total Internal Reflection

Total internal reflection is the fundamental principle upon which fiber optics is based. When light crosses an interface into a medium with a higher index of refraction (n), it bends toward the normal. This is shown in Figure 18–2, where water has a higher index of refraction than air. Conversely, light traveling across an interface from higher n to lower n will bend away from the normal. This means that at some angle (known as the critical angle) light traveling from a medium with higher n to a medium with lower n will be refracted at 90°. In other words, it will be refracted along the interface.

If the light hits the interface at any angle larger than this critical angle, it will not pass through to the second medium at all. Instead, all of it will be reflected back into the first medium, a process known as total internal reflection. This means that light can be confined inside glass or any other transparent substance, provided the material has a higher index of refraction than the material that surrounds it.

In 1841, Swiss physicist Daniel Colladon developed an example of how this refractive index (material density) could contain a light signal. Figure 18–3 shows a drawing of the experiment he designed. The key to making this "total internal reflection" principle work was cladding. Cladding is a method of covering the glass fiber to ensure that all light is internally reflected. If the cladding has a lower refractive index than the carrier, total internal reflection will occur. The first successful glass cladding was accomplished in 1956 by Larry Curtiss of the University of Michigan. Figure 18–4 shows the difference between cladded and uncladded fibers carrying a light signal.

FIGURE 18–3 Total internal reflection.

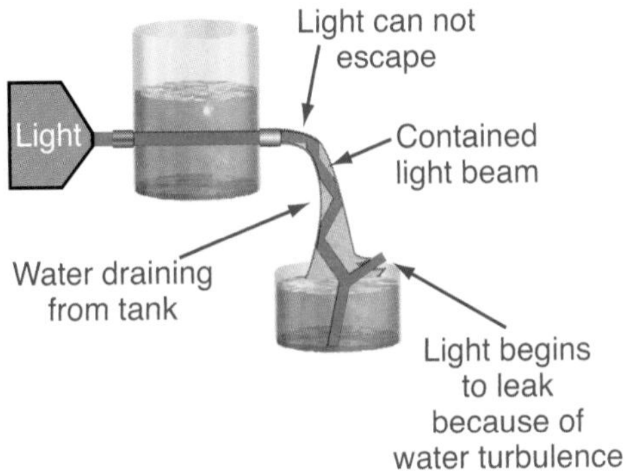

FIGURE 18–4 Cladded and uncladded fibers.

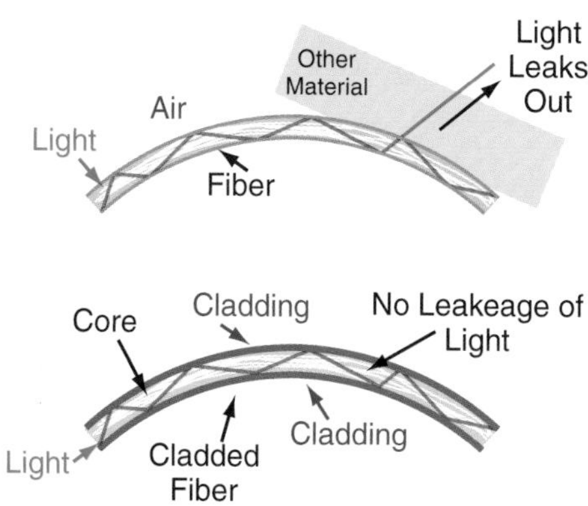

Optical Fiber Construction

The typical fiber-optic cable is made up of several components (see Figure 18–5). The core and the cladding are the components that actually channel the light. Ambient environment protection, moisture protection, and strength are provided by the other components.

FIGURE 18–5 Construction of fiber-optic cable.

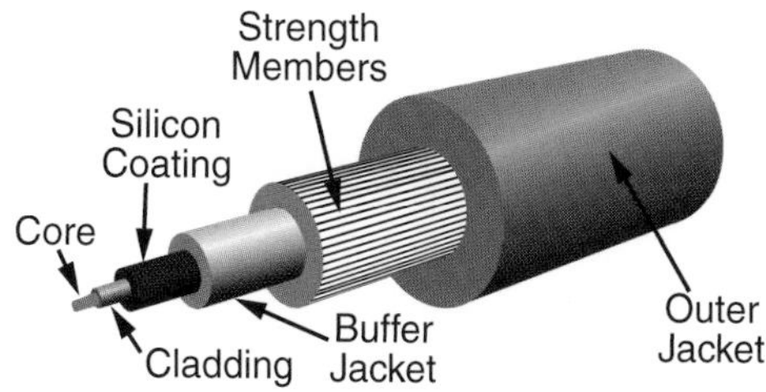

As discussed previously, the glass fiber itself operates on the principle of total internal reflection. Light entering the end of the glass (or plastic) fiber is carried through the glass and is received on the other end. The light is directed down the core. The cladding, which surrounds the core, is also glass but has a lower index of refraction and conducts no useful light.

The light-carrying portion (center) of the cable is very small and is typically measured in micrometers (μm). One common cable size, for example, is 62.5/125, which means that the core diameter is 62.5 μm and the cladding diameter (outside) is 125 μm. By comparison, a human hair is approximately 100 μm. Figure 18–6 shows a diagram of such a cable.

FIGURE 18–6 Internal reflection.

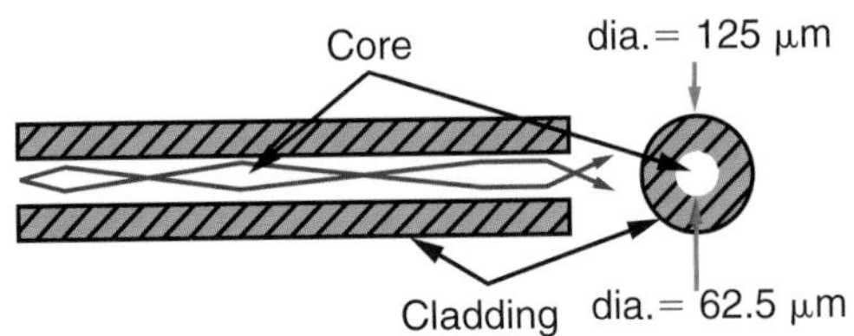

The difference in refraction between the core and the cladding is what allows fiber-optic cable to work. As long as the angle at which the light strikes the boundary (the angle of incidence) is greater than the critical angle, light striking the core-to-cladding boundary is reflected back into the fiber. The angle of incidence is determined by measuring the angle to the axis perpendicular with the surface of the fiber and angle of the incoming light (see Figure 18–7). The critical angle varies among different types of glass (indexes of refraction) and different wavelengths (color) of light. Note that the angle of reflection (∡R) is always equal to the angle of incidence (∡I).

FIGURE 18–7 Critical angle, angle of incidence, angle of reflection.

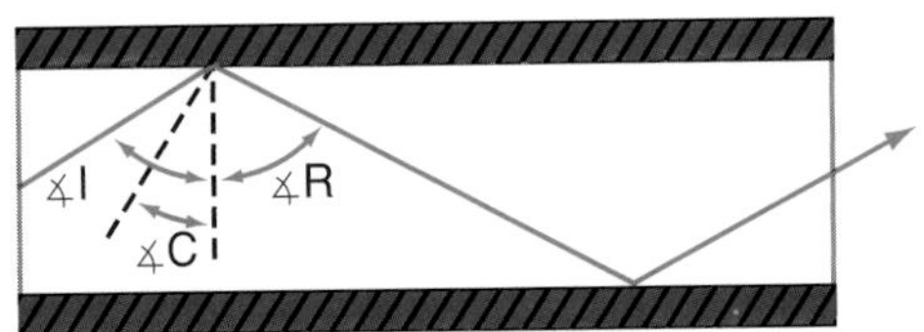

All light striking the core-cladding boundary will be reflected back into the core, as long as the angle of incidence (∡I) is greater than the critical angle (∡C). Optical fiber cables are designed in two broad categories: single mode and multimode. The physical difference between these two is primarily in the diameter of the core. A typical single-mode fiber, for example, is 8/125, where the core diameter is only 8 μm. Single-mode fiber is more expensive to install and requires more sophisticated equipment to communicate. However, single-mode fiber has considerably lower loss at its design frequency.

Normally, single-mode fiber is used only when signals are traveling long distances or when extremely low loss is required. The single-mode fiber core is so small that only a very few modes (rays) of light can travel down the fiber. This largely straight-shot effect results in minimum modal dispersion and a clean undistorted signal at the other end. Single-mode fiber is the most expensive configuration in regard to electronic components and terminations, but it provides the maximum amount of information.

Multimode fiber (which has a larger core) is also subdivided into two categories, depending on how the core-to-cladding boundary is implemented.

One type of fiber, called step-index multimode fiber, has a single transition between the core and the cladding glass. The light in the core reflects from the cladding and propagates (travels) along the fiber.

The drawback is that several light rays enter into the core at different angles, therefore traveling various paths and various distances. As illustrated in Figure 18–8, these different light rays arrive at different times due to their different distances traveled. The different arrival times at the end of the fiber cable lower the ability of the electronic equipment to decipher the signal, resulting in a slower communication rate.

FIGURE 18–8 Step-index multimode fiber.

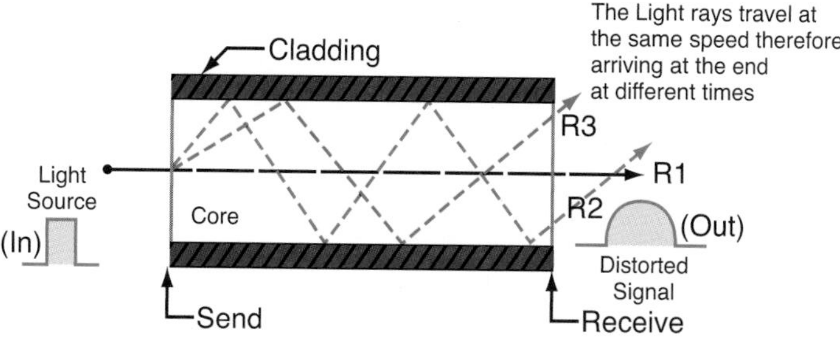

The second type of multimode fiber is the graded-index multimode fiber. Graded-mode fiber has a refractive index that changes gradually from the core to the cladding. As you will study in the next section, light rays will travel at different speeds through different levels of refractive indexes.

As indicated in Figure 18–9, different light rays entering into the core travel at different speeds throughout the length of fiber as they pass through gradual levels of refractive indexes. This has the effect of giving the slower modes a shorter distance to travel. The result is that all rays reach the other end of the fiber at nearly the same time, improving signal quality.

FIGURE 18–9 Graded-index multimode fiber.

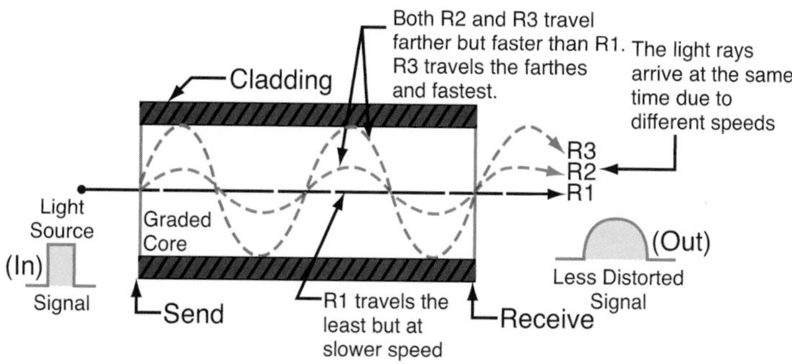

The fibers are made of pure glass. The glass may have small levels of doping called impurities to increase or decrease the refractive index. The glass is usually made of silicon dioxide (silica, SiO_2). Some special fibers are made of plastic. These fibers are less clear than glass but are more flexible and easier to handle.

Fibers usually come in bundles. Bundles are of two types: flexible or rigid. The flexible bundle is usually surrounded by a protective plastic coating, and at the ends of the cable the individual fibers are tied or joined together. In the rigid bundle, the individual fibers are melted together into a single rod and are shaped during the manufacturing process. Figure 18–10 shows a flexible fiber-optic bundle.

FIGURE 18–10 Fiber-optic bundles.

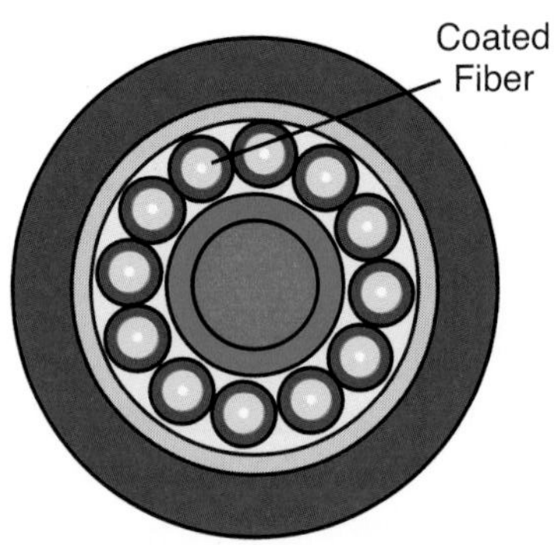

18.2 Optical Fiber Characteristics

To understand how a fiber-optic system operates, you need a fundamental knowledge of three areas: optics, electronics, and communications. In physics, light is treated as either electromagnetic waves or as photons (electromagnetic energy particles). For this discussion, we will concentrate on the electromagnetic wave characteristics of light. The light spectrum (light measured as a wave or electromagnetic frequency) is quite small when compared to the entire spectrum range. Figure 18–11 shows a chart of the electromagnetic spectrum. As you can see, there is only a small area of the spectrum we will consider when dealing with fiber optics, the optical spectrum from infrared to ultraviolet frequencies.

FIGURE 18–11 Electromagnetic spectrum.

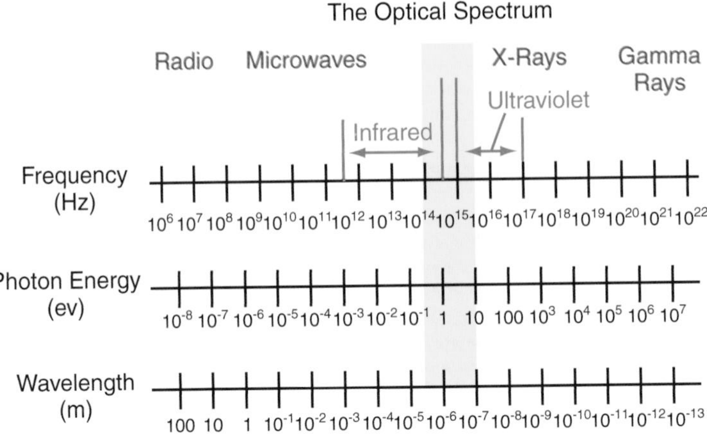

As indicated earlier in this chapter, light bends as it passes from one material to another. The bending is caused by the change in speed. The change in speed is the way the refractive index is calculated, as given by the formula:

$$n = \frac{c_v}{c_m}$$

Where:
c_v is the speed of light in a vacuum
c_m is the speed of light in the other material
This bending is shown in Figure 18–12.

Another critical characteristic of an optical fiber is its acceptance angle, related to the **Cone of Acceptance** (see Figure 18–13). Only signals that enter the core within the Cone of Acceptance will be transmitted through the optical fiber. Optical signals entering from outside the Cone of Acceptance will be reflected back to the source, or will pass through the core into the cladding. The acceptance angle, or **Angle of Acceptance**, is half of the Cone of Acceptance. Another term used for determining the Angle of Acceptance is **Numerical Aperture** (NA). NA is the sine of the Angle of Acceptance. To find the Angle of Acceptance, NA must be determined by using the following formula:

Cone of Acceptance
The angular arc over which an optical signal entering a fiber-optic cable will successfully propagate through the cable.

Angle of Acceptance
One-half the Cone of Acceptance.

Numerical Aperture (NA)
The sine of the Angle of Acceptance.

$$NA = \frac{\sqrt{n_0^2 - n_1^2}}{n_2}$$

Where:
n_0 is the core refractive index
n_1 is the cladding refractive index
n_2 is the input medium refractive index

Note: If the input medium is air, n_2 would have a value of 1 and could be ignored. NA is simply the sine of the Angle of Acceptance θ. Therefore:

$$\theta = \text{ArcSin NA}$$

Typically, NA values vary from 0.2 to 0.5. This results in Angles of Acceptance ranging from 11.5° to 30°. The following example shows how to find NA and Angle of Acceptance for an optical fiber cable.

FIGURE 18–12 Light refraction through glass.

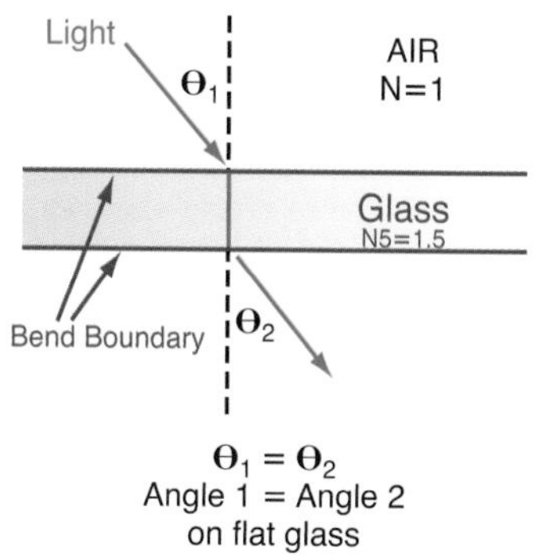

FIGURE 18–13 Optical fiber's acceptance angle.

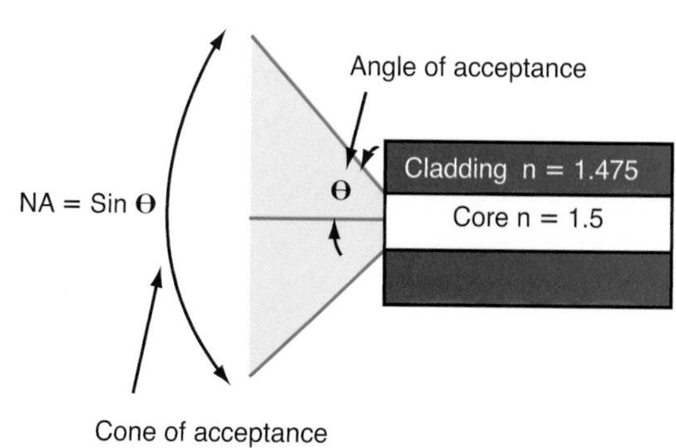

Example

The glass core of a fiber cable has an n of 1.5. Its cladding has an n of 1.475. The input medium is air. Using the previous formulas find the NA and angle of acceptance (θ) for the fiber.

$$NA = \sqrt{n_0^2 - n_1^2} = \sqrt{1.5^2 - 1.475^2} = \sqrt{2.25 - 2.18} = \sqrt{0.7} = 0.26$$

$$\theta = \text{ArcSin } 0.26 = 15°$$

FIBER-OPTIC SIGNALS

18.3 Signal Transmission

From the discussion on NA (numerical aperture), you might guess that light transmission is not 100% efficient. The signal is attenuated because of light loss. This is caused by a number of factors, including:

- Light scattering in the core
- Light leakage from the core to the environment
- Light absorption by the fiber material

Attenuation
Reduction of signal strength, usually through losses.

Recall that **attenuation** is loss of signal strength. To measure the signal strength loss, we use a familiar unit of measurement, the decibel (dB). The two formulas that relate decibels to power levels and power levels to decibels are:

$$dB_{loss} = 10 \times \log_{10}\frac{P_{OUT}}{P_{IN}}$$

$$\frac{P_{OUT}}{P_{IN}} = 10^{\frac{dB}{10}}$$

For instance, if a 1 mile length of fiber cable has a 10 dB/mile loss the amount of energy output compared to the input can be calculated as

$$\frac{P_{OUT}}{P_{IN}} = 10^{\frac{dB}{10}} = 10^{\frac{-10}{10}} = 0.1$$

Note that the dB loss is given a minus sign (−). From this formula, you can see that only 10% of the light entering the cable will come out the other end. If you add another mile to the cable and assume the same loss rate, only 1% of the signal is present at the end of 2 miles. Expressed another way, there was a 20 dB loss over the 2 mile cable length.

Low attenuation or signal loss is possible at frequencies of light near the infrared spectrum (1,300 nm to 1,550 nm). When transmitting a light signal at the high end of this frequency, 1% of the light is still available after 50 miles. Electric wires (copper) and coaxial cable have higher attenuation rates as the frequency of the electrical signal increases. This is not the case for the optical fiber.

Figure 18–14 shows a comparison graph between the coaxial cable and fiber-optic cable. The reason light loss (attenuation) is unaffected by signal frequency changes is that the light is modulated to carry the signal but the light essentially operates within its frequency spectrum and maintains a constant attenuation or loss rate. This is not true of copper or coaxial cable. At higher electrical frequencies, signal loss is caused by counter-electromagnetic fields, heat loss, and increased skin resistance.

FIGURE 18–14 Different cable signal losses.

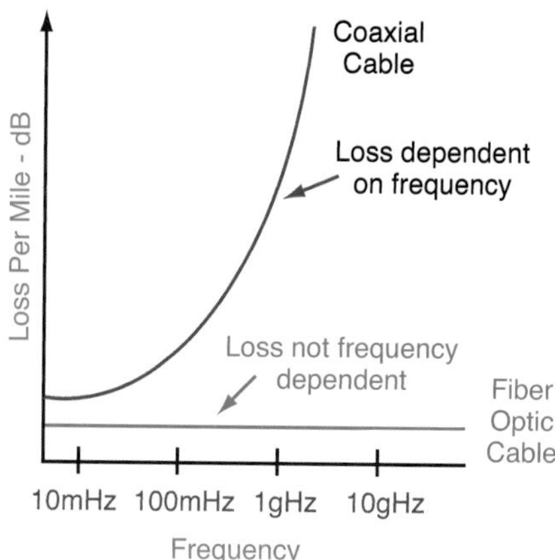

18.4 Signal Modulation

Recall that there are two basic types of signal **modulation.** One type varies the intensity or amplitude of the carrier to encode the signal. The other type varies the frequency of the carrier to encode the signal. For radio or TV, the carrier is typically a single frequency transmitted steadily. For optical signals, it is a beam of light. Both the electrical signal and the light signal can be modulated using an analog or digital signal. Today, most fiber-optic systems use digital modulation.

Modulation
To attach information to a signal by varying its frequency, amplitude, or other characteristic. The information is attached at a transmitter and recovered at the receiver.

To transmit a fiber-optic signal, the intensity of the light source must be modulated. This is done in one of two ways. The input power to the light source can be changed, and thus the intensity of the light beam is changed. Alternatively, the intensity of the light beam can be changed after it leaves the light source. Figure 18–15 shows the difference between an analog and a digital signal.

FIGURE 18–15 Analog and digital signals.

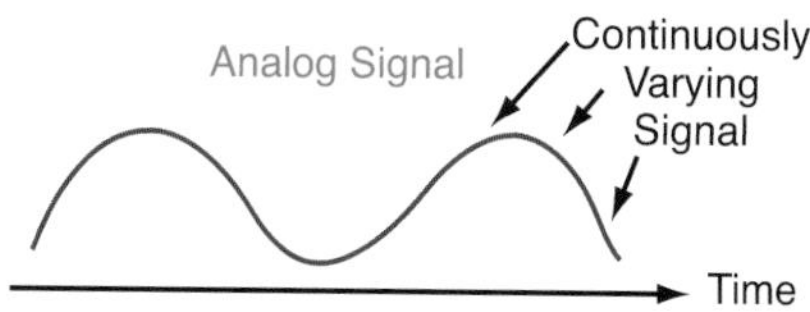

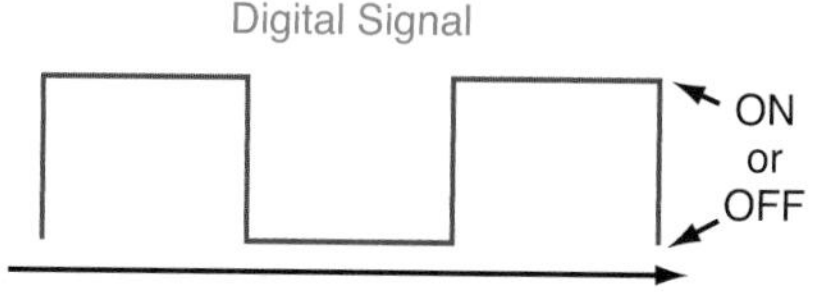

Light Sources

There are three major types of commercially used light sources for fiber-optic signal generation, as outlined in Table 18–1.

TABLE 18–1 Light Sources for Fiber-Optic Systems

Type	Use
Visible light red LEDs	Used with plastic fibers because plastic transmits the visible light wavelengths better than the glass. LEDs are used with systems that transmit short distances at slow to medium speeds.
Near-infrared LEDs and gallium arsenide (GaAs) lasers	These LEDs and lasers use glass fibers and are good for short distances and medium data speeds.
Indium gallium arsenide phosphide (InGaAsP) lasers	These lasers are the most common for telecommunications and transmit at 1,300 nanometers, with a dB loss of only 0.35 dB/km.

Direct Modulation

Direct modulation is very simple and works very well for LEDs and semiconductor lasers. The principle of operation is based on changing the current through the semiconductor. Turning the semiconductor on and off with the digital signal creates a series of light pulses that are identical to the electrical pulses. Figure 18–16 shows how this relationship works.

FIGURE 18–16 Input current versus output light.

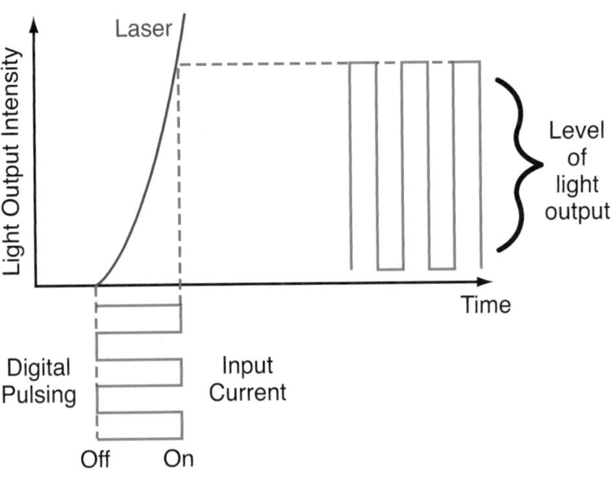

FIBER-OPTIC SYSTEMS

18.5 System Overview

Figure 18–17 shows a block diagram of a simple fiber-optic system. Note that the incoming electrical signal is converted to a light signal, transmitted a distance, received, and converted back to an electrical signal. Fiber-optic cables used for data transmission typically carry light signals at levels of 100 microwatts or less.

FIGURE 18–17 Simple fiber-optic system.

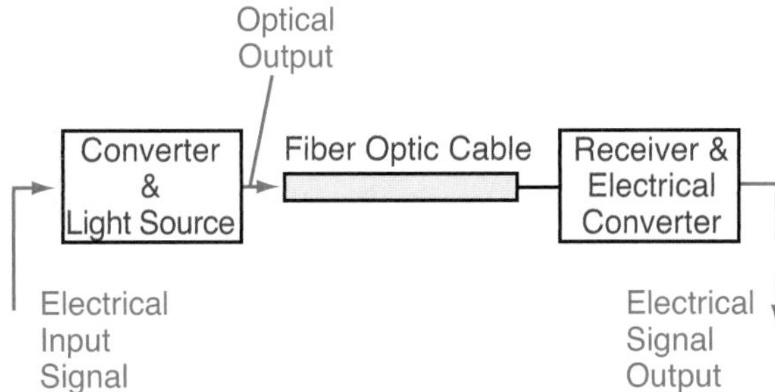

Fiber-optic system components that generate light are called light emitters. LEDs and semiconductor lasers are light emitters. Light receivers and devices that convert light back to electrical energy are called light detectors or photodetectors. Photodiodes, phototransistors (photodarlington), and light-activated SCRs are some of the common photodetectors. Figure 18–18 shows the schematic symbols for some common photodetectors.

FIGURE 18–18 Schematic symbols for photodetectors.

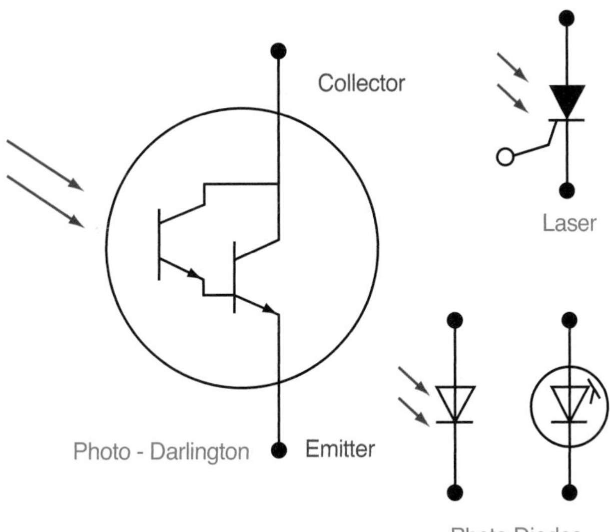

One application of coupling the output of an LED to the input of a phototransistor (optocoupling) is shown in Figure 18–19. Here, the opaque case (opaque means that light is not allowed to penetrate the case) allows the light from the LED to activate the phototransistor only when the LED is forward biased by the incoming electrical signal. When the phototransistor receives light from the forward-biased LED, it becomes forward biased and conducts. The output current is then received by the next amplifier stage and an electric output signal is generated.

FIGURE 18–19 Optocoupling.

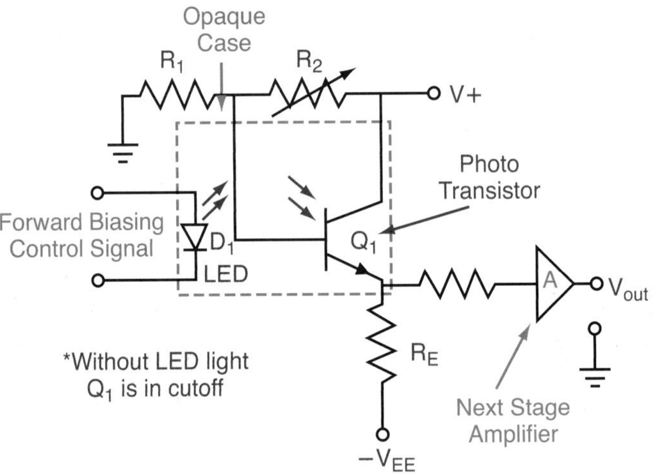

18.6 Optoisolators and Optointerrupters

Two special configurations for optocouplers are optoisolators and the optointerrupters. Optoisolators are six-pin DIPs (dual in-line packages), as shown in Figure 18–20. The circuitry within the optoisolator is shown in Figure 18–19.

FIGURE 18–20 Typical six-pin DIP.

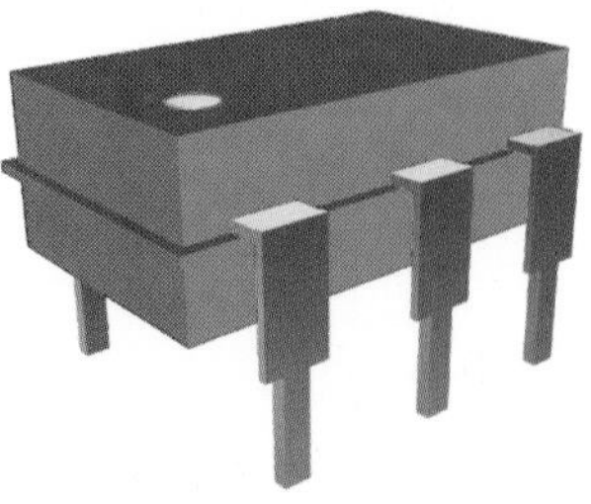

A common application of the optoisolator is that of a solid-state relay. This type of relay uses a DC input voltage to pass or block an AC signal. Because the isolation medium is light, the optoisolator can be designed to attain an equivalent isolation rating of several thousand volts.

The optointerrupter is used as an optical switch. The optointerrupter is designed to have an external object block the light beam path between the photoemitter and the photodetector. A common case or physical construction of the optointerrupter is shown in Figure 18–21a. The write-protect tab on a 3.5-inch floppy drive serves the same purpose. When it is in place, it blocks the light signal. When not in place, the disk is write protected because the sensing light is allowed to activate the detector. Another typical application of the optointerrupter is determination of rotating shaft rpm (see Figure 18–21b).

FIGURE 18–21 (a) Optointerrupter casing (slotted optical switch). (b) Optointerrupter applied to motor speed measurement.

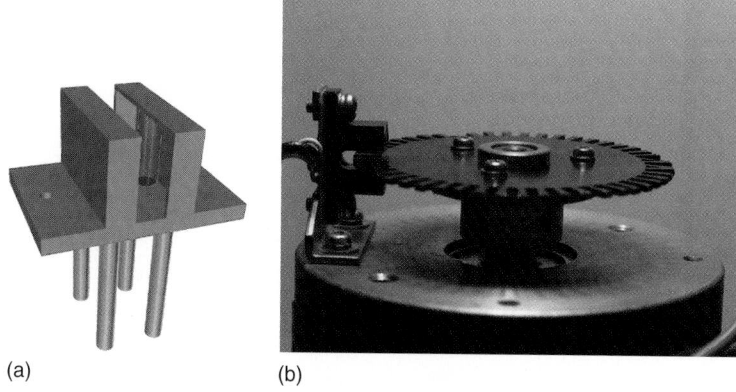

Example

A spinning disk having 10 notches around its perimeter is mounted to a rotating motor shaft. As the disk spins, it makes/breaks the optical path of the optointerrupter. Find the motor rpm if the optointerrupter output pulse frequency is measured to be 1,080 kHz.
RPM = (Frequency ÷ 60 seconds) ÷ Disk slots
RPM = (1,080 kHz ÷ 60 seconds) ÷ 10 slots
RPM = 1,800

FIBER-OPTIC CABLE

18.7 Cable Construction

Optical fiber cable cannot be made from just pure silica (glass). Recall that fibers require at least two refractive indexes, one for the core and one for the cladding. Pure silica has just one refractive index. Doping with impurities is required to get a different cladding index. The most common doping element is germanium. Germanium has very low light absorption and also forms a glass.

The actual fiber is constructed by placing a rod of high-index glass into a tube with a lower refractive index. These two are melted into each other to form one rod. This new rod is called a preform. The rod is heated at one end, and a single very thin fiber is drawn from it. Some plastic fibers are used in communication. Their primary advantages are lighter weight, lower cost, and greater flexibility. Their major disadvantage is higher signal loss. They are used for short communication links such as in office buildings and cars.

Another limitation of plastic fiber-optic cable is the high degree of degradation over time in environments that have high operating temperatures. This is true of sensors and monitoring processes in industrial manufacturing. In reality, fiber-optic cables look much like conventional metal cables. Polyethylene is used on both fiber-optic and metal outdoor cables to protect them against the environment. In theory, fiber cables are stronger than their copper cable counterparts.

When pulled, the fiber cable will not stretch like copper but will extend a little. When released, it will spring back to its original size. A fiber cable will stretch only about 5% before breaking. However, a copper cable can stretch as much as 30% before breaking. Figure 18–22 shows a cross-sectional view of different grades of fiber-optic cable. There are many fiber cable types and specifications. Table 18–2 outlines *NEC*® cable specifications.

FIGURE 18–22 Types of fiber-optic cable construction.

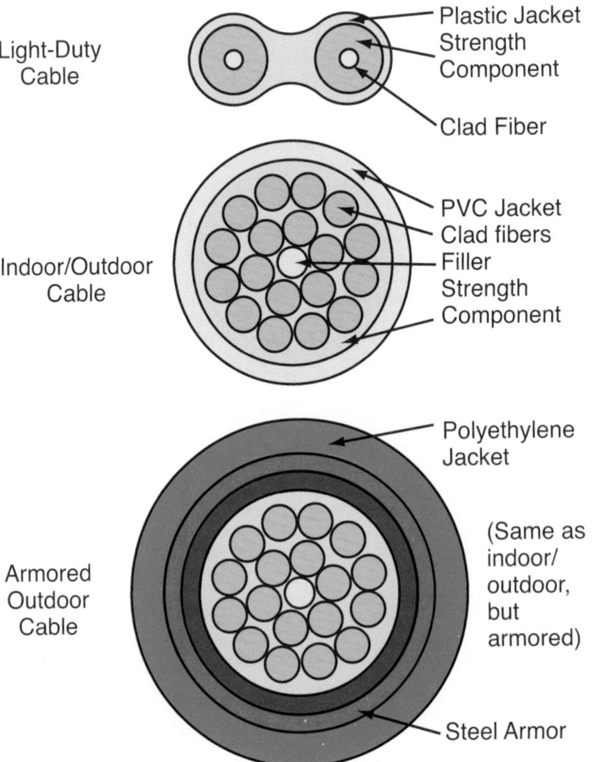

TABLE 18–2 Cable Specifications Under U.S. National Electrical Code

Cable Type	Description	Designation	UL Test
General-purpose (horizontal) fiber only	Nonconductive optical fiber cable	OFN	Tray/1581
General-purpose (horizontal) hybrid (fiber/wire)	Conductive optical fiber cable	OFC	Tray/1581
Riser/backbone fiber only	Nonconductive riser	OFNR	Riser/1666
Riser/backbone hybrid	Conductive riser	OFCR	Riser/1666
Plenum/overhead fiber only	Nonconductive plenum	OFNP	Plenum/910
Plenum/overhead hybrid	Conductive plenum	OFCP	Plenum/910

18.8 Cable Connectors

Single-fiber connections are, of course, easiest. However, seldom do you find just one fiber to connect to another. The main problem with connections is that of alignment. Figure 18–23 illustrates what happens when connecting fibers are misaligned.

FIGURE 18–23 Misaligned fiber-optic cables.

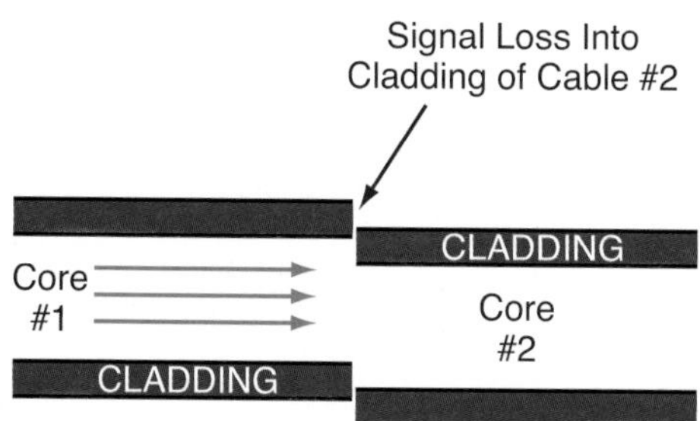

During the past 10 years, many of the industry associations and the telecommunications industry have tried to standardize connector types. The following are some examples of the types of connectors now in use:

- The snap-in single-fiber connector (SC) was developed by Nippon Telegraph and Telephone of Japan. It uses a cylindrical ferrule that holds the fiber and plugs into a coupling receptacle or housing (see Figure 18–24a).
- The twist-on single-fiber connector (ST and FC) was originally designed for copper cable connections. It also uses a cylindrical ferrule. However, the ferrule connects with an interconnection sleeve by twisting into a locked position. The FC type uses threads and is screwed into place instead of twisted into place. See Figure 18–24b.

- Multifiber connectors use an MT ferrule. This connector can align up to 12 fibers. This connector is used with optical fiber ribbon cables (see Figure 18–24c).

FIGURE 18–24 Fiber connectors: (a) snap-in single-fiber connector, (b) twist-on single-fiber connector, (c) multifiber connector.

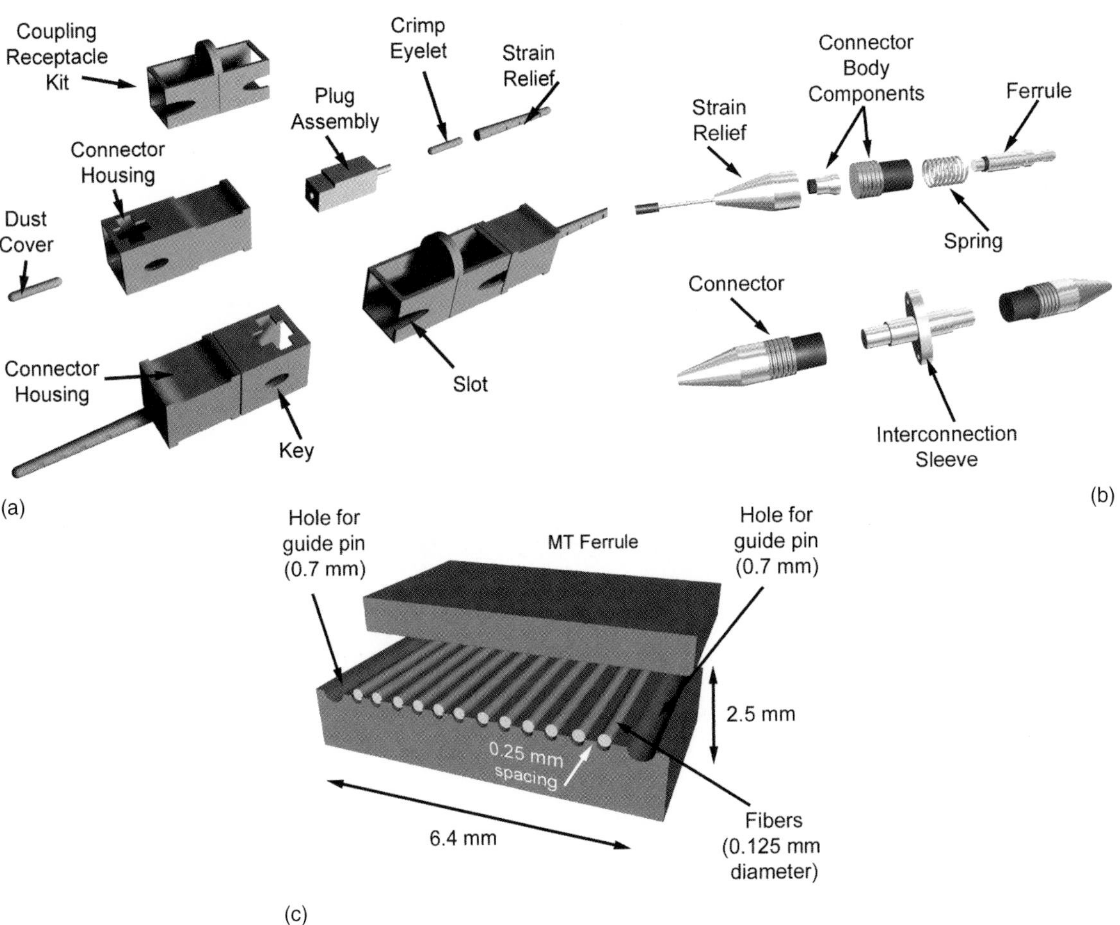

Figure 18–25a shows rack-mounted fiber optic cable connector panels, where simplex fiber (single strand) cables are terminated. Figure 18–25b shows an uninstalled fiber-optic termination panel. Figure 18–25c shows duplex and simplex adapters used for splicing fiber-optic cables.

Figure 18–25d shows the component parts of a PANDUIT duplex LC connector assembly. LC connectors are latched connectors, as opposed to snap, twist, or bayonnet connectors. Figure 18–25e shows a PANDUIT Field Polish Fiber Optic Termination Kit for use in terminating LC multimode and single-mode connectors.

FIGURE 18–25 (a) Assembled Panduit Fiber Optic panel. (b) Panduit Opticon LC fiber optic adapter panel. (c) Panduit LC fiber optic adapters. (d) Panduit Duplex LC connector assembly. (e) Panduit LC field polish optic terminator kit. All photos courtesy of PANDUIT Corp.

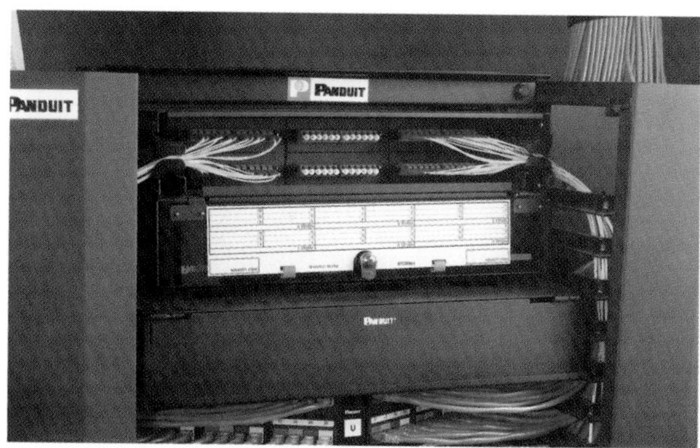

(a)

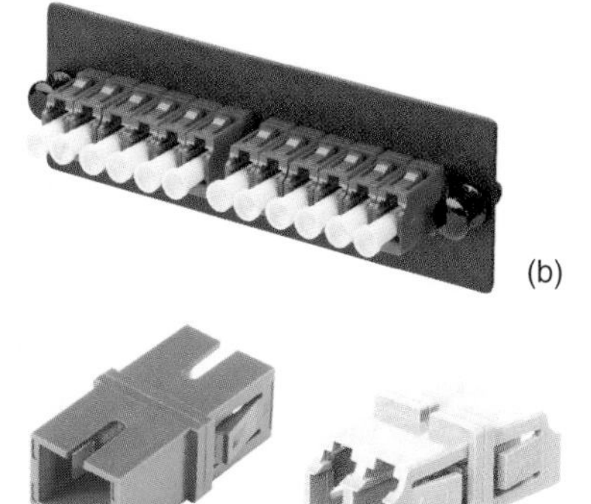
(b)

(c)

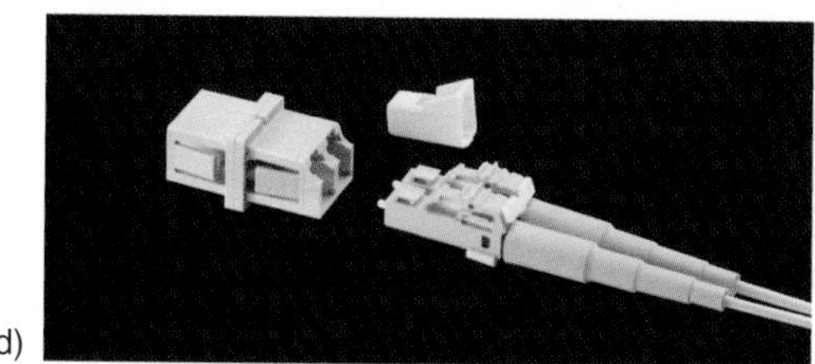
(d)

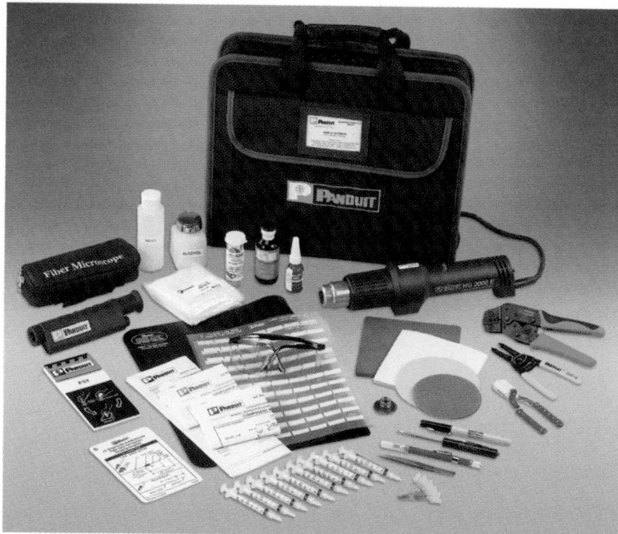

(e)

SUMMARY

The original idea of "piping light" for the purpose of room illumination has given way to the use of light for information transfer. The ability to put information on a light signal and then send it down a fiber-optic channel has started to open so-called wide-bandwidth information channels that provide everything from industrial data and process control to television and communications channels into our homes.

The entire process works because of the fact that light will reflect and/or refract from some surfaces. This allows the light to be "bounced" down a long fiber strand, emerging from the remote end with little or no attenuation compared to other information-transfer methods.

Available in both multimode and single-mode construction, fiber-optic cable has the ability to transmit information for many miles from its source. The original information is captured in electronic form and then converted to modulated light impulses by light sources such as visible red LEDs, near-infrared LEDs, and lasers, either GaAs or In GaAsP semiconductors.

The source then injects the light into the fiber-optic cable, and it travels to the end, where an optocoupler changes the light back into electrical signals for processing. The low attenuation, insensitivity to interference, high data rates, and safer (no electrical shock hazard) installations all combine to make fiber-optic cable the basis of the future of data transmission.

REVIEW QUESTIONS

1. Diagram a cross-section of fiber-optic cable showing:
   a. The core
   b. The outer jacket
   c. The cladding
   d. The strength members
   e. The silicon coating
   f. The buffer jacket
2. Define the following terms:
   a. Angle of incidence
   b. Angle of reflection
   c. Critical angle
   d. Attenuation
3. What is the refractive index? How does it affect the transmission of light down an optical fiber?
4. The acceptance angle of a particular optical fiber is 20°. What will happen to an optical signal that strikes the input of the cable at an angle of 25°?
5. If you are standing on the edge of a swimming pool and see a quarter on the bottom, should you dive straight toward it to recover it? Why or why not?
6. What is modulation?
7. If two fibers are not aligned correctly when they are spliced, what will happen to the signal?
8. In one particular fiber-optic cable, the refractive index of the core (n_0) is 1.25. The refractive index of the cladding (n_1) is 1.225. What is the acceptance angle (NA) of this cable?
9. Discuss the three types of connectors described in the chapter. What are the advantages and disadvantages of each?
10. In electric power distribution systems, fiber-optic cable is often routed along with electric power conductors. Sometimes (in utility overhead line applications, for example), it may be actually wrapped around the energized electric conductor.
    a. What are the advantages to such a practice? Disadvantages?
    b. What safety issues might be involved in such practices?
    c. Why is the use of electric signal cables limited in such applications?

GLOSSARY

Active component Components of an electronic circuit that use a power source to process a signal. The processing usually involves amplification or some other change in the signal that requires additional power. BJTs, FETs, and UJTs are examples of active components.

Active filter A frequency-selective electronic filter utilizing amplification, as opposed to passive frequency-selective filters.

Addend Any of a set of numbers to be added.

A_I Current gain.

Alpha A non-dimensional number representing the ratio of collector current to emitter current.

Amplification The process of increasing the voltage, current, or power of a signal.

Amplifier A device that provides gain without much change in the original signal waveform.

Amplify To make larger or more powerful.

Amplitude The size of a signal. Most commonly, the amplitude is expressed in terms of the signal voltage.

Amplitude modulation A system of attaching information to a single-frequency carrier wave. The information is included by varying the amplitude (magnitude) of the carrier wave.

Analog Referring to a class of circuits in which output varies as a continuous function of input.

Angle of Acceptance One-half the Cone of Acceptance.

Antiparallel Two circuit elements connected in parallel with opposite polarities.

A_P Power gain.

Astable A circuit that alternates automatically and continuously between two unstable states at a frequency dependent on circuit constants; for example, a blocking oscillator.

Attenuation Reduction of signal strength, usually through losses.

Augend The first in a series of addends.

Automatic frequency control (AFC) Circuitry incorporated into most frequency modulation receivers to keep the receiver local oscillator from drifting off frequency.

Automatic gain control (AGC) Circuitry incorporated into most AM and FM receivers that automatically limits incoming signal strength to prevent overloading problems such as variable video contrast and sound volume levels.

Automatic volume control (AVC) See AGC. Only used for voice frequency AM radios.

A_V Voltage gain.

Avalanche voltage See Reverse breakdown voltage.

Base The number of digits or symbols used in a number system.

Beta The ratio of DC collector current to base current.

Bias A current or voltage applied to a semiconductor device to obtain a specific result, such as conduction.

Bipolar junction transistor (BJT) A three-terminal semiconductor device used for the control and amplification of signals in electronic circuitry. The most common of all types of semiconductors.

Bistable Able to operate steadily in either one of two states. Will not leave one state until triggered to do so.

Bistable multivibrator A multivibrator in which either of the two active devices may remain conducting (with the other nonconducting) until the application of an external pulse. Also known as Eccles-Jordan circuit, Eccles-Jordan multivibrator, flip-flop circuit, or trigger circuit.

Bit Single binary digit that can represent 0 or 1.

Byte An 8-bit (XXXXXXXX) binary number (word). Note that this is also the amount of storage space required to store an 8-bit number.

Capture effect The tendency of FM receivers to lock onto the strongest of signals occupying the same carrier frequency to the exclusion of weaker signals.

Carrier wave The signal used to carry the information after it has been modulated.

CD Compact disk. An optical media that stores information using laser-imprinted plastic media. CDs can store up to almost a gigabyte of data.

Central processing unit (CPU) The part of a computer containing the circuits required to interpret and execute instructions.

Choke A coil or inductor used as a filter, often in an AC-to-DC power supply.

Collector feedback Generally, resistance coupling from a transistor collector to its base for the primary purpose of stabilizing collector current.

Commutation The act of turning an "on" thyristor to its "off" state.

Computer A machine into which an operator may apply numerical data and instructions, and from which an operator may retrieve logical numeric results.

Cone of Acceptance The angular arc over which an optical signal entering a fiberoptic cable will successfully propagate through the cable.

Conventional theory of current flow The theory of current flow in which current flows from a positive charge to a negative charge; in other words current flows from positive to negative through a circuit.

Crossover The point(s) in a signal waveform where an amplifier input transitions from positive-to-negative or negative-to-positive polarity.

Crossover distortion The difference in a push-pull (or complementary symmetry) amplifier output compared to its input occurring at the crossover point.

Curve tracer An item of test equipment capable of producing a family of transistor characteristic curves for a transistor under test.

Cutoff frequency The frequency at which output power is half of the peak output power. This point is also called: Half-power point, Critical point, and Filter cutoff frequency.

Damping The gradual reduction of an oscillation, usually caused by resistance in an electric circuit.

DC working voltage The voltage rating of a capacitor based upon average applied voltage.

Demodulator An electronic circuit that recaptures the information signal from the received signal.

Depletion layer The layer that forms between the P and N material in a PN junction. Also called the depletion region.

Detector See Demodulator.

Dielectric The insulating material separating capacitor plates, where electric energy is stored.

Digital Referring to a class of circuits in which output operates like a switch (on or off), with no intermediate values.

Diode A two-terminal semiconductor device that passes current of one polarity and blocks current of the opposite polarity.

Direct coupling The transfer of current from the output of one stage directly to another by way of a conductor; capacitance and inductance are not employed.

Dividend A quantity to be divided. In 45 / 3 = 15, 45 is the dividend.

Divisor The quantity by which another quantity, the dividend, is to be divided. In 45 / 3 = 15, 3 is the divisor.

Doping material A material added to a semiconductor to cause either an N-type material (electron excess) or a P-type material (electron deficiency).

Dual conversion The use of two intermediate frequencies in a communication receiver. These are known as the first IF and second IF.

Duty cycle Usually, duty cycle is that percentage of the input cycle period when a load is driven by its source.

DVD Digital video disk or digital versatile disk. A modern optical storage media that can store multiple gigabytes of data on a disk.

Dynamic testing A procedure for determining operation characteristics of an operating circuit.

Effective value The magnitude of a DC waveform that generates as much power as a measured AC waveform.

Electron theory of current flow The theory of current flow in which current flows from a negative change to a positive change; in other words current flows from negative to positive through a circuit.

Emitter follower Another term for a common collector amplifier.

Endoscopy A minimally invasive diagnostic medical procedure used to evaluate the interior surfaces of an organ by inserting a small scope in the body, usually through a natural body opening.

Feedback The transfer of a portion of the output signal back to the input.

Filter cutoff frequency [Please supply definition.]

Flip-flop See bistable multivibrator.

Flywheel effect The maintenance of oscillation in an LC circuit resulting from the alternate charging of the capacitor, while the inductor field collapses, then increasing inductor field, as the capacitor discharges. Flywheeling dissipates as resistive factors, radiation, and coupling remove energy from the circuit.

Frequency modulation A system of attaching information to a single-frequency carrier wave. The information is included by varying the frequency of the carrier wave.

Gain The amplification of current, voltage, or power by a transistor per selected circuit values (page 80); the ratio of the output signal to the input signal of an active component (page 115).

GND (or Gnd) A common abbreviation for the term "Ground."

Hertz The SI unit of frequency. It replaced cycles per second.

Heterodyne (1) To mix one signal frequency with another in a nonlinear device to produce their sum and/or difference. (2) The sum or difference signal resulting from the mixing of two or more signals in a nonlinear device.

Hot-carrier diode A diode exhibiting a thermal difference across the PN junction, having a very fast off-to-on and on-to-off transition.

Image interference Interference produced by a signal present on the image frequency.

Impedance coupling A coupling technique similar to capacitive (or RC) coupling wherein an inductor is used in lieu of a resistor.

Input impedance The total of all electrical opposition, both resistive, and reactive, to current flow offered by a circuit input element.

Inverting input The differential op-amp input that drives the output 180° out-of-phase.

Isolation diffusion Part of the IC fabrication process that creates insulating barriers.

Isolation transformers A power transformer having turns ratio of 1:1.

Knee voltage The voltage at which current (other than leakage) begins to flow across a diode junction.

Least significant digit (LSD) The digit in any number that carries the least weight (i.e., represents the smallest part). For example, in the number 4,256.79 the numeral 9 is the LSD.

Lower sideband (LSB) The carrier frequency minus the modulating frequencies.

Low-pass The filter type which passes low frequencies more readily than higher frequencies.

Luminous flux (light flux) The rate at which visible light energy is produced or received at a source. It is measured in lumen (lm).

Luminous intensity (light intensity) The luminous flux per unit area, measured as lumens/meter^2 (lm/m^2)

Mainframe A large computer. Older models are made of discrete components and often require cabinetry that fills several rooms.

Microcomputer A microprocessor combined with input/output interface devices, some type of external memory, and the other elements required to form a working computer system. It is smaller, lower in cost, and usually slower than a minicomputer. Also known as a micro.

Microcontroller A small electronic device, usually containing a microprocessor, which based on a fixed set of input data and instructions produces a limited number of predetermined outputs for the accomplishment of specific tasks.

Microprocessor An integrated circuit that today normally contains a CPU, processor clock, some memory, buffered address, data input/output lines, and control signals.

Minicomputer A relatively small general-purpose digital computer intermediate in size between a microcomputer and a mainframe.

Minuend The quantity from which another quantity, the subtrahend, is to be subtracted. In the equation $50 - 16 = 34$, the minuend is 50.

Modulation In electronics and communications, the process of attaching information to a signal or carrier wave (page 273); to attach information to a signal by varying its frequency, amplitude, or other characteristic. The information is attached at a transmitter and recovered at the receiver (page 409).

Modulator The electronic circuit that impresses the information signal onto the carrier frequency.

Monostable Having only one stable state.

Most significant digit (MSD) The digit in any number that carries the most weight (i.e., represents the biggest part). For example, in the number 4,256.79, the numeral 4 is the MSD.

Multiplexer An electrical or electronic device that allows the transmission of two or more signals along a smaller number of pathways.

Multiplicand The number to be multiplied by another. In 8×32, the multiplicand is 8.

Multiplier The number by which another number is multiplied. In 8×32, the multiplier is 32.

Negative feedback Any method used, whether intentional or unintentional, to couple a portion of amplifier output back to its input, 180° out-of-phase (page 152); Feedback that decreases input (page 226).

Negative resistance A special purpose diode in which current reaches a peak. Further voltage increase results in a rapid current decrease.

Negative resistance region With reference to a UJT, range of emitter current depicted on the VEB1/1E characteristic curve, where VEB declines as emitter current increases.

Nibble Four-bit (XXXX) binary value.

Noninverting input The differential op-amp input that drives the output in-phase.

Numerical Aperture (NA) The sine of the Angle of Acceptance.

Octave An octave is a range of frequencies in which the highest frequency in the range is double the lowest frequency in the range. For example, 3,000 Hertz to 6,000 Hertz is one octave.

Optical spectrum The range of electromagnetic frequencies that fall into the region we call light.

Oscillation The movement from one point to another in a smooth rhythmic manner. This term can refer to either electrical or mechanical motion.

Oscillator An electronic circuit that creates an alternating (AC) output waveform. The waveform may be sinusoidal or any other regular shape, such as a square wave or triangular wave.

Passive circuit element An electric device that does not add energy to an electric circuit. Examples include resistors, inductors, and capacitors.

Peak average The magnitude of a waveform as measured from the zero value to the peak value.

Peak inverse voltage See Reverse breakdown voltage.

Peak-to-peak The peak-to-peak magnitude of an AC signal.

Peak value The magnitude of a waveform as measured from the zero value to the peak value.

Peta (P) The unit prefix indicating 10^{15}. For example, 30 PHz $= 30 \times 10^{15}$ Hz.

Photolithography Creation of ICs using a photographic system.

Piezoelectric effect The effect exhibited by some crystals in which pressure causes electric charge, and electric charge causes a change in shape.

Pinchoff Voltage (V_P) The drain-to-source voltage, with a given gate voltage, beyond which drain current will increase no further.

PN junction The interface formed when a P-type material is conjoined with an N-type material.

Positive feedback Signal returned from circuit output to circuit input in such a way that the returned signal is in phase (additive) with other signal input.

Positive temperature coefficient A condition whereby resistance, capacitance, length, or other physical characteristic of a material increases as its temperature rises.

Power supply An electric circuit used to convert electric voltages and currents from one form to another form suitable for a particular application. Example: change the AC main voltage into a DC voltage suitable for powering an electronic computer.

Q The measure of the quality of an inductor, or capacitor $\left(\frac{X}{R}\right)$. The determining factor in selectivity of a resonant circuit.

Quiescent point (Q-point) Collector current present when no input signal is applied.

Quotient The number obtained by dividing one quantity by another. In 45 / 3 = 15, 15 is the quotient.

Radix The base of a system of numbers, such as 2 in the binary system and 10 in the decimal system. Also called the base.

Rectify To change alternating current (AC) to direct current (DC).

Reflect To throw or bend back (light, for example) from a surface.

Reflection The act of reflecting or the state of being reflected.

Refraction The turning or bending of any wave, such as a light or sound wave, when it passes from one medium into another of different density.

Regenerative feedback See positive feedback.

Reverse breakdown voltage The reverse bias voltage required to cause a PN junction to fail (page 8); the reverse-bias voltage beyond which current in excess of leakage current begins to flow. This is usually a destructive condition (page 36).

RMS value The same as effective value. RMS stands for "root mean square."

Semiconductor Any of various solid crystalline substances, such as germanium or silicon, having electric conductivity greater than insulators but less than good conductors. (Excerpted from the American Heritage Talking Dictionary. Copyright © 1997 The Learning Company, Inc. All rights reserved.)

Static switch A switch that uses semiconductor devices such as thyristors to perform the on and off operations.

Stimulated emission The emission of radiation from a system as an excited electron energy level transitions to a lower energy level.

Subtrahend A quantity or number to be subtracted from another. In the equation 50 – 16 = 34, the subtrahend is 16.

Surface-barrier diode See Hot-carrier diode.

Tera (T) The unit prefix indicating 10^{12}. For example, 30 THz = 30×10^{12} Hz.

Thermal instability Change in bias as the result of a change in ambient temperature.

Thermal runaway A circuit condition where an increase in transistor temperature results in higher current, and higher temperature, continuously, until saturation or a circuit breakdown occurs.

Thermistor A two-wire, temperature sensitive semiconducting device that exhibits resistance that varies inversely with temperature.

Thyristor A semiconductor device with three or more junctions that has the ability to turn on and/or off by application of an external signal.

Transient A temporary change in a circuit, usually used when referring to a momentary high-energy pulse.

TTL A standard abbreviation for the solid-state transistor family of logic devices known as Transistor–Transistor Logic.

Turn-off time Time that elapses after anode current stops, and before forward bias can be applied without turning the device on.

Upper sideband (USB) The carrier frequency plus the modulating frequencies.

Virtual ground A point in a circuit which, although not hardwired to a ground point, maintains a zero-volt potential.

Voltage-capacitance curve A graph upon which Voltage on the x-axis, is plotted against capacitance on the y-axis (or reverse).

Zener region Voltages equal to, or greater than Zener voltage.

Zener voltage The Zener diode's rated reverse breakdown voltage. The regulating voltage of a zener diode.

INDEX

Q

R

S